Conductive Polymers

Series in Materials Science and Engineering

Other books in the series:

Conductive Polymers: Electrical Interactions in Cell Biology and Medicine
Ze Zhang, Mahmoud Rouabhia, Simon E. Moulton (Eds)

Advanced Thermoelectrics: Materials, Contacts, Devices, and Systems
Zhifeng Ren, Yucheng Lan, Qinyong Zhang (Eds)

Physical Methods for Materials Characterisation, Third Edition
Peter E J Flewitt, Robert K Wild

Multiferroic Materials: Properties, Techniques, and Applications
J Wang (Ed)

Computational Modeling of Inorganic Nanomaterials
S T Bromley, M A Zwijnenburg (Eds)

Automotive Engineering: Lightweight, Functional, and Novel Materials
B Cantor, P Grant, C Johnston

Strained-Si Heterostructure Field Effect Devices
C K Maiti, S Chattopadhyay, L K Bera

Spintronic Materials and Technology
Y B Xu, S M Thompson (Eds)

Fundamentals of Fibre Reinforced Composite Materials
A R Bunsell, J Renard

Novel Nanocrystalline Alloys and Magnetic Nanomaterials
B Cantor (Ed)

3-D Nanoelectronic Computer Architecture and Implementation
D Crawley, K Nikolic, M Forshaw (Eds)

Computer Modelling of Heat and Fluid Flow in Materials Processing
C P Hong

High-K Gate Dielectrics
M Houssa (Ed)

Metal and Ceramic Matrix Composites
B Cantor, F P E Dunne, I C Stone (Eds)

High Pressure Surface Science and Engineering
Y Gogotsi, V Domnich (Eds)

Physical Methods for Materials Characterisation, Second Edition
P E J Flewitt, R K Wild

Topics in the Theory of Solid Materials
J M Vail

Solidification and Casting
B Cantor, K O'Reilly (Eds)

Conductive Polymers

Electrical Interactions in Cell Biology and Medicine

Editors: Ze Zhang
Mahmoud Rouabhia
Simon E. Moulton

CRC Press
Taylor & Francis Group
Boca Raton London New York

CRC Press is an imprint of the
Taylor & Francis Group, an **informa** business

CRC Press
Taylor & Francis Group
6000 Broken Sound Parkway NW, Suite 300
Boca Raton, FL 33487-2742

CRC Press is an imprint of Taylor & Francis Group, an Informa business

Printed on acid-free paper

International Standard Book Number-13: 978-1-4822-5928-5 (Hardback)

Library of Congress Cataloging-in-Publication Data

Names: Zhang, Ze, 1957- editor. | Rouabhia, Mahmoud, 1957- editor. | Moulton, Simon E., editor.
Title: Conductive polymers : electrical interactions in cell biology and medicine / edited by Ze Zhang, Mahmoud Rouabhia, Simon E. Moulton.
Description: Boca Raton, FL : CRC Press, Taylor & Francis Group, 2017.
Identifiers: LCCN 2016042529| ISBN 9781482259285 (hardback ; alk. paper) | ISBN 1482259281 (hardback ; alk. paper)
Subjects: LCSH: Conducting polymers. | Polymers in medicine. | Biomedical materials. | Organic conductors. | Polymers--Electric properties. | Cell culture.
Classification: LCC QD382.C66 C686 2017 | DDC 610.28/4--dc23
LC record available at https://lccn.loc.gov/2016042529

Visit the Taylor & Francis Web site at
http://www.taylorandfrancis.com

and the CRC Press Web site at
http://www.crcpress.com

To my parents, Li Zhang and Yanjun Tai,
my wife, Jin, and my children, Issan and Roger,
for their love, support, and understanding.

Ze Zhang

I would like to dedicate this book to my wife, Jamila,
to my daughter, Dounia, and my son, Réda, for their continued love, support,
patience, etc., and having made this edition possible and exciting.
I would also like to dedicate this book to my father and
my father-in-law for inspiration and guidance.
I am terribly missing both of you, but your inspiration
and guidance are with me forever.

Mahmoud Rouabhia

I wish to dedicate this book to my wife, Louise, and
my children, Aleida and Liam, who encourage me to aim high and
achieve my goals.
A special mention to Max, Stella, and Tess
for their silent but ever-present support.

Simon E. Moulton

Contents

Series Preface

The Series in Materials Science and Engineering publishes cutting-edge monographs and foundational textbooks for interdisciplinary materials science and engineering. It is aimed at undergraduate and graduate-level students, as well as practicing scientists and engineers.

Its purpose is to address the connections between properties, structure, synthesis, processing, characterization, and performance of materials. The subject matter of individual volumes spans fundamental theory, computational modeling, and experimental methods used for design, modeling, and practical applications. The series encompasses thin films, surfaces, and interfaces, and the full spectrum of material types, including biomaterials, energy materials, metals, semiconductors, optoelectronic materials, ceramics, magnetic materials, superconductors, nanomaterials, composites, and polymers.

New books in the series are commissioned by invitation. Authors are also welcome to contact the publisher (Luna Han: luna.han@taylorandfrancis.com) to discuss new title ideas.

Foreword

The dramatic developments in microelectronics technology over the past generation or so have been nothing less than transformational for science, engineering, and society as a whole. The ability to rapidly design and manufacture on a large-scale solid-state components with ever-smaller, faster, inexpensive, and complex computational circuitry has now put simply incredible amounts of processing power into the hands of essentially everyone on the planet. What is perhaps most remarkable is that the platform substrates that drove this revolution are single crystals of semiconducting silicon, locally doped with different electron-rich or electron-poor elements (such as phosphorous or boron) to provide controlled variations in electron transport performance. These are then layered up with appropriate insulators (typically oxides) and metals (such as gold, copper, and tungsten) to provide the necessary means for interconnecting these ever-shrinking components with the external world.

In many instances, there is a need to directly integrate these engineered, typically inorganic electronic devices with living systems. This might be necessary to provide support for a function that no longer exists (such as for an auditory or visual prosthesis), or to directly communicate with the central or peripheral nervous system. In such instances, it is important to maintain efficient, stable interactions that will allow for the patient to establish and maintain an intimate interface with the technology. However, traditional electronic components are not typically designed to work well in this environment, since they are composed of solid, stiff, essentially flat, inorganic, and relatively inert metals, semiconductors, or dielectrics. Living tissue, on the other hand, is wet, soft, dynamic, articulated, mostly organic, and conducts charge predominantly by ionic transport.

Conjugated polymers have recently emerged as a class of organic materials that can work well at the interface between living systems and engineered components. Examples of these materials include functionalized poly(pyrroles) and poly(thiophenes) that have interesting chemical similarities to melanin, a natural conjugated polymer found in skin, hair, and certain electrically active organs, such as the ear and the brain. The mechanical and electrical properties of conjugated polymers are typically intermediate to those of the tissue and the solid-state devices, and they have the ability to efficiently transmit charge as both electrons and holes in the solid state, as well as ionically through their precisely controlled counterion and side-group chemistry.

This monograph focuses on this important biology-conducting materials interface. There have been considerable efforts to create mechanically stable, biocompatible interactions between implanted devices and living tissue. However, for an electrically active system, it is absolutely critical to maintain long-term, facile charge transport between the implanted device and the cells in the tissue of interest. Professors Zhang, Rouabhia, and Moulton have assembled a group of investigators who are working on issues ranging from materials synthesis to device characterization to analytical measurements of performance. Of particular interest and value are several reports from clinically inclined investigators that describe recent studies of electrically mediated cell response. These areas represent opportunities for future developments and collaborations between chemists, materials scientists, biomedical engineers, and physicians. Taken together, these chapters provide a comprehensive overview of issues related to the interface between active devices and biological systems, and emphasize the need to consider future opportunities in this area.

We of course never know for sure what the future will bring. However, by looking at the progress that is being made by the international collection of research groups included in this book, it is clear that there are many opportunities for creating ever-more intricate, sophisticated components that will effectively integrate advanced microelectronics technology with living systems. I am hopeful that this book will not only serve as a useful snapshot of the state of the art at this point in time but also help to guide future investigators as the field rapidly evolves into the unknown frontier.

David C. Martin
The University of Delaware
Newark, Delaware

Preface

Research in the fields of conductive polymers (defined as intrinsically conducting polymers for most of this book) and cellular electrophysiology has been disconnected until the 1990s, when Masuo Aizawa's lab published electrochemical modulation of enzyme activities of yeast cells (Haruyama et al. 1993) and Robert Langer's lab published work on the noninvasive modulation of the shape and function of endothelial cells (Wong et al. 1994). In these two works, however, cells were found to be affected through the secondary effect of the electrical field, that is, because of redox-induced chemical or physical changes at the interface of the polypyrrole and cells. The first report on conductive polymer–mediated electrical stimulation to directly enhance mammalian cell performance is probably the work of Christine Schmidt et al. (1997), who found that neurite outgrowth in PC12 cells was doubled in the presence of electrical current. The work of Wong and Schmidt also brought tissue engineering into the context of electrical stimulation and conducive polymers.

Bioelectricity was first documented by Aristotle (350 BCE), who wrote, "The torpedo narcotizes the creatures that it wants to catch, … and the torpedo is known to cause a numbness even in human beings." This phenomenon was used to release pain in humans as early as 46 AD by Scribonius Largus, a court physician to the Roman Emperor Claudius (Kane and Taub 1975). Since Luigi Galvani, bioelectricity has been studied extensively in biosciences and medicine with many modern medical diagnostics and treatments based on electrical and electromagnetic phenomena in the human body, such as electrocardiography, magnetic resonance imaging, and pacemaker technology. Despite the wide recognition of bioelectrical phenomena in humans at the organ, tissue, and cellular levels, electrical and electromagnetic fields are yet to be widely accepted as effective tools to induce cell growth and treat diseases. One major obstacle is the lack of specificity. The size or range of a field generated by instruments currently used in most labs, either electrical or magnetic, is much larger than the size of cells, not to mention the much smaller membrane receptors. When the same field is applied to different cell populations, diseased and normal alike, the absence of specificity is expected. This is just an example of how cell biologists, material researchers, and electrical engineers can work together to address the outstanding challenges. Of course, how electrical or electromagnetic fields initiate cell signaling cascades remains the central issue that requires the collaboration of those with different expertise, including physicists, computational chemists, and cell biologists. Solving such issues will bring about truly exciting science and engineering advances in areas such as the brain–machine interface and field-assisted tissue engineering and regenerative medicine.

Compared with the long research records of bioelectricity in human history, research on conductive polymers has a much shorter registry, taking off only in the 1970s and peaking again following the awarding of the Nobel Prize in Chemistry in 2000 to the founders of conducting polymers. Most current industrial research on conductive polymers has focused on electrochromic, photovoltaic, energy storage, and anticorrosion applications. In fact, as a class of material, conductive polymers have natural advantages over most synthetic substances when it comes to their interaction with living systems. First, they are electrically conductive and ionically active, similar to biological tissues. All biological systems are made up of electrolytes, proteins, sugars, lipids, and so forth, which carry electrical charges and react to electrical and electromagnetic fields. In fact, when we talk about the basic force in molecular interactions, it boils down to electrical and electromagnetic fields. Conductive polymers can carry electrical current and generate electrical fields, facilitating their ease of "communicating" with living tissues (compared with insulating synthetic polymers). The second advantage of conductive polymers in interacting with living tissues stems from their functionality. As synthetic materials, conductive polymers can be modified to carry functional groups or biologically active molecules, making conductive polymers "friendlier" with living tissues. Biologically functional conductive polymers may confine electrical impact to specific molecular interactions such as adhesion. Third, with respect to metals and inorganic conductors, conductive polymers can be fabricated or modified to acquire mechanical properties similar to those of living tissue. Evidently, these advantages do not mean that conductive polymers can be used to substitute for natural tissues, but they do open a new dimension in conductive polymer research.

This book is intended to provide readers with a relatively comprehensive picture regarding conductive polymers and electrical modulation of cellular activities in the context of medicine. Different from traditional electrophysiology, which is more about diagnosis and recording, our objective is to treat, cure, and communicate using electrical and electromagnetic fields, with the help of conductive polymers as the interface, scaffold, substrate, guidance channel, and so forth. To achieve this objective, we need to see several changes, including better materials allowing us to perform electrical field intervention with high specificity; a more thorough understanding about how electrical and electromagnetic fields interact with electrolytes, ligands, and receptors; and improved awareness of safety issues related to electrically activated cells. Obviously, these challenges cannot be met without collaboration among scientists and engineers of different disciplines. Hopefully, the information and core messages in this book will contribute to the success of our objective.

Ze Zhang
Laval University, Quebec, Canada

Mahmoud Rouabhia
Laval University, Quebec, Canada

Simon E. Moulton
Swinburne University of Technology, Melbourne, Victoria, Australia

MATLAB® is a registered trademark of The MathWorks, Inc. For product information, please contact:

The MathWorks, Inc.
3 Apple Hill Drive
Natick, MA 01760-2098 USA
Tel: 508-647-7000
Fax: 508-647-7001
E-mail: info@mathworks.com
Web: www.mathworks.com

REFERENCES

Aristotle. *The History of Animals*. 350 BCE.

Haruyama, T., E. Kobatake, Y. Ikariyama, and M. Aizawa. Stimulation of acid phosphatase induction in *Saccharomyces cerevisiae* by electrochemical modulation of effector concentration. *Biotechnology and Bioengineering* 42 (1993): 836–842.

Kane, K., and A. Taub. A history of local electrical analgesia. *Pain* 1 (1975): 125–138.

Schmidt, C. E., V. R. Shastri, J. P. Vacanti, and R. Langer. Stimulation of neurite outgrowth using an electrically conducting polymer. *Proceedings of the National Academy of Sciences of the United States of America* 94 (1997): 8948–8953.

Wong, J. Y., R. Langer, and D. E. Ingber. Electrically conducting polymers can noninvasively control the shape and growth of mammalian cells. *Proceedings of the National Academy of Sciences of the United States of America* 91 (1994): 3201–3204.

Editors

Dr. Ze Zhang is a full professor of the Department of Surgery at Laval University and a senior researcher in the Division of Regenerative Medicine of the University Hospital Center in Quebec City. He received his bachelor's and master's degrees in engineering from Chengdu University of Science & Technology (now Sichuan University), China, in 1982 and 1984, and then a PhD degree in experimental medicine from Laval University in 1993. After a postdoctoral training in Japan, he returned to Laval University in 1995. Dr. Ze Zhang's main research focuses are cardiovascular implants and tissue repair using synthetic polymers and electrical stimulation. He has published more than 100 peer-reviewed papers and 4 book chapters.

Dr. Mahmoud Rouabhia is a full professor at the Faculty of Dentistry of Laval University, Quebec City. He is a senior scientist in the field of immunology, cell biology, and tissue engineering. He got his PhD in France, followed by a postdoctoral training for four years in Canada. Dr. Rouabhia's research interests include studying the interaction between human cell biomaterials and electrical stimulation for better wound healing. Dr. Rouabhia has more than 130 peer-reviewed scientific publications. He also published more than 15 book chapters and review articles and holds 2 patents. He is the editor or coeditor of two books in the field of tissue engineering and wound healing.

Dr. Simon E. Moulton is a full professor of biomedical electromaterials science in the Faculty of Science, Engineering and Technology at Swinburne University of Technology, Melbourne, Victoria, Australia. He completed his PhD at the University of Wollongong, Wollongong, New South Wales, Australia, in 2002, and has developed a substantial research track record in the synthesis and fabrication of organic conducting materials for use in a variety of biomedical applications. He has a strong focus in materials chemistry research, with an emphasis in developing composite biomaterials through the integration of electroactive materials with conventional biomaterials. He has published 1 book, 4 book chapters, and 95 journal papers with an h-index of 27.

Contributors

Ardeshir Bayat
Plastic and Reconstructive Surgery Research
Faculty of Biology, Medicine and Health
The University of Manchester
Manchester, United Kingdom

and

Centre for Dermatology
Institute of Inflammation and Repair
The University of Manchester
Manchester, United Kingdom

and

University Hospital of South Manchester NHS Foundation Trust
Manchester, United Kingdom

and

Institute of Inflammation and Repair
Faculty of Medical and Human Sciences
Manchester Academic Health Science Centre
The University of Manchester
Manchester, United Kingdom

Peter R. Bergethon
Department of Anatomy and Neurobiology
Boston University School of Medicine
Boston, Massachusetts

and

Pfizer Neuroscience and Pain Research Unit
Pfizer, Inc.
Cambridge, Massachusetts

Miina Björninen
ARC Centre of Excellence for Electromaterials Science
Intelligent Polymer Research Institute
Australian Institute of Innovative Materials
Innovation Campus
University of Wollongong
Wollongong, New South Wales, Australia

Justin L. Bourke
Department of Medicine
St. Vincent's Hospital Melbourne
University of Melbourne
Fitzroy, Victoria, Australia

and

ARC Centre of Excellence for Electromaterials Science
Intelligent Polymer Research Institute
Australian Institute of Innovative Materials
Innovation Campus
University of Wollongong
Wollongong, New South Wales, Australia

and

Clinical Neurosciences,
St. Vincent's Hospital Melbourne,
Fitzroy, Victoria, Australia

Jun Chen
ARC Centre of Excellence for Electromaterials Science
Intelligent Polymer Research Institute
Australian Institute of Innovative Materials
Innovation Campus
University of Wollongong
Wollongong, New South Wales, Australia

Jeremy M. Crook
ARC Centre of Excellence for Electromaterials Science
Intelligent Polymer Research Institute
Australian Institute of Innovative Materials
Innovation Campus
University of Wollongong
Wollongong, New South Wales, Australia

and

Illawarra Health and Medical Research Institute
University of Wollongong,
Wollongong, New South Wales, Australia

and

Department of Surgery
St. Vincent's Hospital Melbourne
The University of Melbourne
Fitzroy, Victoria, Australia

María B. Decca
Centro de Investigaciones en Química Biológica de Córdoba (CIQUIBIC)
CONICET
Departamento de Química Biológica
Facultad de Ciencias Químicas
Ciudad Universitaria
Universidad Nacional de Córdoba
Córdoba, Argentina

Tessa Gordon
Division of Plastic Reconstructive Surgery
Department of Surgery
Peter Gilgan Centre for Research and Learning
Hospital for Sick Children (SickKids)
Toronto, Ontario, Canada

Michelle Griffin
Centre for Nanotechnology and Regenerative Medicine
UCL Division of Surgery and Interventional Science
University College London
London, United Kingdom

Suvi Haimi
Department of Biomaterials Science and Technology
University of Twente
Enschede, The Netherlands

and

Institute of Biosciences and Medical Technology
University of Tampere
Tampere, Finland

Michael J. Higgins
ARC Centre of Excellence for Electromaterials Science
Intelligent Polymer Research Institute
Australian Institute of Innovative Materials
Innovation Campus
University of Wollongong
Wollongong, New South Wales, Australia

Andreas Hildebrandt
Department of Scientific Computing and Bioinformatics
Johannes Gutenberg University Mainz
Mainz, Germany

Anna Katharina Hildebrandt
Max Planck Institute for Informatics
Saarbrücken, Germany

Peter C. Innis
ARC Centre of Excellence for Electromaterials Science
Intelligent Polymer Research Institute
Australian Institute of Innovative Materials
Innovation Campus
University of Wollongong
Wollongong, New South Wales, Australia

Trenton A. Jerde
Department of Psychology
New York University
New York, New York

Robert M. I. Kapsa
ARC Centre of Excellence for Electromaterials Science
Intelligent Polymer Research Institute
Australian Institute of Innovative Materials
Innovation Campus
University of Wollongong
Wollongong, New South Wales, Australia

and

Department of Medicine
St. Vincent's Hospital Melbourne
University of Melbourne
Fitzroy, Victoria, Australia

and

Clinical Neurosciences,
St. Vincent's Hospital Melbourne,
Fitzroy, Victoria, Australia

Thomas Kemmer
Department of Scientific Computing and Bioinformatics
Johannes Gutenberg University Mainz
Mainz, Germany

Lichun Lu
Department of Orthopaedic Surgery

and

Department of Physiology and Biomedical Engineering
Mayo Clinic College of Medicine
Rochester, Minnesota

Yizhong Luo
Key Laboratory of Polymer Ecomaterials
Changchun Institute of Applied Chemistry
Chinese Academy of Sciences
Changchun, China

Shiyun Meng
College of Environment and Resources
Chongqing Technology and Business University
Chongqing, China

A. Lee Miller II
Department of Orthopaedic Surgery

and

Department of Physiology and Biomedical Engineering
Mayo Clinic College of Medicine
Rochester, Minnesota

Guillermo G. Montich
Centro de Investigaciones en Química Biológica de Córdoba (CIQUIBIC)
CONICET
Departamento de Química Biológica
Facultad de Ciencias Químicas
Ciudad Universitaria
Universidad Nacional de Córdoba
Córdoba, Argentina

Simon E. Moulton
Faculty of Science, Engineering and Technology
Swinburne University of Technology
Melbourne, Victoria, Australia

David L. Officer
ARC Centre of Excellence for Electromaterials Science
Intelligent Polymer Research Institute
Australian Institute of Innovative Materials
Innovation Campus
University of Wollongong
Wollongong, New South Wales, Australia

Trisha M. Pfluger
Juno Biomedical, Inc.
Mountain View, California

Anita F. Quigley
ARC Centre of Excellence for Electromaterials Science
Intelligent Polymer Research Institute
Australian Institute of Innovative Materials
University of Wollongong
Wollongong, New South Wales, Australia

and

Department of Medicine
St. Vincent's Hospital Melbourne
University of Melbourne
Fitzroy, Victoria, Australia

and

Clinical Neurosciences,
St. Vincent's Hospital Melbourne,
Fitzroy, Victoria, Australia

Seth C. Rasmussen
Department of Chemistry and Biochemistry
North Dakota State University
Fargo, North Dakota

Mahmoud Rouabhia
Faculté de Médecine Dentaire
Université Laval
Québec, Canada

Darren Svirskis
Faculty of Medical and Health Sciences
The University of Auckland
Auckland, New Zealand

Sara Ud-Din
Plastic and Reconstructive Surgery Research
The University of Manchester
Faculty of Biology, Medicine and Health
The University of Manchester
Manchester, United Kingdom

Klaudia Wagner
ARC Centre of Excellence for Electromaterials Science
Intelligent Polymer Research Institute
Australian Institute of Innovative Materials
University of Wollongong
Wollongong, New South Wales, Australia

Pawel Wagner
ARC Centre of Excellence for Electromaterials Science
Intelligent Polymer Research Institute
Australian Institute of Innovative Materials
Innovation Campus
University of Wollongong
Wollongong, New South Wales, Australia

Huan Wang
Department of Neurologic Surgery
Mayo Clinic College of Medicine
Rochester, Minnesota

Xianhong Wang
Key Laboratory of Polymer Ecomaterials
Changchun Institute of Applied Chemistry
Chinese Academy of Sciences
Changchun, China

Natalia Wilke
Centro de Investigaciones en Química Biológica de Córdoba (CIQUIBIC)
CONICET
Departamento de Química Biológica
Facultad de Ciencias Químicas
Ciudad Universitaria
Universidad Nacional de Córdoba
Córdoba, Argentina

Michael J. Yaszemski
Department of Orthopaedic Surgery

and

Department of Physiology and Biomedical Engineering
Mayo Clinic College of Medicine
Rochester, Minnesota

Ze Zhang
Département de Chirurgie
Axe Médecine Régénératrice–CHU
Université Laval
Québec, Canada

Siwei Zhao
Department of Biomedical Engineering
Tufts University
Medford, Massachusetts

Wen Zheng
Department of Mechanical Engineering
Shanghai Jiao Tong University
Shanghai, China

1 Early history of conductive organic polymers

Seth C. Rasmussen
North Dakota State University
Fargo, North Dakota

Contents

1.1 INTRODUCTION

1.1.1 COMMONLY STUDIED PARENT CONDUCTIVE ORGANIC POLYMERS

The most common and successful examples of conductive organic polymers are doped conjugated organic polymers. While the total number of such polymers is now easily in the thousands, all known conjugated polymers may be considered derivatives of a number of parent polymers, the most important of which are shown in Figure 1.1. Unlike the more common saturated organic polymers, conjugated polymers are a class of organic semiconducting materials that exhibit enhanced electronic conductivity in their oxidized or reduced states (Perepichka and Perepichka 2009; Skotheim and Reynolds 2007). As such, these materials combine the conductivity of classical inorganic systems with many of the desirable properties of organic plastics, including mechanical flexibility and low production costs. This combination has led to considerable fundamental and technological interest over the last few decades, resulting in the current field of organic electronics and the development of a variety of modern technological applications, including sensors, electrochromic devices, organic photovoltaics, field effect transistors, and organic light-emitting diodes (Perepichka and Perepichka 2009; Skotheim and Reynolds 2007).

The oxidation of conjugated organic polymers generates positive charge carriers (i.e., holes) (Figure 1.2) and an increase of p-type character (MacDiarmid and Epstein 1994; MacDiarmid 2001a). As such, these polymers in their oxidized form are referred to as p-doped in analogy to p-doped inorganic semiconductors such as gallium-doped silicon. This oxidation process can be accomplished either by treating the polymer with an oxidizing agent or via electrochemical oxidation. As this p-doping process results in the generation of a polycationic material, the material must incorporate anions in order to maintain charge neutrality. If accomplished via an oxidizing agent, the anions generated by the redox process then become the counterions incorporated into the polymer.

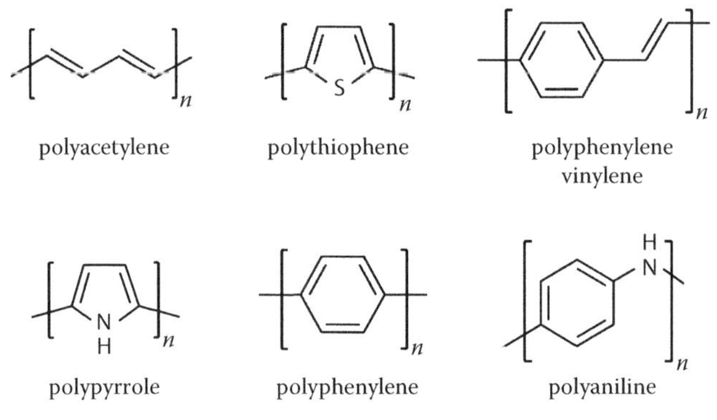

Figure 1.1 Commonly studied parent conjugated organic polymers.

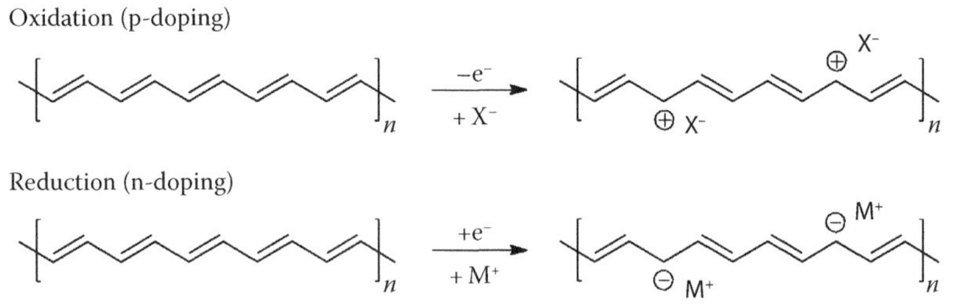

Figure 1.2 Doping of polyacetylene.

If accomplished via electrochemical oxidation, the anions come from the supporting electrolyte utilized during the electrochemical process.

For some conjugated polymers, reduction or n-doping is also possible, resulting in the addition of negative charge carriers (i.e., electrons) (Figure 1.2) and an increase of n-type character (MacDiarmid and Epstein 1994; MacDiarmid 2001a, 2002). As with p-doping, this can be accomplished either electrochemically or by treatment of the polymer with a reducing agent. In either case, the material must incorporate cationic species in order to maintain charge neutrality. The counterions incorporated into these polymers during either p-doping or n-doping are often referred to as "dopants," which can be somewhat misleading as the counterion itself does not cause the enhanced conductivity, although they are necessary to allow the formation of oxidized or reduced forms that do provide the resulting conductive materials.

1.1.2 2000 NOBEL PRIZE IN CHEMISTRY

The impact and importance of these organic conductors was recognized by the awarding of the 2000 Nobel Prize in Chemistry to Professors Alan J. Heeger, Alan G. MacDiarmid, and Hideki Shirakawa "for the discovery and development of conductive polymers" (Rasmussen 2011, 2014, 2015). This award was in acknowledgment of their early contributions to the field of conjugated polymers, particularly their collaborative work on conducting polyacetylene beginning in the mid-to-late 1970s. As illustrated by this Nobel Prize, the discovery of conducting polymers via doping is most often attributed to Heeger, MacDiarmid, and Shirakawa, although reports of electrically conductive conjugated polymers date back to the early 1960s. The most notable of these reports was the investigations of Donald Weiss and coworkers on conducting polypyrrole, as well as that of René Buvet and Marcel Jozefowicz on conducting polyaniline. That these previous studies are overlooked in most discussions of the history of conjugated polymers is unfortunate and results in the fact that the majority of researchers in this expanding field of materials science are unaware of these previous contributions.

Recently, the current author has attempted to educate the community with a series of publications detailing the early history of conjugated polymers and the discovery of their conductivity (Rasmussen 2011, 2014, 2015). Along with these efforts, two additional historical accounts have been published during this time frame that have also tried to shed light on some of these previous contributions (Elschner et al. 2011; Inzelt 2008). In continuing these collective efforts, this chapter provides an overview of the history of the first three primary conducting organic polymers from their origins in the early nineteenth century up through the polyacetylene work recognized by the Nobel Prize in 2000.

1.1.3 CARBON BLACK AND ORIGIN OF CONDUCTIVE ORGANIC POLYMERS

Scientists began to speculate about the possibility that electronic conduction might be observed in organic materials as early as the 1930s. Nevertheless, it was not until the 1950s that significant experimental efforts were undertaken in attempts to produce organic conductors. Of the various materials investigated in these efforts, it was graphite and the carbon blacks (material from the partial burning or pyrolysis of organic matter) that gave the most significant electrical conductivity (up to 50 Ω^{-1} cm^{-1}) (Weiss and Bolto 1965; Gutmann and Lyons 1967).

Although the modern description of carbon blacks is a particulate, quasi-graphitic material, these materials were initially proposed to be three-dimensional, cross-linked organic polymers in which the specific structural nature of the carbon black was dependent on the method of its production (Riley 1947; Weiss and Bolto 1965). As the chemical structures of these "carbonaceous polymers" were considered to be too complex and ill-defined, efforts turned to the production of related organic polymers with more defined and controllable compositions. Such efforts typically utilized either the high-temperature pyrolysis of various synthetic organic polymers or the direct polymerization of conjugated precursors in order to produce potential model systems (Weiss and Bolto 1965). Examples of conjugated polymers in these early efforts include polyenes, polyvinylenes, and polyphenylenes. For the most part, however, these initially investigated materials exhibited high resistivities and could nearly be considered insulators. The first real successful synthesis of an organic polymer with significant conductivity was reported in 1963 by Donald Weiss and coworkers with polypyrrole materials (McNeill et al. 1963). As such, our discussion of the history of conjugated polymers begins with that of polypyrrole.

1.2 POLYPYRROLE

1.2.1 ANGELI AND PYRROLE BLACK

The history of polypyrrole dates back to 1915 with the work of Angelo Angeli (1864–1931) (Figure 1.3) at the University of Florence (Rasmussen 2015). At that time, Angeli began studying the treatment of pyrrole with mixtures of hydrogen peroxide and acetic acid, which produced a black precipitate that he named *nero di pirrolo* or "pyrrole black" (Angeli 1915; Angeli and Alessandri 1916). This was typically accomplished by adding 50% H_2O_2 to pyrrole dissolved in a sufficient amount of acetic acid. After a short period of time, the solution turned greenish-brown, ultimately turning a black-brown color over the space of a couple of days. The product could be isolated as a thin black powder, via either spontaneous precipitation, dilution of the final solution with water, or addition of aqueous sodium sulfate. The product was insoluble in everything but basic solutions, and thus purification was typically accomplished by dissolving the powder in base, followed by precipitation with either acetic acid or dilute sulfuric acid. The purified solid was then filtered and dried at 120°C to give a fine dark brown to black powder.

Angeli went on to find that pyrrole blacks could be obtained using a variety of additional oxidizing agents, including nitrous acid (Angeli and Cusmano 1917), potassium dichromate (Angeli 1918) or chromic acid (Angeli and Lutri 1920), lead oxide (Angeli 1918), potassium permanganate (Angeli and Pieroni 1918), and various quinones (Angeli and Lutri 1920). Oxygen could also be used as the oxidant when used in combination with either light or ethylmagnesium iodide (Angeli and Cusmano 1917). Comparison of

Figure 1.3 Angelo Angeli (1864–1931). (Courtesy of the "Ugo Schiff" Chemistry Department, University of Florence, Italy.)

the results obtained from these various oxidants ultimately led Angeli (1918, p. 22; translated to English) to conclude:

> These facts are of special interest because it shows that the formation of pyrrole blacks is most likely preceded by a process of polymerization of the pyrrole molecule, which takes place more or less rapidly depending on the reagents that are used.

Attempts to probe the structure of pyrrole black were limited by its insoluble nature, but analysis via oxidative degradation revealed cleavage products consistent with pyrrole and indole derivatives, thus leading to the conclusion that the pyrrole ring was retained within the structure of pyrrole black. Angeli then extended the scope of his study to functionalized pyrroles, finding that the treatment of various functionalized pyrroles with oxidants produced colored products but did not result in the typical precipitate characteristic of pyrrole black (Angeli and Cusmano 1917; Angeli 1918). Ultimately, these combined studies led Angeli to propose that the structure of pyrrole black contained units consisting of direct carbon–carbon bonds between pyrroles, as shown in Figure 1.4a (Angeli 1918). It should be noted that this proposed structure is very similar to the currently accepted structure for oxidized portions of the polypyrrole backbone (Figure 1.4b).

Angeli ultimately concluded his work with pyrrole black in the early 1920s in order to move on to other research topics, although he did return to the subject with a final paper in 1930. At about that same time, however, another Italian scientist, Riccardo Ciusa (1877–1965), began investigating the polymerization of heterocycles in efforts to generate graphitic analogues from pyrrole, thiophene, and furan (Rasmussen 2015).

1.2.2 CIUSA AND GRAPHITE FROM PYRROLE

Starting in the early 1920s at the University of Bologna, Riccardo Ciusa began investigating the thermal polymerization of tetraiodopyrrole as a potential route to materials that could be considered a type of graphite generated from pyrrole (Ciusa 1921, 1922, 1925).

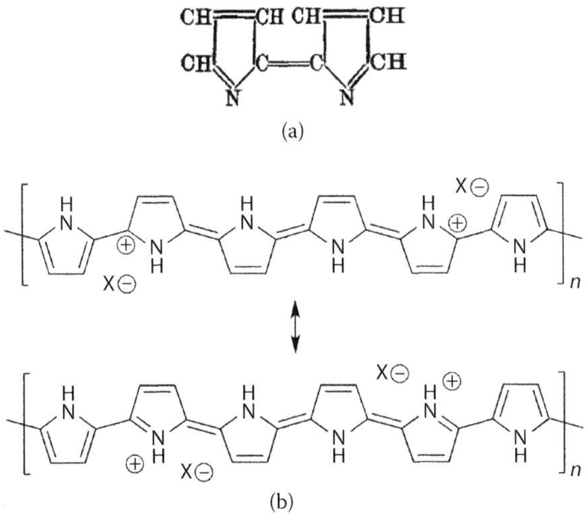

Figure 1.4 (a) Angeli's proposed basic unit for the structure of pyrrole black. (From Angeli, A., *Gazz. Chim. Ital.* 48[II], 21–25, 1918. With permission.) (b) Modern resonance structures describing oxidized (p-doped) polypyrrole.

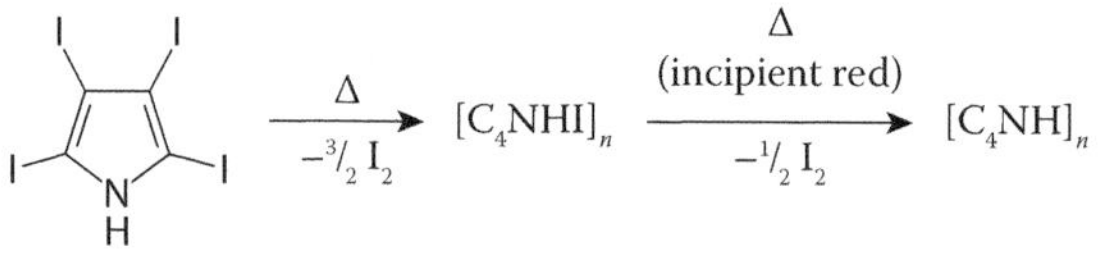

Figure 1.5 Thermal polymerization of tetraiodopyrrole.

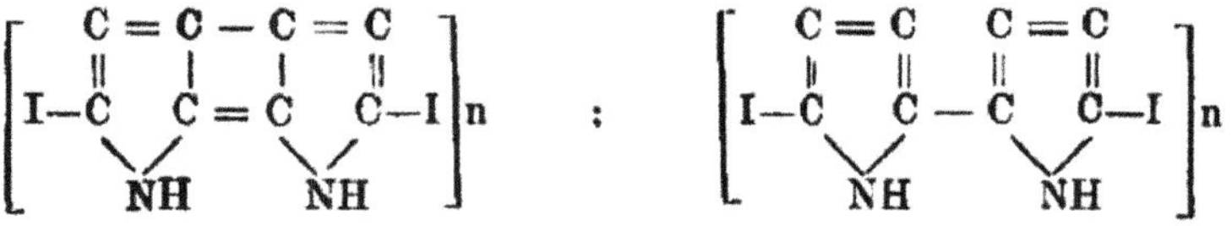

Figure 1.6 Ciusa's proposed structures for $[C_4NHI]_n$. (Reprinted from Ciusa, R., *Gazz. Chim. Ital.* 55, 385–389, 1925.)

By heating tetraiodopyrrole under vacuum at 150°C–200°C, a black material with a graphitic appearance was produced, which gave an elemental composition corresponding to $[C_4NHI]_n$ (Figure 1.5). Ciusa concluded this to be an intermediate in the formation of the desired pyrrole "graphite" and later proposed the two structures given in Figure 1.6 as possible representations of this intermediate species (Ciusa 1925).

Ciusa then reheated the material at a higher temperature (incipient red), which liberated the final iodine atom to give a black material with an appearance similar to that of graphite flakes. The determined elemental composition was consistent with $[C_4NH]_n$ (Ciusa 1922). Ciusa repeated these methods with thiophene and furan to obtain similar results before finally investigating the thermal polymerization of hexaiodobenzene to produce a graphite material that he described to be similar to ordinary graphite. Comparing the resistivity of the synthesized material to an authentic sample of graphite, however, showed that while his material did exhibit a low resistivity, it was approximately six times more resistive than graphite (Ciusa 1925).

Unfortunately, Ciusa did not report the resistivity of the heterocyclic graphites, and thus it is unclear how they might have compared. In reality, he did not actually report any other characterization of these materials beyond their graphite-like appearances and elemental compositions. Nearly 40 years later, however, Ciusa's thermal polymerization of tetraiodopyrrole became the foundation of efforts to produce conductive organic polymers by Donald Weiss and coworkers in Melbourne, Australia (Rasmussen 2011, 2015).

Figure 1.7 Donald E. Weiss (1924–2008). (Courtesy of Robert Weiss.)

1.2.3 WEISS AND CONDUCTING POLYPYRROLE

In the late 1950s, a group of Council for Scientific and Industrial Research (CSIR) researchers led by Donald Weiss (1924–2008) (Figure 1.7) began studying semiconducting organic polymers as potential electrically activated and easily regenerated adsorbents for a proposed electrical process for water desalination (Rasmussen 2011). These efforts began with the preparation of xanthene polymers in 1959, and while these materials did exhibit p-type semiconductor behavior, the resulting resistivity was still fairly high (McNeill and Weiss 1959; Rasmussen 2011). During these initial efforts, however, Weiss came across Ciusa's reports on pyrrole graphite, which suggested this might provide a new type of conducting organic material.

As Ciusa had not reported any electrical properties, Weiss and coworkers began with reproducing Ciusa's pyrrole graphite in order to study the material's structure and potential conductivity. Using modifications of Ciusa's conditions, Weiss prepared a series of polymers by heating tetraiodopyrrole in a rotating flask under a flow of nitrogen at temperatures as low as 120°C. The flow of nitrogen was used to both provide an inert atmosphere and transfer iodine vapor away from the reaction. The products of these reactions were reported to be black, insoluble powders, which Weiss described as "polypyrroles" consisting of (McNeill et al. 1963, p. 1062)

> a three-dimensional network of pyrrole rings cross-linked in a nonplanar fashion by direct carbon to carbon bonds.

Analysis of the products revealed that the materials contained both "adsorbed molecular iodine" and nonreactive iodine that was concluded to be "iodine of substitution" (McNeill et al. 1963). A hypothetical structure of the polymeric material, based on the various descriptions given by Weiss and coworkers, is given in Figure 1.8.

The resistivity (R) of the polypyrrole powders as pressed pellets was measured under a stream of nitrogen to give values of 11–200 Ω cm at 25°C, which correspond to conductivities (1/R) of 0.005–0.09 Ω^{-1} cm^{-1} (Bolto et al. 1963; Weiss and Bolto 1965). Further study of the resistivity at variable temperatures also revealed a temperature profile consistent with a standard semiconductor. While the measured conductivities were still below that of carbon black, they were drastically better than those of the previous xanthene polymers and represented the highest reported conductivities

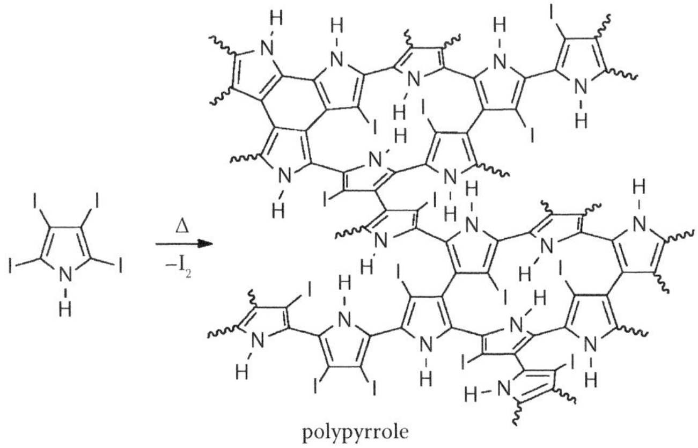

Figure 1.8 Hypothetical structure of Weiss's polypyrrole.

to date for a nonpyrolyzed organic polymer. Weiss described the nature of this conductivity as follows (Bolto et al. 1963, p. 1091):

> However it is apparent that the polymers are relatively good conductors of electricity. Since no polarization was observed during the measurement of the electrical resistance, even over substantial periods of time, it is assumed that the conductivity is of electronic origin.

Of particular note was the discovery that the removal of adsorbed molecular iodine from the polymer via solvent extraction (Bolto et al. 1963), chemical or electrochemical reduction (Bolto and Weiss 1963; McNeill et al. 1963), or thermal vacuum treatment (Bolto et al. 1963) resulted in a corresponding increase in resistance. Study of this relationship via electron spin resonance (ESR) revealed evidence for the formation of a strong charge–transfer complex between the polymer and iodine (Bolto and Weiss 1963), leading to the following conclusion (Bolto et al. 1963, p. 1090):

> Charge-transfer complexes of strength sufficient to cause partial ionization induce extrinsic [semiconductor] behaviour by changing the ratio of the number of electrons to the number of holes.

Thus, they understood that the presence of iodine, and in its absence oxygen, facilitated oxidation of the polymer (McNeill et al. 1963). Of course, this oxidative process (Figure 1.8) is now referred to as p-doping of the polymer and was ultimately determined to be the key in producing highly conductive organic polymers (Chiang et al. 1977b, 1978a; Skotheim and Reynolds 2007). Weiss admitted, however, that the full role of this oxidation in determining conductivity was not realized at the time (Rasmussen 2011, 2015).

1.2.4 PYRROLE BLACK AT THE UNIVERSITY OF PARMA

As Weiss and coworkers were wrapping up their work with the polypyrrole–iodine materials, a new resurgence in the study of Angeli's pyrrole black was occurring at the University of Parma in northern Italy, due to the research of Luigi Chierici (d. 1967), Gian Piero Gardini (d. 2001), and Vittorio Bocchi (Rasmussen 2015). While the specifics of their collaborations are unclear, Chierici appears to have been the guiding force in these efforts, as he was studying pyrrole black as early as 1953, before the others had come to Parma. However, the three did not work together for long, as Chierici died in 1967, leaving Gardini and Bocchi to continue the research.

The majority of their research focused on identifying the intermediates and byproducts formed during the oxidative polymerization of pyrrole via peroxide (Bocchi et al. 1967, 1970; Chierici and Gardini 1966), but the most significant results came via collaboration with Parma's Institute of Physics. These efforts focused on ESR studies of pyrrole black, with the first of these studies utilizing pyrrole blacks produced via Angeli's initial H_2O_2–acetic acid conditions (Dascola et al. 1966). The second study, however, utilized a polymeric material obtained via electrolysis (Dall'Olio et al. 1968), thus representing the first example of an electropolymerized polypyrrole. These polymeric samples were obtained by applying a constant current of 100 mA to a Pt electrode in a H_2SO_4 solution of pyrrole. This resulted in the production of a laminar film on the electrode over a period of 2 h, which was then rinsed with distilled water and dried under vacuum (Dall'Olio et al. 1968). X-ray analysis of the film indicated an essentially amorphous material, and conductivity measurements gave a value of 7.54 Ω^{-1} cm^{-1}, considerably higher than that reported by Weiss and coworkers for the thermally produced polypyrrole–iodine materials (Bolto et al. 1963).

Chierici and Bocchi went on to analyze the composition of the electropolymerized material via oxidative degradation (Chierici et al. 1968). As for the traditional pyrrole blacks produced via H_2O_2 oxidation, the major degradation product was pyrrole-2,5-dicarboxylic acid, although some additional products of unknown composition were also detected in the electropolymerized materials. These results led to the conclusion that all of the pyrrole blacks studied consisted of chains of α,α'-linked pyrroles (Figure 1.1) (Chierici et al. 1968). In 1975, Gardini then started the first of several stays as a visiting scientist at the IBM Research Laboratory in San Jose, California (Rasmussen 2015), where he began working with Arthur Diaz.

1.2.5 DIAZ AND ELECTROPOLYMERIZED POLYPYRROLE FILMS

Upon arriving at IBM in the mid-1970s, Arthur F. Diaz (b. 1938) was given the task of developing a new project of significant impact, preferably with a focus in electrochemistry, as this was a subject in which IBM was interested in building capabilities (Rasmussen 2015). This ultimately led to work in modified electrodes, and as conducting polymers were a hot topic at the time, he considered the use of such materials but was unsure how to successfully modify electrodes with polyacetylene. Gardini, who was currently visiting IBM, provided the solution to this problem when he mentioned to Diaz about the pyrrole black work being done at Parma, particularly the most recent success in electropolymerization (Rasmussen 2015).

The combination of the material's intractability and conductivity was attractive to Diaz, and he thus began investigating electropolymerized polypyrrole films. In time, he was able to perform the electropolymerization under controlled conditions, allowing the repeatable generation of strongly adhered films onto electrode surfaces (Diaz et al. 1979). It was found that the use of deoxygenated aprotic solvents resulted in better material properties (Diaz et al. 1979; Kanazawa et al. 1980) than the aqueous conditions previously utilized at Parma (Dall'Olio et al. 1968). Under optimum conditions, polypyrrole films were produced galvanostatically on Pt from pyrrole in a 99:1 CH_3CN–H_2O mixture with Et_4NBF_4 as a supporting electrolyte (Diaz et al. 1979; Kanazawa et al. 1980). The water content of the solution was found to affect the film adherence to the substrate, with the absence of water resulting in poorly adhering, nonuniform films. In contrast, increased water content improved film adherence (Kanazawa et al. 1980).

Elemental analysis of the films was consistent with a composition comprising mainly coupled pyrrole units, as well as BF_4^- anions, in a ~4:1 ratio (Diaz et al. 1979; Kanazawa et al. 1980). It was concluded that the polypyrrole backbone carried a partial positive charge balanced by the BF_4^- ions (Kanazawa et al. 1979; Diaz et al. 1981), and the pyrrole-linked structure was confirmed by Raman and reflective infrared (IR) analysis. Lastly, electron diffraction indicated the films to be not very crystalline, showing only diffuse rings corresponding to a 3.4 Å lattice spacing (Kanazawa et al. 1979).

Thicker free-standing polypyrrole films (5–50 μm) were evaluated via four-point probe to give room temperature conductivities of 10–100 Ω^{-1} cm^{-1}, much higher than the previous Parma results (Diaz et al. 1979, 1981; Kanazawa et al. 1979, 1980). This improvement in conductivity was thought to be at least partially due to higher quality films resulting from slow growth and their very thin nature (Diaz et al. 1981). This is consistent with the modern understanding that the structural order of the electropolymerized film decreases with the corresponding film thickness, which also results in a decrease in film conductivity. As with the materials of Wiess and coworkers (Bolto et al. 1963), the electropolymerized films were characterized via temperature-dependent conductivity measurements to reveal a temperature profile consistent with a classical semiconductor (Diaz et al. 1981; Kanazawa et al. 1979).

1.3 POLYANILINE

1.3.1 EARLY REPORTS OF THE OXIDATION OF ANILINE

Although the conductivity of polyaniline was not studied until after that of polypyrrole, aniline materials are the oldest example of conjugated polymers and have an extensive history with observations of colored precipitates resulting from the oxidation of aniline dating to the early 1800s (Inzelt 2008). The earliest such observations were reported by Friedlieb F. Runge (1794–1867) (Figure 1.9) in 1834, who treated aniline nitrate with copper oxide in hydrochloric acid to produce a dark green–black color (Runge 1834). He then went on to show that the treatment of either aniline nitrate or aniline hydrochloride with a variety of copper salts resulted in the same reaction and noted that if enough of the aniline salts could be prepared, this reaction could provide a practical use. Later, in 1840, Carl Julius Fritzsche (1808–1871) showed that the treatment of aniline salts with chromic acid gave similar results (Fritzsche 1840).

This oxidative process was then applied to the production of a black dye by John Lightfoot (1831–1872) in 1859, who applied aniline hydrochloride to cotton in the presence of potassium chlorate and a copper salt (Travis 1994, 1995). The resulting dye was developed by Lightfoot in the early 1860s (Travis 1995) and became known as aniline black, a term that was later used to refer to polyanilines in general. During this same time period, a second such species was produced by Heinrich Caro (1834–1910) in 1860 (Travis 1991, 1994). This second aniline black was a byproduct from the production of aniline purple (mauve), which was prepared by aniline oxidation with copper salts. After alcoholic extraction of the desired purple dye, a black residue remained that could be printed with wooden hand blocks. Caro's aniline black was then commercialized by Roberts, Dale & Co. for sale to printers in 1862 (Travis 1994).

By far the most commonly referenced early report of aniline oxidation is the work of Henry Letheby (1816–1876), who is often incorrectly credited with the first production of polyaniline. In 1862, Letheby investigated the treatment of acidic solutions of aniline with various oxidizing agents to produce blue to purple colors (Letheby 1862). Continuing his studies, he then electrochemically oxidized a sulfuric acid solution of aniline via a Pt electrode at the positive pole of a battery to generate a deep blue to bluish-green pigment that adhered to the

Figure 1.9 Friedlieb F. Runge (1794–1867). (From the Edgar Fahs Smith Collection, University of Pennsylvania libraries, Philadelphia. With permission.)

electrode as a fine powder. This could be removed from the electrode, washed with water, and dried to give a bluish-black powder that was only soluble in sulfuric acid. Dilution of the resulting acid solution with water then resulted in the precipitation of a dirty emerald green powder, which could be made blue with concentrated ammonia or blue to purple with concentrated sulfuric acid. While this was not the first report of aniline oxidation, it was the first example of its electrochemical oxidation and the earliest report of this method for the production of conjugated materials.

1.3.2 WILLSTÄTTER, GREEN, WOODHEAD, AND THE IDENTIFICATION OF OXIDATION PRODUCTS

It is important to point out that in all the aniline examples above, the identity of the resulting oxidation products was completely unknown, and it was not until the early 1900s that significant efforts to determine the structure and identity of such products were reported. These efforts began with the characterization of the aniline oxidation products by Richard Willstätter (1872–1942), first reported between 1907 and 1911, which concluded that these materials comprised linear octameric species existing in various degrees of oxidation (Willstätter and Dorogi 1909a, 1909b; Willstätter and Moore 1907). In a separate series of studies starting in 1910, Arthur G. Green (1864–1941) and Arthur E. Woodhead came to some of the same conclusions but found the conclusions of Willstätter and coworkers to be oversimplified. In the process, they reinterpreted the previously reported results and provided additional data to produce a more detailed and complete structural model of aniline materials (Elschner et al. 2011; Green and Wolff 1911; Green and Woodhead 1910, 1912a, 1912b).

As with Willstätter, they viewed the initial oxidation products as various octameric species. From their work, these species included the fully reduced, colorless base leucoemeraldine, along with the sequential oxidized analogues protoemeraldine (violet base, forming yellowish-green salts), emeraldine (violet-blue base, forming green salts), nigraniline (dark blue base, forming blue salts), and pernigraniline (purple base, forming purple salts), as shown in Figure 1.10. Unlike Willstätter and coworkers, however, Green and Woodhead felt that these species did not represent true aniline blacks, but were only intermediates to the final, highly stable aniline black.

1.3.3 BUVET, JOZEFOWICZ, AND CONDUCTING POLYANILINE

The first detailed characterization of the electronic properties of polyaniline was carried out in the laboratory of Rene Buvet (1930–1992) at the Ecole Supérieure de Physique et de Chimie Industrielles de la ville de Paris (ESPCI ParisTech) in the mid-1960s. The majority of this work was carried out by Marcel Jozefowicz (b. 1934), working under Buvet (Rasmussen 2011). Their initial work focused on optimizing reproducible methods for the preparation of polyaniline samples in order to carry out detailed studies of their electronic properties (Constantini et al. 1964). These materials were prepared by the oxidative polymerization of aniline using persulfate in sulfuric acid solutions to afford the emeraldine sulfate. Finally, controlling the level of protonation and the nature of the counterion was investigated, including the neutralization of the initial emeraldine sulfate to produce the emeraldine base, followed by generation of emeraldine salts of either chloride or formate counterions. This was then followed with a report of the redox properties of the resulting polyaniline materials in 1965 (Jozefowicz et al. 1965).

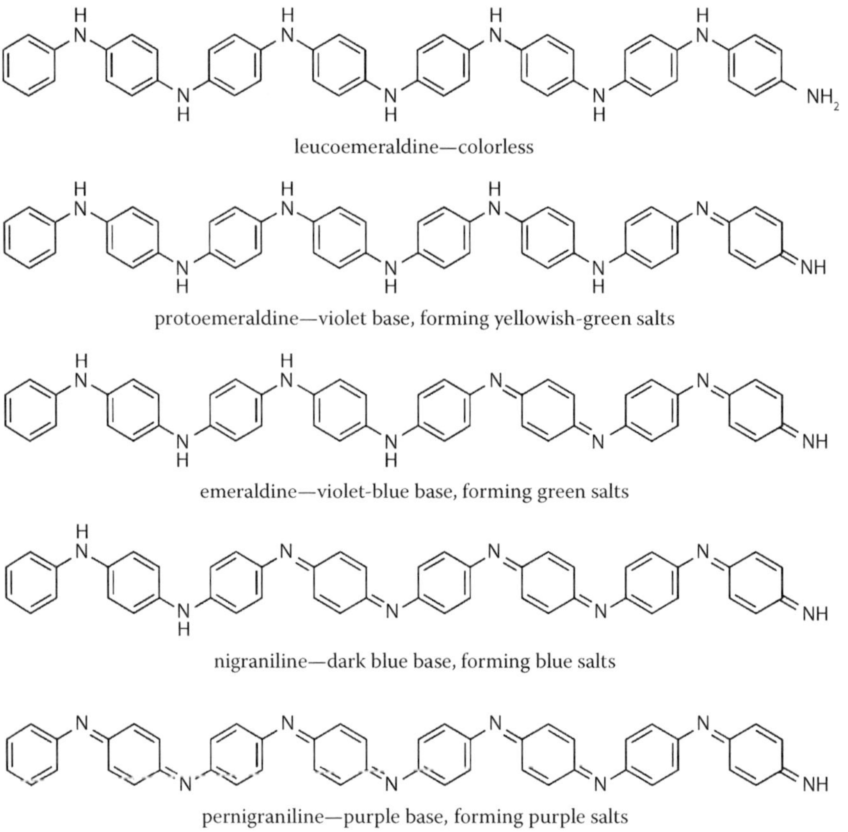

Figure 1.10 Names and structures of aniline oxidation products as reported by Green and Woodhead.

By 1966 (Combarel et al. 1966; Jozefowicz and Yu 1966; Yu and Jozefowicz 1966), they had developed optimized methods for the production of emeraldine sulfates of controlled compositions and began studies of the resulting conductive properties. Pressed pellets of the resulting polyaniline materials were indeed found to be quite conductive, with Jozefowicz stating (Jozefowicz and Yu 1966, p. 1011; translated to English):

> The conductivity of the polyanilines is very high and classifies these compounds among the best known organic conductors. This conductivity is, without possible dispute, electronic.

Continued studies showed that the conductivity of the oxidized polyaniline was dependent on both the extent of protonation and the water content of the resulting sample (De Surville et al. 1968; Jozefowicz and Yu 1966; Yu and Jozefowicz 1966). Of particular interest was the pH dependence of the material, in which it was found that the conductivity increased linearly with decreasing pH to give conductivities that ranged from 10^{-9} to 30 S cm^{-1}. In contrast, the effect of water was not as great, with the conductivity increasing with increasing water content. During a lecture presented in April 1967 at the 18th meeting of Comité International de Thermodynamique et Cinétique Electrochimiques (CITCE), Buvet reviewed the electronic characteristics of polyaniline and concluded that its conductivity was electronic in nature and not due to any ion transport (De Surville et al. 1968; Inzelt 2008).

1.4 POLYACETYLENE

1.4.1 NATTA AND THE POLYMERIZATION OF ACETYLENE

Although polyacetylene is often presented as the first conducting polymer, its history is actually fairly recent in comparison with both polypyrrole and polyaniline. In fact, the production of polyacetylene by the direct polymerization of acetylene was not reported until the mid-1950s, when Giulio Natta (1903–1979) began to apply his previously successful catalytic methods for α-olefin and diolefin polymerizations to the polymerization of acetylenes. These efforts generated an initial Italian patent in 1955 (Natta et al. 1955), followed by a 1958 publication detailing the successful catalytic polymerization of acetylene via triethylaluminum (Et_3Al)–titanium alkoxide combinations (Natta et al. 1958). The optimized conditions utilized Et_3Al and titanium(IV) propoxide at 75°C, with a catalyst molar ratio (Al:Ti) of 2.5 (Figure 1.11). These conditions resulted in 98.5% conversion of monomeric acetylene to produce a dark, crystalline polymer.

The products were completely insoluble in organic solvents, but powder samples were characterized by X-ray diffraction, which were found to have a crystalline content of ~90%–95%. These data were consistent with linear chains of polyacetylene in which the double-bond configuration was thought to be predominantly *trans* (Natta et al. 1958). The combination of the black color, metallic luster, and relatively low electrical resistivity (~10^{10} Ω cm, compared with 10^{15}–10^{18} Ω cm for typical polyhydrocarbons) led to the conclusion that the polyacetylene product was structurally identical to a very long conjugated polyene, although Shirakawa later stated that this conclusion was not accepted widely at the time (Shirakawa 2001a, 2002).

Although the samples exhibited poor solubility, Natta's polyacetylene was found to be fairly reactive, particularly with oxidants such as O_2 and Cl_2 (Natta et al. 1958). Reaction with chlorine resulted in the production of a white solid that was found to be amorphous by X-ray characterization. It was found that heating the white product at 70°C–80°C resulted in a rapid loss of HCl accompanied by a darkening of the polymer. Alternately, treatment with potassium metal in hot ethanol resulted in removal of the majority of the chlorine to give a black amorphous solid.

Although Natta stated that his 1958 paper represented only an initial communication and that additional publications were planned (Natta et al. 1958), no additional studies on polyacetylenes were ever published. Other groups, however, did not hesitate to continue the work Natta began. As such, the modern term *polyacetylene* gradually replaced the term *polyene* as more studies began to utilize Natta's methods (Shirakawa 2001a, 2002).

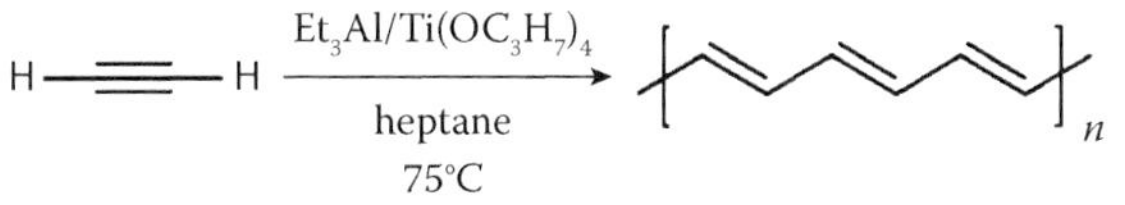

Figure 1.11 Catalytic polymerization of acetylene.

Conductive polymers

1.4.2 SHIRAKAWA AND POLYACETYLENE FILMS

One such research group that continued Natta's efforts was that of Sakuji Ikeda at the Tokyo Institute of Technology. Starting in the mid-1960s, his group began studying the mechanism of acetylene polymerization using Ziegler–Natta catalysts, as well as developing new polymerization catalysts (Rasmussen 2014). As a part of these investigations, it was found that benzene was produced as a byproduct of the polymerization process, and that the ratio of benzene to polymer varied with the catalyst used. These mechanistic investigations were then continued by a new research associate, Hideki Shirakawa (b. 1936) (Figure 1.12), who joined Ikeda's group in April of 1966 (Shirakawa 2001a, 2001b, 2002).

Figure 1.12 Hideki Shirakawa (1936–). (Reproduced from Hall, N., *Chem. Commun.*, 1–4, 2003. With permission of the Royal Society of Chemistry.)

In the fall of 1967 (Hall 2003; Shirakawa 2001a, 2001b, 2002), a visiting Korean coworker, Hyung Chick Pyun, was assisting Shirakawa to produce polyacetylene using conditions nearly identical to those previously reported by Natta (Shirakawa and Ikeda 1971). However, instead of generating the typical powder samples as expected, these efforts produced ragged pieces of a polymer film (Shirakawa 2001b). Upon reviewing the experimental conditions used by Pyun, Shirakawa found that the catalyst concentration used had been 1000 times higher than intended (Hall 2003; Shirakawa 2001a, 2001b, 2002). Shirakawa posed the following as explanations for the mistake (Shirakawa 2001b, p. 214):

> I might have missed the "m" for "mmol" in my experimental instructions, or the visitor might have misread it.

In contrast, MacDiarmid (2001b, p. 187) gives a quite different account, stating,

> he [Shirakawa] replied that this occurred because of a misunderstanding between the Japanese language and that of a foreign student who had just joined his group.

However, it has been pointed out that Pyun spoke fluent Japanese, which casts doubt on MacDiarmid's statement (Hargittai 2010). Regardless of reason, the high catalyst content accelerated the rate of polymerization such that acetylene polymerization occurred at the air–solvent interface, rather than in solution, as was typical under normal conditions (Shirakawa 2001b; Shirakawa and Ikeda 1971). Using these new conditions, Shirakawa was now able to reproducibly produce silvery plastic films of polyacetylene via polymerization at the surface of unstirred, concentrated catalyst solutions (Ito et al. 1974, 1975; Shirakawa and Ikeda 1971; Shirakawa et al. 1973, 1978).

The resulting polyacetylene films exhibited strongly temperature-dependent backbone configurations (Figure 1.13) due to an irreversible isomerization of the *cis* to *trans* forms above 145°C. Polyacetylenes with all-*cis* structures were copper colored and gave conductivities of 10^{-9} to 10^{-8} S cm^{-1}, while all-*trans* samples were silver in color and exhibited higher conductivities (10^{-5} to 10^{-4} S cm^{-1}) (Shirakawa et al. 1978).

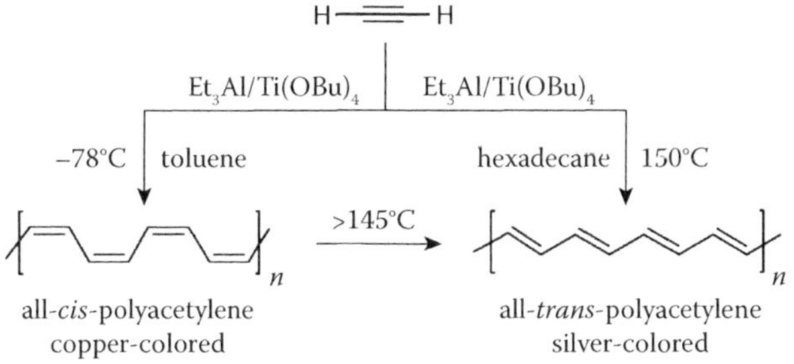

Figure 1.13 Temperature dependence of polymerization.

Surprisingly, the values of the latter samples are essentially the same as those previously reported for highly crystalline polyacetylene powders. As it had been previously shown that conductivity increased with crystallinity, one could expect increased order in the films with a corresponding rise in conductivity, but this was not the case (Shirakawa 2001a). X-ray diffraction of the polyacetylene films (Ito et al. 1974) gave data nearly identical to the previous studies of Natta (Natta et al. 1958).

1.4.3 SMITH AND DOPED POLYACETYLENE

About the same time that the first polyacetylene films were produced, effects of additives on the conductivity of polyacetylene powders were being studied by Dorian S. Smith (1933–2010) and Donald J. Berets (d. 2003) at the American Cyanamid Company (Berets and Smith 1968). Initially, their efforts focused on the effect of oxygen impurities on the conductivity of polyacetylene pressed pellets, finding that samples with lower oxygen content gave lower resistivity. Along the way, however, an interesting phenomenon was observed (Berets and Smith 1968, pp. 825–826):

> On admission of 150 mm pressure of oxygen to the measuring apparatus (normally evacuated or under a few cm pressure of He gas), the resistivity of polyacetylene decreased by a factor of 10. If the oxygen was pumped off within a few minutes and evacuation continued at 10^{-4} mm pressure for several hours, the original electrical properties of the specimen were restored.

They went on to conclude that the polymer first adsorbed oxygen in a reversible manner, causing a reduction in resistivity. Ultimately, however, the oxygen causes an irreversible chemical reaction resulting in the typically observed increase in resistivity.

An investigation of the effects of various gases on the conductivity was then performed to find that the addition of various electron acceptors (BF_3, BCl_3, Cl_2, SO_2, NO_2, O_2, etc.) all resulted in decreases in resistivity, although oxidizing gases ultimately resulted in chemical reaction with the polymer. Electron donors (NH_3, CH_3NH_2, H_2S, etc.), however, had the opposite effect. The most dramatic results were obtained using BF_3, giving an increase in conductivity of three orders of magnitude (to ~0.0013 S cm^{-1}). These trends were then explained as follows (Berets and Smith 1968, p. 827):

> The effect on conductivity of the adsorbed electron-donating and electron-accepting gases is consistent with the p-type nature .... If holes are the dominant carriers, electron donation would be expected to compensate them and reduce conductivity; electron acceptors would be expected to increase the concentration of holes and increase conductivity; this is observed.

Although the effect of the gaseous additions was not completely understood, they quite clearly state (Berets and Smith 1968) that the electrical conductivity depended on the extent of oxidation of the samples. The results of this study did not seem to generate much interest, however, and the authors never followed this initial report with any additional studies.

1.4.4 MacDIARMID, HEEGER, AND POLY(SULFUR NITRIDE)

A few years after the Smith and Berets report, Alan G. MacDiarmid (1927–2007) and Alan J. Heeger (b. 1936) (Figure 1.14) began a related study with the addition of gaseous Br_2 to the inorganic polymer poly(sulfur nitride) at the University of Pennsylvania (Chiang et al. 1976, 1977a; MacDiarmid et al. 1975; Mikulski et al. 1975). This collaboration between the Penn colleagues began in 1975, after Heeger had become intrigued by reports of the metallic nature of poly(sulfur nitride), $(SN)_x$ (Hall 2003; Heeger 2001; MacDiarmid 2001b). Heeger then approached MacDiarmid about working together to study this new polymer after learning that MacDiarmid had some experience with sulfur nitride chemistry.

In order to obtain a reliable sample of the material, they first developed a method to prepare the polymer via the solid-state polymerization of S_2N_2. This gave a lustrous golden material and represented the first reproducible preparation of analytically pure $(SN)_x$ (MacDiarmid et al. 1975; Mikulski et al. 1975). With the polymer in hand, they then reported its electronic properties the following year, giving conductivities of $1.2–3.7 \times 10^3\ \Omega^{-1}\ cm^{-1}$ (Chiang et al. 1976). Finally, following up on previous reports that $(SN)_x$ reacted with halides, they treated the material with Br_2 vapor to produce the derivative $(SNBr_y)_x$. In comparison with pristine $(SN)_x$, this derivative exhibited a 10-fold increase in conductivity (Hall 2003; Heeger 2001; MacDiarmid 2001b).

1.4.5 DOPED POLYACETYLENE FILMS

Not long after starting the collaboration with Heeger (2001), MacDiarmid spent time as a visiting professor at Kyoto University (MacDiarmid 2001b). During his time in Japan, he was invited to speak at the Tokyo Institute of Technology where he met Shirakawa at tea after his lecture (Hall 2003; MacDiarmid 2001b). After MacDiarmid showed him

(a) (b)

Figure 1.14 (a) Alan G. MacDiarmid (1927–2007) and (b) Alan J. Heeger (1936–). (Reproduced from Hall, N., *Chem. Commun.*, 1–4, 2003. With permission of the Royal Society of Chemistry.)

a sample of his golden $(SN)_x$ film, Shirakawa mentioned that he had a similar material and retrieved a sample of his silver polyacetylene film to show MacDiarmid (Rasmussen 2014). MacDiarmid was quite interested in Shirakawa's result, and after returning to the states, he arranged for funding to bring Shirakawa to Penn to work on polyacetylene (MacDiarmid 2001b). As a result, Shirakawa began working with MacDiarmid and Heeger as a visiting scientist in September 1976 (Shirakawa 2001b).

Upon arriving at Penn, the focus of Shirakawa and MacDiarmid was to improve the purity of the polyacetylene films in an effort to increase its conductivity (MacDiarmid 2001b). As discussed earlier, Smith and Berets had shown that decreased oxygen content increased conductivity, and thus limiting other impurities could potentially further increase the polymer conductivity. As a result, they were ultimately able to make films with purities as high as ca. 98.6% but found that conductivity actually decreased with increasing film purity (MacDiarmid 2001b; Rasmussen 2014). Based on the observed trends between purity and conductivity, it was proposed that perhaps the film impurities were acting as dopants, which increased the polyacetylene conductivity (MacDiarmid 2001b) in a manner similar to that previously seen in the addition of Br_2 to $(SN)_x$ (Chiang et al. 1977a). Previous *in situ* IR measurements by Shirakawa and Ikeda also supported this reasoning, in which a dramatic decrease in IR transmission was observed during the treatment of polyacetylene films with halide vapors (Shirakawa 2001a). This decrease in transmission suggested that the initial halogen-treated material might have unusual electronic properties, and therefore it was decided to determine the conductivity of the films upon Br_2 addition.

On November 23, 1976 (Shirakawa 2001a), Shirakawa and Dr. Chwan K. Chiang, a postdoctoral fellow of Heeger, measured the conductivity of a *trans*-polyacetylene film by four-point probe while being exposed to Br_2 vapor (Hall 2003; Rasmussen 2014; Shirakawa et al. 1977). Upon the addition of 1 Torr of Br_2, the film's conductivity increased rapidly, rising from 10^{-5} to 0.5 S cm^{-1} within only 10 min. This experiment was then repeated using iodine in place of bromine, resulting in even greater increases in conductivity (up to 38 S cm^{-1}) (Shirakawa et al. 1977). Optimization of this iodine treatment later that same year showed that conductivities up to 160 S cm^{-1} could be obtained, although use of AsF_5 as the oxidizing agent was found to produce even higher conductivities (Chiang et al. 1977b, 1978a). Treatment of *trans*-polyacetylene films with AsF_5 produced conductivities of 220 S cm^{-1}, while treatment of *cis*-polyacetylene gave even higher values (560 S cm^{-1}). The magnitude of the values for the AsF_5-doped *cis*-polyacetylene films then led them to revisit the iodine treatments, and they then treated *cis*-polyacetylene with iodine in 1978 to produce conductivity values above 500 S cm^{-1} (Chiang et al. 1978c). In that same year, it was also demonstrated that doping of polyacetylene with electron-donating species such as sodium resulted in conductivities of 8 S cm^{-1} (Chiang et al. 1978a). In a final 1978 paper, Heeger and MacDiarmid reported values as high as 200 S cm^{-1} for polyacetylene films doped with electron donors (Chiang et al. 1978b).

1.5 COMPARISONS AND THE GROWTH OF THE FIELD OF CONDUCTIVE POLYMERS

In terms of impact, the polyacetylene work of Heeger, MacDiarmid, and Shirakawa was the first demonstration of an organic polymer that exhibited conductivities in the metallic range, and it is quite clear that these dramatic results sparked the significant growth in the study of conjugated and conducting polymers that followed. However, considering that some of the previously reported results were also quite impressive for the time, it might be asked what factors of the polyacetylene studies contributed to such widespread interest in comparison with the previous reports of polypyrrole and polyaniline (Rasmussen 2011). For example, the conductivity of polyaniline was quite similar to the initially reported polyacetylene values (30 vs. 38 S cm^{-1}), yet it has been said that reports of the electronic nature of polyaniline's conductivity did not give rise to great excitement at the time (Inzelt 2008).

One important factor may have been in how the published results were disseminated. For example, the polypyrrole papers of Weiss and coworkers were published only in Australian journals, which Weiss believed he had a strong duty to support. In hindsight, however, Weiss believed that this contributed to the fact that the work received little attention (Rasmussen 2011). In a similar manner, the majority of the polyaniline work of Buvet and Jozefowicz was limited to the French literature. In comparison, the polyacetylene papers were broadly published in various high-profile, international journals, and thus more widely read by the scientific community.

Another factor that could have played an equally important role was the physical form of the materials investigated. With the exception of the electropolymerized polypyrrole films produced at Parma, all the previous studies were the study of dark powders that had to be pressed into pellets to investigate the resulting electronic properties. In comparison, the conductive organic polyacetylene samples of Shirakawa were plastic, free-standing films. In addition, these samples were not just simple plastic films, but *silvery, metallic-looking films*, which easily captivated spectators, just as MacDiarmid was initially captivated when Shirakawa first showed a sample to him.

Ultimately, the field of conjugated organic materials owes much to Heeger, MacDiarmid, and Shirakawa, whose early work initiated the rise of a niche area of scientific interest into the wide community of organic electronics today. As such, it was a joyous moment in the fall of 2000 when the field received the news that they were being justly recognized by the Nobel committee for their contributions. Of course, it is just as important for all working in this field to understand how far back this history really stretches and all the early important contributions that helped shape its origins.

REFERENCES

Angeli, A. Sopra il nero del pirrolo. Nota preliminare. *Atti della Accademia Nazionale dei Lincei, Classe di Scienze Fisiche, Matematiche e Naturali, Rendiconti* 24 (1915): 3–6.

Angeli, A. Sopra i neri di pirrolo. Nota. *Gazzetta Chimica Italiana* 48(II) (1918): 21–25.

Angeli, A., and L. Alessandri. Sopra il nero pirrolo II. Nota. *Gazzetta Chimica Italiana* 46(II) (1916): 283–300.

Angeli, A., and G. Cusmano. Sopra i neri di nitropirrolo. Nota. *Atti della Accademia Nazionale dei Lincei, Classe di Scienze Fisiche, Matematiche e Naturali, Rendiconti* 26(I) (1917): 273–278.

Angeli, A., and C. Lutri. Nuove ricerche sopra i neri di pirrolo. Nota. *Atti della Accademia Nazionale dei Lincei, Classe di Scienze Fisiche, Matematiche e Naturali, Rendiconti* 29(I) (1920): 14–22.

Angeli, A., and A. Pieroni. Sopra un nuovo modo di formazione del nero di pirrolo. Nota. *Atti della Accademia Nazionale dei Lincei, Classe di Scienze Fisiche, Matematiche e Naturali, Rendiconti* 27(II) (1918): 300–304.

Berets, D. J., and D. S. Smith. Electrical properties of linear polyacetylene. *Transactions of the Faraday Society* 68 (1968): 823–828.

Bocchi, V., L. Chierici, and G. P. Gardini. Structure of the oxidation product of pyrrole. *Tetrahedron* 23 (1967): 737–740.

Bocchi, V., L. Chierici, G. P. Gardini, and R. Mondelli. Pyrrole oxidation with hydrogen peroxide. *Tetrahedron* 26 (1970): 4073–4082.

Bolto, B. A., R. McNeill, and D. E. Weiss. Electronic conduction in polymers. III. Electronic properties of polypyrrole. *Australian Journal of Chemistry* 16 (1963): 1090–1103.

Bolto, B. A., and D. E. Weiss. Electronic conduction in polymers. II. The electrochemical reduction of polypyrrole at controlled potential. *Australian Journal of Chemistry* 16 (1963): 1076–1089.

Chiang, C. K., M. J. Cohen, A. F. Garito, A. J. Heeger, C. M. Mikulski, and A. G. MacDiarmid. Electrical conductivity of $(SN)_x$. *Solid State Communications* 18 (1976): 1451–1455.

Chiang, C. K., M. J. Cohen, D. L. Peebles, A. J. Heeger, M. Akhtar, J. Kleppinger, A. G. MacDiarmid, J. Milliken, and M. J. Moran. Transport and optical properties of polythiazyl bromides: $(SNBr_{0.4})_x$. *Solid State Communications* 23 (1977a): 607–612.

Chiang, C. K., M. A. Druy, S. C. Gau, A. J. Heeger, E. J. Louis, A. G. MacDiarmid, Y. W. Park, and H. Shirakawa. Synthesis of highly conducting films of derivatives of polyacetylene, $(CH)_x$. *Journal of the American Chemical Society* 100 (1978a): 1013–1015.

Chiang, C. K., C. R. Fincher Jr., Y. W. Park, A. J. Heeger, H. Shirakawa, B. J. Louis, S. C. Gau, and A. G. MacDiarmid. Electrical conductivity in doped polyacetylene. *Physics Review Letters* 39 (1977b): 1098–1101.

Chiang, C. K., S. C. Gau, C. R. Fincher Jr., Y. W. Park, A. G. MacDiarmid, and A. J. Heeger. Polyacetylene, $(CH)_x$: n-type and p-type doping and compensation. *Applied Physics Letters* 33 (1978b): 18–20.

Chiang, C. K., Y. W. Park, A. J. Heeger, H. Shirakawa, E. J. Louis, and A. G. MacDiarmid. Conducting polymers: Halogen doped polyacetylene. *Journal of Chemical Physics* 69 (1978c): 5098–5104.

Chierici, L., G. C. Artusi, and V. Bocchi. Sui neri di ossipirrolo. *Annales de Chimie* 58 (1968): 903–13.

Chierici, L., and G. P. Gardini. Structure of the product of oxidation of pyrrole $C_8H_{10}N_2O$. *Tetrahedron* 22 (1966): 53–56.

Ciusa, R. Sulla scomposizione dello iodolo. *Atti della Accademia Nazionale dei Lincei, Classe di Scienze Fisiche, Matematiche e Naturali, Rendiconti* 30(II) (1921): 468–469.

Ciusa, R. Sulle grafiti da pirrolo e da tiofene (Nota preliminare). *Gazzetta Chimica Italiana* 52(II) (1922): 130–131.

Ciusa, R. Su alcune sostanze analoghe alla grafite. *Gazzetta Chimica Italiana* 55 (1925): 385–389.

Combarel, M. F., G. Belorgey, M. Jozefowicz, L. T. Yu, and R. Buvet. Conductivité en courant continu des polyanilines oligomères: Influence de l'état acide-base sur la conductivité électronique. *Comptes Rendus des Séances de l'Académie des Science, Série C: Sciences Chimiques* 262 (1966): 459–462.

Constantini, P., G. Belorgey, M. Jozefowicz, and R. Buvet. Préparation, propriétés acides et formation de complexes anioniques de polyanilines oligomères. *Comptes Rendus des Séances de l'Académie des Science* 258 (1964): 6421–6424.

Dall'Olio, A., G. Dascola, V. Varacca, and V. Bocchi. Resonance paramagnètique èlectronique et conductiviè d'un noir d'oxypyrrol èlectrolytique. Note. *Comptes Rendus des Séances de l'Académie des Science, Série C: Sciences Chimiques* 267 (1968): 433–435.

Dascola, G., D. C. Giori, V. Varacca, and L. Chierici. Rèsonance paramagnètique èlectronique des radicaux libres crèès lors de la formation des noirs d'oxypyrrol. Note. *Comptes Rendus des Séances de l'Académie des Science, Série C: Sciences Chimiques* 267 (1966): 433–435.

De Surville, R., M. Jozefowicz, L. T. Yu, J. Perichon, and R. Buvet. Electrochemical chains using protolytic organic semiconductors. *Electrochimica Acta* 13 (1968): 1451–1458.

Diaz, A. F., K. K. Kanazawa, and G. P. Gardini. Electrochemical polymerization of pyrrole. *Journal of the Chemical Society, Chemical Communications* (1979): 635–636.

Diaz, A. F., A. Martinez, K. K. Kanazawa, and M. Salmon. Electrochemistry of some substituted pyrroles. *Journal of Electroanalytical Chemistry* 130 (1981): 181–187.

Elschner, A., S. Kirchmeyer, W. Lovenich, U. Merker, and K. Reuter. PEDOT: Principles and applications of an intrinsically conductive polymer. Boca Raton, FL: CRC Press, 2011, 1–20.
Fritzsche, J. Ueber das Anilin, ein neue, Zersetzungsproduct des Indigo. *Journal fuer Praktische Chemie* 20 (1840): 453–459.
Green, A. G., and S. Wolff. Anilinschwarz und seine Zwischenkörper, I. *Berichte der deutschen chemischen Gesellschaft* 44 (1911): 2570–2582.
Green, A. G., and A. E. Woodhead. Aniline-black and allied compounds. Part I. *Journal of the Chemical Society Transactions* 97 (1910): 2388–2403.
Green, A. G., and A. E. Woodhead. Anilinschwarz und seine Zwischenkörper, II. *Berichte der deutschen chemischen Gesellschaft* 45 (1912a): 1955–1959.
Green, A. G., and A. E. Woodhead. Aniline-black and allied compounds. Part II. *Journal of the Chemical Society Transactions* 101 (1912b): 1117–1123.
Gutmann, F., and L. E. Lyons. *Organic Semiconductors.* New York: Wiley, 1967, vii, 448–484.
Hall, N. Twenty-five years of conducting polymers. *Chemical Communications* (2003): 1–4.
Hargittai, I. *Drive and Curiosity: What Fuels the Passion for Science.* Amherst, NY: Prometheus Books, 2010, 173–190.
Heeger, A. J. Semiconducting and metallic polymers: The fourth generation of polymeric materials (Nobel lecture). *Angewandte Chemie International Edition* 40 (2001): 2591–2611.
Inzelt, G. *Conducting Polymers: A New Era in Electrochemistry*, ed. F. Scholz. Monographs in Electrochemistry. Berlin: Springer-Verlag, 2008, 265–269.
Ito, T., H. Shirakawa, and S. Ikeda. Simultaneous polymerization and formation of polyacetylene film on the surface of concentration soluble Ziegler-type catalyst solution. *Journal of Polymer Science: Polymer Chemistry Edition* 12 (1974): 11–20.
Ito, T., H. Shirakawa, and S. Ikeda. Thermal cis-trans isomerization and decomposition of polyacetylene. *Journal of Polymer Science: Polymer Chemistry Edition* 13 (1975): 1943–1950.
Jozefowicz, M., G. Belorgey, L. T. Yu, and R. Buvet. Oxydation et réduction de polyanilines oligomères. *Comptes Rendus des Séances de l'Académie des Science* 260 (1965): 6367–6370.
Jozefowicz, M., and L. T. Yu. Relations entre propriétés chimiques et électrochimiques de semi-conducteurs macromoléculaires. *Revue Générale de l'Électricité*, 75 (1966): 1008–1013.
Kanazawa, K. K., A. F. Diaz, R. H. Geiss, W. D. Gill, J. F. Kwak, J. A. Logan, J. F. Rabolt, and G. B. Street. 'Organic metals': polypyrrole, a stable synthetic 'metallic' polymer. *Journal of the Chemical Society, Chemical Communications* (1979): 854–855.
Kanazawa, K. K., A. F. Diaz, W. D. Gill, P. M. Grant, G. B. Street, G. P. Gardini, and J. F. Kwak. Poly-pyrrole: An electrochemically synthesized conducting organic polymer. *Synthetic Metals* 1 (1980): 329–336.
Letheby, H. On the production of a blue substance by the electrolysis of sulphate of aniline. *Journal of the Chemical Society* 15 (1862): 161–163.
MacDiarmid, A. G. "Synthetic metals": A novel role for organic polymers (Nobel lecture). *Angewandte Chemie International Edition* 40 (2001a): 2581–2590.
MacDiarmid, A. G. Alan G. MacDiarmid. In *Les Prix Nobel: The Nobel Prizes 2000*, ed. T. Frängsmyr. Stockholm: Nobel Foundation, 2001b, 183–190.
MacDiarmid, A. G. Synthetic metals: A novel role for organic polymers. *Synthetic Metals* 125 (2002): 11–22.
MacDiarmid, A. G., and A. J. Epstein. Conducting polymers: Past, present, and future. *Materials Research Society Symposium Proceedings* 328 (1994): 133–144.
MacDiarmid, A. G., C. M. Mikulski, P. J. Russo, M. S. Saran, A. F. Garito, and A. J. Heeger. Synthesis and structure of the polymeric metal, $(SN)_x$, and its precursor, S_2N_2. *Journal of the Chemical Society, Chemical Communications* (1975): 476–477.
McNeill, R., R. Siudak, J. H. Wardlaw, and D. E. Weiss. Electronic conduction in polymers. *Australian Journal of Chemistry* 16 (1963): 1056–1075.
McNeill, R., and D. E. Weiss. A xanthene polymer with semiconducting properties. *Australian Journal of Chemistry* 12 (1959): 643–656.

Mikulski, C. M., P. J. Russo, M. S. Saran, A. G. MacDiarmid, A. F. Garito, and A. J. Heeger. Synthesis and structure of metallic polymeric sulfur nitride, $(SN)_x$, and its precursor, disulfur dinitride, S2N2. *Journal of the American Chemical Society* 97 (1975): 6358–6363.

Natta, G., G. Mazzanti, and P. Corradini. Polimerizzazione stereospecifica dell'acetilene. *Atti della Accademia Nazionale dei Lincei, Classe di Scienze Fisiche, Matematiche e Naturali, Rendiconti* 25 (1958): 3–12.

Natta, G., P. Pino, and G. Mazzanti. Polimeri ad elevato peso molecolore degli idrocarburi acetilenici e procedimento per la loro preparozione. Italian Patent 530,753 (July 15, 1955); *Chemical Abstracts* 52 (1958): 15128b.

Perepichka, I. F., and D. F. Perepichka, eds. *Handbook of Thiophene-Based Materials.* Hoboken, NJ: John Wiley & Sons, 2009.

Rasmussen, S. C. Electrically conducting plastics: Revising the history of conjugated organic polymers. In *100+ Years of Plastics: Leo Baekeland and Beyond*, ed. E. T. Strom and S. C. Rasmussen. ACS Symposium Series 1080. Washington, DC: American Chemical Society, 2011, 147–163.

Rasmussen, S. C. The path to conductive polyacetylene. *Bulletin for the History of Chemistry* 39 (2014): 64–73.

Rasmussen, S. C. Early history of polypyrrole: The first conducting organic polymer. *Bulletin for the History of Chemistry* 40 (2015): 45–55.

Riley, H. L. Amorphous carbon and graphite. *Quarterly Review of the Chemical Society* 1 (1947): 59–72.

Runge, F. F. Ueber einige Producte der Steinkohlen-destillation. *Annalen der Physik und Chemie* 31 (1834): 513–524.

Shirakawa, H. The discovery of polyacetylene film: The dawning of an era of conducting polymers (nobel lecture). *Angewandte Chemie International Edition* 40 (2001a): 2574–2580.

Shirakawa, H. Hideki Shirakawa. In *Les Prix Nobel: The Nobel Prizes 2000*, ed. T. Frängsmyr. Stockholm: Nobel Foundation, 2001b, 213–216.

Shirakawa, H. The discovery of polyacetylene film. The dawning of an era of conducting polymers. *Synthetic Metals* 125 (2002): 3–10.

Shirakawa, H., and S. Ikeda. Infrared spectra of poly(acetylene). *Polymer Journal* 2 (1971): 231–244.

Shirakawa, H., T. Ito, and S. Ikeda. Raman scattering and electronic spectra of poly(acetylene). *Polymer Journal* 4 (1973): 460–462.

Shirakawa, H., T. Ito, and S. Ikeda. Electrical properties of polyacetylene with various cis-trans compositions. *Makromolekulare Chemie* 179 (1978): 1565–1573.

Shirakawa, H., E. J. Louis, A. G. MacDiarmid, C. K. Chiang, and A. J. Heeger. Synthesis of Electrically Conducting Organic Polymers: Halogen Derivatives of Polyacetylene, $(CH)_x$. *Journal of the Chemical Society, Chemical Communications* (1977): 578–580.

Skotheim, T. A., and J. R. Reynolds, eds. *Handbook of Conducting Polymers.* 3rd ed. Boca Raton, FL: CRC Press, 2007.

Travis, A. S. *Heinrich Caro at Roberts, Dale & Co.* Ambix 38 (1991): 113–134.

Travis, A. S. From Manchester to Massachusetts via Mulhouse: The transatlantic voyage of aniline black. *Technology and Culture* 35 (1994): 70–99.

Travis, A. S. Artificial dyes in John Lightfoot's Broad Oak Laboratory. *Ambix* 42 (1995): 10–27.

Weiss, D. E., and B. A. Bolto. Organic polymers that conduct electricity. In *Physics and Chemistry of the Organic Solid State.* Vol. II. New York: Interscience Publishers, 1965, 67–120.

Willstätter, R., and S. Dorogi. Über Anilinschwarz. II. *Berichte der deutschen chemischen Gesellschaft* 42 (1909a): 2147–2168.

Willstätter, R., and S. Dorogi. Über Anilinschwarz. III. *Berichte der deutschen chemischen Gesellschaft* 42 (1909b): 4118–4151.

Willstätter, R., and C. W. Moore. Über Anilinschwarz. I. *Berichte der deutschen chemischen Gesellschaft* 40 (1907): 2665–2689.

Yu, L. T., and M. Jozefowicz. Conductivité et constitution chimique pe semi-conducteurs macromoléculaires. *Revue Générale de l'Électricité* 75 (1966): 1014–1018.

2 Synthesis of biomedically relevant conducting polymers

Simon E. Moulton
Swinburne University of Technology
Hawthorn, Victoria, Australia

Darren Svirskis
University of Auckland
Auckland, New Zealand

Contents

2.1 INTRODUCTION

This chapter describes the synthesis of conducting polymers (CPs) and outlines factors of synthesis that have a significant impact on the properties of the formed CPs. The vast majority of publications detailing research undertaken in this area of biomedical uses of CPs focus on the CPs, such as polypyrrole (PPy) and poly(3,4-ethylenedioxythiophene) (PEDOT). Therefore, in order to cover such an extensive area of literature, the factors that influence CP growth are highlighted using PPy as an example. Where appropriate, polyaniline (PANi) and polythiophene are discussed to highlight variations between the three most common CPs. In addition, CPs that form an interface with biological systems are typically synthesized using electrochemical techniques; therefore, this chapter only focuses on the electrochemical polymerization approaches. Detailed information regarding chemical polymerization can be found for pyrrole (Vernitskaya and Efimov 1997; Malinauskas 2001;

Stejskal 2001; Wang et al. 2001; Cosnier 2005; Jang and Springer 2006; Pomogailo 2006), aniline (Ameen et al. 2010; Boeva et al. 2010; Stejskal et al. 2010), and thiophene (Bhattacharyya et al. 2012; Higashihara and Ueda 2013; Khalid et al. 2014) in the listed references. This chapter also describes methods used to impart micro- and nanostructures to CPs and associated effects on cellular interactions and responses.

2.2 POLYMERIZATION PROCESS

2.2.1 POLYPYRROLE

Among other CPs, PPy and its derivatives are of particular interest, owing to their excellent biocompatibility, high conductivity, stability in the oxidized state, and interesting redox properties. The simplicity of the synthetic procedures and availability of the initial monomers are also attractive features of PPy (Wallace et al. 2003). Electrochemical polymerization occurs without the use of a chemical oxidant, as it is the application of a suitable voltage or current that drives oxidation of the appropriate monomer (Ren and Pickup 1993) that undergoes radical–radical coupling to form polymer chains, with the process simplified in Figure 2.1.

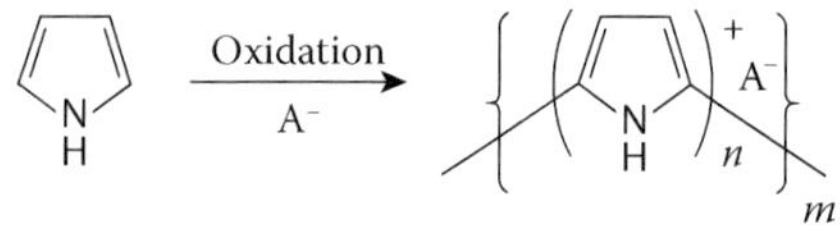

Figure 2.1 Schematic of pyrrole oxidation indicating the requirement of a dopant (A^-) in the polymerization process.

This simplistic reaction belies the complexity of the polymerization process, which can be broken down into multiple steps, with Figure 2.2 showing the complex mechanism of pyrrole oxidation to form PPy. Polymerization is believed to proceed via a radical–radical coupling mechanism (Andrieux et al. 1991), wherein the natural repulsion of the radicals is assumed to be negated by the solvent, the counterion (A^-), and even the monomer. Chain growth then continues until the charge on the chain is such that a counterion is incorporated. Eventually, as the polymer chain exceeds a critical length (molecular weight), the solubility limit is exceeded and polymer deposits onto the electrode surface. While polymer formation occurs predominantly on the anode, some polymerization does occur in solution (Beck and Oberst 1989), with the amount depending on the polymerization conditions.

Dopant counterions (A^-) are incorporated during synthesis to balance the positive charge formed on the polymer backbone during growth. A wide range of dopants can be incorporated using this approach, such as biological molecules (Thompson et al. 2006, 2010b) and even drugs (Moulton et al. 2008; Stevenson et al. 2010; Esrafilzadeh et al. 2013). When polymerization is performed using an electrochemical approach, it is necessary to incorporate a dopant counterion into the monomer solution, such as *para*-toluenesulfonate sodium salt ($C_7H_8S_2O_8Na$) (Quigley et al. 2009) or dodecyl benzene sulfonate sodium salt ($C_{18}H_{30}S_2O_8Na$) (Baldissera et al. 2015; Mane et al. 2015). The ability to incorporate molecules of choice into the monomer solution provides the ability to use functional molecules, such as drugs and biological molecules (provided they possess the necessary negative charge) to dope the polymer.

Electrochemical polymerization typically occurs in a three-electrode cell (Figure 2.3) comprising a reference electrode (i.e., Ag/AgCl), counterelectrode (i.e., platinum mesh),

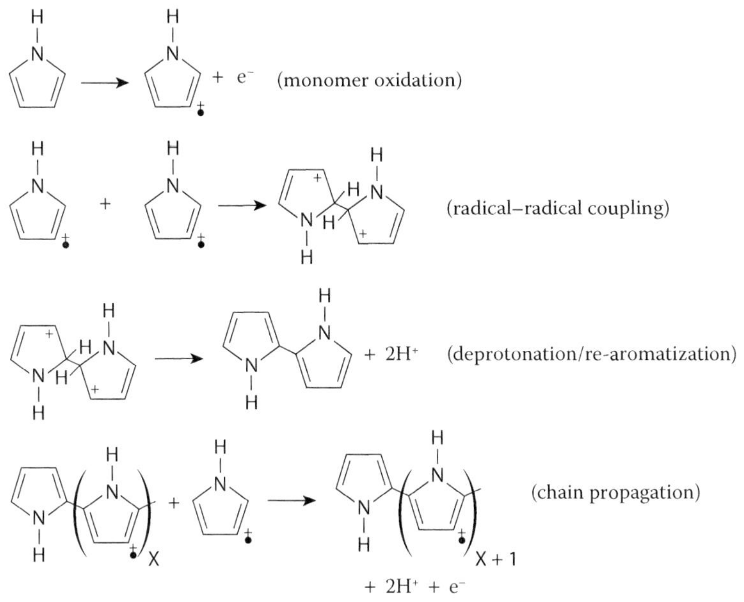

Figure 2.2 Detailed reaction mechanism for polymerization of pyrrole into PPy showing the intermediate reaction process.

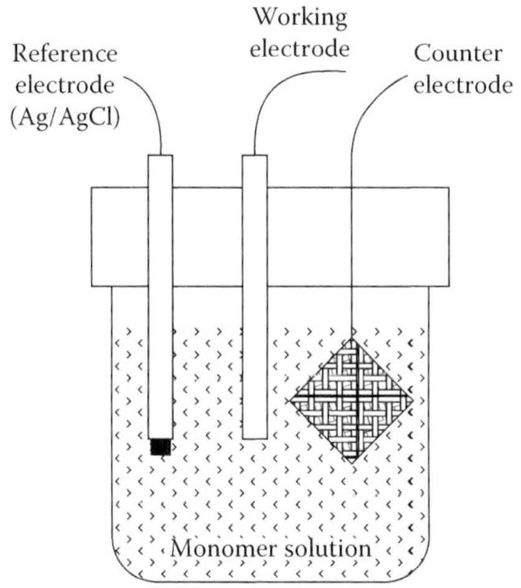

Figure 2.3 Three-electrode cell setup showing the working, counter, and reference electrodes used in electrochemical polymerization of CPs.

and working electrode (anode) (i.e., gold Mylar [Higgins et al. 2011; Hwang et al. 2011] and platinum [Adeloju and Hussain 2016; Jafari et al. 2016; Yu et al. 2016]). The electrodes are then connected to a power source that can facilitate polymerization through either potentiometric (voltage) or galvanostatic (current) techniques. The positioning of the electrodes shown in Figure 2.3 is critical in that it determines the electrical field generated, which can influence the quality and evenness of the polymer deposited. With respect to cell design, the working electrode geometry, the working counterelectrode separation, and the nature of the working electrode all influence the nature of the polymer formed. Given the steps involved in the polymerization process, the hydrodynamics and temperature control of the cell design are important. Not only does the temperature influence the rates of transport in the cell, but it is also important in determining the extent to which unwanted side reactions occur (Wallace et al. 2003).

2.3 ELECTRODE MATERIAL

For all CPs, the working electrode substrate should be considered an integral part of the polymerization process, particularly in the initial stage of growth. The nature of the working electrode influences the ease of the monomer oxidation, with deposition of the polymer being dependent on the surface energy of the electrode and its hydrophobic or hydrophilic nature (Gvozdenovic et al. 2014). The major limitations of electrochemical synthesis of electroconductive polymers on numerous metals and alloys are the relatively high potentials required for monomer oxidation (ranging between +0.6 and +1.2 V vs. Ag/AgCl reference). At these potentials, most metals either dissolve (e.g., iron, steel, and copper) or form low or even nonconductive passive layers (aluminum and its alloys) (Popovic and Grgur 2004; Biallozor and Kupniewska 2005; Grgur et al. 2006; Gvozdenovic and Grgur 2009). It has been shown, however, that if the rate of polymerization is faster than the rate of oxide formation, it is possible to deposit CPs on metals such as aluminum (Tallman et al. 2002). However, when more inert materials such as platinum (El-Rahman et al. 2000; Yalcinkaya et al. 2010), gold (Xue et al. 2005; Kannan et al. 2012; Komarova et al. 2015), or carbon (Jo et al. 2012; Xu et al. 2015) are used as the working electrode, synthesis of CPs occurs easily.

2.4 ELECTROLYTE

The composition of electrolytes used for electrochemical synthesis of CPs involves, besides the selected monomer, solvent and acid, which serve as sources of dopants ion and in some cases may contain some additional compounds (Innis et al. 1998; Thompson et al. 2010b, 2011). Apart from being capable of dissolving the monomer, the solvent has to be as pure as possible and stable at potentials of interest for the polymerization. For example, the presence of dissolved oxygen may be problematic due to reaction with radical intermediates, and it can also be reduced at the counterelectrode and form hydroxide (Wallace et al. 2003). Interaction of the solvent, monomers, and electrode materials cannot be neglected even before the polymerization is initiated, since they will all have an impact on monomer adsorption onto the electrode (Gvozdenovic et al. 2014). Once the polymerization is started, the properties of solvent will have an influence on the solubility of the polymer. It has also been shown that the concentration of the counterion in the electrolyte has an effect not only on the amount of anion incorporated into the polymer but also on its structure and morphology (Demoustier-Champagne and Stavaux 1999; Eftekhari et al. 2006). The ability for the counterion to ion pair with the oligomers (that possess a positive charge) during polymerization will also have an effect on the kinetics of the polymerization, as well as the final polymer film properties. Sulfonated molecules are preferential counterions for PPy growth that generally result in CPs with improved conductivities compared with those grown using other molecules, such as phosphates (Warren and Anderson 1987). This improved conductivity arises due to sulfonates inducing a degree of crystallinity to the CP (Wallace et al. 2003). However, it is proposed that an increase in the number of sulfonate groups in a doped molecule causes localization of positive charge carriers in PPy films, resulting in an observed decrease of the film conductivity (Kuwabata et al. 1987).

In the case of PANi, the electropolymerization is almost always conducted in strong aqueous acids (e.g., sulfuric and hydrochloric acid) at a low pH (<3.0) (Syed and Dinesan 1991; Daprano et al. 1995); in this situation, the sulfur and chloride from the acids dope the PANi. This low pH is required to solubilize the monomer, as well as to generate the PANi/HA (HA = acid) emeraldine salt as the only conducting form of PANi (Wallace et al. 2003; Bhadra et al. 2009). While most of the literature reports the use of aqueous acids for PANi formation, the use of organic acids has also been reported (Osaka et al. 1991; Abdiryim et al. 2007). It may also be necessary to employ a cosolvent system comprising aqueous and organic solvents mixed at various ratios to achieve optimal solubility of the dopant and monomer. The cosolvent system is advantageous when the dopant molecule is sparingly soluble in the monomer solvent, or vice versa. Stevenson et al. (2010) employed a cosolvent system to incorporate the aqueous-soluble anti-inflammatory drug dexamethasone phosphate (Dex) with the organic-soluble monomer terthiophene to successfully electropolymerize polyterthiophene doped with dexamethasone (PTTh–Dex). They confirmed the inclusion of the Dex molecule and demonstrated the ability to control the release of the drug from the polymer (Figure 2.4).

Another attractive feature of the cosolvent system is that it provides a means of lowering the oxidation potential onset, which facilitates the use of some metals that would normally start to dissolve or generate an insulating oxide layer at oxidation potentials generated in pure solvent. Hu et al. (1999) demonstrated that the use of a suitable ratio of water to acetonitrile could lower the onset oxidation potential of bithiophene to about 0.6 V (vs. Ag/AgCl) in the aqueous–organic mixed solution. This means that bithiophene can be electropolymerized at very low potential in such solutions, which is usually unachievable in pure organic media.

2.5 ELECTROCHEMICAL TECHNIQUE

Generally, electrochemical techniques used for the synthesis of CPs can be classified as galvanostatic, potentiostatic, or potentiodynamic. The potential at which oxidation occurs depends heavily on the monomer type (i.e., pyrrole, aniline, or thiophene) and also on whether these monomers possess any additional functional group. The solvent–electrolyte composition also plays a significant role in determining the oxidation at which polymerization occurs. The potential required to oxidize the monomer is greater than the oxidation potential of the formed polymer, suggesting that as monomer oxidation continues to occur, there is a possibility of overoxidizing the deposited polymer. When the oxidizing potential greatly exceeds the oxidation potential of the polymer for an extended period of time, irreversible damage is done to the polymer due to overoxidation, resulting in the CP typically exhibiting inferior conductivity and electroactivity (Krische and Zagorska 1989).

The morphology of the resulting film depends in particular on the nature of the supporting electrolyte, the crystallographic structure of the underlying working electrode, the kinetics of the process (related to the electrode material), the potential used

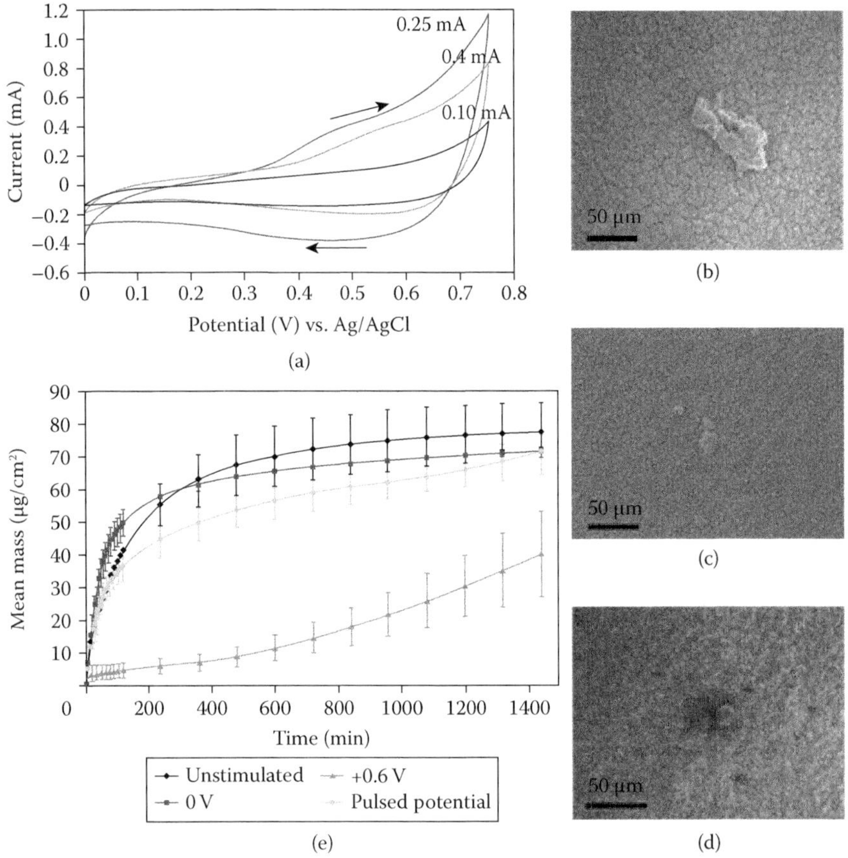

Figure 2.4 (a) Cyclic voltammetry of PTTh–Dex films grown for 15 min at 0.1, 0.25, and 0.4 mA. The cyclic voltammograms (CVs) were recorded in phosphate-buffered saline at a scan rate of 50 mV s^{-1}. The arrows on the cyclic voltammogram indicate the direction of the potential scan. Characterization SEM images for films grown at (b) 0.40 mA, (c) 0.25 mA, and (d) 0.10 mA. Cumulative Dex release over 24 h period from PTTh–Dex films stimulated with constant potential (0 V) (■), constant potential (+0.6 V) (▲), and pulsed potential (0 to +0.6 V at 1 Hz) (●). (e) Passive release from unstimulated films (♦) is included for comparison. Error bars indicate the standard error of the mean (n = 3). (Reprinted from Stevenson, G., et al., *Synthetic Metals*, 160, 1107–1114, 2010. With permission.)

for deposition, the nature of the dopant, and the concentration of the original monomer solution (Ateh et al. 2006). Temperature and pH also have an effect on the ensuing film. Besides composition of the reaction solution, that is, electrolyte and temperature, the electrochemical synthesis is subjected to the influence of the electrode material and selected electrochemical technique (Pringle et al. 2004; Pornputtkul et al. 2006).

Having the ability to choose the appropriate electrochemical polymerization method affords the researcher the ability to judiciously select the method that produces a CP based on the synthesis environment (i.e., electrolyte, electrode material, pH, etc.). The research team lead by Professor David Martin have utilized the facile electrochemical approach to synthesize the biocompatible-functionalized thiophene PEDOT directly within living

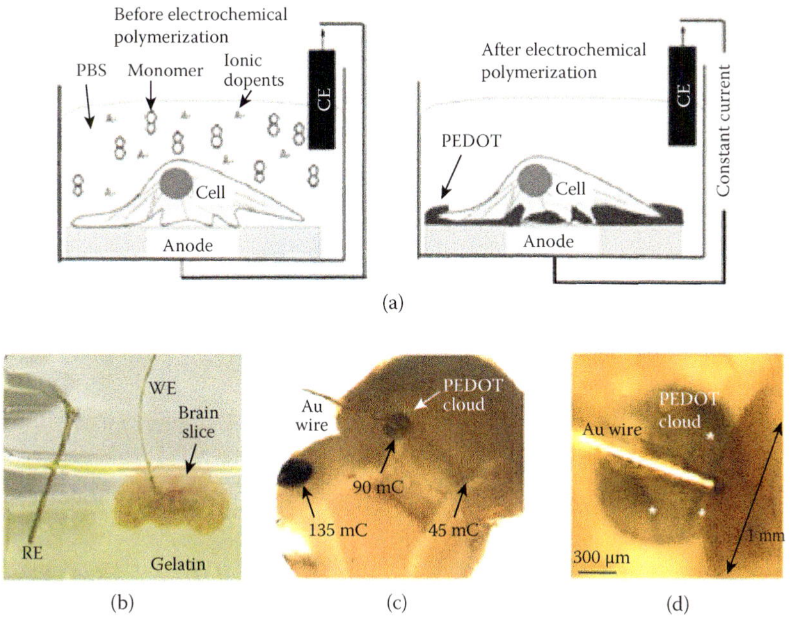

Figure 2.5 (a) Diagram representing the process of electrochemical polymerization in the presence of living cells cultured on an electrode substrate. A network of conducting PEDOT filaments can be polymerized directly within brain tissue from an implanted electrode. (b) Image of the setup used to polymerize PEDOT directly within a mouse brain slice. The tissue slice is physically stabilized within gelatin, and then a microwire electrode is inserted into the tissue and a platinum wire reference or counterelectrode is inserted into the gelatin. (c) Mouse brain slice (2 mm thick) within which three distinct PEDOT networks were polymerized from a Au microwire using increasing polymer deposition charges of 45, 90, and 135 mC. (d) A higher magnification image reveals that the PEDOT cloud integrates directly within the tissue and appears to intensify near and follow white matter fiber tracts (see white stars). (Reprinted from Richardson-Burns, S. M., et al., *J. Neural Eng.*, 4, L6–L13, 2007. With permission.)

neural tissue, resulting in an electrically conductive network that is integrated within the tissue (Richardson-Burns et al. 2007) (Figure 2.5). Martin and his team (Richardson-Burns et al. 2007) utilized the interplay between the surrounding electrode environment, strategic orientation of the counter- and working electrodes, and ability to utilize an electrochemical synthesis method that facilitated the accurate control of the polymerization rate (i.e., the oxidation potential generated during growth).

2.6 MICRO- AND NANOSTRUCTURED CPs AND BIOMEDICAL APPLICATIONS

Research efforts have explored the fabrication of CPs with a micro- or nanostructure to exaggerate properties and extend the applications of these versatile materials. CPs with a micro- or nanostructure have an increased surface area compared with unstructured CPs. As many CP applications rely on the redox altering properties of the CP and associated movement of charged ions into and out of the polymer, the enhanced surface area in micro- and nanostructured materials facilitates ion movement with increased redox responses observed. The enhanced effective electrochemical surface area of these materials is also useful when reduced impedance is required, such as at a bioelectrode interface.

A wide range of micro- and nanostructures can be prepared by polymerizing CPs within and over existing template structures. Such templates to direct CP growth can be "hard," such as physical membranes, or "soft," such as liquid crystals (LCs). To direct the electrochemical growth of CPs, the template must be in physical contact with a conductive substrate from which polymerization is initiated. The final structure of the CP is dictated by the template used. A broad range of templates have been used to direct CP growth, with new reports appearing frequently. Meanwhile, so-called template-free approaches have been explored where the nature and concentration of the reactants, the media used, and the electro-oxidative conditions can all be modified to polymerize CPs with the desired structure.

2.6.1 HARD TEMPLATES

The most widely explored templates are solid materials, and include fibers, membranes, colloidal arrangements, and carbon nanotubes. These templates are arranged on a conductive substrate, or are made conductive themselves. The polymerization media occupies available volume between the conductive substrate and template; when polymerization is initiated, the template guides growth. Following polymerization, the template can remain or be removed.

PEDOT has been electropolymerized through and over the top of poly(lactic-co-glycolic acid) nanofibers, which were electrospun onto a conductive substrate (Abidian et al. 2006). Following electropolymerization, the PLGA fibers could be selectively removed to leave hollow PEDOT fibers (Figure 2.6). Alternately, by leaving the PLGA in place, a controlled release delivery system was fabricated. PLGA is widely used for drug delivery purposes, with drug release via both diffusion through the PLGA bulk and polymer degradation. Using the PEDOT-coated, drug-loaded PLGA nanofibers, Abidian et al. (2006) demonstrated electrochemically modulated drug release, presumably with the PEDOT coating limiting drug diffusion and exposure of the PLGA to degrading biological fluids. When electrically stimulated, the PEDOT coating had altered drug permeability and the appearance of cracks, promoting the movement of drug and biological fluids.

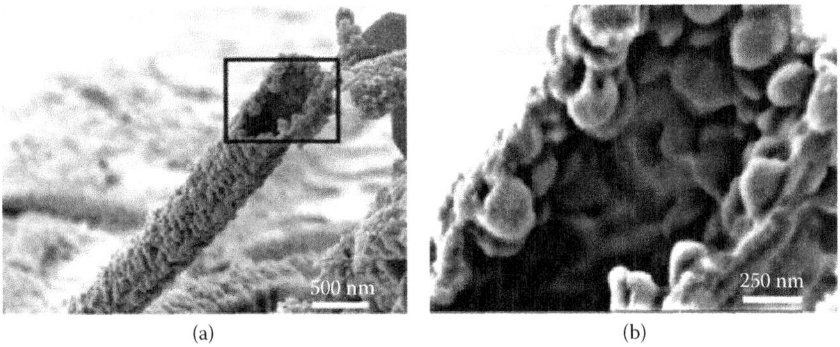

Figure 2.6 (a) PEDOT electropolymerized over electrospun PLGA fibers with (b) subsequent dissolution of PLGA, leaving hollow PEDOT fibers. (From Abidian, M. R., et al., *Adv. Mater.*, 18, 405–409, 2006. With permission.)

Colloids can also be used as hard templates to guide the growth of CPs. Colloids can be arranged on conductive substrates through a variety of methods. Sedimentation is perhaps the simplest method to prepare the colloidal template and has been used to deposit polystyrene beads of 500 or 750 nm diameter onto a gold working electrode (Bartlett et al. 2011). While simple and effective, the colloids do not tend to pack with a high degree of order (Figure 2.7a), which may or may not be important, depending on the final application. A higher degree of order can be obtained by arranging the colloids by flow-controlled vertical deposition (Waterhouse and Waterland 2007) (Figure 2.7c), electrophoretic deposition (Pokki et al. 2012), or even using capillary forces inside a microfluidic channel (Gorey et al. 2012). CP films can then be obtained by electropolymerizing the polymer through the interstitial voids of a colloidal arrangement (Figure 2.7b and 2.7d). The template can subsequently be removed by selective chemical etching to reveal a porous CP structure with a high surface area. The size of the colloids is then represented by the size of the resulting pores, as is the degree of order.

Porous CP structures prepared using colloidal templates have been used for drug delivery, with the drug incorporated either predominantly within the pores due to surface interactions, or as a dopant within the polymer bulk. Porous and nonporous structures were incubated with the small hydrophobic molecule rhodamine B (Pokki et al. 2012). Considerably more drug was spontaneously released from the porous structures, which could be attributed to the increased CP surface area available to interact with the model drug compared with nonporous films. Similarly, risperidone can be loaded onto CP films with porous films having higher drug loading levels with faster and

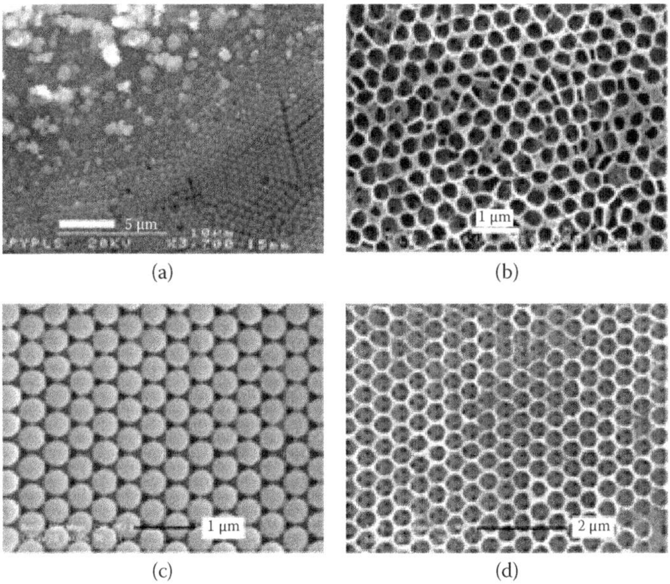

Figure 2.7 CP films prepared over colloidal templates. (a) A hard template formed sedimented colloids displaying a moderate degree of order and (b) the resulting CP structure produced. (From Bartlett, P. N., et al., *J. Mater. Chem.*, 11, 849–853, 2011. With permission.) (c) A hard template formed through vertical deposition displaying a high degree of order and (d) the resulting CP structure produced. (From Sharma, M., et al., *Int. J. Pharm.* 443, 163–168, 2013. With permission.)

more responsive release than nonporous structures (Sharma et al. 2013). Drug can also be included as a dopant molecule into the polymer bulk; recently, porous PPy structures were polymerized using Dex as a dopant (Seyfoddin et al. 2015). Release was highly responsive to electrical stimulation. In a combination approach, fluorescein was incorporated as a dopant within the porous CP structure, with dexamethasone base loaded within the pores themselves (Luo and Cui 2009). Through this approach, two different drugs could be loaded and released in response to an electrical stimulus.

Membranes with precise pores can be used to direct the growth of CPs with tightly controlled dimensions. Polycarbonate membranes, with pores between 20 and 220 nm, and a gold layer deposited on one side have been used to initiate and guide the growth of electrochemically polymerized PANI (Delvaux et al. 2000). The membrane pore size dictated the diameter of the PANI tubes produced, while different-length tubes were prepared by controlling the charge passed during polymerization. The membrane could be selectively removed by soaking in dichloromethane, leaving behind nanorods of PANI.

Meanwhile, membranes with CPs partially filling the pores can function as systems with electrically tunable pore size and thus permeability (Figure 2.8) (Jeon et al. 2011; Abelow et al. 2014). Jeon et al. (2011) used anodized aluminum oxide membranes as a starting template with an initial pore size of 410 nm. The pore size decreased to 380 nm following the deposition of a gold layer to make the membrane conductive, facilitating the electropolymerization of PPy. Once PPy was polymerized, the pore size altered between 190 and 140 nm in the oxidized and reduced states, respectively. Such membranes with electrically modifiable pore size could find use in a range of applications, including controlled drug release and in separation devices.

The versatile electrical, optical, mechanical, and thermal properties of carbon nanotubes can be blended with the properties of CPs when composite materials are formed. In 2000, Gao et al. reported PANI electropolymerized over well-aligned carbon nanotubes (Figure 2.9). More recently, these composite materials have been used for the guidance and stimulation of various cell types. Nanostructures of PPy electrodeposited over multiwalled carbon nanotubes have been used to influence the orientation of primary myoblasts *in vitro* (Quigley et al. 2012). Cell differentiation and division could subsequently be influenced by electrical stimulation transmitted through these structures.

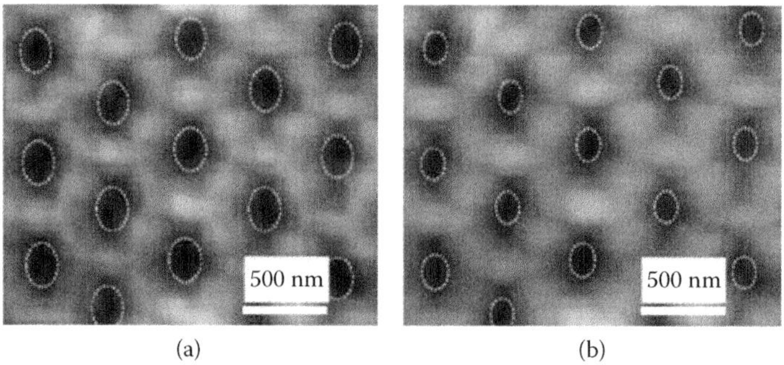

Figure 2.8 Pores of a membrane that are partially filled with PPy–DBS demonstrate electrically controllable actuation. (a) Oxidized state with more open pores. (b) Reduced state with the pores in the closed state. (From Delvaux, M., et al., *Synth. Metals*, 113, 275–280, 2000. With permission.)

Figure 2.9 PANI nanotubes electrochemically polymerized over carbon nanotubes. (From Gao, M., et al., *Angew. Chem. Int. Ed.*, 39, 3664–3667, 2000. With permission.)

Composite materials consisting of carbon nanotubes and CPs have also found use as amperometric biosensors. Multiwalled carbon nanotubes were coated with PANI, with dendrimer-encapsulated Pt nanoparticles adsorbed and glucose oxidase cross-linked (Xu et al. 2009). These structures demonstrated a high sensitivity of 42.0 mA M^{-1} cm^{-2} to glucose. Meanwhile, the versatility of such fabrications was demonstrated when multiwalled carbon nanotubes were electrochemically coated with PPy and cross-linked to both glucose oxidase and horseradish peroxidase to yield an amperometric biosensor with greater sensitivity than when glucose oxidase was linked alone (Singh et al. 2012).

2.6.2 SOFT TEMPLATES

Soft templates are typically in the form of structures that self-assemble from amphiphilic molecules. Such templates can be used to direct the growth of CPs. LCs exist in different phases, depending on the amphiphilic materials used, the water content, and external conditions, including temperature and pressure (Boyd et al. 2011). Therefore, it is possible to select a phase with an appropriate long-range order and internal nanostructure to prepare CPs with desirable properties. An exciting possibility is that the same materials can be used as a template, and the structure controlled by temperature or other external factors. An additional benefit of LCs as soft templates is the relative ease and gentle conditions required to remove these templates, compared with hard templates.

Hulvat and Stupp (2003, 2004) synthesized PEDOT through a poly(oxyethylene)*n*-oleyl ether (*n*~10) LC. When an anisotropic phase of LC was used, the resulting PEDOT was anisotropic, as demonstrated by cross-polarized optical micrographs, even after the LC template was removed (Figure 2.10a and b). Using the same materials, the LC was transformed into the reverse hexagonal phase by adjusting the temperature; electropolymerization then produced PEDOT fibers. These one-dimensional PANI nanowires were electropolymerized within the aqueous channels of a reverse hexagonal LC (Figure 2.10c) (Huang et al. 2002). Ordered PEDOT has also been electropolymerized through a chitosan-based lyotropic LC (Meng et al. 2013).

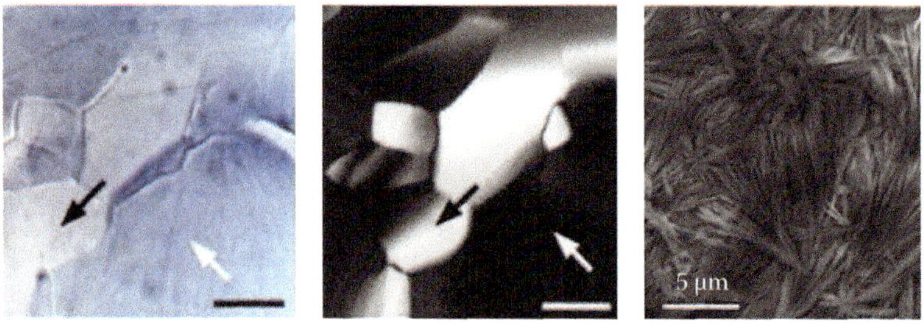

(a) (b) (c)

Figure 2.10 Optical micrographs of PEDOT electropolymerized through an LC template. (a) Nonpolarized light and (b) polarized light. Anisotropy demonstrated through observed birefringence. (From Hulvat, J. F., and Stupp, S. I., *Angew. Chem. Int. Ed.*, 42, 778–781, 2003. With permission.) PANI nanowires (c) prepared through the LC template in the reverse hexagonal phase. (From Huang, L., et al., *J. Mater. Chem.*, 12, 388–391, 2002. With permission.)

Self-assembled structures based on surfactants can be used as soft templates. PPy micro- and nanowires were prepared electrochemically using the surfactant pyrenesulfonic acid as the dopant. The surfactant molecules form either spherical micelles or rod-like structures, depending on the surfactant concentration. High-surface-area PPy fibers could be produced through a single-step electrochemical reaction with diameters ranging from 120 to 500 nm.

2.6.3 TEMPLATE-FREE SYNTHESIS

The so-called template-free methods have focused on manipulating polymerization conditions to form CPs with desired structures. Such approaches avoid any constraints presented by the hard or soft template, including internal and overall dimensions.

PPy fibers can be electropolymerized by appropriate control over monomer and dopant concentrations, alongside set polymerization conditions (Palod and Singh 2015). The resulting high-surface-area PPy fiber mats were applied to glucose biosensing (Figure 2.11). Meanwhile, large arrays of PANI particles and fibers can be formed using higher than typical current densities (0.08 mA cm^{-2}) to form nucleation sites (Liang et al. 2002). These structures could subsequently be used to guide the polymerization of more PANI prepared at more typical and lower current densities.

Alternately, an elegant approach has seen a reducing potential being applied to the working electrode, resulting in hydrogen bubble formation; the bubbles were stabilized on the electrode using anionic surfactants (Bajpai et al. 2004). When the bias is switched, the anionic surfactant-coated bubbles are attracted to the working electrode from which the CP is electropolymerized. The resulting PPy microcontainers were used to seal fluorescein, which was contained until the containers were mechanically ruptured and their payload released (Bajpai et al. 2004). In related work, the resulting high surface area of the produced microstructures was applied to glucose sensing following the incorporation of glucose oxidase (Bajpai et al. 2006). This approach of "bubble templating" was recently advanced by Diaz-Orellana and Roberts 2015 (2015), who present a scalable method to prepare PPy microtubes. Hydrogen bubbles, reduced at the counterelectrode, nucleated at the joins in the stainless steel mesh working electrode. On switching polarity, PPy microtubes could subsequently be formed (Figure 2.12).

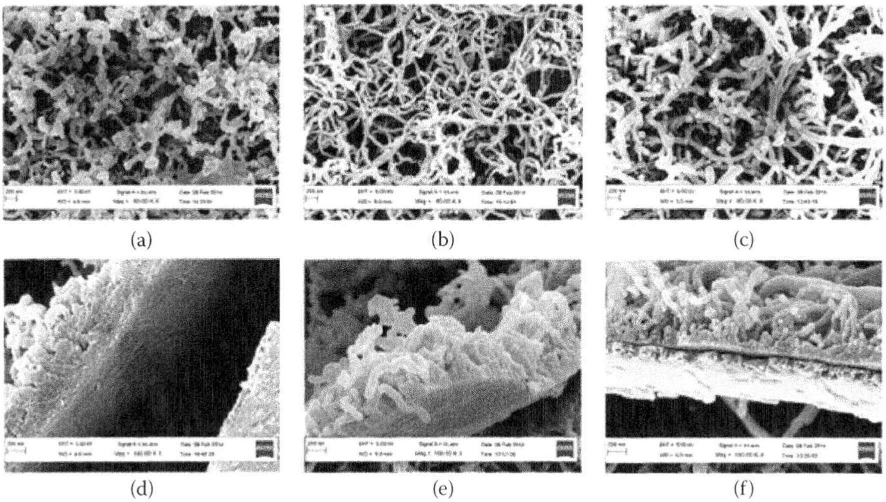

Figure 2.11 PPy nanofibers prepared without a hard or soft template from an aqueous solution containing 0.15 M pyrrole with 20 mM $LiClO_4$ and 200 mM Na_2HPO_4 polymerized at 0.85 V (a and d), 0.8 V (b and e), and 0.7 V (c and f). (From Palod, P. A., and Singh, V., *Sensors Actuators B*, 209, 85–93, 2015. With permission.)

Figure 2.12 PPy microtubes formed over hydrogen bubbles electrochemically produced on a stainless steel mesh. (From Diaz-Orellana, K. P., and Roberts, M. E., *RSC Adv.*, 5, 25504–25512, 2015. With permission.)

The tube shape, diameter, and density were controllable using different mesh densities, polymerizing current densities, and total charges passed.

2.7 CONCLUSION

Attention to synthesis procedures is essential to form CPs appropriate to perform at biological interfaces. Electrochemical polymerization is frequently preferred, with the final product influenced by geometries of the electrochemical cells, electrode materials, electrolyte composition, and electrochemical technique itself. The polymerization method can be carefully designed to prepare CPs with desirable properties, within environments compatible with living tissue. Micro- and nanostructures can be imparted into CPs following polymerization over and through templating materials. This approach is used to impart order into the CP and increase the surface area of the polymer–media interface. By doing so, impedance can be reduced and the reversible properties that make CPs attractive can be exaggerated with the speed of response increased. CPs with deliberate micro- and nanostructures have demonstrated benefits when applied to drug delivery, for separation devices, for biosensors, and to influence cellular interactions.

REFERENCES

Abdiryim, T., R. Jamal, and I. Nurulla. Doping effect of organic sulphonic acids on the solid-state synthesized polyaniline. *Journal of Applied Polymer Science* 105 (2007): 576–584.

Abelow, A. E., K. M. Persson, E. W. H. Jager, M. Berggren, and I. Zharov. Electroresponsive nanoporous membranes by coating anodized alumina with poly(3,4-ethylenedioxythiophene) and polypyrrole. *Macromolecular Materials and Engineering* 299 (2014): 190–197.

Abidian, M. R., D. H. Kim, and D. C. Martin. Conducting-polymer nanotubes for controlled drug release. *Advanced Materials* 18 (2006): 405–409.

Adeloju, S. B., and S. Hussain. Potentiometric sulfite biosensor based on entrapment of sulfite oxidase in a polypyrrole film on a platinum electrode modified with platinum nanoparticles. *Microchimica Acta* 183 (2016): 1341–1350.

Ameen, S., M. S. Akhtar, and M. Husain. Polyaniline and its nanocomposites: Synthesis, processing, electrical properties and applications. *Science of Advanced Materials* 2 (2010): 441–462.

Andrieux, C. P., P. Audebert, P. Hapiot, and J. M. Savéant. Identification of the first steps of the electrochemical polymerization of pyrroles by means of fast potential step techniques *Journal of Physical Chemistry* 95 (1991): 10158–10164.

Ateh, D. D., H. A. Navsaria, and P. Vadgama. Polypyrrole-based conducting polymers and interactions with biological tissues. *Journal of the Royal Society Interface* 3 (2006): 741–752.

Bajpai, V., P. He, and L. Dai. Conducting-polymer microcontainers: Controlled synthesis and potential applications. *Advanced Functional Materials* 14 (2004): 145–151.

Bajpai, V., P. He, L. Goettler, J. H. Dong, and L. Dai. Controlled syntheses of conducting polymer micro- and nano-structures for potential applications. *Synthetic Metals* 156 (2006): 466–469.

Baldissera, A. F., K. L. de Miranda, C. Bressy, C. Martin, A. Margaillan, and C. A. Ferreira. Using conducting polymers as active agents for marine antifouling paints. *Materials Research: Ibero-American Journal of Materials* 18 (2015): 1129–1139.

Bartlett, P. N., P. R. Birkin, M. A. Ghanem, and C.-S. Toh. Electrochemical syntheses of highly ordered macroporous conducting polymers grown around self-assembled colloidal templates. *Journal of Materials Chemistry* 11 (2011): 849–853.

Beck, F., and M. Oberst. Electrocatalytic deposition and transformation of polypyrrole layers. *Synthetic Metals* 28 (1989): 43–50.

Bhadra, S., D. Khastgir, N. K. Singha, and J. H. Lee. Progress in preparation, processing and applications of polyaniline. *Progress in Polymer Science* 34 (2009): 783–810.

Bhattacharyya, D., R. M. Howden, D. C. Borrelli, and K. K. Gleason. Vapor phase oxidative synthesis of conjugated polymers and applications. *Journal of Polymer Science Part B: Polymer Physics* 50 (2012): 1329–1351.

Biallozor, S., and A. Kupniewska. Conducting polymers electrodeposited on active metals. *Synthetic Metals* 155 (2005): 443–449.

Boeva, Z. A., O. A. Pyshkina, A. A. Lezov, G. E. Polushina, A. V. Lezov, and V. G. Sergeyev. Matrix synthesis of water-soluble polyaniline in the presence of polyelectrolytes. *Polymer Science Series C* 52 (2010): 35–43.

Boyd, B. J., T.-H. Nguyen, and A. Müllertz. Lipids in oral controlled release drug delivery. In *Controlled Release in Oral Drug Delivery*, ed. C. G. Wilson and P. J. Crowley. Berlin: Springer, 2011, 299–327.

Cosnier, S. Affinity biosensors based on electropolymerized films. *Electroanalysis* 17 (2005): 1701–1715.

Daprano, G., M. Leclerc, G. Zotti, and G. Schiavon. Synthesis and characterisation of polyaniline derivatives—Poly(2-alkoxyanilines) and poly(2,5-dialkoxyanilines). *Chemistry of Materials* 7 (1995): 33–42.

Delvaux, M., J. Duchet, P.-Y. Stavaux, R. Legras, and S. Demoustier-Champagne. Chemical and electrochemical synthesis of polyaniline micro- and nano-tubules. *Synthetic Metals* 113 (2000): 275–280.

Demoustier-Champagne, S., and P. Y. Stavaux. Effect of electrolyte concentration and nature on the morphology and the electrical properties of electropolymerized polypyrrole nanotubules. *Chemistry of Materials* 11 (1999): 829–834.

Diaz-Orellana, K. P., and M. E. Roberts. Scalable, template-free synthesis of conducting polymer microtubes. *RSC Advances* 5 (2015): 25504–25512.

Eftekhari, A., M. Kazemzad, and M. Keyanpour-Rad. Significant effect of dopant size on nanoscale fractal structure of polypyrrole film. *Polymer Journal* 38 (2006): 781–785.

El-Rahman, H. A. A., A. A. Hathoot, M. El-Bagoury, and M. Abdel-Azzem. Electroactive polymer films formed from the Schiff base product of 1,8-diaminonaphthalene and dehydroacetic acid I. Preparation and characterization. *Journal of the Electrochemical Society* 147 (2000): 242–247.

Esrafilzadeh, D., J. M. Razal, S. E. Moulton, E. M. Stewart, and G. G. Wallace. Multifunctional conducting fibres with electrically controlled release of ciprofloxacin. *Journal of Controlled Release* 169 (2013): 313–320.

Gao, M., S. Huang, L. Dai, G. G. Wallace, R. Gao, and Z. Wang. Aligned coaxial nanowires of carbon nanotubes sheathed with conducting polymers. *Angewandte Chemie International Edition* 39 (2000): 3664–3667.

Gorey, B., J. Galineau, B. White, M. Smyth, and A. Morrin. Inverse-opal conducting polymer monoliths in microfluidic channels. *Electroanalysis* 24 (2012): 1318–1323.

Grgur, B. N., P. Zivkovic, and M. M. Gvozdenovic. Kinetics of the mild steel corrosion protection by polypyrrole-oxalate coating in sulfuric acid solution. *Progress in Organic Coatings* 56 (2006): 240–247.

Gvozdenovic, M. M., and B. N. Grgur. Electrochemical polymerization and initial corrosion properties of polyaniline-benzoate film on aluminum. *Progress in Organic Coatings* 65 (2009): 401–404.

Gvozdenovic, M. M., B. Z. Jugovic, J. S. Stevanovic, and B. N. Grgur. Electrochemical synthesis of electroconducting polymers. *Hemijska Industrija* 68 (2014): 673–684.

Higashihara, T., and M. Ueda. Precision synthesis of tailor-made polythiophene-based materials and their application to organic solar cells. *Macromolecular Research* 21 (2013): 257–271.

Higgins, T. M., S. E. Moulton, K. J. Gilmore, G. G. Wallace, and M. I. H. Panhuis. Gellan gum doped polypyrrole neural prosthetic electrode coatings. *Soft Matter* 7 (2011): 4690–4695.

Hu, X., G. Wang, and T. K. S. Wong. Effect of aqueous and organic solvent ratio on the electropolymerization of bithiophene in the mixed solutions. *Synthetic Metals* 106 (1999): 145–150.

Huang, L., Z. Wang, H. Wang, X. Cheng, A. Mitra, and Y. Yan. Polyaniline nanowires by electropolymerization from liquid crystalline phases. *Journal of Materials Chemistry* 12 (2002): 388–391.

Hulvat, J. F., and S. I. Stupp. Liquid-crystal templating of conducting polymers. *Angewandte Chemie International Edition* 42 (2003): 778–781.

Hulvat, J. F., and S. I. Stupp. Anisotropic properties of conducting polymers prepared by liquid crystal templating. *Advanced Materials* 16 (2004): 589–592.

Hwang, J., I. Schwendeman, B. C. Ihas, R. J. Clark, M. Cornick, M. Nikolou, A. Argun, J. R. Reynolds, and D. B. Tanner. In situ measurements of the optical absorption of dioxythiophene-based conjugated polymers. *Physical Review B* 83 (2011): 12.

Innis, P. C., I. D. Norris, L. A. P. Kane-Maguire, and G. G. Wallace. Electrochemical formation of chiral polyaniline colloids codoped with (+)- or (–)-10-camphorsulfonic acid and polystyrene sulfonate. *Macromolecules* 31 (1998): 6521–6528.

Jafari, M., R. Sedghi, and H. Ebrahimzadeh. A platinum wire coated with a composite consisting of poly pyrrole and poly(E >-caprolactone) for solid phase microextraction of the antidepressant imipramine prior to its determination via ion mobility spectrometry. *Microchimica Acta* 183 (2016): 805–812.

Jang, J., and V. Springer. (2006). Conducting polymer nanomaterials and their applications. *Emissive Materials: Nanomaterials* 199: 189–259.

Jeon, G., S. Y. Yang, J. Byun, and J. K. Kim. Electrically actuatable smart nanoporous membrane for pulsatile drug release. *Nano Letters* 11 (2011): 1284–1288.

Jo, S. H., Y. K. Lee, J. W. Yang, W. G. Jung, and J. Y. Kim. Carbon nanotube-based flexible transparent electrode films hybridized with self-assembling PEDOT. *Synthetic Metals* 162 (2012): 1279–1284.

Kannan, B., D. E. Williams, C. Laslau, and J. Travas-Sejdic. A highly sensitive, label-free gene sensor based on a single conducting polymer nanowire. *Biosensors & Bioelectronics* 35 (2012): 258–264.

Khalid, H., H. J. Yu, L. Wang, W. A. Amer, M. Akram, N. M. Abbasi, Z. ul-Abdin, and M. Saleem. Synthesis of ferrocene-based polythiophenes and their applications. *Polymer Chemistry* 5 (2014): 6879–6892.

Komarova, E., M. Aldissi, and A. Bogomolova. Design of molecularly imprinted conducting polymer protein-sensing films via substrate-dopant binding. *Analyst* 140 (2015): 1099–1106.

Krische, B., and M. Zagorska. Overoxidation in conducting polymers. *Synthetic Metals* 28 (1989): C257–C262.

Kuwabata, S., K. I. Okamoto, O. Ikeda, and H. Yoneyama. Effect of organic dopants on electrical-conductivity of polypyrrole films. *Synthetic Metals* 18 (1987): 101–104.

Liang, L., J. Liu, C.F. Windisch Jr., G. J. Exarhos, and Y. Lin. Direct assembly of large arrays of oriented conducting polymer nanowires. *Angewandte Chemie International Edition* 41 (2002): 3665–3668.

Luo, X., and X. T. Cui. Sponge-like nanostructured conducting polymers for electrically controlled drug release. *Electrochemistry Communications* 11 (2009): 1956–1959.

Malinauskas, A. Chemical deposition of conducting polymers. *Polymer* 42 (2001): 3957–3972.

Mane, A. T., S. D. Sartale, and V. B. Patil. Dodecyl benzene sulfonic acid (DBSA) doped polypyrrole (PPy) films: Synthesis, structural, morphological, gas sensing and impedance study. *Journal of Materials Science: Materials in Electronics* 26 (2015): 8497–8506.

Meng, X., Z. Wang, L. Wang, M. Pei, W. Guo, and X. Tang. Electrosynthesis of pure poly(3,4-ethylenedioxythiophene) (PEDOT) in chitosan-based liquid crystal phase. *Electronic Materials Letters* 9 (2013): 605–608.

Moulton, S. E., M. D. Imisides, R. L. Shepherd, and G. G. Wallace. Galvanic coupling conducting polymers to biodegradable Mg initiates autonomously powered drug release. *Journal of Materials Chemistry* 18 (2008): 3608–3613.

Osaka, T., T. Nakajima, K. Shiota, and T. Momma. Electroactive polyaniline film deposited from nonaqueos media. 3. Effects of mixed organic-solvent on polyaniline deposition and its battery performance. *Journal of the Electrochemical Society* 138 (1991): 2853–2858.

Palod, P. A., and V. Singh. Facile synthesis of high density polypyrrole nanofiber network with controllable diameters by one step template free electropolymerization for biosensing applications. *Sensors and Actuators B* 209 (2015): 85–93.

Pokki, J., O. Ergeneman, K. M. Sivaraman, B. Özkale, M. A. Zeeshan, T. Lühmann, B. J. Nelson, and S. Pané. Electroplated porous polypyrrole nanostructures patterned by colloidal lithography for drug-delivery applications. *Nanoscale* 4 (2012): 3083–3088.

Pomogailo, A. D. Synthesis and intercalation chemistry of hybrid organo-inorganic nanocomposites. *Polymer Science Series C* 48 (2006): 85–111.

Popovic, M. M., and B. N. Grgur. Electrochemical synthesis and corrosion behavior of thin polyaniline-benzoate film on mild steel. *Synthetic Metals* 143 (2004): 191–195.
Pornputtkul, Y., L. A. P. Kane-Maguire, and G. G. Wallace. Influence of electrochemical polymerization temperature on the chiroptical properties of (+)-camphorsulfonic acid-doped polyaniline. *Macromolecules* 39 (2006): 5604–5610.
Pringle, J. M., J. Efthimiadis, P. C. Howlett, J. Efthimiadis, D. R. MacFarlane, A. B. Chaplin, S. B. Hall, D. L. Officer, G. G. Wallace, and M. Forsyth. Electrochemical synthesis of polypyrrole in ionic liquids. *Polymer* 45 (2004): 1447–1453.
Quigley, A. F., J. M. Razal, M. Kita, R. Jalili, A. Gelmi, A. Penington, R. Ovalle-Robles, R. H. Baughman, G. M. Clark, G. G. Wallace, and R. M. I. Kapsa. Electrical stimulation of myoblast proliferation and differentiation on aligned nanostructured conductive polymer platforms. *Advance Healthcare Materials* 1 (2012): 801–808.
Quigley, A. F., J. M. Razal, B. C. Thompson, S. E. Moulton, M. Kita, E. L. Kennedy, G. M. Clark, G. G. Wallace, and R. M. I. Kapsa. A conducting-polymer platform with biodegradable fibers for stimulation and guidance of axonal growth. *Advance Materials* 21 (2009): 4393–4397.
Ren, X., and P. G. Pickup. Ion transport in polypyrrole and a polypyrrole/polyanion composite. *Journal of Physical Chemistry* 97 (1993): 5356–5362.
Richardson-Burns, S. M., J. L. Hendricks, and D. C. Martin. Electrochemical polymerization of conducting polymers in living neural tissue. *Journal of Neural Engineering* 4 (2007): L6–L13.
Seyfoddin, A., A. Chan, W.-T. Chen, I. D. Rupenthal, G. I. N. Waterhouse, and D. Svirskis. Electro-responsive macroporous polypyrrole scaffolds for triggered dexamethasone delivery. *European Journal of Pharmaceutics and Biopharmaceutics* 94 (2015): 419–426.
Sharma, M., G. Waterhouse, S. Loader, S. Garg, and D. Svirskis. High surface area polypyrrole scaffolds for tunable drug delivery. *International Journal of Pharmaceutics* 443 (2013): 163–168.
Singh, K., B. P. Singh, R. Chauhan, and T. Basu. Fabrication of amperometric bienzymatic glucose biosensor based on MWCNT tube and polypyrrole multilayered nanocomposite. *Journal of Applied Polymer Science* 125 (2012): E235–E246.
Stejskal, J. Colloidal dispersions of conducting polymers. *Journal of Polymer Materials* 18 (2001): 225–258.
Stejskal, J., I. Sapurina, and M. Trchova. Polyaniline nanostructures and the role of aniline oligomers in their formation. *Progress in Polymer Science* 35 (2010): 1420–1481.
Stevenson, G., S. E. Moulton, P. C. Innis, and G. G. Wallace. Polyterthiophene as an electrostimulated controlled drug release material of therapeutic levels of dexamethasone. *Synthetic Metals* 160 (2010): 1107–1114.
Syed, A. A., and M. K. Dinesan. Polyaniline—A novel polymeric material. *Talanta* 38 (1991): 815–837.
Tallman, D. E., C. Vang, G. G. Wallace, and G. P. Bierwagen. Direct electrodeposition of polypyrrole on aluminum and aluminum alloy by electron transfer mediation. *Journal of the Electrochemical Society* 149 (2002): C173–C179.
Thompson, B. C., J. Chen, S. E. Moulton, and G. G. Wallace. Nanostructured aligned CNT platforms enhance the controlled release of a neurotrophic protein from polypyrrole. *Nanoscale* 2 (2010a): 499–501.
Thompson, B. C., S. E. Moulton, J. Ding, R. Richardson, A. Cameron, S. O'Leary, G. G. Wallace, and G. M. Clark. Optimising the incorporation and release of a neurotrophic factor using conducting polypyrrole. *Journal of Controlled Release* 116 (2006): 285–294.
Thompson, B. C., S. E. Moulton, R. T. Richardson, and G. G. Wallace. Effect of the dopant anion in polypyrrole on nerve growth and release of a neurotrophic protein. *Biomaterials* 32 (2011): 3822–3831.
Thompson, B. C., R. T. Richardson, S. E. Moulton, A. J. Evans, S. O'Leary, G. M. Clark, and G. G. Wallace. Conducting polymers, dual neurotrophins and pulsed electrical stimulation—Dramatic effects on neurite outgrowth. *Journal of Controlled Release* 141 (2010b): 161–167.
Vernitskaya, T. V., and O. N. Efimov. Polypyrrole: A conducting polymer (synthesis, properties, and applications). *Uspekhi Khimii* 66 (1997): 489–505.

Wallace, G. G., G. M. Spinks, L. A. P. Kane-Maguire, and P. R. Teasdale. *Conductive Electroactive Polymers: Intelligent Material Systems*. Boca Raton, FL: CRC Press, 2003.

Wang, L. X., X. G. Li, and Y. L. Yang. Preparation, properties and applications of polypyrroles. *Reactive & Functional Polymers* 47 (2001): 125–139.

Warren, L. F., and D. P. Anderson. Polypyrrole films from aqueous electrolytes—The effect of anions upon order. *Journal of the Electrochemical Society* 134 (1987): 101–105.

Waterhouse, G. I. N., and M. R. Waterland. Opal and inverse opal photonic crystals: Fabrication and characterization. *Polyhedron* 26 (2007): 356–368.

Xu, G. Y., W. T. Wang, B. B. Li, Z. L. Luo, and X. L. Luo. A dopamine sensor based on a carbon paste electrode modified with DNA-doped poly(3,4-ethylenedioxythiophene). *Microchimica Acta* 182 (2015): 679–685.

Xu, L., Y. Zhu, X. Yang, and C. Li. Amperometric biosensor based on carbon nanotubes coated with polyaniline/dendrimer-encapsulated Pt nanoparticles for glucose detection. *Materials Science and Engineering C* 29 (2009): 1306–1310.

Xue, F. L., Y. Su, and K. Varahramyan. Modified PEDOT-PSS conducting polymer as S/D electrodes for device performance enhancement of P3HT TFTs. *IEEE Transactions on Electron Devices* 52 (2005): 1982–1987.

Yalcinkaya, S., C. Demetgul, M. Timur, and N. Colak. Electrochemical synthesis and characterization of polypyrrole/chitosan composite on platinum electrode: Its electrochemical and thermal behaviors. *Carbohydrate Polymers* 79 (2010): 908–913.

Yu, Y., Q. W. Tang, B. L. He, H. Y. Chen, Z. M. Zhang, and L. M. Yu. Platinum alloy decorated polyaniline counter electrodes for dye-sensitized solar cells. *Electrochimica Acta* 190 (2016): 76–84.

3 Properties and characterization of conductive polymers

David L. Officer, Klaudia Wagner, and Pawel Wagner
University of Wollongong
Wollongong, New South Wales, Australia

Contents

3.1 INTRODUCTION TO CONDUCTIVE POLYMER PROPERTIES AND CHARACTERIZATION

3.1.1 GENERAL

As has been described in the previous chapters, organic conductive polymers (OCPs) exhibit the behavior of metals or semiconductors. Their backbones have a delocalized bond structure that can be modified through "oxidation (doping) or reduction" processes, as shown in Figure 3.1 for oxidation. By changing the charge carriers

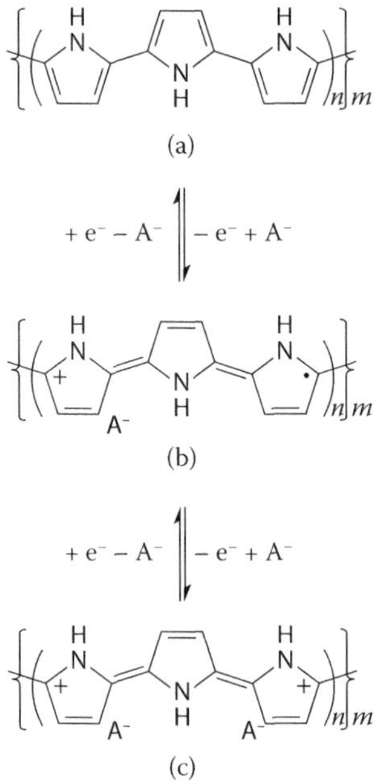

Figure 3.1 Conversion of (a) neutral polypyrrole into (b) its oxidized polaron form and (c) oxidized bipolaron form. A^- denotes anion; e^- denotes electron; *n* determines degree of doping; and *m* determines molecular weight.

(addition or removal of electrons) in the conduction or valence bands, the conductivity of the polymer can be tuned. Therefore, the most significant properties of OCPs are related to their electronic (due to charge carriers), optical (from their highly conjugated backbone), and ionic (whether oxidized or reduced) characteristics (Wallace et al. 2009). These major characteristics are briefly described below and further elaborated on within the more detailed discussion of the methods that are used to characterize OCPs, particularly when they are associated with biological systems. These characterization methods include electrochemical and spectroelectrochemical methods and electronic, vibrational, X-ray, and nuclear magnetic resonance (NMR) spectroscopies.

The polymeric nature of the conducting polymers is, of course, critical to their interaction with biological materials, and the salient polymeric characteristics are also discussed below. In this regard, physical characterization techniques useful for studying biological interactions such as microscopies, light scattering, and microbalance techniques are also considered, along with molecular weight determination.

3.1.2 ELECTRONIC PROPERTIES

One of the unique features of conducting polymers is their ability to act as both semiconductors and conductors, depending on whether the polymer backbone is oxidized (positively charged), reduced (negatively charged), or not (neutral) (Kar 2013). In the neutral state (Figure 3.1a), conducting polymers are semiconductors that become hole conductors or p-type materials when electrons are chemically or electrochemically removed from the semiconductor state, or electron conductors when electrons are added to the semiconductor. As shown in Figure 3.1, the initial removal of electrons (oxidation) leads to the formation of radical cations or polarons (Figure 3.1b) along the polymer backbone, and further oxidation gives bipolarons (Figure 3.1c) that make up the metallic state of the conductive polymer. Each of these states has a completely different spectral signature such that it is easy to distinguish between them, as well as observe the transition from one state to the next.

Changes in the redox state of OCPs will affect the OCP physical, chemical, and electrical properties, including changes in surface charge, wettability, local pH, ion gradients, conductivity, surface roughness, porosity, and elasticity, resulting in an ability to fine tune interactions with biomaterials that can then be probed using electrochemical and spectroscopic techniques. While this is typically done electrochemically or using optical spectroscopy, or spectroelectrochemically, a combination of the two, other spectroscopic and microscopy techniques can also provide a wealth of information about the electronic state of the polymer and how that is affected by materials associated with the OCP. This is particularly important since a biomaterial associated with the OCP or the conditions used to prepare such a composite can frequently lead to a decrease in the stability or complete loss of the oxidized state, reducing the OCP conductivity and therefore the point of making the composite.

As a result, measuring conductivity is important in order to gauge the impact of a biomaterial. This can be carried out in a way similar to that used for inorganic semiconductors. However, the conduction mechanism in OCPs is somewhat different from other conducting materials. The charged centers are mobile and can move along the conjugated polymeric chain by the rearrangement of the double and single bonds in the conjugated system that occurs in an electric field. The movement of polarons and bipolarons carriers is the main mechanism for the conduction of electricity by the doped conjugated polymer. Therefore, the conductivity of OCPs is directly related to the concentration and mobility of the charged species inside the polymer.

3.1.3 IONIC PROPERTIES

The ionic properties of conducting polymers primarily result from the presence of charge on the polymer backbone and its associated dopant, although charged groups can also be introduced into the polymer by way of functionality on the backbone. Doping and dedoping of conducting polymers allow anions to move into and out of the polymer film, as illustrated in Figure 3.1 (Kar 2013). However, if the anion is large, it may become trapped in the polymer on polymer dedoping, resulting in the ingress of a cation into the neutral polymer to balance the charge. Whenever ions move into the polymer, the polymer volume must increase to accommodate them. Therefore, in a typical doping–dedoping process, the OCP swells and contracts, leading to significantly different polymer morphology that can be monitored by surface characterization techniques. Impedance spectroscopy has been used to measure ionic conductivities in OCPs. In addition, since the weights of the oxidized and neutral OCPs are different, the quartz crystal microbalance (QCM) has proved particularly useful for monitoring the polymer dopant ions. Given that the presence or absence of dopant ions is associated with different electronic states of the OCP, most of the other spectroscopic techniques also provide information about the ionic properties of the OCP.

Since many biomolecules or biopolymers are charged, they can be used as dopants for OCPs, and once they are in the polymer, their size often precludes them being easily released from the neutral polymer. Dopant composites such as these can be easily characterized as discussed above.

The ionic nature of doped OCPs also allows the surface energy of polymer films to be varied to suit particular biomolecules or enhance cell adhesion and growth. Contact angle measurements have been essential for determining polymer surface energies. With the development of a range of scanning probe techniques, in particular atomic force microscopy (AFM), highly detailed information about biomaterial polymer interactions can also now be obtained.

3.1.4 OPTICAL PROPERTIES

The conjugated nature of OCPs means that they absorb visible light and are highly colored (Patil et al. 1988). The absorption by the neutral polymer will depend on the length of the conjugation, which will be determined by not only the regioregularity of the polymer but also the twisting of the backbone as a result of the monomer unit interactions. In addition, substituents can cause steric effects and change the absorption wavelength. Consequently, optical spectroscopy provides a great deal of information about the neutral OCP.

Of equal significance are the changes that can be observed in the optical properties of the oxidized or reduced OCP. Oxidation of the polymer drives its absorption into the infrared (IR), affording a significantly different absorption spectrum that clearly shows the degree of oxidation of the OCP. As the polymer oxidation is influenced by the dopant, functionality, or associated materials, the biomaterials will have a significant effect on the optical properties that will be reflected in the optical spectroscopy of the oxidized polymer biomaterial composite.

Highly oxidized OCPs are metallic in appearance and, as a result, are highly reflecting. Therefore, they are often characterized via reflectance from films.

3.1.5 PHYSICAL PROPERTIES

The physical characteristics of OCPs have a significant influence on their interaction with other materials such as biomaterials, and it is therefore important to be able to probe the relevant characteristics, such as surface energy, film morphology, and porosity (Pethrick 2014). The conjugated nature of the OCPs generally means that long-chain regioregular polymers form insoluble rigid rods unless appropriate functionality is introduced onto the polymer backbone to promote solubilization in a suitable solvent. The majority of OCPs are not regioregular, however, due to polymerization taking place at more than one position on the monomer. This can be commonly seen in the morphology of the OCP film. For example, scanning electron micrographs of electrochemically polymerized polypyrrole (PPy) show "cauliflower-like" structures on the electrode surface as a result of irregular polymer growth (Pringle et al. 2004). This type of morphology will be influenced by an associated biomaterial, impacting other polymer properties, such as surface energy and polymer porosity.

3.2 ELECTROCHEMICAL METHODS

Electrochemistry has played a key role in the preparation and characterization of conductive polymers. Electrochemical techniques are especially beneficial in the case of the controlled synthesis of conducting polymers and for the tuning of their oxidation–reduction states. Electrochemistry combined with other characterization methods (spectroscopic, microscopic, or microgravimetric) can help with understanding the mechanisms of charge transfer and transport processes occurring within these materials, as well as provide insights into the interactions with other species present in the system. Thus, when conducting polymers are utilized in biomedical applications, electrochemical methods can provide a significant amount of information about the biomaterial–polymer interactions. The basic electrochemical techniques are introduced below, and more detailed descriptions can be found in a variety of books, such as that of Bard and Faulkner (2001).

3.2.1 POTENTIAL CONTROLLED TECHNIQUES

The electrochemical setup for potential controlled techniques typically comprises a three-electrode cell consisting of working, auxiliary, and reference electrodes (Figure 3.2). The potential applied to the working electrode can be varied with time in a defined manner or can be held constant while the current is measured as a function of potential or time. The basic experimental system involves a potentiostat that controls the voltage across the working electrode and counter electrode, and it adjusts this voltage to maintain the potential difference between the working and the reference electrode. No current should flow through the reference electrode to maintain a constant potential over all applied conditions.

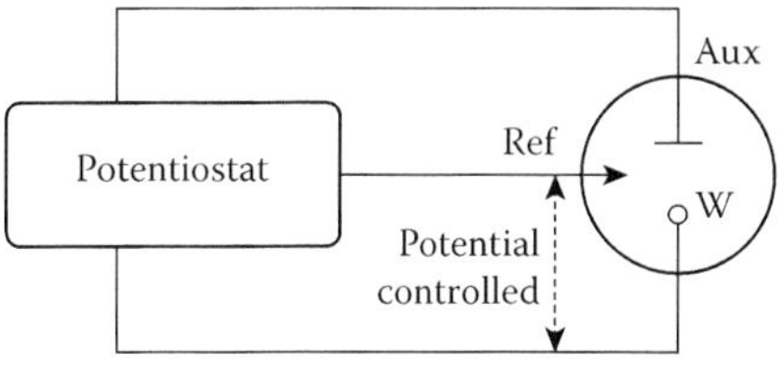

Figure 3.2 Experimental system for controlled-potential measurements. Aux, auxiliary or counter electrode; Ref, reference electrode; W, working electrode.

3.2.1.1 Cyclic voltammetry

Cyclic voltammetry (CV) is a potentiodynamic electrochemical technique that allows the measurement of current as a function of varied applied voltage. In CV, a linear potential sweep is applied to the electrode (Figure 3.3a), and after reaching a switching potential (E_s), the sweep is reversed to its initial value. This sweep time can be widely varied with the scan rate. The recorded curve in the time–potential–current domain is known as a cyclic voltammogram. For a diffusion-controlled electron transfer reaction on an inert planar electrode, the cyclic voltammogram is shown in Figure 3.3b. When the potential is applied to the electrode (Figure 3.3a), the current increases from a background (capacitance) current to a limiting current (Figure 3.3b) that is restricted by the rate of diffusion of the redox-active species to the electrode. Because the limiting current can only occur if the thickness of the diffusion layer remains constant (e.g., stirring the solution when a rotating disc electrode is used, spherical diffusion in the case of microelectrodes, or using a thin layer cell), scanning the potential above the (discharge) potential (E_{ox}) causes depletion of the concentration of the electroactive species close to the electrode surface since diffusion is not fast enough in this overpotential condition. This reduced concentration in the diffusion layer (close to the electrode) decreases the current, which results in the peak shape of the voltammogram (Figure 3.3b). The characteristic shapes of the voltammetric waves and their position on the potential scale fingerprint the individual electrochemical properties of the redox system being investigated.

An example of the cyclic voltammogram of a conducting polymer film on the surface of an electrode is shown in Figure 3.3c. A large increase in capacitive current is observed upon switching from the semiconducting or neutral (reduced) state to the conductive or charged (oxidized) state. The presence of the peaks is characteristic of the typical doping and dedoping processes of conductive polymers, which result from the transitions between the neutral, polaron, bipolaron, and metallic states of the conducting polymer and involve structural reorganization and charging of polymer fragments of different chain lengths (Chen and Inganaes 1996; Skompska 1998; Skotheim and Reynolds 2007).

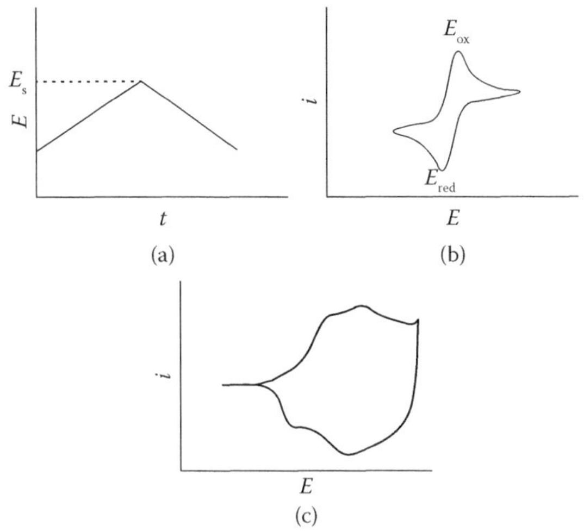

Figure 3.3 (a) Linear potential sweep. (b) Cyclic voltammogram resulting from a linear potential sweep. (c) Postpolymerization cyclic voltammogram of PTh.

CV is usually the first choice of electrochemistry technique used because it provides a quick visualization of the effects of potential and timing on the current in a single experiment. Consequently, any biomaterial influence on the conducting polymer would be detected by way of redox peak numbers, height and position, and capacitance response.

3.2.1.2 Chronoamperometry

In chronoamperometry (CA), the potential step is applied to the working electrode (Figure 3.4a) and the current is recorded as a function of time (Figure 3.4b). The current response (Figure 3.4b) is controlled by diffusion, and the current magnitude is determined by the geometric area of the electrode. Upon application of each potential step (Figure 3.4a), there is a current "spike" (Figure 3.4b) due to potential-induced reorganization of the molecules adjacent to the electrode surface. Once the electrochemical process starts, the rate of reaction is measured as a current that is controlled by the rate of diffusion of the electroactive species to the electrode surface. The measured current is proportional to the concentration gradient and the continued flux, causing the depletion of electroactive species near the working electrode. This results in a concentration (current) profile decreasing with time. CA is an excellent technique for measuring diffusion coefficients (Galus et al. 1994; Yap and Doane 1982).

The electropolymerization of conductive polymers can be achieved using CA. Comparisons of the chronoamperograms recorded during 3,4-ethylenedioxythiophene (EDOT) electropolymerization in the two different electrolytes and two conventional acetonitrile-based electrolytes are shown in Figure 3.4c and d (Wagner et al. 2005). The current transient presented in Figure 3.4c suggests that the polymer growth process is initially much slower in the solution containing Bu_4NClO_4 than in the acetonitrile–lithium bis(trifluoromethylsulphonyl)imide (LiTFSA) solution (Figure 3.4d). Moreover, the shape of the curve (Figure 3.4c) suggests a progressive nucleation mechanism of deposition (Randriamahazaka et al. 1999) in contrast to the instantaneous nucleation suggested by Figure 3.4d (Vico et al. 1999).

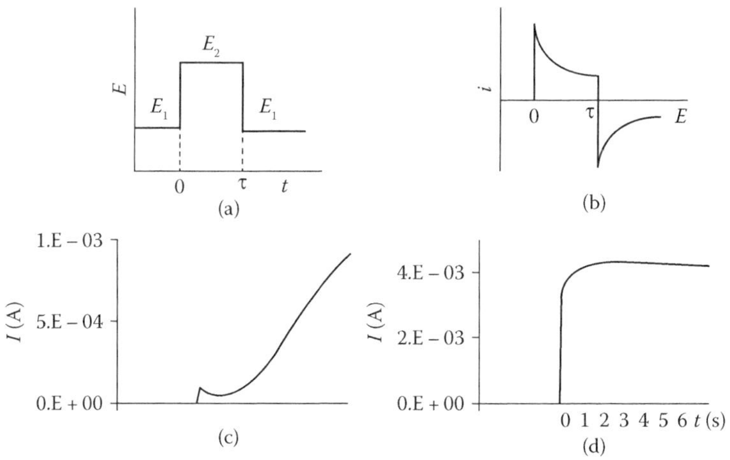

Figure 3.4 (a) Double potential step and (b) current response in chronoamperometry. Current–time responses to potential step from 0 to 1.4 V for the electropolymerization of EDOT in different electrolytes: 0.1 M Bu_4NClO_4 in acetonitrile (c) and in 0.1 M LiTFSA in acetonitrile (d). (From *Synthetic Metals*, 153 (1–3), Wagner, K., et al., Investigation of the electropolymerization of EDOT in ionic liquids, 257–260, Copyright 2005, with permission from Elsevier.)

3.2.1.3 Chronocoulometry

In the chronocoulometry (CC) experiment, the potential step is applied in a manner similar to that for chronamperometry (Figure 3.4a), and the resulting curve is plotted as charge (coulombs) as a function of time (Figure 3.5). As in chronamperometry, the response to a potential step originates from the current. The current (i) is equal to the change in charge (Q) with time (Equation 3.1):

$$i = \frac{dQ}{dt} \tag{3.1}$$

CC is useful for measuring electrode surface areas, diffusion coefficients, the time window of an electrochemical cell, and the mechanisms and rate constants for chemical reactions coupled to electron transfer reactions. One of the major applications of CC is the study of species adsorbed to the surface of the working electrode, which was the reason the technique was originally devised (Anson and Epstein 1968; Anson and Osteryoung 1983). In the case of conductive polymers, CC can be useful for the estimation of the polymerization yield and the doping level of the polymer during its formation, as well as to evaluate the reversibility of the polymer charge processes (Simonet and Rault-Berthelot 1991).

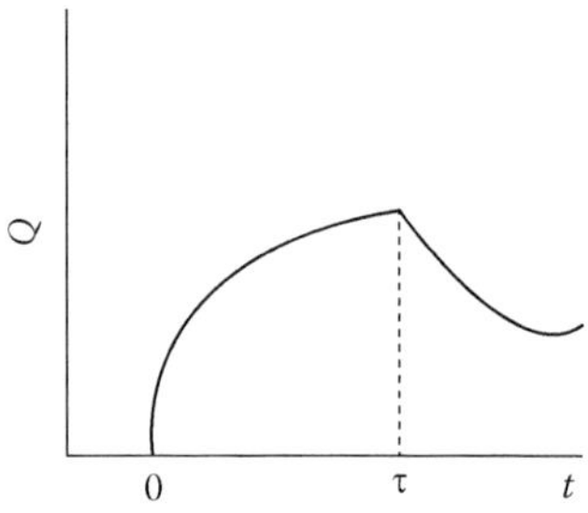

Figure 3.5 Charge versus time response in chronocoulometry.

3.2.2 CURRENT-CONTROLLED TECHNIQUES

In Section 3.2.1, the methods that were described, like CV, CA, and CC, involve controlling the potential while the resulting current is measured as a function of time. The contrary techniques, in which the current is controlled (applied as constant or modulated) and the potential variation measured, are called chronopotentiometric (or galvanostatic) since the potential is determined as a function of time. The controlled current is applied with a current source (galvanostat) between the working and auxiliary electrodes, recording the potential between the working and reference electrodes (Figure 3.6).

Current is an easier parameter to control than potential, because only two elements of the cell, the working and counter electrodes, are involved in the control circuit.

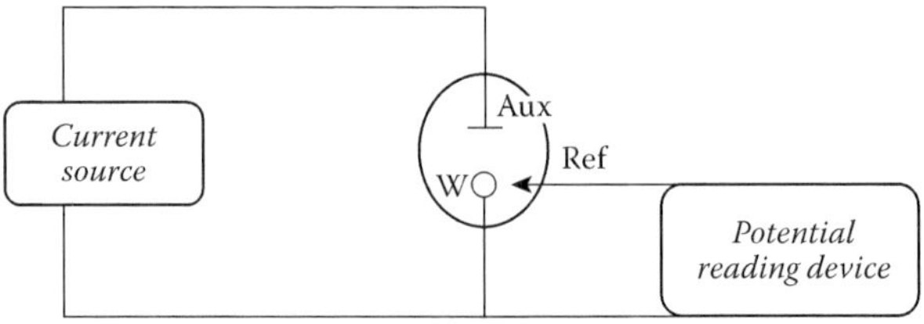

Figure 3.6 Experimental systems for controlled-current measurements. Aux, auxiliary or counter electrode; Ref, reference electrode; W, working electrode.

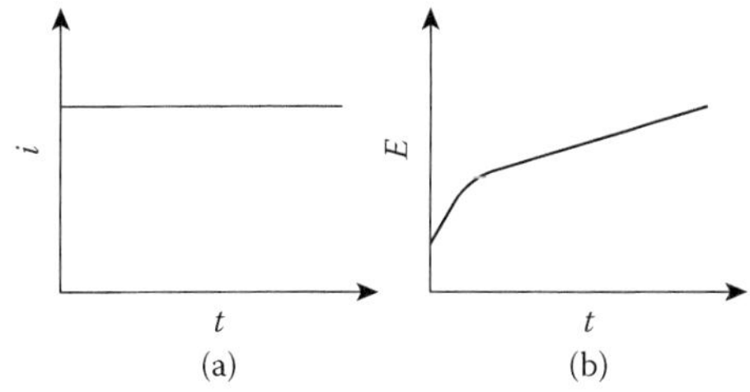

Figure 3.7 (a) Constant current (i) applied during time (t). (b) Example of a potential–time (E–t) curve response that is governed by the reversibility or the kinetics of the electrode reaction.

In chronopotentiometric experiments, the potential is monitored as a function of time (Figure 3.7b), while the current is applied to the working electrode (Figure 3.7a).

In the case of conductive polymer films, parameters like concentration (the number of electroactive sites in the polymer films) or diffusion coefficient refer to the polymer film properties, not solution characteristics, and the information about these values can be estimated from the chronocoulometric plots (Skotheim and Reynolds 2007).

The choice of the electropolymerization method in general has an impact on the structure and properties of the electrogenerated films (Roncali 1992). For example, for polythiophenes (PThs), the most homogeneous and conducting films are generally obtained under galvanostatic conditions, at potentials ~0.5 V more positive than the oxidation potential of the monomer (Yassar et al. 1989). Applying an overpotential initially allows the instantaneous creation of a larger number of nucleation sites, which improves the compactness and thus conductivity of the polymer (Roncali 1992). The choice of counterions has a similar effect, influencing not only the morphology of the films but also their conductivity.

3.2.3 APPLICATION OF ELECTROCHEMICAL METHODS TO CONDUCTING POLYMER–BIOMOLECULE INTERACTIONS

The integration of biomolecules with conductive polymers can be achieved using electrochemical methods in a variety of different ways, such as electrochemical polymerization with the biomolecule as dopant (Garner et al. 1999), covalent binding to the monomer and subsequent polymerization (including using the biomolecule for cross-linking), or polymer modifications (physical entrapment, adsorption, or covalent attachment) (Balint et al. 2014). It should be noted that the term *entrapment* is commonly used in the literature to mean both doping (electrochemical entrapment) and physical entrapment (Cosnier 1999). Ensuring that the properties and functions of the biomolecules, such as enzymes, oligonucleotides, or antibodies, remain after immobilization on conductive surfaces or in polymers is quite often a crucial challenge.

All the electrochemical techniques discussed above can be used to create biomaterial–CP composites. For example, incorporation of biomolecules into CPs through doping (electrochemical entrapment) can be achieved by constant current (galvanostatic) or constant voltage (potentiostatic) deposition or by cycling the potential through the polymer oxidation point in the presence of the biomaterial as they are charged (Ahuja et al. 2006).

Entrapment of biomolecules in electropolymerized films is the most popular approach. This is achieved by mixing the biomolecule with the monomer in the electropolymerization medium prior to synthesis. If the biomolecules possess charges and act as a dopant, electrochemical entrapment takes place, whereas for neutral biomolecules, physical entrapment occurs (Cosnier 1999). Since the pioneering work done by Foulds and Lowe (1986) and Umana and Waller (1986) on enzyme entrapment in OCPs, many other papers have been published on this approach.

However, introducing bioactive molecules through entrapment provides poor control of the amount of biomolecule in the polymer film. The other disadvantage is that a compromise on the ideal electropolymerization conditions for the monomer must be made in order to prevent degradation of the biomaterial. Incorporating a bulky group as a dopant also has a negative effect on the polymer's conductivity (Bendrea et al. 2011; Kar 2013). Nonetheless, the advantage of the electrochemical polymerization is that biocomposite films can be prepared in a one-step procedure, with the possibility of precise deposition of the polymer coating over a conductive surface having a complex geometry. It also offers precise control of the thickness of the polymer film based, for example, on the measurement of the electrical charge passed during the electrochemical polymerization.

A key driver for the incorporation of biomaterials into OCPs has been the development of biosensors. In clinical diagnostics, the need to easily and cheaply monitor metabolites such as glucose, cholesterol, or urea, both quantitatively and qualitatively, has inspired this development. A biosensor is a device having a biological sensing element either connected to or integrated with a transducer. A transducer converts the biochemical signal to an electronic signal. A conducting polymer modified with a biomolecule can directly act as an electronic device that transduces and amplifies a signal, enhancing speed, sensitivity, and biosensor versatility. An OCP chemical or biosensor based on conducting polymers relies on changes in the optical and electrical properties of these materials. The choice of the mediator, conducting polymer, and method of immobilization have a strong influence on properties of the biosensor.

An amperometric biosensor measures the current produced during the oxidation or reduction of a product or reactant, usually at a constant applied potential. Such sensors have fast response times and good sensitivity. However, the excellent specificity of the biological component can be compromised by the partial selectivity of the electrode. This lack of specificity requires sample preparation, separation, or some compensation for interfacing signals. Conducting polymers have attracted much interest as a suitable matrix for the entrapment of enzymes (Adeloju and Wallace 1996).

Potentiometric biosensors relate electrical potentials to the concentration of analyte, and they are selective. The measurements can be carried out in two ways: either by an amperometric measurement (at the potential for which the peak occurs) and the comparison of the obtained result with the calibration curve (amperometric biosensor), or by carrying out a voltammetric measurement whose result is compared with voltammograms obtained earlier. Interaction of NH_3 with PPy was utilized for a design of a potentiometric biosensor of urea with urease immobilized on an electrodeposited PPy layer (Pandey and Mishra 1988; Trojanowicz et al. 1993).

A change in redox potential or pH of the biological environment on the conjugated polymer matrix can change the response of electronic conductivity of conjugated polymers over several orders of magnitude. Biosensors based on the conductivity response of conjugated polymers have been demonstrated for polyaniline (PANI) and the detection of glucose, urea, lipids, and hemoglobin (Contractor et al. 1994; Nishizawa et al. 1992).

3.2.4 CONDUCTIVITY MEASUREMENTS

3.2.4.1 Electrical conductivity

3.2.4.1.1 Introduction to electrical conductivity

The electrical conductivity of conducting polymers is typically expressed in terms of volume or bulk conductivity (σ). This is determined as siemens per meter (or commonly in siemens per centimeter) and is defined as the ratio of the current density (J) to the electric field strength (E) (Equation 3.2)

$$\sigma = \frac{J}{E} \tag{3.2}$$

It is the inverse of the electrical volume or bulk resistivity (ρ) that has units of $\Omega \cdot m$, which describes the resistance of a given material regardless of the shape or size.

The determination of conductivity is practically done by determining the resistivity using Ohm's law (Equation 3.3):

$$R = \frac{V}{I} \tag{3.3}$$

where V is the voltage (V) and I is the current (A).

The resistance (R) is obtained by measuring the voltage when a known current is applied. The bulk resistivity ρ ($\Omega \cdot m$) is then calculated by including the geometry of the material (Equation 3.4)

$$\rho = R\frac{A}{l} \tag{3.4}$$

where R is the resistance (Ω), l is the length of the conductor (m), and A is the cross-sectional area of the conductor (m^2).

However, since thin films of conducting polymers are commonly produced, the term *sheet resistance* (R_s) is also commonly used. R_s may be thought of as a material property of a thin film with uniform thickness, which is essentially two-dimensional. It is nothing more than a restatement of volume resistivity without the film thickness being specified as calculated by Equation 3.5

$$R_s = \frac{\rho}{d} \tag{3.5}$$

where ρ is the bulk resistivity and d is the film thickness (m).

Sheet resistance (R_s) has units of ohms but is frequently given as ohms per square ($\Omega/\square$). This was introduced in 1968 by Berry et al. (1968), where the material dimension term (length/width) was called the number of "squares" in the resistor and was a pure number, having no dimensions. The authors claim that the sheet resistance has the unit of ohms, but it is convenient to refer to it as "ohms per square ($\Omega/\square$)" since the sheet resistance produces the resistance of the resistor when multiplied by the number of squares.

The issue is further confused by the frequent use of the term *surface resistivity*, which assumes that current can only flow at the surface of a material. The surface resistivity description will be used, for example, for highly resistive material with a conductive surface where the current flows almost entirely in the surface. Since no significant current is carried through the resistive bulk of the material, the thickness of the bulk material does not affect the resistance measurement.

3.2.4.1.2 Conventional resistivity measurement

For general purposes, two-point probe electrical measurements are normally used. However, a problem that occurs when using a two-point setup is that the voltage that is measured arises not only from the material resistance but also from the resistance of the leads and contacts. In practice, the method is generally restricted to resistivities below ~10^6 Ω·m; otherwise, currents become too small to measure accurately and voltmeter resistances become significant (Blythe 1984). When the resistance being measured is relatively low, or the resistance of the probes or the contacts is relatively high, a four-point probe will provide more accurate results. Four-point probe instruments measure the average resistance of a thin layer or sheet by passing current through the two outer points of the probe and measuring the voltage across the inside two points; this allows the probes that supply the current to be separate from the probes measuring the voltage. This separation of current and voltage eliminates the impedance contribution of the wiring and contact resistances (Figure 3.8).

There are many challenges in measuring the conductivity of conducting polymer films since the film must be measured on an insulating surface, be thick enough to allow contact with the measurement probes, and be hard enough such that the probes make good electrical contact without piercing the film. Soluble chemically prepared OCPs can be cast onto an insulating surface providing suitable thick films. However, electrochemically polymerized OCPs must be grown on a suitably large electrode, and then the film removed and placed on the insulating surface. The brittleness of OCP films makes this problematic.

Four-point probe measurements are usually the better choice for determining the resistivity of biologically modified conductive polymers whose conductivity is quite often compromised by the incorporation of biomaterials. Generally, conductive polymers doped with organic dopants show lower conductivity than polymers doped with inorganic ions, due to restricted ion mobility within the PPy polymer chain (Kar 2013).

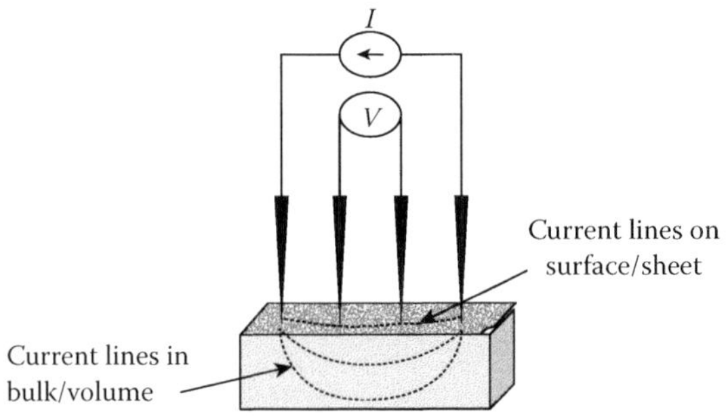

Figure 3.8 Representation of a four-point probe resistivity measurement showing the current (*I*) and voltage (*V*) sources, and current lines.

The effect of incorporating a bulky group, such as a protein or peptide, therefore often has a negative effect on the polymer's conductivity (Bendrea et al. 2011). For example, it has been reported that the conductivity of electrochemically polymerized hyaluronic acid (HA)–doped PPy films is greatly reduced compared with the conductivity of PPy films doped with polystyrene sulfonate (PSS) (Collier et al. 2000). However, despite the low conductivity of most bioconductive materials, they are still able to pass the low currents that are required in many bioapplications, such as cell stimulation (typically in the range of 10–100 μA) (Rowlands and Cooper-White 2008).

3.2.4.1.3 In situ *conductivity measurement*

In the late 1980s, Schiavon et al. (1989) developed a special electrochemical setup that allows the deposition of a polymer via anodic electropolymerization onto two electrodes and the measuring of its conductivity *in situ*. Conductivity measurements were performed with the setup illustrated in Figure 3.9, following the method used by Wrighton et al. (Kittlesen et al. 1984; Paul et al. 1985; Thackeray et al. 1985). The key advantage of using such an *in situ* technique is that it minimizes the risk of the film damage that invariably occurs and affects conductivity measurements when trying to remove electrochemically polymerized films from electrodes.

In situ conductivity measurements have been performed on PPys (Vercelli et al. 2002), PThs (Aubert et al. 2002; Vercelli et al. 2002), and PANIs (MacDiarmid and Epstein 1994).

3.2.4.2 Ionic conductivity

While the electronic conductivity measurement is a standard and noncomplicated technique, ionic conductivity is difficult to quantify, and there have been few attempts to achieve this. Pickup and his coworkers have utilized impedance spectroscopy

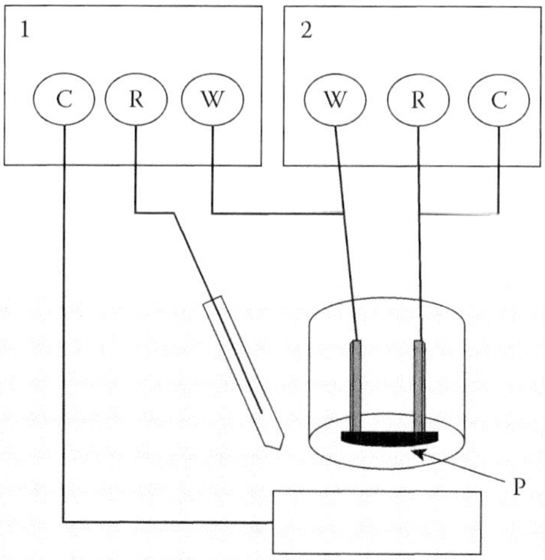

Figure 3.9 Schematic representation of the electrochemical setup for *in situ* conductivity measurements: 1, potentiostat for changing the oxidation level of polymer P; 2, potentiostat for DC conductivity measurements; C, counter electrode; R, reference electrode; and W, working electrode.

(see Section 3.2.5) to investigate ionic conductivities in a number of OCPs (Ren and Pickup 1993a, 1993b). More recently, an electrophoretic-like methodology to measure cation mobility into the conducting polymer during dedoping was established (Stavrinidou et al. 2013).

3.2.5 ELECTROCHEMICAL IMPEDANCE SPECTROSCOPY

Electrochemical impedance spectroscopy (EIS) is a common technique for studying charging and transport phenomena in conjugated polymers (Stavrinidou et al. 2014). In EIS, the electrochemical cell behavior in general is viewed as a network of resistors and capacitors (Figure 3.10a). The model displayed in Figure 3.10a is a modification of the Randles model (Randles 1947) and is valid for diffusion driven by a gradient and not by an electric field. Since this simple model only considers a pure diffusional transport of the single charge carrier and a uniform film, more advanced models have been developed for conductive polymers (Rubinson and Kayinamura 2009), such as the dual-transmission-line model comprising ionic and electronic resistance rails connected in parallel with a capacitance, as proposed by Albery and coworkers (1989).

Electrical impedance (Z), like resistance, describes the ability of a circuit to resist the flow of electrical current, and this is represented by the term *real impedance* (z'). It also refers to the ability of the circuit to store electrical energy in the "imaginary impedance" (z'') component. The measurements are performed under conditions in

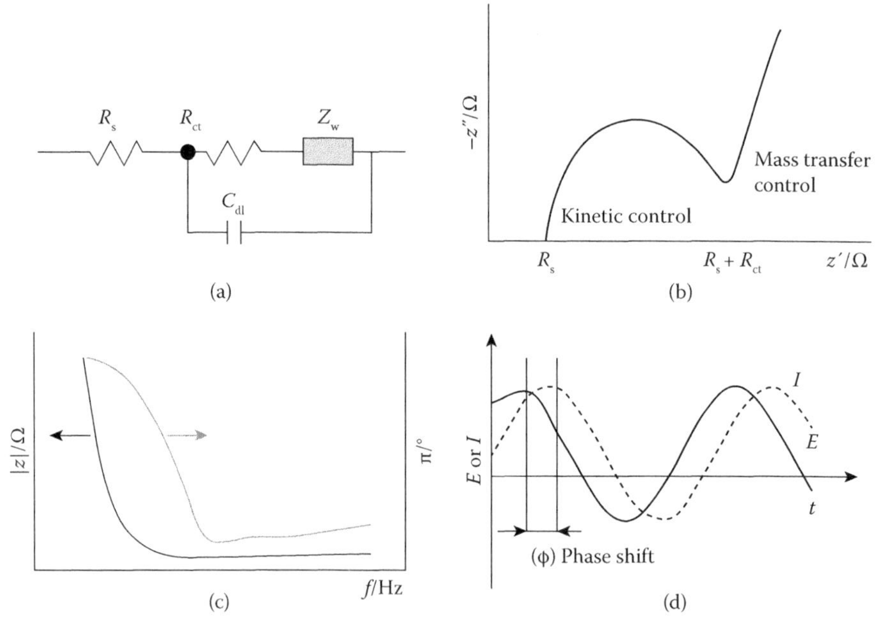

Figure 3.10 (a) Randles circuit where R_s is the solution resistance, C_{dl} is the double-layer capacitance, R_{ct} is the charge transfer resistance, and Z_w is the Warburg impedance, which accounts for the diffusion of ions from the bulk electrolyte to the electrode interface. (b) Nyquist plot representative of the previous circuit. (c) Example of a Bode plot. (d) Time difference between the applied voltage (V) and resulting current (I) response in reaching the same amplitude resulting in the phase shift (ϕ).

which the main potential is usually fixed at the working electrode (the applied direct current [DC] potential can be constant or scanned) with a small perturbation of alternating current (AC) voltage (usually 5–10 mV) imposed on the constant potential.

In an electrochemical cell, electrode kinetics, redox reactions, diffusion phenomena, and molecular interactions at the electrode surface can impede the flow of electrons in an AC circuit. The measured change in current describes the way in which the system, being in a steady-state condition, follows the perturbation. At these conditions, the feedback of the electrochemical systems can be considered to be linear (Rangarajan 1974). Working close to equilibrium offers a high-precision measurement, with the response steady and averaged over a long period. This differs from the electrochemical methods described in Sections 3.2.1 and 3.2.2, in which the application of a potential or current to the electrode gives a response that is usually a transient signal as a consequence of driving the electrode to a condition far from equilibrium.

Electrical impedance is defined as the ratio of an incremental change in voltage to the resulting change in current, that is, $V(t)/I(t)$. It is measured over a spectrum of frequency, which not only provides the information about magnitude of resistance and capacitance but also indicates the system behavior in this frequency domain. To capture responses from processes with different time constants (kinetic), the measurement of impedance can be performed over a wide range of frequencies (10^{-4} to 10^{6} Hz). The electrode processes are controlled by electrochemical reactions at high frequencies and by mass transfer at low frequencies.

The most popular format for evaluating electrochemical impedance data is the Nyquist plot. In the Nyquist plot (Figure 3.10b), the imaginary impedance (z'', Y-axis) is plotted against the real impedance (z', X-axis) at each excitation frequency. The data from the Nyquist plot are analyzed by fitting to an equivalent electrical circuit model, such as Figure 3.10a, and provide solution resistance (R_s), which can be found by reading the real axis value at the high-frequency intercept; charge transfer resistance (R_{ct}); double-layer capacitance (C_{dl}); and ion diffusion inside the polymer (usually the Warburg element, Z_w). An alternative but less common interpretation of electrochemical impedance data is the Bode plot, which shows more clearly the general frequency dependence of impedance and phase angle variation, and has been used to correlate polymer morphology and impedance results (Sarac et al. 2008). In the Bode plot (Figure 3.10c), both the logarithm of the absolute impedance ($|Z|$) and the phase shift (φ) are plotted against the logarithm of the excitation frequency (f); φ, as shown in Figure 3.10d, is the time difference between the applied voltage (V) and resulting current (I) response in reaching the same amplitude.

EIS is an excellent tool for obtaining highly resolved kinetic data related to thin films of electroactive polymers using a small-amplitude perturbation signal (Levi and Aurbach 2002; Macdonald 2006). In the most common configuration, a polymer film is deposited on a conductive substrate (usually metal) and a DC bias is applied to the electrode in the electrochemical cell. This leads to ion injection from the electrolyte into the polymer and sets the doping level of the conductive polymer. For example, during oxidation processes, anions, cations, and solvents can simultaneously cross the polymer–electrolyte interface where, for example, anions can be inserted, and cations, as well as solvent, can be simultaneously expelled to compensate for the positive charges created during the redox reaction. Once the desired doping level is reached, a small AC modulation is applied, causing a movement of charge carriers. The interpretation of the obtained impedance data can be complex given the presence of both ionic and electronic carriers, which can move at the same time upon applying this perturbation (Vorotyntsev et al. 1999).

EIS-based sensors (impedimetric sensors or immunosensors in the biomedical field) are ideal tools for observing the direct dynamics of biomolecule interactions (Prodromidis 2010). The resistive component of the biomolecule recognition element arises when charges transfer between the electrode and solution, whereas capacitance arises when charges are bonded in the layer, causing, for example, a change in permittivity or dielectric properties. Both components can be used as the detecting parameters (Tagawa et al. 2011). For example, impedance results showed that the resistance of a functionalized bithiophene polymer film changed due to binding of the target oligonucleotide (Sosnowska et al. 2013).

EIS measurements have shown that the charge transport characteristics of polymer coatings are extremely sensitive to the surface morphology, a polymer property critically important for biomaterial interaction. The rough, fuzzy structure of the electrochemically deposited conducting polymer leads to a significant (~2 orders of magnitude) decrease in the sample impedance (Yang and Martin 2006). In addition, conducting polymer coatings lowers the impedance of electrodes and provides a mechanical buffer between a hard device and soft tissue. Thus, two synthetic anionically modified laminin peptides were used for doping poly(3,4-ethylenedioxythiophene) (PEDOT) electrodeposited on platinum electrodes (Green et al. 2009) and compared with conventional PEDOT-pTS films. It was found that the use of the bulky synthetic peptides as anionic dopants produced a softer interface with improved impedance characteristics, especially in the low-frequency, biologically significant region.

There are also many examples from the literature of the use of EIS with other techniques to characterize conductive polymers in the area of bioapplications. For example, EIS measurements on a polymer biosensor combined with microscopic techniques were used for the detection of viral infection of human cells (Kiilerich-Pedersen et al. 2011). Changes in the electrochemical impedance responses measured on PEDOT films cultured with cells were good indicators of cell morphology changes.

EIS-based sensors have been miniaturized and integrated into more complex diagnostic tools (Prodromidis 2010). However, despite such advances, a lack of reliability of this type of instrumentation has limited the usage of EIS techniques to academic studies.

3.3 SPECTROSCOPIC METHODS

Spectroscopic studies are critical for understanding the change of properties of conducting polymers on doping, dedoping, and their interaction with other materials. Common spectroscopic methods, for example, ultraviolet–visible (UV–Vis), IR, and Raman spectroscopy, provide direct evidence for OCP properties, such as the oxidation level of conjugated polymers in either solution or the solid state. In contrast, NMR spectroscopy is typically limited to soluble neutral OCPs and X-ray spectroscopy to film properties. The key aspects of these spectroscopic techniques are briefly discussed in the following sections.

3.3.1 ELECTRONIC ABSORPTION SPECTROSCOPY

The conjugated nature of OCPs is the primary influence on their electronic absorption, with increasing conjugation length reducing the absorption energy, that is, shifting it to the red. However, the change in the electronic structure caused by oxidation (Figure 3.1) has a profound influence on the electronic absorption spectrum of an OCP such that polaron and bipolaron bands are readily identified. This is nicely illustrated in the electronic absorption spectrum of PEDOT:DS (Figure 3.11a), a biomaterial OCP composite in which PEDOT is doped with the polysaccharide dextran sulfate (Harman et al. 2015). The broad absorption band at λ_{max} = 600 nm (A, Figure 3.11a) results from the π–π^* energy absorption of the conjugated polymer backbone, often referred to as the transition from the valence band to the conduction band. It is indicative of the mean effective conjugation length in the polymer since the polymer regioregularity and twisting of the backbone affects the planarity of the conjugated π system, limiting the absorption energy (wavelength). Consequently, any shift of this absorption to the red (absorption band shifting toward longer wavelength) indicates an extension of the conjugation. The energy band gap, E_g, of the polymer can be calculated from the long-wavelength onset of this absorption band (Gierschner et al. 2007).

The formation of polarons (radical cations) upon oxidation of the PEDOT:DS, which corresponds to point B in the cyclic voltammogram (Figure 3.11b), is readily observed (B, Figure 3.11a). Further removal of electrons leads to the appearance of the so-called free carrier tail (C, Figure 3.11a and b) that results from the presence of bipolarons (dications) along the polymer chains (Figure 3.1). In its most highly doped

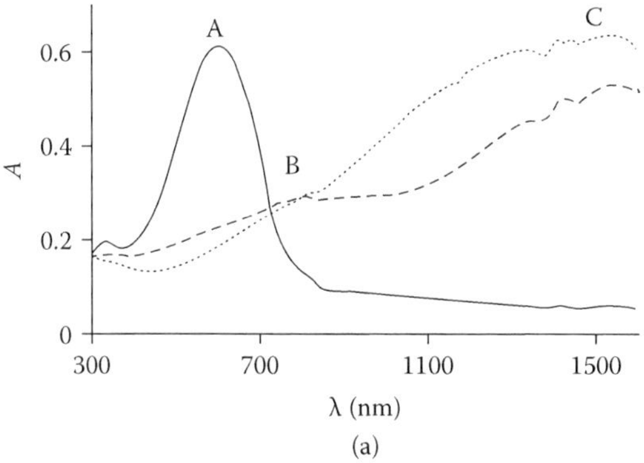

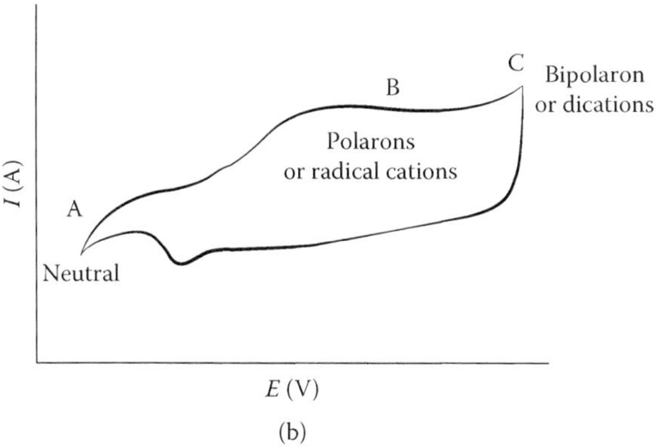

Figure 3.11 (a) UV-vis absorption spectra for reduced (solid line, A), partially oxidized (dashed line, B) and oxidized (dotted line, C) PEDOT:DS polymer film. (b) Cyclic voltammogram for PEDOT:DS. (From *Acta Biomaterialia*, 14, Wagner, K., et al., Poly(3,4-ethylenedioxythiophene): dextran sulfate (PEDOT:DS)—A highly processable conductive organic biopolymer, 33–42, Copyright 2015, with permission from Elsevier.)

ordered state, the conductive polymer shows "metallic" absorption (C) and reflection behavior (Wang et al. 1991).

Numerous examples of the absorption characterization of OCP biomaterial composites have been reported (Bendrea et al. 2011; Ravichandran et al. 2010).

3.3.2 EMISSION SPECTROSCOPY

The emission, usually fluorescence, of conductive polymers can be based on two unique features: superquenching and the dependence of fluorescence emission on the chain conformation. The process of quenching, or deactivation of the excited state, decreases the fluorescence intensity. As noted by Swager (1998), the fluorescence emission of conjugated polymers can be made to respond to very small quantities of analytes. Changes in emission intensity of the polymer can be achieved simply by interaction of a quencher (which could be a biomolecule) with the polymer units, as illustrated in Figure 3.12.

The delocalized electronic structure of conductive polymers promotes efficient coupling between polymer segments, allowing excitons to move to lower electron or energy acceptor sites over long distances. As a result of this exciton mobility, greater amplification of fluorescence quenching or superquenching is achievable. Such a process is reversible if a polymer and a quencher are free to dissociate. With an increase in chain

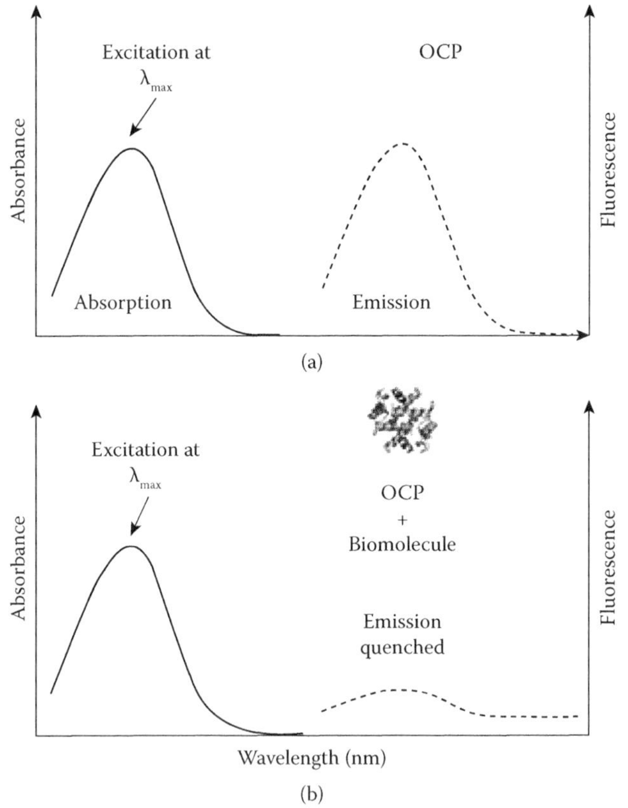

Figure 3.12 Representation of typical absorption and fluorescence emission spectra of (a) an OCP and (b) an OCP quenched by a biomolecule.

length, the fluorescence intensity of conductive polymers becomes stronger, as does the effect of external quenchers, although this is limited by exciton mobility and lifetime. Consequently, the binding of a single-analyte molecule can quench the fluorescence of the whole polymer chain. Therefore, the presence of only one such molecule or group interacting with one element of the polymer chain can provide enough signal to classify conductive polymers as high-sensitivity transducers. Transducers are molecular or supermolecular structures, which, when activated by the signal from one system, provide this signal, often in another form, to a different system.

In most conductive polymer cases, the quenching of polymer fluorescence can be considered a sensing mechanism, and this was initially carried out in organic solvents (Yang and Swager 1998). However, this greatly limits their application in the biosensor field. To overcome this problem, the inherently hydrophobic nature of conjugated polymer backbones must be modified by the addition of water solubilizing groups, usually in the form of ionized side chains (Harrison et al. 2000). The signal transduction mechanism in this system is reliant on electrostatic interaction between the biomolecular quencher and the charged groups of the conductive polymers. For example, extremely efficient fluorescence quenching of conjugated polyelectrolyte fluorescence has been observed with positively charged heme-containing proteins and anionic polyelectrolyte (Fan et al. 2002). The groups of Leclerc (Dore et al. 2004; Ho et al. 2002) and Nilsson (Nilsson and Inganaes 2003) have also made use of changes in the photophysical properties of cationic PThs on complexation to anionic DNA to create highly sensitive DNA sensors.

The emission of conductive polymers also depends on chain conformation (Demchenko 2010). Planar conformations extend the chain conjugation and favor the fluorescence quantum yield, and may also cause a shift of emission wavelength (Stokes shift). When the planarity is destroyed as a result of conformational change, the peaks shift to shorter wavelengths. Modification of the polymer with side groups and further interaction with them can determine the conformational states of the backbone, and thus the fluorescence signal. Of special interest is the modification with a charged side group, which transforms a low polarity polymer into a polyelectrolyte and allows target binding based on electrostatic interactions. For example, with this kind of modification, the polymer can specifically recognize charged molecules such as adenosine triphosphate (ATP) through electrostatic and hydrophobic cooperative interactions, which can induce changes in polymer conformation (Li et al. 2005). Ordering of electronic polymers through interactions with biological macromolecules was demonstrated after complexation with DNA (Nilsson and Inganaes 2003), proteins (Nilsson 2005; Nilsson and Inganaes 2004), and synthetic polypeptides (Nilsson et al. 2003, 2004).

3.3.3 INFRARED SPECTROSCOPY

Vibrational spectroscopy is advantageous as an analytical tool for polymers because it provides information about composition, structures, conformations, and intermolecular interactions. This is because the vibrational frequencies of a molecule depend on the nature of the polymer motion, the mass of the atoms, detailed geometric arrangements, the nature of the chemical bonding, and the polymer's chemical and physical environment (Koenig 2001). Vibrational spectra result from the interactions of the vibrational motions of a molecule with the energy at which any peak in an absorption or transmission spectrum appears corresponding to the frequency of vibration of a part of the molecule (Stuart 2004). Both IR and Raman are complementary techniques in vibrational spectroscopy. In IR spectroscopy, absorption of IR radiation not only results in an absorption or transmission spectrum that represents a molecular fingerprint of the sample but also allows identification of specific groups on the molecule.

In conducting polymers, the incident IR radiation interacts with vibrational excitations of the material and also with free carriers in their electronic structure. These interactions create phenomena such as free-carrier absorption, excitation across the energy gap, exciton transitions, or light scattering by free electrons, and are reflected in the IR spectrum. On oxidation and reduction, the conductive polymers undergo structural transitions that give rise to spectroscopic changes in the mid- and near-IR regions.

In mid-IR, the absorptions are strongly enhanced due to the coupling of the ring vibrations to the motion of the charge carriers arising from the oxidation. These absorptions are called doping-induced bands or infrared-active vibrations (IRAVs) (Neugebauer 2004). The near-IR absorptions are associated directly with charge carriers.

IR spectroscopy can be most useful for the characterization of the chemical composition of conductive polymers and biomolecule coatings. In the work of Tam et al. (2010), IR spectroscopy was used to verify the existence of PPy and DNA sequences on the electrode surface—a sensor developed for rapid herpes virus detection. IR is also a good tool to determine the type of intermolecular interactions, for example, to distinguish between physical and chemical protein adsorption on the polymer surface (Zhang et al. 2006). Thus, greater bovine serum albumin (BSA) adsorption occurs on PPy around the isoelectric point of the protein due to the fewer net protein charges reducing the electrostatic repulsion between proteins, and thereby enabling incoming proteins to more easily adsorb in the presence of the existing adsorbed proteins.

3.3.4 RAMAN SPECTROSCOPY

Raman spectroscopy is a powerful technique for structural investigation, able to provide the needed information on the molecular level. The Raman effect occurs when a sample is irradiated by monochromatic light, causing a small fraction of the scattered radiation to exhibit shifted frequencies that correspond to the sample's vibrational transitions. Ground-state molecules produce lines shifted to energies lower than the source, while the slightly weaker lines at higher frequency are due to molecules in excited vibrational states. These lines, the result of the inelastic scattering of light by the sample, are called Stokes and anti-Stokes lines, respectively. The chemical structure of electrically conducting PPy films doped with *p*-toluene sulfonate and dodecyl sulfate was studied by Fourier transform (FT)–Raman spectroscopy. The spectra were compared with those from the corresponding reduced polymers after dedoping and found to be consistent with polaron and bipolaron descriptions of the electron transport mechanism in PPy (Jenden et al. 1993).

3.3.5 SPECTROELECTROCHEMISTRY

Spectroelectrochemistry (SEC) is the coupling of a variety of spectroscopic techniques with electrochemical methods, with the most common spectroscopic techniques used for OCPs being optical and vibrational. SEC techniques allow *in situ* spectroscopic investigations of electrogenerated species. Thus, coupling optical spectroscopy to electrochemistry allows the optical changes of OCPs at a range of different oxidation and reduction potentials to be determined. As discussed in Section 3.3.1 for dextran sulfate–doped PEDOT (Figure 3.11), this allows a correlation between the polymer structure and the electronic nature of the polymer. In a similar manner, IR and Raman measurements at a range of potentials can be used to identify the effects of charge carriers on the polymer structure.

However, the combination of spectroscopy with electrochemistry requires changes in the way the measurements are carried out, leading to less than optimal data; the change in the configuration of the electrochemical cell in order to allow electromagnetic radiation to pass through the polymer sample on the working electrode creates nonideal conditions for electrochemical measurements. Other problems also arise for particular techniques. For example, a thin-layer cell is required for IR spectroscopic measurements in order to minimize strong solvent absorption; however, this creates a problem with uncompensated solution resistance, leading to nonuniform potential across the electrode surface. A conventional reference electrode cannot be used in this kind of cell configuration, and as a result, potential drift when using a pseudoreference electrode made from silver or platinum wire (or foil) has to be taken into account (Kaim and Klein 2008).

For optical measurements, the transparent electrodes are required, such as glass with a thin film of an optically transparent, conducting material, for example, indium-doped tin oxide (ITO).

Typical plots of spectroelectrochemical measurements will show absorbance or transmittance as a function of

1. Wavelength at a constant potential
2. Potential at a fixed wavelength
3. Time after a potential perturbation, usually a potential step, or alternatively a potential scan

For example, large and reverse hypochromic and bathochromic shifts on either polaronic or bipolaronic bands have been determined from electrochromic PPy films by "*in situ*" spectroelectrochemical measurements in the visible region during oxidation–reduction switching and linked to the conformational changes of the polymeric chains during the electrochemically induced swelling–shrinking processes (Otero and Bengoechea 1999).

SEC measurements have also been used to identify conformational changes in amorphous structures of conducting polymers able to mimic ion channels in membranes, enzymatic processes, and conformational changes in muscles (Otero 2013).

SEC has also been used to determine the effects of the position of amine functionality within an OCP structure for glucose biosensor platforms constructed with covalently immobilized glucose oxidase on polythienylpyrrole derivatives (Ayranci et al. 2015).

"*In situ*" measurements are highly important to gain some information regarding doping processes or the dynamics of the electrochemical processes accompanied by ionic fluxes. By coupling electrochemistry with IR (Yohannes et al. 2009) or electron paramagnetic resonance (EPR) (Domagala et al. 2008) spectroscopy, a more complete description of the molecular aspects of charge transfer mechanisms and surface equilibria is achievable.

3.3.6 NUCLEAR MAGNETIC RESONANCE SPECTROSCOPY

Over the past 50 years, NMR spectroscopy has become the preeminent technique for determining the structure of organic compounds. The technique can be used for conducting polymer characterization in solution and solid state. The data obtained provide information about functionality, regioregularity, packing, and the doped state of the material.

Solution-based ^{1}H NMR spectroscopy can be used to determine the regioregularity of poly(alkylthiophene)s (Chen and Rieke 1992). The regioregular polymer displays a much simpler spectrum than the regiorandom or regioirregular material. Solid-state ^{13}C analysis has been used to establish the molecular structure of poly(isothianaphthene) (Ottenbourgs et al. 1997). This technique can be used for all insoluble undoped materials. Apart from molecular structure determination, NMR spectral analysis can also be used for determining macrostructural arrangements, such as packing of the polymer chain in the semicrystalline phase (Dudenko et al. 2012).

Solid- and sometimes liquid-state NMR techniques can also be used to determine the doping levels of oxidized polymers. For example, ^{1}H NMR spectroscopy was utilized for investigating the doping level of PANI (Wang et al. 2010) and a poly(phenylene vinylene) derivative (Lee et al. 1995). Solid-state, cross-polarization magic-angle spinning (CMPAS) ^{13}C NMR spectroscopy was used to characterize doping effects on molecular structure and the doping level of PPy and its N-substituted derivatives (Forsyth et al. 1994), as well as poly(acetylene) (Clarke and Scott 1983; Vainrub et al. 1993).

3.3.7 ELECTRON SPIN RESONANCE SPECTROSCOPY

Electron spin resonance (ESR) or EPR spectroscopy is a technique for studying unpaired electrons, and consequently, it is very useful for studying doped conducting polymers, their doping level, charge delocalization, and other aspects of OCPs associated with unpaired electrons. The technique is discussed in detail in Chapter 4.

3.4 X-RAY-BASED TECHNIQUES

There are a number of techniques that use X-rays to characterize solid-state materials. Some of these techniques have proved to be useful tools for characterizing conducting polymers.

3.4.1 X-RAY PHOTOELECTRON SPECTROSCOPY

X-ray photoelectron spectroscopy (XPS) is a surface-sensitive quantitative spectroscopic technique that can be used to determine elemental composition, empirical formula, and the chemical and electronic states of the elements that exist within a material. XPS has been used to determine the degree of doping in PPy (Naudin et al. 2002) and PEDOT (Massonnet et al. 2015). However, because of the surface sensitiveness of this technique, the results from the bulk of the film can be obscured by the surface information (Neoh et al. 1997).

XPS can be used for monitoring the surface chemistry of conducting polymers, for example, click chemistry (Daugaard et al. 2008) or degradation (Rannou et al. 1999). The technique is also useful for observing the adsorption and interaction of biological molecules, such as proteins and DNA, with a conducting polymer surface (Saoudi et al. 2004; Zhang et al. 2011).

3.4.2 SMALL-ANGLE X-RAY SCATTERING AND WIDE-ANGLE X-RAY SCATTERING TECHNIQUES

The following techniques, in contrast to XPS, are designed to give information about the macrostructural arrangement of the investigated materials, such as level of crystallinity and porosity, rather than the elemental composition. Both the small-angle X-ray scattering (SAXS) and wide-angle X-ray scattering (WAXS) techniques are complementary and usually performed together.

SAXS is recorded at a very low angular range (typically 0.1°–10°), which contains information about the shape and size of macromolecules, characteristic distances of partially ordered materials, pore sizes, and other data. SAXS is capable of delivering macromolecular structural information between 5 and 25 nm, with repeat distances in partially ordered systems of up to 150 nm. WAXS is also used to determine the crystalline structure of polymers, specifically for the analysis of Bragg peaks scattered to wide angles that are caused by sub-nanometer-sized structures (Kostorz 1988; Rabiej and Wlochowicz 1990). Because of the crystallinity, crystal domain size, and size of the pores influence conducting polymer properties such as conductivity or adhesiveness, those techniques are useful for their characterization (Balko et al. 2013; Massonnet et al. 2015).

3.4.3 X-RAY DIFFRACTION

X-ray diffraction (XRD) is a tool primarily used for identifying the atomic and molecular structure of a crystal; however, the analysis can be used to investigate any regular structures in a material, such as aligned polymer chains. Like the two techniques described before (SAXS and WAXS), it uses the same physical principles and gives insight into the macromolecular arrangements of the conducting polymer. Polymers rarely crystallize completely but frequently have some ordered or crystalline domains.

XRD can be used for studying the influence of the dopant on the macromolecular structure of OCPs. It has been found that poly(alkylthiophene) and PPy doped with alkylphenyl sulfonate anion can make helical-like structures (Davidson et al. 1996; Watanabe et al. 2012). Lu et al. (2014) used this technique to determine poly(3-hexylthiophene) chain alignment in polymer nanofibers. McCullough et al. (1993) were able to find distances between poly(alkylthiophene) chains and established the correlation between polymer structure and electronic properties. Fischer et al. (1994) studied the temperature-induced nucleation of PANI and its influence on the electronic properties of the material.

Crystallization of conducting polymers has a huge impact on charge transport properties. In general, transport is more efficient the more regular the polymer, and so the more crystalline (Hotta and Waragai 1993).

XRD can also be used to investigate intermolecular interactions. This technique, along with solid-state NMR techniques, has provided insight into the intermolecular interactions between a conjugated polymer and fullerene derivative (Miller et al. 2012).

3.5 SURFACE CHARACTERIZATION

The interaction between a conducting polymer and another material, such as a biomolecule, is often dependent on the polymer surface or interface. Therefore, characterization techniques that provide insight into the types of surface characteristics that determine biomolecular interactions, such as chemical composition, heterogeneity, morphology, and potential of the polymer, are important. In order to understand the properties of such polymer biomaterial interfaces, it is necessary to consider the balance of forces that exists at the interface (Pethrick 2014).

3.5.1 CONTACT ANGLE MEASUREMENT

The simplest method of assessment of the energy of the polymer surface involves the study of the contact angle. This is a direct measurement of the tangent between a liquid drop and a solid. The contact angle depends very much on the properties of the surface: roughness, uniformity (defects), whether it absorbs solvents, and so forth. The wetting properties at the molecular level will result from intermolecular interactions between the liquid and solid, such that a variety of liquids can be used to measure a polymer film contact angle. However, given the charged nature of many biomaterials, it is the hydrophilic characteristics of the polymer that are important in many polymer biomaterial composite studies. Therefore, a water droplet is typically used to assess the contact angle of polymer films for biomaterial applications.

For example, contact angle measurements of PTh polymers modified with various backbone functionalities and dopants provided insights into muscle cell adhesion and differentiation on the polymer films, demonstrating that even a hydrophobic surface can support cell growth (Breukers et al. 2010; Quigley et al. 2013).

3.5.2 SCANNING PROBE TECHNIQUES

The fundamental feature of all scanning probe microscopies is the interaction between a sharp tip and the surface of a sample to measure its local physical properties. The stylus profilometer, which draws a stylus across a surface to record vertical deviation, is one of the oldest techniques that uses a scanning probe to visualize the roughness of a surface, being developed in 1929 (Whitehouse 2010). This mechanical process enables the roughness to be measured in terms of peak and valley deviation of the surface profile. In contrast to this contact profilometer, the optical profilometer is a noncontact method that images the surface irregularities by mapping distortions resulting from laser reflection. These techniques are similar to the more recent scanning probe technique, AFM, but the fundamental difference is the attainable resolution. AFM is a powerful imaging technique for the characterization of the morphology of materials on the nanoscale (Haugstad 2012). Given the nondestructive nature of AFM, it is now one of the most commonly used scanning probe techniques for polymer biomaterial composites and is discussed in more detail next.

3.5.3 ATOMIC FORCE MICROSCOPY

AFM is not only a high-resolution microscope probe capable of capturing a nanoscale topography of OCPs but also an excellent tool for directly measuring forces at the single-molecule level. For example, the correlation between the topography of PPy, electrosynthesized under different conditions, and its local surface potential was achieved by using AFM equipped with a mode called electric force microscopy (EFM), commonly referred to in the literature as Kelvin probe force microscopy (KFM) (Barisci et al. 2000).

In the biomaterial area, the direct measurement of intermolecular surface forces between complex biomolecules and conducting polymers is a relatively new area, with AFM being one of the main tools used. AFM characterization includes adhesion, electrostatic and entropic forces of single-polymer chains, and specific binding of proteins (Baró and Reifenberger 2012). Furthermore, the forces are often measured as a function of the polymer redox state since the surface properties, such as charge, energy, and their related interactions, are determined by the oxidation state of the polymer and physiochemical interactions of the dopants (Higgins and Wallace 2013).

The principle of AFM force measurement involves measuring the change in deflection of a flexible cantilever with sharp tip while bringing the tip into contact and then withdrawing it from a surface. Force measurements can be performed where a known potential is also applied to the AFM tip and polymer. For example, the interaction between a gold-coated tip with PPy–hyaluronic acid films produces interactions that are dependent on prior charging of the polymer (Pelto et al. 2013). The AFM tip has also been modified with PPy to study the forces and local interactions between the polymer interfaces and cells at the nanoscale (Knittel et al. 2014). Chemical modification of AFM tips and surfaces enables the interactions of functional groups such as $-COOH$, $-NH_2$, $-OH$, and $-CH_3$. This approach has been applied to model surfaces (e.g., self-assembled monolayers) (Butt et al. 2005) and more recently to conducting polymers. For example, a series of functional groups such as those used for gluteraldehyde cross-linking of proteins have been introduced onto a silicon AFM tip and force measurements performed after each functionalization step to assess their involvement in the interaction with the conducting polymer (Gelmi et al. 2013).

The main advantage of this modern microscopy lies in the variety of different auxiliary scanning techniques combined simultaneously with the topography. Some of these techniques include current-sensing AFM (CS-AFM), which can determine local electrical properties and, specifically, local conductivity; KFM (as discussed at the beginning of this section), which can be used to assess the local work function of materials and, through it, the chemical composition and the degree of oxidation; and phase imaging AFM (PI-AFM), which can determine local mechanical properties and, in particular, local crystallinity.

3.5.4 SCANNING ELECTRON MICROSCOPY

Scanning electron microscopy (SEM) is frequently used for the surface visualization of OCPs (Reimer 1998). By scanning over a surface with a focused beam of electrons, an image is created as a result of the interaction of the electrons with the specimen of interest. This method requires high vacuum (10^{-7} to 10^{-9} Torr), a conductive coating film if an insulator is being analyzed, and a dry sample—conditions quite often challenging for some types of materials. However, SEM has been successfully used for imaging conducting polymer biocomposite films, such as a virus-doped PEDOT developed for biosensing (Donavan et al. 2012).

Biological samples, typically wet materials, are often destroyed during standard SEM sample preparation. To overcome this limitation, the environmental scanning electron microscope (ESEM), which works at low vacuum (0.1–50 Torr) and in hydrated conditions, is more useful for the characterization of biological samples (McGregor et al. 2013). For example, ESEM has been used in the characterization of the polymerization of PEDOT around living cells (Richardson-Burns et al. 2007).

3.6 NONSPECTROSCOPIC METHODS

There are a number of other important nonspectroscopic techniques that are commonly used for characterizing OCPs. Polymer chain length and molecular weight are key parameters in characterizing OCPs, especially when trying to understand interactions between OCPs and other materials, such as biomaterials. Given also that OCP chain lengths will vary greatly depending on the polymerization method, as well as within the same sample, it is important to determine the effect of this on the polymer biomaterial composite properties. In addition, molecular weight or chain length has a significant impact on the mechanical properties of the polymer.

3.6.1 MOLECULAR WEIGHT DETERMINATION

Polymer chains very rarely have the same molecular weight. Normally, it is a wide or narrow distribution of different chain lengths described by the term *polydispersity*. Polydispersity is represented by the symbol Đ (pronounced D-stroke), which can refer to either molecular mass or degree of polymerization. It can be calculated using Equation 3.6:

$$Đ_M = \frac{M_w}{M_n} \tag{3.6}$$

where M_w is the weight-average molar mass and M_n is the number-average molar mass. It can also be calculated according to degree of polymerization:

$$Đ_X = \frac{X_w}{X_n} \tag{3.7}$$

where X_w is the weight-average degree of polymerization and X_n is the number-average degree of polymerization. In certain limiting cases, where $Đ_M = Đ_X$, it is simply referred to as Đ. The International Union of Pure and Applied Chemistry (IUPAC) has also deprecated the terms *monodisperse*, which is considered to be self-contradictory, and *polydisperse*, which is considered redundant, preferring the terms *uniform* and *nonuniform* instead (Stepto et al. 2009). There are three standard methods to measure those values: gel permeation chromatography (GPC), mass spectrometry (MS), and light scattering, as briefly discussed below.

3.6.1.1 Gel permeation chromatography

GPC is a type of size-exclusion chromatography (SEC) that separates analytes on the basis of size. As a technique, GPC is widely used in the characterization of polymers. However, the measurement requires an internal standard—normally a polystyrene of known molecular weight. Because conducting polymers are more rigid compared with the flexible saturated chains of the standards, the measured values are much higher than in reality (Holdcroft 1991). In addition, the conducting polymer needs to be soluble, thus restricting the method to a limited number of OCPs. The technique can, however, be used successfully for comparative studies (Hontis et al. 2003).

3.6.1.2 Mass spectrometry

MS is an analytical technique that helps identify the amount and type of chemicals present in a sample by measuring the mass-to-charge ratio. There several techniques to do that, but there are two used for conducting polymer characterization: electrospray ionization mass spectrometry (ESI-MS) and matrix-assisted laser-induced mass spectrometry (MALDI).

3.6.1.2.1 Electrospray ionization mass spectrometry

ESI-MS is an important, sensitive, robust, and reliable technique to determine the molecular weight of nonvolatile and thermally labile molecules and biomolecules. Because of the technical limitations, this technique can be used only for short-oligomer characterization, such as oligoaniline (Dolan and Wood 2004).

3.6.1.2.2 *Matrix-assisted laser-induced mass spectrometry*

MALDI has been used successfully to determine the molecular weight of large molecules such as proteins.

However, the analysis of the soluble conducting polymers showed that the values seem to be lower than in reality and the polydispersity is difficult to determine (Folch et al. 2000). A use of a special ionization technique (cationization salt) allowed the successfully analysis of a conducting polymer (P3HT). However, the values were lower than measured by GPC (Liu et al. 1999).

3.6.2 DYNAMIC LIGHT SCATTERING

Dynamic light scattering (DLS) is a technique that can be used to determine the size distribution profile of small particles in suspension or polymers in solution. Most light scattering experiments are performed on transparent suspensions. Solutions that absorb the incident light are subject to several problems that make the interpretation of light scattering data difficult. However, the data can still be interpreted by employing mathematical tools (Sehgal and Seery 1999). It has been shown that data obtained from GPC measurements are typically of higher value than that obtained from DLS (Yamamoto et al. 1996).

DLS is commonly used for the determination of the size of conducting polymer nanoparticles such as PANI (Riede et al. 1998) and their combination with biomolecules like DNA (Dawn and Nandi 2005).

3.6.3 QUARTZ CRYSTAL MICROBALANCE

Piezoelectric microgravimetry using a QCM with nanogram sensitivity is an important technique for monitoring mass change. As a result, it is a very useful tool for studying polymer biomolecule interactions (Marx 2003). The principal behind the microbalance relies on the piezoelectric effect of the quartz crystal. When a piezoelectric material is subjected to an AC potential, a mechanical oscillation of the material is effected. The frequency change of such an oscillation can be correlated to the mass change on the quartz crystal surface.

In the QCM, the adsorption of target analytes onto the quartz surface causes the shift of resonant frequency that can be used to derive information for quantitative analysis. By coupling electrochemistry to the QCM (EQCM), a comparison of the change of the surface mass with the charge consumed in the course of redox transformations can shed light on the mechanism of electrochemical reactions. It may also supply information on the sorption of neutral species and protonation equilibria. Conclusions can be drawn from QCM or EQCM data concerning the relative contribution of different anions and cations to the overall ion-exchange processes, as well as the rate of the mass transport processes. The analysis of the growth mechanism of conductive polymers, doping–dedoping process, and charge transfer properties of the polymer can be undertaken using EQCM.

For example, the effects of counterions and growth methods on PANI have been studied using this technique and showed that the polymer growth rates and the morphology of polymer surfaces are very different depending on the electrolytes and growth methods used, and the reversibility of doping–dedoping processes was poor for thick films (Choi and Park 2002). The assembly and disassembly of electroactive multilayer films based on water-soluble PThs could be monitored by EQCM (Mawad et al. 2011, 2012).

Quartz crystal microgravimetry with dissipation monitoring (QCM-D) has been used to study the fundamental physical properties of PEDOT films polymerized with the biological dopants dextran sulfate, chondroitin sulfate, and alginic acid, as well as their ability to interact with extracellular matrix proteins and neural cells (Molino et al. 2014).

The interaction of OCPs with biomolecules has also been studied using QCM-D. An investigation of the adsorption of the proteins bovine serum albumin and fibronectin to PPy demonstrated that modification of the polymer physicochemical and redox condition alters the nature of protein–polymer interactions (Molino et al. 2012).

REFERENCES

Adeloju, S. B., and G. G. Wallace. Conducting polymers and the bioanalytical sciences: New tools for biomolecular communications. A review. *Analyst (Cambridge, UK)* 121 (6) (1996): 699–703.

Ahuja, T., I. A. Mir, D. Kumar, and Rajesh. Biomolecular immobilization on conducting polymers for biosensing applications. *Biomaterials* 28 (5) (2006): 791–805.

Albery, W. J., Z. Chen, B. R. Horrocks, et al. Spectroscopic and electrochemical studies of charge transfer in modified electrodes. *Faraday Discussions of the Chemical Society* 88 (1989): 247–259.

Anson, F. C., and B. Epstein. Chronocoulometric study of the adsorption of anthraquinone monosulfonate on mercury. *Journal of the Electrochemical Society* 115 (11) (1968): 1155–1158.

Anson, F. C., and R. A. Osteryoung. Chronocoulometry: A convenient, rapid and reliable technique for detection and determination of adsorbed reactants. *Journal of Chemical Education* 60 (4) (1983): 293–296.

Aubert, P. H., L. Groenendaal, F. Louwet, et al. In situ conductivity measurements on polyethylenedioxythiophene derivatives with different counter ions. *Synthetic Metals* 126 (2–3) (2002): 193–198.

Ayranci, R., T. Soganci, M. Guzel, et al. Comparative investigation of spectroelectrochemical and biosensor application of two isomeric thienylpyrrole derivatives. *RSC Advances* 5 (65) (2015): 52543–52549.

Balint, R., N. J. Cassidy, and S. H. Cartmell. Conductive polymers: Towards a smart biomaterial for tissue engineering. *Acta Biomaterialia* 10 (6) (2014): 2341–2353.

Balko, J., R. H. Lohwasser, M. Sommer, M. Thelakkat, and T. Thurn-Albrecht. Determination of the crystallinity of semicrystalline poly(3-hexylthiophene) by means of wide-angle x-ray scattering. *Macromolecules (Washington, DC)* 46 (24) (2013): 9642–9651.

Bard, A. J., and L. R. Faulkner. *Electrochemical Methods: Fundamentals and Applications.* 2nd ed. New York: John Wiley & Sons, 2001.

Barisci, J. N., R. Stella, G. M. Spinks, and G. G. Wallace. Characterization of the topography and surface potential of electrodeposited conducting polymer films using atomic force and electric force microscopies. *Electrochimica Acta* 46 (4) (2000): 519–531.

Baró, A. M., and R. G. Reifenberger. 2012. *Atomic Force Microscopy in Liquid: Biological Applications.* Weinheim: Wiley-VCH Verlag.

Bendrea, A.-D., L. Cianga, and I. Cianga. Review paper: Progress in the field of conducting polymers for tissue engineering applications. *Journal of Biomaterial Applications* 26 (1) (2011): 3–84.

Berry, R. W., P. M. Hall, and M. T. Harris. *Thin Film Technology.* New York: D. Van Nostrand Company, 1968.

Blythe, A. R. Electrical resistivity measurements of polymer materials. *Polymer Testing* 4 (2–4) (1984): 195–209.

Breukers, R. D., K. J. Gilmore, M. Kita, et al. Creating conductive structures for cell growth: Growth and alignment of myogenic cell types on polythiophenes. *Journal of Biomedical Materials Research Part A* 95A (1) (2010): 256–268.

Butt, H.-J., B. Cappella, and M. Kappl. Force measurements with the atomic force microscope: Technique, interpretation and applications. *Surface Science Reports* 59 (1–6) (2005): 1–152.

Chen, T. A., and R. D. Rieke. The first regioregular head-to-tail poly(3-hexylthiophene-2,5-diyl) and a regiorandom isopolymer: Nickel versus palladium catalysis of 2(5)-bromo-5(2)-(bromozincio)-3-hexylthiophene polymerization. *Journal of the American Chemical Society* 114 (25) (1992): 10087–10088.

Chen, X., and O. Inganaes. Three-step redox in polythiophenes: Evidence from electrochemistry at an ultramicroelectrode. *Journal of Physical Chemistry* 100 (37) (1996): 15202–15206.

Choi, S.-J., and S.-M. Park. Electrochemistry of conductive polymers XXVI. Effects of electrolytes and growth methods on polyaniline morphology. *Journal of the Electrochemical Society* 149 (2) (2002): E26–E34.

Clarke, T. C., and J. C. Scott. Magic angle spinning NMR of conducting polymers. *IBM Journal of Research and Development* 27 (4) (1983): 313–320.

Collier, J. H., J. P. Camp, T. W. Hudson, and C. E. Schmidt. Synthesis and characterization of polypyrrole-hyaluronic acid composite biomaterials for tissue engineering applications. *Journal of Biomedical Materials Research* 50 (4) (2000): 574–584.

Contractor, A. Q., T. N. Sureshkumar, R. Narayanan, et al. Conducting polymer-based biosensors. *Electrochimica Acta* 39 (8–9) (1994): 1321–1324.

Cosnier, S. Biomolecule immobilization on electrode surfaces by entrapment or attachment to electrochemically polymerized films. A review. *Biosensors & Bioelectronics* 14 (5) (1999): 443–456.

Daugaard, A. E., S. Hvilsted, T. S. Hansen, and N. B. Larsen. Conductive polymer functionalization by click chemistry. *Macromolecules* 41 (12) (2008): 4321–4327.

Davidson, R. G., L. C. Hammond, T. G. Turner, and A. R. Wilson. An electron and x-ray diffraction study of conducting polypyrrole/dodecyl sulfate. *Synthetic Metals* 81 (1) (1996): 1–4.

Dawn, A., and A. K. Nandi. Biomolecular hybrid of a conducting polymer with DNA: Morphology, structure, and doping behavior. *Macromolecular Bioscience* 5 (5) (2005): 441–450.

Demchenko, A. P., ed. *Advanced Fluorescence Reporters in Chemistry and Biology II: Molecular Constructions, Polymers and Nanoparticles.* Berlin: Springer-Verlag, 2010.

Dolan, A. R., and T. D. Wood. Synthesis and characterization of low molecular weight oligomers of soluble polyaniline by electrospray ionization mass spectrometry. *Synthetic Metals* 143 (2) (2004): 243–250.

Domagala, W., B. Pilawa, and M. Lapkowski. Quantitative in-situ EPR spectroelectrochemical studies of doping processes in poly(3,4-alkylenedioxythiophene)s. Part 1: PEDOT. *Electrochimica Acta* 53 (13) (2008): 4580–4590.

Donavan, K. C., J. A. Arter, G. A. Weiss, and R. M. Penner. Virus-poly(3,4-ethylenedioxythiophene) biocomposite films. *Langmuir* 28 (34) (2012): 12581–12587.

Dore, K., S. Dubus, H.-A. Ho, et al. Fluorescent polymeric transducer for the rapid, simple, and specific detection of nucleic acids at the zeptomole level. *Journal of the American Chemical Society* 126 (13) (2004): 4240–4244.

Dudenko, D., A. Kiersnowski, J. Shu, et al. A strategy for revealing the packing in semicrystalline π-conjugated polymers: Crystal structure of bulk poly-3-hexyl-thiophene (P3HT). *Angewandte Chemie International Edition* 51 (44) (2012): 11068–11072.

Fan, C., K. W. Plaxco, and A. J. Heeger. High-efficiency fluorescence quenching of conjugated polymers by proteins. *Journal of the American Chemical Society* 124 (20) (2002): 5642–5643.

Fischer, J. E., Q. Zhu, X. Tang, et al. Polyaniline fibers, films, and powders: X-ray studies of crystallinity and stress-induced preferred orientation. *Macromolecules* 27 (18) (1994): 5094–5101.

Folch, I., S. Borros, D. B. Amabilino, and J. Veciana. Matrix-assisted laser desorption/ionization time-of-flight mass spectrometric analysis of some conducting polymers. *Journal of Mass Spectrometry* 35 (4) (2000): 550–555.

Forsyth, M., T. T. Van, and M. E. Smith. Structural characterization of conducting polypyrrole using 13C cross-polarization/magic-angle spinning solid-state nuclear magnetic resonance spectroscopy. *Polymer* 35 (8) (1994): 1593–1601.

Foulds, N. C., and C. R. Lowe. Enzyme entrapment in electrically conducting polymers. Immobilization of glucose oxidase in polypyrrole and its application in amperometric glucose sensors. *Journal of the Chemical Society, Faraday Transactions 1: Physical Chemistry in Condensed Phases* 82 (4) (1986): 1259–1264.

Galus, Z., R. A. Chalmers, and W. A. J. Bryce. *Fundamentals of Electrochemical Analysis*. 2nd ed. New York: Ellis Horwood Ltd, 1994.

Garner, B., A. J. Hodgson, G. G. Wallace, and P. A. Underwood. Human endothelial cell attachment to and growth on polypyrrole-heparin is vitronectin dependent. *Journal of Materials Science: Materials in Medicine* 10 (1) (1999): 19–27.

Gelmi, A., M. J. Higgins, and G. G. Wallace. Resolving sub-molecular binding and electrical switching mechanisms of single proteins at electroactive conducting polymers. *Small* 9 (3) (2013): 393–401.

Gierschner, J., J. Cornil, and H.-J. Egelhaaf. Optical bandgaps of π-conjugated organic materials at the polymer limit: Experiment and theory. *Advanced Materials (Weinheim, Germany)* 19 (2) (2007): 173–191.

Green, R. A., N. H. Lovell, and L. A. Poole-Warren. Cell attachment functionality of bioactive conducting polymers for neural interfaces. *Biomaterials* 30 (22) (2009): 3637–3644.

Harman, D. G., R. Gorkin III, L. Stevens, et al. Poly(3,4-ethylenedioxythiophene):dextran sulfate (PEDOT:DS)—A highly processable conductive organic biopolymer. *Acta Biomaterialia* 14 (2015): 33–42.

Harrison, B. S., M. B. Ramey, J. R. Reynolds, and K. S. Schanze. Amplified fluorescence quenching in a poly(p-phenylene)-based cationic polyelectrolyte. *Journal of the American Chemical Society* 122 (35) (2000): 8561–8562.

Haugstad, G. *Atomic Force Microscopy*. Hoboken, NJ: Wiley, 2012.

Higgins, M. J., and G. G. Wallace. Surface and biomolecular forces of conducting polymers. *Polymer Reviews (Philadelphia, PA)* 53 (3) (2013): 506–526.

Ho, H.-A., M. Boissinot, M. G. Bergeron, et al. Colorimetric and fluorometric detection of nucleic acids using cationic polythiophene derivatives. *Angewandte Chemie International Edition* 41 (9) (2002): 1548–1551.

Holdcroft, S. Determination of molecular weights and Mark-Houwink constants for soluble electronically conducting polymers. *Journal of Polymer Science Part B: Polymer Physics* 29(13) (1991): 1585–1588.

Hontis, L., V. Vrindts, D. Vanderzande, and L. Lutsen. Verification of radical and anionic polymerization mechanisms in the sulfinyl and the gilch route. *Macromolecules* 36(9)(2003): 3035–3044.

Hotta, S., and K. Waragai. Crystal structures of oligothiophenes and their relevance to charge transport. *Advanced Materials (Weinheim, Germany)* 5 (12) (1993): 896–908.

Jenden, C. M., R. G. Davidson, and T. G. Turner. A Fourier transform-Raman spectroscopic study of electrically conducting polypyrrole films. *Polymer* 34 (8) (1993): 1649–1652.

Kaim, W., and A. Klein, eds. *Spectroelectrochemistry*. Cambridge: Royal Society of Chemistry, 2008.

Kar, P. *Doping in Conjugated Polymers*. Beverly, MA: Scrivener Publishing, 2013.

Kiilerich-Pedersen, K., C. R. Poulsen, T. Jain, and N. Rozlosnik. Polymer based biosensor for rapid electrochemical detection of virus infection of human cells. *Biosensors & Bioelectronics* 28 (1) (2011): 386–392.

Kittlesen, G. P., H. S. White, and M. S. Wrighton. Chemical derivatization of microelectrode arrays by oxidation of pyrrole and N-methylpyrrole: Fabrication of molecule-based electronic devices. *Journal of the American Chemical Society* 106 (24) (1984): 7389–7396.

Knittel, P., M. J. Higgins, and C. Kranz. Nanoscopic polypyrrole AFM-SECM probes enabling force measurements under potential control. *Nanoscale* 6 (4) (2014): 2255–2260.

Koenig, J. L. Infrared and Raman spectroscopy of polymers. *Rapra Review Reports* 12 (2) (2001): 1–143.

Kostorz, G. Small-angle scattering in materials science. *Makromolekulare Chemie–Macromolecular Symposia* 15 (1988): 131–151.

Lee, C. E., C. H. Lee, K. H. Yoon, and J. I. Jin. NMR study of the PPV conducting polymers. *Synthetic Metals* 69 (1–3) (1995): 427–428.

Levi, M. D., and D. Aurbach. The behavior of polypyrrole-coated electrodes in propylene carbonate solutions. II. Kinetics of electrochemical doping studied by electrochemical impedance spectroscopy. *Journal of the Electrochemical Society* 149 (6) (2002): E215–E221.

Li, C., M. Numata, M. Takeuchi, and S. Shinkai. A sensitive colorimetric and fluorescent probe based on a polythiophene derivative for the detection of ATP. *Angewandte Chemie International Edition* 44 (39) (2005): 6371–6374.

Liu, J., R. S. Loewe, and R. D. McCullough. Employing MALDI-MS on poly(alkylthiophenes): Analysis of molecular weights, molecular weight distributions, end-group structures, and end-group modifications. *Macromolecules* 32 (18) (1999): 5777–5785.

Lu, G., J. Chen, W. Xu, S. Li, and X. Yang. Aligned polythiophene and its blend film by direct-writing for anisotropic charge transport. *Advanced Functional Materials* 24 (31) (2014): 4959–4968.

MacDiarmid, A. G., and A. J. Epstein. The concept of secondary doping as applied to polyaniline. *Synthetic Metals* 65 (2–3) (1994): 103–116.

Macdonald, D. D. Reflections on the history of electrochemical impedance spectroscopy. *Electrochimica Acta* 51 (8–9) (2006): 1376–1388.

Marx, K. A. Quartz crystal microbalance: A useful tool for studying thin polymer films and complex biomolecular systems at the solution–surface interface. *Biomacromolecules* 4 (5) (2003): 1099–1120.

Massonnet, N., A. Carella, A. de Geyer, J. Faure-Vincent, and J.-P. Simonato. Metallic behaviour of acid doped highly conductive polymers. *Chemical Science* 6 (1) (2015): 412–417.

Mawad, D., K. Gilmore, P. Molino, et al. An erodible polythiophene-based composite for biomedical applications. *Journal of Materials Chemistry* 21 (15) (2011): 5555–5560.

Mawad, D., P. J. Molino, S. Gambhir, et al. Electrically induced disassembly of electroactive multilayer films fabricated from water soluble polythiophenes. *Advanced Functional Materials* 22 (23) (2012): 5020–5027.

McCullough, R. D., S. Tristram-Nagle, S. P. Williams, R. D. Lowe, and M. Jayaraman. Self-orienting head-to-tail poly(3-alkylthiophenes): New insights on structure-property relationships in conducting polymers. *Journal of the American Chemical Society* 115 (11) (1993): 4910–4911.

McGregor, J. E., L. T. L. Staniewicz, S. E. Guthrie, and A. M. Donald. 2013. Environmental scanning electron microscopy in cell biology. In *Methods in Molecular Biology*. Secaucus, NJ: Springer, 493–516.

Miller, N. C., E. Cho, M. J. N. Junk, et al. Use of x-ray diffraction, molecular simulations, and spectroscopy to determine the molecular packing in a polymer-fullerene bimolecular crystal. *Advanced Materials (Weinheim, Germany)* 24 (45) (2012): 6071–6079.

Molino, P. J., M. J. Higgins, P. C. Innis, R. M. I. Kapsa, and G. G. Wallace. Fibronectin and bovine serum albumin adsorption and conformational dynamics on inherently conducting polymers: A QCM-D study. *Langmuir* 28 (22) (2012): 8433–8445.

Molino, P. J., Z. Yue, B. Zhang, et al. Influence of biodopants on PEDOT biomaterial polymers: Using QCM-D to characterize polymer interactions with proteins and living cells. *Advanced Material Interfaces* 1 (3) (2014): 1300122/1–1300122/12.

Naudin, E., P. Dabo, D. Guay, and D. Belanger. X-ray photoelectron spectroscopy studies of the electrochemically n-doped state of a conducting polymer. *Synthetic Metals* 132 (1) (2002): 71–79.

Neoh, K. G., E. T. Kang, and K. L. Tan. Limitations of the x-ray photoelectron spectroscopy technique in the study of electroactive polymers. *Journal of Physical Chemistry B* 101 (5) (1997): 726–731.

Neugebauer, H. Infrared signatures of positive and negative charge carriers in conjugated polymers with low band gaps. *Journal of Electroanalytical Chemistry* 563 (1) (2004): 153–159.

Nilsson, K. P. R., A. Herland, P. Hammarstroem, and O. Inganaes. Conjugated polyelectrolytes: Conformation-sensitive optical probes for detection of amyloid fibril formation. *Biochemistry* 44 (10) (2005): 3718–3724.

Nilsson, K. P. R., and O. Inganaes. Chip and solution detection of DNA hybridization using a luminescent zwitterionic polythiophene derivative. *Nature Materials* 2 (6) (2003): 419–424.

Nilsson, K. P. R., and O. Inganaes. Optical emission of a conjugated polyelectrolyte: Calcium-induced conformational changes in calmodulin and calmodulin-calcineurin interactions. *Macromolecules* 37 (24) (2004): 9109–9113.

Nilsson, K. P. R., J. Rydberg, L. Baltzer, and O. Inganaes. Twisting macromolecular chains: Self-assembly of a chiral supermolecule from nonchiral polythiophene polyanions and random-coil synthetic peptides. *Proceedings of the National Academy of Sciences of the United States of America* 101 (31) (2004): 11197–11202.

Nilsson, K. P. R., J. Rydberg, L. Baltzer, and O. Inganäs. Self-assembly of synthetic peptides control conformation and optical properties of a zwitterionic polythiophene derivative. *Proceedings of the National Academy of Sciences of the United States of America* 100 (18) (2003): 10170–10174.

Nishizawa, M., T. Matsue, and I. Uchida. Penicillin sensor based on a microarray electrode coated with pH-responsive polypyrrole. *Analytical Chemistry* 64 (21) (1992): 2642–2644.

Otero, T. F. Biomimetic conducting polymers: Synthesis, materials, properties, functions, and devices. *Polymer Reviews (Philadelphia, PA)* 53 (3) (2013): 311–351.

Otero, T. F., and M. Bengoechea. UV–visible spectroelectrochemistry of conducting polymers. Energy linked to conformational changes. *Langmuir* 15 (4) (1999): 1323–1327.

Ottenbourgs, B., H. Paulussen, P. Adriaensens, D. Vanderzande, and J. Gelan. Characterization of poly(isothianaphthene) derivatives and analogs by using solid-state 13C NMR. *Synthetic Metals* 89 (2) (1997): 95–102.

Pandey, P. C., and A. P. Mishra. Conducting polymer-coated enzyme microsensor for urea. *Analyst (London)* 113 (2) (1988): 329–331.

Patil, A. O., A. J. Heeger, and F. Wudl. Optical properties of conducting polymers. *Chemical Reviews* 88 (1) (1988): 183–200.

Paul, E. W., A. J. Ricco, and M. S. Wrighton. Resistance of polyaniline films as a function of electrochemical potential and the fabrication of polyaniline-based microelectronic devices. *Journal of Physical Chemistry* 89 (8) (1985): 1441–1447.

Pelto, J. M., S. P. Haimi, A. S. Siljander, et al. Surface properties and interaction forces of biopolymer-doped conductive polypyrrole surfaces by atomic force microscopy. *Langmuir* 29 (20) (2013): 6099–6108.

Pethrick, R. A. *Polymer Structure Characterization: From Nano to Macro Organization in Small Molecules and Polymers*. 2nd ed. Cambridge: Royal Society of Chemistry, 2014.

Pringle, J. M., J. Efthimiadis, P. C. Howlett, et al. Electrochemical synthesis of polypyrrole in ionic liquids. *Polymer* 45 (5) (2004): 1447–1453.

Prodromidis, M. I. Impedimetric immunosensors—A review. *Electrochimica Acta* 55 (14) (2010): 4227–4233.

Quigley, A. F., K. Wagner, M. Kita, et al. In vitro growth and differentiation of primary myoblasts on thiophene based conducting polymers. *Biomaterials Science* 1 (9) (2013): 983–995.

Rabiej, S., and A. Wlochowicz. SAXS and WAXS investigations of the crystallinity in polymers. *Angewandte Makromolekulare Chemie* 175 (1990): 81–97.

Randles, J. E. B. Kinetics of rapid electrode reactions. *Discussions of the Faraday Society* 1 (1947): 11–19.

Randriamahazaka, H., V. Noe, and C. Chevrot. Nucleation and growth of poly(3,4-ethylene-dioxythiophene) in acetonitrile on platinum under potentiostatic conditions. *Journal of Electroanalytical Chemistry* 472 (2) (1999): 103–111.

Rangarajan, S. K. High amplitude periodic signal theory. *Journal of Electroanalytical Chemistry and Interfacial Electrochemistry* 56 (1) (1974): 55–71.

Rannou, P., D. Rouchon, Y. F. Nicolau, M. Nechtschein, and A. Ermolieff. Chemical degradation of aged CSA-protonated PANI films analyzed by XPS. *Synthetic Metals* 101 (1–3) (1999): 823–824.

Ravichandran, R., S. Sundarrajan, J. R. Venugopal, S. Mukherjee, and S. Ramakrishna. Applications of conducting polymers and their issues in biomedical engineering. *Journal of the Royal Society Interface* 7 (Suppl. 5) (2010): S559–S579.

Reimer, L. *Scanning Electron Microscopy.* 2nd ed. Berlin: Springer-Verlag, 1998.

Ren, X., and P. G. Pickup. Coupling of ion and electron transport during impedance measurements on a conducting polymer with similar ionic and electronic conductivities. *Journal of the Chemical Society, Faraday Transactions* 89 (2) (1993a): 321–326.

Ren, X., and P. G. Pickup. Ion transport in polypyrrole and a polypyrrole/polyanion composite. *Journal of Physical Chemistry* 97 (20) (1993b): 5356–5362.

Richardson-Burns, S. M., J. L. Hendricks, B. Foster, et al. Polymerization of the conducting polymer poly(3,4-ethylenedioxythiophene) (PEDOT) around living neural cells. *Biomaterials* 28 (8) (2007): 1539–1552.

Riede, A., M. Helmstedt, V. Riede, and J. Stejskal. Polyaniline dispersions. 9. Dynamic light scattering study of particle formation using different stabilizers. *Langmuir* 14 (23) (1998): 6767–6771.

Roncali, J. Conjugated poly(thiophenes): Synthesis, functionalization, and applications. *Chemical Reviews (Washington, DC)* 92 (4) (1992): 711–738.

Rowlands, A. S., and J. J. Cooper-White. Directing phenotype of vascular smooth muscle cells using electrically stimulated conducting polymer. *Biomaterials* 29 (34) (2008): 4510–4520.

Rubinson, J. F., and Y. P. Kayinamura. Charge transport in conducting polymers: Insights from impedance spectroscopy. *Chemical Society Reviews* 38 (12) (2009): 3339–3347.

Saoudi, B., N. Jammul, M. M. Chehimi, et al. XPS study of the adsorption mechanisms of DNA onto polypyrrole particles. *Spectroscopy (Amsterdam, Neth.)* 18 (4) (2004): 519–535.

Sarac, A. S., S. Sezgin, M. Ates, and C. M. Turhan. Electrochemical impedance spectroscopy and morphological analyses of pyrrole, phenylpyrrole and methoxyphenylpyrrole on carbon fiber microelectrodes. *Surface & Coatings Technology* 202 (16) (2008): 3997–4005.

Schiavon, G., S. Sitran, and G. Zotti. A simple two-band electrode for in situ conductivity measurements of polyconjugated conducting polymers. *Synthetic Metals* 32 (2) (1989): 209–217.

Sehgal, A., and T. A. P. Seery. Anomalous dynamic light scattering from solutions of light absorbing polymers. *Macromolecules* 32 (23) (1999): 7807–7814.

Simonet, J., and J. Rault-Berthelot. Electrochemistry: A technique to form, to modify and to characterize organic conducting polymers. *Progress in Solid State Chemistry* 21 (1) (1991): 1–48.

Skompska, M. Alternative explanation of asymmetry in cyclic voltammograms for redox reaction of poly(3-methylthiophene) films in acetonitrile solutions. *Electrochimica Acta* 44 (2–3) (1998): 357–362.

Skotheim, T. A., and J. R. Reynolds, eds. *Handbook of Conducting Polymers: Conjugated Polymers, Theory, Synthesis, Properties, and Characterization.* 3rd ed. Boca Raton, FL: CRC Press, 2007.

Sosnowska, M., P. Pieta, P. S. Sharma, et al. Piezomicrogravimetric and impedimetric oligonucleotide biosensors using conducting polymers of biotinylated bis(2,2′-bithien-5-yl)methane as recognition units. *Analytical Chemistry* 85 (15) (2013): 7454–7461.

Stavrinidou, E., P. Leleux, H. Rajaona, et al. Direct measurement of ion mobility in a conducting polymer. *Advanced Materials (Weinheim, Germany)* 25 (32) (2013): 4488–4493.

Stavrinidou, E., M. Sessolo, B. Winther-Jensen, S. Sanaur, and G. G. Malliaras. A physical interpretation of impedance at conducting polymer/electrolyte junctions. *AIP Advances* 4 (1) (2014): 017127/1–017127/6.

Stepto, R. F. T., R. G. Gilbert, M. Hess, et al. Polydispersity in polymer science. *Pure and Applied Chemistry* 81 (2009): 351–353.

Stuart, B. H. *Infrared Spectroscopy: Fundamentals and Applications.* Chichester: Wiley, 2004.

Swager, T. M. The molecular wire approach to sensory signal amplification. *Accounts of Chemical Research* 31 (5) (1998): 201–207.

Tagawa, T., T. Tamura, and P. Å. Öberg. *Biomedical Sensors and Instruments.* 2nd ed. Boca Raton, FL: CRC Press, 2011.

Tam, P.-D., M.-A. Tuan, T.-Q. Huy, A.-T. Le, and N.-V. Hieu. Facile preparation of a DNA sensor for rapid herpes virus detection. *Materials Science and Engineering C* 30 (8) (2010): 1145–1150.

Thackeray, J. W., H. S. White, and M. S. Wrighton. Poly(3-methylthiophene)-coated electrodes: Optical and electrical properties as a function of redox potential and amplification of electrical and chemical signals using poly(3-methylthiophene)-based microelectrochemical transistors. *Journal of Physical Chemistry* 89 (23) (1985): 5133–5140.

Trojanowicz, M., W. Matuszewski, B. Szczepanczyk, and A. Lewenstam. 1993. Clinical application of biosensing with amperometric detection of ammonia-nitrogen. In *Uses of Immobilized Biological Compounds (Proceedings of the NATO Advanced Research Workshop on Uses of Immobilized Biological Compounds for Detection, Medical, Food and Environmental Analysis, Brixen, Italy, May 9–14, 1993)*, ed. G. G. Guilbault and M. Mascini. Amsterdam: Kluwer Academic Publishers, 577.

Umana, M., and J. Waller. Protein-modified electrodes. The glucose oxidase/polypyrrole system. *Analytical Chemistry* 58 (14) (1986): 2979–2983.

Vainrub, A., I. Heinmaa, E. Lippmaa, and G. I. Kozub. Comparative high-resolution NMR studies of n- and p-type doped polyacetylene films. *Synthetic Metals* 55 (1) (1993): 660–665.

Vercelli, B., S. Zecchin, N. Comisso, et al. Solvoconductivity of polyconjugated polymers: The roles of polymer oxidation degree and solvent electrical permittivity. *Chemistry of Materials* 14 (11) (2002): 4768–4774.

Vico, S., V. Carlier, and C. Buess-Herman. Spectroelectrochemical study of the influence of anions on the behaviour of poly(N-vinylcarbazole) films. *Journal of Electroanalytical Chemistry* 475 (1) (1999): 1–8.

Vorotyntsev, M. A., J.-P. Badiali, and G. Inzelt. Electrochemical impedance spectroscopy of thin films with two mobile charge carriers: Effects of the interfacial charging. *Journal of Electroanalytical Chemistry* 472 (1) (1999): 7–19.

Wagner, K., J. M. Pringle, S. B. Hall, et al. Investigation of the electropolymerization of EDOT in ionic liquids. *Synthetic Metals* 153 (1–3) (2005): 257–260.

Wallace, G. G., G. M. Spinks, L. A. P. Kane-Maguire, and P. R. Teasdale. *Conductive Electroactive Polymers: Intelligent Polymer Systems*. 3rd ed. Boca Raton, FL: CRC Press, 2009.

Wang, X., T. Sun, C. Wang, et al. 1H NMR determination of the doping level of doped polyaniline. *Macromolecular Chemistry and Physics* 211 (16) (2010): 1814–1819.

Wang, Z. H., C. Li, E. M. Scherr, A. G. MacDiarmid, and A. J. Epstein. Three dimensionality of "metallic" states in conducting polymers: Polyaniline. *Physical Review Letters* 66 (13) (1991): 1745–1748.

Watanabe, K., I. Osaka, S. Yorozuya, and K. Akagi. Helically π-stacked thiophene-based copolymers with circularly polarized fluorescence: High dissymmetry factors enhanced by self-ordering in chiral nematic liquid crystal phase. *Chemistry of Materials* 24 (6) (2012): 1011–1024.

Whitehouse, D. J. *Handbook of Surface and Nanometrology*. 2nd ed. Boca Raton, FL: CRC Press, 2010.

Yamamoto, T., D. Oguro, and K. Kubota. Viscometric and light scattering analyses of CHCl3 solutions of poly(3-alkylthiophene-2,5-diyl)s. *Macromolecules* 29 (5) (1996): 1833–1835.

Yang, J., and D. C. Martin. Impedance spectroscopy and nanoindentation of conducting poly(3,4-ethylenedioxythiophene) coatings on microfabricated neural prosthetic devices. *Journal of Materials Research* 21 (5) (2006): 1124–11232.

Yang, J.-S., and T. M. Swager. Porous shape persistent fluorescent polymer films: An approach to TNT sensory materials. *Journal of the American Chemical Society* 120 (21) (1998): 5321–5322.

Yap, W. T., and L. M. Doane. Determination of diffusion coefficients by chronoamperometry with unshielded planar stationary electrodes. *Analytical Chemistry* 54 (8) (1982): 1437–1439.

Yassar, A., J. Roncali, and F. Garnier. Conductivity and conjugation length in poly(3-methylthiophene) thin films. *Macromolecules* 22 (2) (1989): 804–809.

Yohannes, T., S. Lattante, H. Neugebauer, N. S. Sariciftci, and M. Andersson. In situ FTIR spectroelectrochemical characterization of n- and p-dopable phenyl-substituted polythiophenes. *Physical Chemistry Chemical Physics* 11 (29) (2009): 6283–6288.

Zhang, X., R. Bai, and Y. W. Tong. Selective adsorption behaviors of proteins on polypyrrole-based adsorbents. *Separation and Purification Technology* 52 (1) (2006): 161–169.

Zhang, Z., Y. Liang, P. Liang, C. Li, and S. Fang. Protein adsorption materials based on conducting polymers: Polypyrrole modified with ω-(N-pyrrolyl)-octylthiol. *Polymer International* 60 (4) (2011): 703–710.

4 Mechanism in charge transfer and electrical stability

Wen Zheng
Shanghai Jiao Tong University
Shanghai, China

Jun Chen and Peter C. Innis
University of Wollongong, Wollongong
New South Wales, Australia

Contents

4.1 INTRODUCTION

In recent decades, a series of new polymeric organic conductors have been developed with conductivities tailorable both during and after polymerization. Unlike traditional conductive polymers, which have conducting fillers loaded into an insulating polymeric matrix, the conductivity of the intrinsically conducting polymer originates as a consequence of mobile charge carriers delocalized across a conjugated electronic structure. The addition of conducting fillers to an insulating polymer has the limitation that the observed conductivity eventually plateaus, at a low filler volume fraction, with an associated decrease in mechanical properties with increasing filler content.

In the case of conducting polymers, conductivity originates from a combination of the polymer molecular structure and doping additives. In the case of the conductive polymer, the filler additive is solely the conductive medium (Roth 1989). This effectively limits the onset of conductivity to a high percolation threshold of 10%–15% conductive filler. Conversely, since the conductivity of the conducting polymers originates from within their molecular structure, a much higher conductivity can be realized by selective design of the molecular architecture. Additional enhancement and stabilization of the electrical property of conducting polymers are achieved by the incorporation of dopant molecules, which have a dual role of charge stabilization on the polymer backbone and electrostatic balancing of charges within the polymer matrix.

A number of intrinsically conducting polymers are known. The most successful of these polymeric systems are the homopolymers of polyacetylene (PA), polypyrrole (PPy), polythiophene (PT), polyaniline (PAn), poly(*para*-phenylene) (PPP), and poly(*para*-phenylene vinylene) (PPV), as shown in Figure 4.1, which can be synthesized into a range of polymeric derivatives and copolymers.

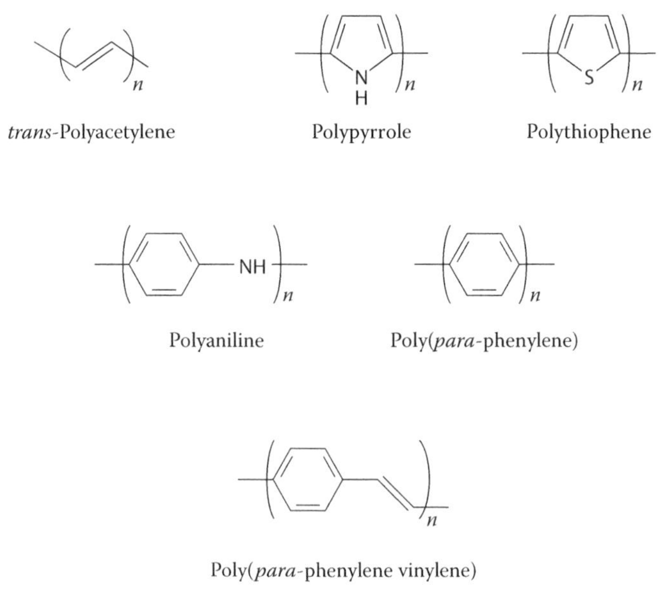

Figure 4.1 Conducting polymer types.

The most conductive, but least stable, conducting polymer is PA, which is the *trans* isomer, especially when highly crystalline and doped with iodine. The instability problems of PA in air are the major limiting factors to the applications of this polymer. Even though the polymer has serious environmental instability, some of the most promising electronic devices have been fabricated using highly crystalline *trans*-PA.

The most significant development in the field of conducting polymers occurred in 1979 when Diaz et al. electrochemically formed a PPy film with conductivities of up to 100 S cm^{-1} (Diaz et al. 1979). However, the conductivities achieved with PPy were significantly lower than those of *trans*-PA. The significant development was that PPy was thermally and chemically stable, although unprocessable. This lack of processability has since been offset by the ability to manipulate the properties of the polymer at the synthesis stage to yield the desired conductivity characteristics. The initial formation of PPy, otherwise called as a polyheterocycle, opened up an entire field of conducting polyheterocyclic materials, such as PT, and a broad range of substitute heterocyclic polymers.

4.2 BASIC PRINCIPLE: MECHANISM IN CHARGE TRANSFER

Conductivity in materials has been defined through the band theory, which provides a simplistic way to visualize the electronic transitions that occur from metallic to semiconductor and insulating states. In the band theory, electrons are characterized into bands of electrons rather than discrete quantized states. Crucially, conductivity is considered to arise when electrons can transition an energy barrier, or band gap (E_g), between the valence band electrons (or the highest occupied molecular orbital [HOMO]) and the conduction band (or the lowest unoccupied molecular orbital [LUMO]) in organic conductors. In an insulating material, the conduction and valence bands are separated by a sufficiently large energy barrier such that electrons will not jump or transition between them. As a consequence, there are no free electrons to conduct electricity and the material is classified as an insulator with band gaps above 3.0 eV. In conductors, the valence–conduction band gaps are extremely low (band gap <0.001 eV) or are overlapped, permitting free electron mobility. For semiconductors, the energy difference between the valence and conduction bands is relatively small (with band gaps between 0.01 and 3.0 eV), such that thermal excitation of electrons can bridge the gap and increase the conductivity. The Fermi level is defined as the energy level halfway between the valence and conduction bands. An effective method of excitation for semiconductors is doping, which inserts additional energy levels to reduce the materials band gap. Such doping leads to two types of semiconductors, which are called n-type and p-type.

The n-type silicon semiconductors are doped intrinsic semiconductors in which the dopant is a pentavalent element such as arsenic (As), antimony (Sb), or phosphorus (P) substituted for the silicon atom. These substitutional impurities provide a supplementary electron, owing to their ns^2np^3 electronic configuration, which contains five valence electrons, rather than the four outer-shell electrons seen in silicon. As a consequence, the density of electrons in the conduction band exceeds the density of holes in the valance band. The holes in the valance band act like a positive charge that is also being driven by the electric field. The pentavalent element is called the donor, as it supplies the electrons without distortion of the crystal structure. As in this type of semiconductor, electrons are the majority carriers and holes are the minority carriers; they are called n-type

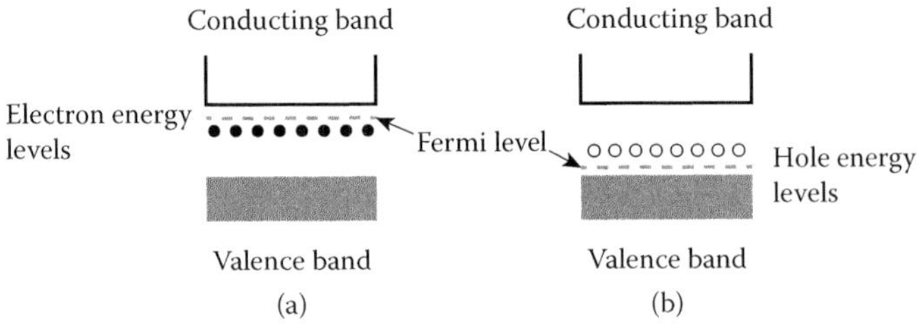

Figure 4.2 (a) n-Type and (b) p-type semiconductors.

semiconductors. The energy level of donor electrons is just below that of the conduction band (Figure 4.2a).

The p-type semiconductors are doped semiconductors in which the dopant is a trivalent element, such as boron (B), aluminum (Al), and gallium (Ga). The substitutional impurities will trap an electron owing to its ns^2np^1 electron configuration containing three, rather than four, outer-shell electrons. Therefore, the density of electrons in the conduction band is exceeded by the density of holes in the valence band. The trivalent element behaves as an electron acceptor as it provides holes without distortion of the crystal structure. In these types of semiconductors, the majority carriers are holes, so they are referred to as p-type. The energy of the accepter electron is located just above the valence band (Figure 4.2b).

Conjugated polymers, also known as inherently conducting polymers (ICPs), are a type of organic polymer that can conduct electricity. It is widely recognized that the conductivity range is similar to that of semiconductors, as its typical band gap is in the range of 1–3 eV.

4.3 REQUIREMENTS FOR CONDUCTIVITY

4.3.1 CONJUGATION AND STRUCTURE

For a polymer to become intrinsically conducting, it has been shown that a number of characteristics are essential for conductivity. The most important aspect for these types of polymers is the presence of overlapping π-bond molecular orbitals and a high degree of π-bond conjugation. When a highly conjugated polymer is observed in its pristine non-oxidized state, the conducting polymer is essentially an insulating material. It is only by the removal of a π-bond electron from the conjugated polymer chain, forming a radical cation defect (or polaron), that conductivity can arise.

4.3.2 POLYMER CHAIN LENGTH

Generally, the longer the effective polymer conjugation length, the higher the polymer conductivity. Ideally, charge injection into one end of a conducting polymer chain of infinite length would result in a simple charge transfer to its end due to the delocalization of the conjugated bonds on the polymer chain. Given that a conducting polymer must have a finite molecular weight, and coupled with the fact that conjugation defects may be prevalent, the concept of a continuous "molecular wire" is too simplistic. What must occur is an interchain charge transfer process, the mechanism of which has been an area of intense fundamental research. A brief overview of this process is discussed in Section 4.3.6.

4.3.3 CONDUCTION MECHANISM: THE POLARON AND BIPOLARON

Upon oxidation of a conducting polymer, there is an initial formation of a radical cation or polaron, as mentioned previously. Investigations using electron spin resonance (ESR) spectroscopy have shown the presence of a non-spin-paired electron ($s = \pm\frac{1}{2}$) due to the formation of the radical cation (Waller et al. 1989). As the level of oxidation is increased within the polymer, the ESR signal is observed to decrease rather than increase, as would be expected from an increasing concentration of polaronic structures. The disappearance of the ESR signal has been attributed to two polaron structures interacting and recombining to form a dicationic structure or bipolaron (Figure 4.3). The recombination of the polaron to form the bipolaron results in zero free electron spin ($s = 0$) due to the coupling of the un-spin-paired polaron electron. This effect has been observed to occur when the dopant concentration exceeds 1% in PPy (Street 1986). An equilibrium between the polaron and the bipolaron structures is said to exist and is dependent on the level of polymer doping (Bertho et al. 1988) where

$$P^{\bullet+} + P^{\bullet+} \Leftrightarrow BP^{2+}$$

$$(P^{\bullet+} = \text{Polaron}, BP^{2+} = \text{Bipolaron})$$

At the lowest levels of doping and polymer oxidation, polarons predominate, and conversely, bipolarons predominate at higher levels on the basis of spin, as observed by ESR measurements (Dennany et al. 2010, 2011). Even though an equilibrium exists between the two spin states, pure spinless bipolaron conductivity is not apparent—hence, a combination of the two radical species must exist.

The mechanism of charge transfer from one adjacent chain to the next under an electrical field has been described as being preferentially a bipolaronic process when viewed on an energetic basis (Nalwa 1989). This implies that in the lower conductivity states, or at lower doping levels, a bipolaronic structure must precede any charge hop to an adjacent chain. On completion of the bipolaron hop, the bipolaron is free to dissociate into its component polaronic states. The level of anion doping has been determined for PPy using elemental analysis as approximately one dopant molecule per four monomer

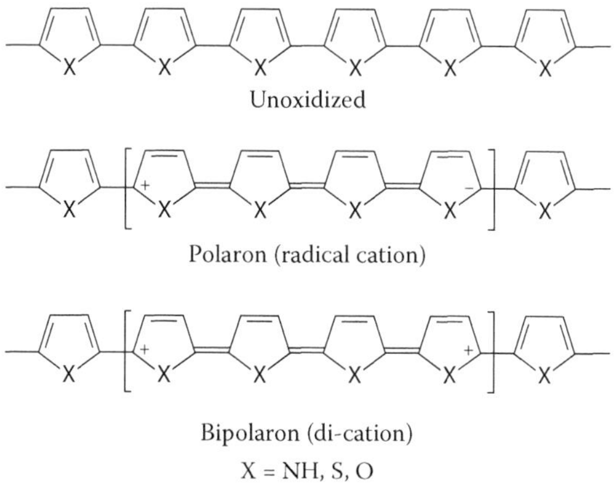

Figure 4.3 Polaron (radical cation) and bipolaron (dication) structures.

residue units (0.25–0.33 anions per pyrrole unit) and has been reviewed as being similar for other polyheterocycles (Bakhshi 1988; Diaz et al. 1986).

Conductivity in ICPs is attributed to the presence of charge carriers, which are typically p-type polarons, bipolarons in nondegenerate ICPs, or solitons in degenerate PA (Bredas and Street 1985). The observed electrical conductivity can be described by semiconductor theory incorporating electron–lattice or electron–electron interactions (Baeriswyl et al. 1992), which are free to move along the conjugated polymer backbone of the ICP as a result of π-bond delocalization. In chemical terminology, polarons and bipolarons correspond to stable un-spin-paired radical cations and spinless divalent cations, respectively. Solitons arise from bond alternation defects or misfits in polyenes and only exist in degenerate polymer systems such as PA (Figure 4.4). Under an electrical field, the movement of polarons and bipolarons conducts charge.

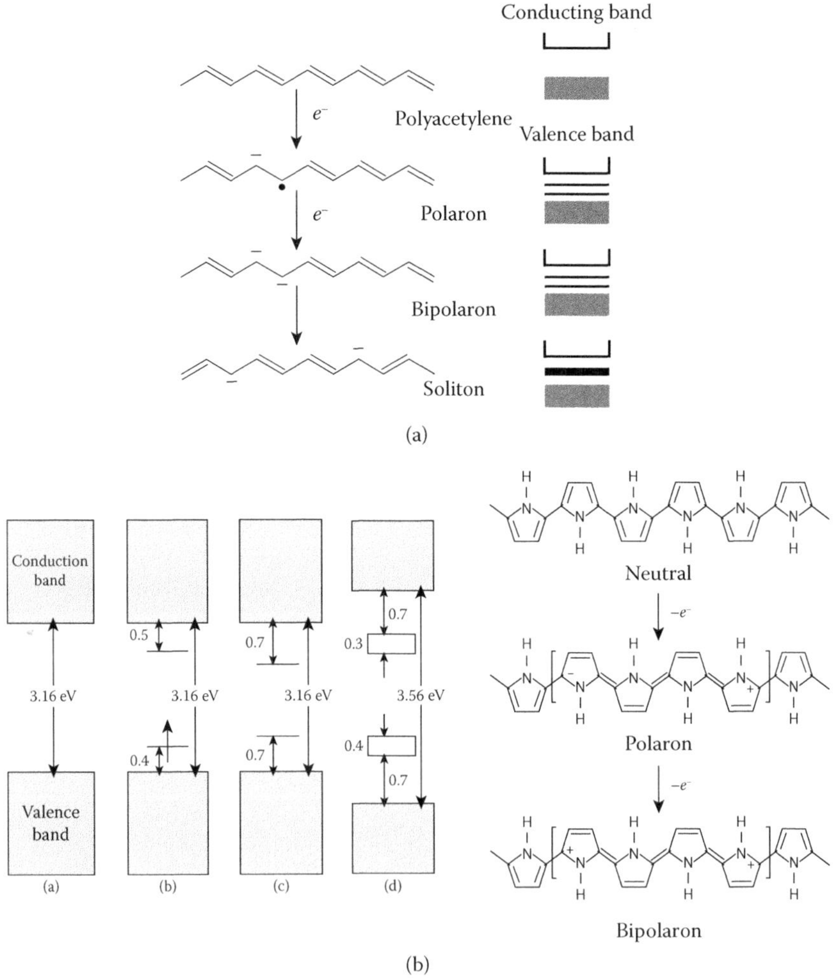

Figure 4.4 Schematic formation of polaron, bipolaron, and soliton. (a) Polyacetylene and (b) polypyrrole are used as examples. (From Bredas, J. L., and Street, G. B., *Acc. Chem. Res.*, 18 [10], 309–315, 1985. With permission.)

4.3.4 DOPING OF CONJUGATED POLYMERS

The basic principle of doping in conjugated polymers is equivalent to that of doping in inorganic semiconductors. Impurities (dopant) with appropriate electronic properties can be added, via reductive processes, to donate an electron to the LUMO levels for n-type doping or to remove an electron, via oxidative processes, from the HOMO levels for p-type doping. The measured conductivities of conjugated polymers are enhanced with increasing doping, but only up to a discrete point (Yim et al. 2008). Unlike inorganic semiconductors, p- or n-type doping disrupts conjugation along the polymer backbone. Excessive disruption to the polymer inhibits the mobility of the charge carrier and therefore reduces conductivity. Density function theory (DFT) confirms that band gap is reduced at higher doping levels in both doping formats (Ullah et al. 2014).

The doping processes of conjugated polymers are not comparable to classic doping processes of semiconductors. Rather than replacing an atom within an atomic lattice structure that is electron rich or deficient, as in the process of doping in inorganic semiconductors, doping of conjugated polymers is an oxidation and reduction process.

Oxidative (p-type) doping in conjugated polymers is commonly achieved through chemical or electrochemical approaches. However, this has also been demonstrated through photodoping or charge injection doping (MacDiarmid 2001). Chemical doping of conjugated polymer can be facilitated by simple exposure of the conjugated polymer to halogens such as Br_2 or I_2. For electrochemical doping of conjugated polymers, these processes correspond to electrochemical reactions that are either oxidative, in the case of p-doping, or reductive, in the case of n-doping (Figure 4.5).

Removal of the π-bond electron causes the remaining π electrons to become delocalized along the length of conjugation, permitting free mobility of the radical cation or polaron. Molecular distortions that inhibit or hinder charge carrier mobility will result

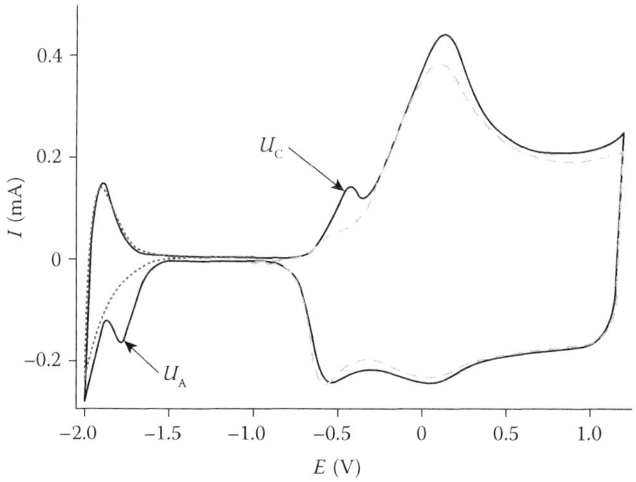

Figure 4.5 Cyclic voltammetry of a poly(3,4-ethylenedioxythiophene) (PEDOT) film in 0.1 M $TEABF_4/CH_3CN$: the response of the film cycled through both n- and p-doping regions (solid line), while only p-doping is shown in dash and n-doping in dot lines. (From Hillman, A. R., et al., *Electrochim. Acta*, 53 [11], 3763–3771, 2008. With permission.)

in a lower conductivity (Schwartz 2003). The chain conformation, degree of interchain contact, and rate of energy transfer can be controlled by factors such as choice of polymerization parameters (Sabouraud et al. 2000), solvent (Nguyen et al. 1999; Xu et al. 2010), thermal annealing (Nguyen et al. 2000), and side-chain types and their distributions (Hu et al. 2013). Side-chain engineering has been recognized as an effective method to tune the electrical properties of conjugated polymers (Lei et al. 2014). Mai et al. (2014) demonstrated that by using π-conjugated cyclopenta-[2,1-b;3,4-b′]-dithiophene-alt-4,7-(2,1,3-benzothiadiazole) backbones, shorting the side chains leads to an increased doping level, higher crystallinity, and more molecular orientation, all of which help to enhance electrical conductivity.

4.3.5 STERIC FACTORS

Free mobility of the polaron can be hindered by twisting out of plane the overlapping π-bonds. Excessive twisting effectively reduces conjugation length and limits conductivity. This effect is routinely observed when PPy or PT is β substituted, as shown in Figures 4.6 and 4.7. A general review article of PT was prepared by Roncali (1992) covering PT structure, 3-substitution, and applications. The review covered topics such as synthesis, functionalization of the monomer, and potential application of PT. Roncali (1992) gave a critical review of the reasons for functionalization of the thiophene monomer. The initial use of substitution was to force an ideal planar polymer structure, which is α–α′ in character. Typically, the reactivity of PT is 95% α and 5% β in character. Due to the β site reactivity, the α–β′ type mislinkage is significant and, as a result, causes disruption to the mean polymer conjugation length and limits conductivity. One approach used to limit these defects is to functionalize the β site to sterically force the ideal α–α′ structure. Typical substituents are C1–C18 alkyl chains, which have the additional role of inducing polymer solubility to common organic solvents. The major drawback with this approach is that such substitution destroys the conductivity maximum due to steric distortion of the π-bond conjugation, as discussed in Section 4.3.1. This type of substitution, with an electron-donating substituent, lowers the oxidation potential but induces polymer chain distortion and generally results in lower conductivities.

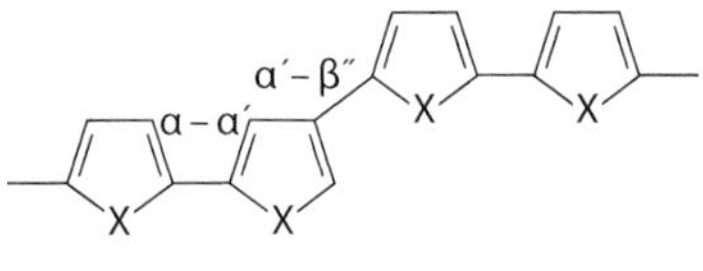

Figure 4.6 α-β′ linkages in polyheterocyclics.

Structure effects can also be observed for PA, where the *trans* structure has a much higher conductivity than the *cis* (Figure 4.8). The *trans* structure has also been shown to be the more thermodynamically stable conformation, and a transformation between

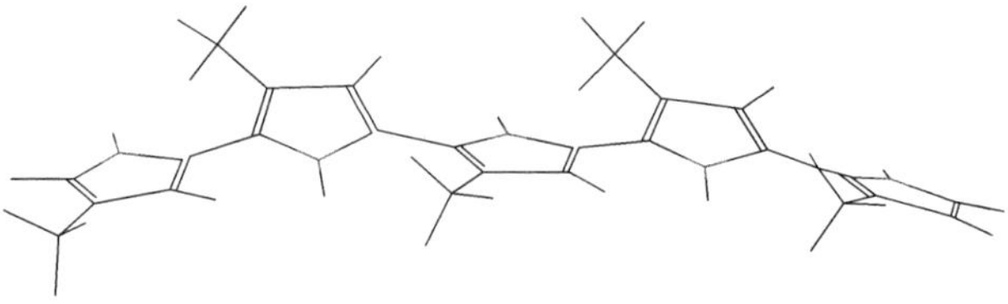

Figure 4.7 Polyheterocyclic showing out-of-plane twisting due to β substitution.

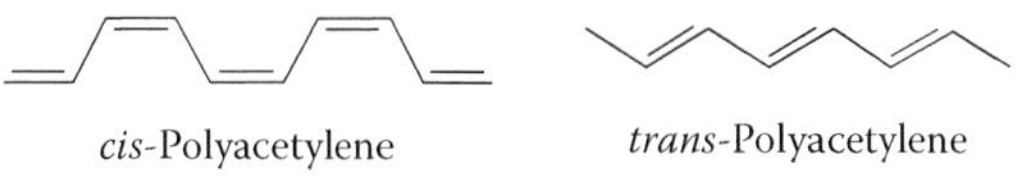

Figure 4.8 *cis*- and *trans*-polyacetylene structures.

cis and *trans* can be achieved readily by heating the polymer to 150°C for a few minutes.

Another approach is to directly synthesize α–α' bonded PPy or PT via a chemical Grignard synthesis technique (Galal et al. 1989). Although this method yields an ideally structured polymer, the material synthesized is nonconducting and requires secondary oxidation and doping to obtain conductivity. The process of postoxidation is limited to a small range of oxidants that must also function in the dual role of dopant. The chemical route of synthesis is limited as the polymer requires postprocessing to obtain film or pellets, and hence can be of poorer quality than electrochemically synthesized polymer.

A more elegant approach to forcing pristine structure has been developed for PT using activating silane leaving groups (Lemaire et al. 1990; Roncali et al. 1991), as opposed to the blocking substituent strategy. The approach adopted was to substitute functional groups in the position to act as leaving and activating groups. The most successful monomer developed was tetrathienylsilane, which had an oxidation potential similar to that of the unsubstituted thiophene monomer but produced a conductivity of >200 S cm^{-1}. The major advantage of this synthesis method is that PT can be electropolymerized and doped in a simple one-step process to yield higher conductivities than achievable with the chemical Grignard synthetic approach.

4.3.6 PROCESS OF CHARGE TRANSFER

The variable-range (Mott) hopping (VRH) theory was first developed by Mott and Davis (2012) for the study of amorphous silicon semiconductors. Applications of this theory to conducting polymers have been made on the basis that the polymers are amorphous, and that transport originates from localized or fixed states within the materials' structure (Plocharski 1989). In the case of the polyheterocycle, the localized state originates from an assumption that there is short conjugation length within the polymer due to the formation of missed linkages or low polymer molecular weight. Charge transfer from one polymer chain to the adjacent one occurs via an electron hop, referred to as a phonon-assisted hop between two localized states. The Mott VRH model predicts the following relationship:

$$\ln \sigma \propto T^{-1/n}$$

where σ is the electrical conductivity (S cm^{-1}) and T refers to temperature (K). The nature of the dimensionality of the conductivity is defined by the term n, where for a one-dimensional conductor (or molecular wire), $n = 2$. For a two-dimensional conduction (planar sheet), $n = 3$. For a three-dimensional conductor, $n = 4$. The dimensionality of the conductor determined by a plot of $\ln \sigma$ versus $T^{-1/n}$ should yield a straight line. The VRH model holds and has been shown to hold for conducting PPy above 70 K (Maddison et al. 1988) with an n-dimensionality term between 3 and 4, indicating that the conduction mode is typically a mixture of both two- and three-dimensional charge carrier mobility.

4.4 SPECTROSCOPIC EVIDENCE FOR ICP CONDUCTION MECHANISMS

Commonly employed spectroscopic techniques used to reveal the chemical structure and chain conformation of conjugated polymers are ultraviolet–visible (UV–Vis), Fourier transform infrared (FTIR), ESR, and Raman spectroscopy. Details on how these tools can be utilized are given in the next sections.

4.4.1 UV–VISIBLE SPECTROSCOPY

Most neutral (nonoxidized) conjugated polymers show the π–π^* electronic absorptions in the region of UV to visible. Depending on the nature of the dopant and doping level, in the doped conjugated polymer, an extra energy state or energy band is formed between the valence band and conducting band, creating additional absorption bands of lower energy. Due to the presence of these intermediate bands, new absorptions associated with the presence of the midstate polaronic and bipolaronic charge carriers are observed in the visible to near-IR or IR region of the visible spectrum. The optical absorption as a function of doping level for poly(3,4-ethylenedioxythiophene) (PEDOT) is shown in Figure 4.9. The maximum absorption occurs around 620 nm (1.99 eV) with a band gap of 1.63 eV, and the absorption band shows a hardly noticeable shoulder around 675 nm, which corresponds to the reduced deeply blue-colored polymer, which is insulating in character. In the oxidized p-doped state of the polymer, the electrons are delocalized along the conjugated chains through overlapped π-orbitals. Further oxidation leads to the formation of polarons and bipolarons and introduces subband energy levels (Scrosati 1988). PEDOT doping shifts the polaron or bipolaron band into the near IR. Consequently, an increase in optical transparency in the visible domain is then observed, as well as additional absorption bands in the near-IR region due to the polaronic-type conduction (Monk et al. 2007).

4.4.2 FOURIER TRANSFORM INFRARED SPECTROSCOPY

Peak shifts in FTIR spectra can also be observed for doped conjugated polymer compared with undoped ones. If the electronic conjugation is extended, the corresponding

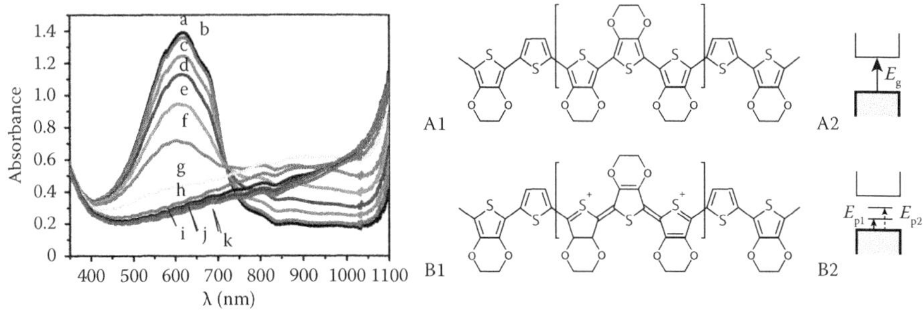

Figure 4.9 Optical absorption spectra of PEDOT depending on the applied potential. Experiment carried out in 0.03 M Lithium bis(trifluoromethylsulphonyl)imide (LiTFSI) in 1-butyl-3-methylimidazolium bis(trifluoromethanesulfonyl)imide (BMITFSI). Ag/AgCl is used as a reference: (a) −1.3 V, (b) −1 V, (c) −0.7 V, (d) −0.6 V, (e) −0.5 V, (f) −0.4 V, (g) −0.3 V, (h) −0.2 V, (i) 0 V, (j) 0.2 V, and (k) 0.4 V. Molecular structure of undoped PEDOT (A1) and its band gap structure (A2). Bipolaronic species obtained via two-electron oxidation (p-doping) and band gap structure. (From Duluard, S., et al., *J. Phys. Chem. B*, 114 [22], 7445–7451, 2010. With permission.)

vibrational peak will shift toward a lower wavenumber (referred to as bathochromic shift). Furthermore, the structure differences can be detected in vibrational spectra based on changes in position and shape of FTIR bonds, such as N–H and C–H. Typical bands for PAn are ~1560 cm^{-1} (C=N quinoid ring [Q] stretching), ~1490 cm^{-1} (C–C benzenoid ring [B] stretching), ~1300 cm^{-1} (C–N stretching of secondary amine), ~1115 cm^{-1} (C–H in-plane bending of aromatic rings), ~817 cm^{-1} (C–H deformation in the Q ring), and ~796 cm^{-1} (C–H out-of-plane bending of aromatic rings) (Coskun et al. 2012). The region of 1400–1000 cm^{-1} is of particular interest because it corresponds to the stretching of intermediate CN bonds (i.e., the intermediate CN bonds between the single and double CN bonds). As shown in Figure 4.10, single and double CN stretching peaks are presented in 1224 and 1517 cm^{-1}, respectively; peaks that emerge between those indicate the delocalization of π-electrons along the conjugated chain, which also reveals an increase in the conjugated chain length.

4.4.3 RAMAN SPECTROSCOPY

Raman spectroscopy is a powerful method for studying the structure of conjugated polymers (Furukawa 1996). While the nature and evolution of polarons and zero-spin diacationic bipolarons in conducting polymers can be examined effectively using ESR spectroscopy (Furukawa 1995; Furukawa et al. 1992; Mu et al. 1998), Raman spectroscopy is a complementary tool to characterize the presence of radical cations and dications in conducting polymers such as PAn and PPy (Pereira Da Silva et al. 1999; Fukuda et al. 1995; Bernard and Hugot-Le 2006a, 2006b). The degree of

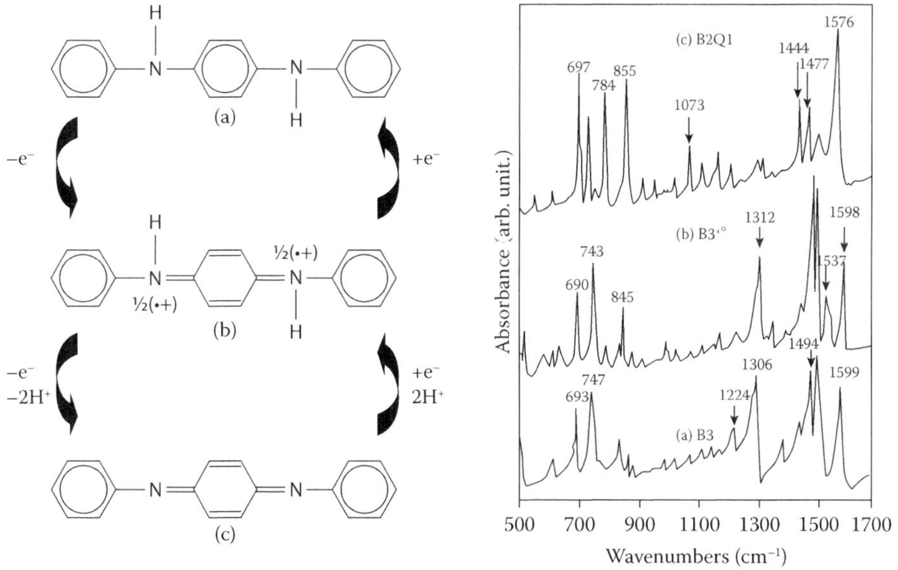

Figure 4.10 Schematic representation of (a) *N,N′*-diphenyl-1,4-phenylenediamine (B3), (b) the $FeCl_3$-doped dimer, and (c) *N,N′*-diphenyl-1,4-benzoquininediimime (B2Q1) (left) and corresponding infrared spectra (right). (From Boyer, M. I., et al., *J. Phys. Chem. B*, 104 [38], 8952–8961, 2000. With permission.)

conjugation can lead to broadening of the Raman bands, indicating a considerable degree of disorder in the polymeric material, resulting in shifts in the vibrational frequencies of the Raman band.

The Raman spectrum is produced by irradiating samples with laser at a certain frequency and analyzing the scattered radiation by Stokes shift. Similar to UV–Vis and FTIR techniques, Raman spectroscopy can also give information on the structure of conjugated polymers in the doped and neutral states, as shown in Figure 4.11. A freshly prepared PPy has three important pairs of peaks: ~933 and ~973 cm^{-1} are attributed to the ring deformation associated with polarons and bipolarons; ~1085 and ~1051 cm^{-1} arise from C–H in-plane bending vibration due to the oxidation and reduction states of PPy, respectively (Furukawa et al. 1988); and ~1607 and ~1581 cm^{-1} are believed to be the C=C backbone stretching in the oxidation and reduction states of PPy (Santos et al. 2007, 2013). With the increase in the applied potential, the Raman peaks at 933 and 973 cm^{-1} keep nearly constant during both oxidation and reduction processes, indicating that the ring structures of the polaron and bipolaron are very stable and do not change significantly during such processes. However, the peaks at 1051, 1085, 1581, and 1607 cm^{-1} shift to higher frequencies, revealing that the bond strengths of C–H and C=C increase. During the discharge process, the Raman peaks at 1051 and 1085 cm^{-1} gradually shift back to low frequencies with the decrease of applied potential, demonstrating that the reduction reaction of PPy gives rise to the molecular chain stretching mainly by the relaxation of C–H bonds.

PAns are also an important class of conducting polymers. The different modes of vibration have been assigned as C–H bending between 1100 and 1700 cm^{-1}, C–N stretching modes (amines, imines, and polarons) between 1210 and 1520 cm^{-1}, and C=C ring quinoid stretching at ca. 1600 cm^{-1} (Bernard and Hugot-Le 2006a, 2006b; Niaura et al. 2004; Louarn et al. 1996). An important spectral feature at 1333 and 1386 cm^{-1}

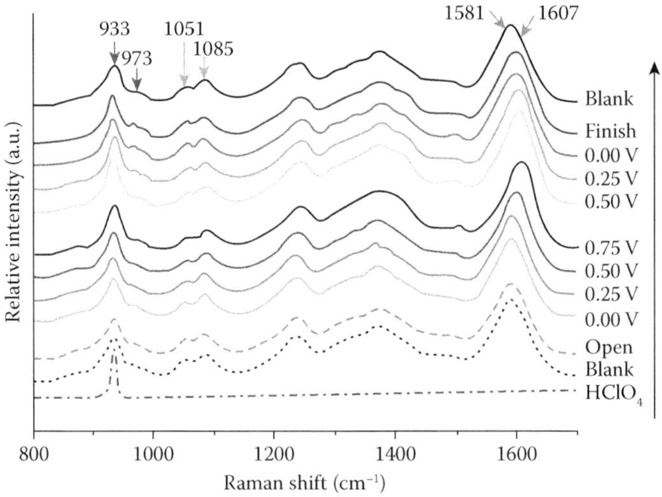

Figure 4.11 *In situ* Raman spectroscopy of PPy/ClO_4 under different voltages; excitation laser at 632.8 nm. (From Hou, Y., et al., *Phys. Chem. Chem. Phys.*, 16 [8], 3523–3528, 2014. With permission.)

has been reported as a broad polaron doublet resulting from the presence of the C–N$^{•+}$ polaron (Bernard and Hugot-Le 2006a, 2006b).

4.4.4 ELECTRON SPIN RESONANCE SPECTROSCOPY

ESR (or electron paramagnetic resonance) spectroscopy in conducting polymer studies is commonly used to identify the nature of the charge carrier (Pratt et al. 1997; Salafsky 1999; Chakrabarti et al. 1999; Chipara et al. 2003; Mizoguchi 2001), such as excitons (Salafsky 1999), solitons, polarons (stable radical cations) (Pratt et al. 1997), and bipolarons (dications) (Chakrabarti et al. 1999), to elucidate the physical properties of conducting polymers. ESR is used to probe the nature of the quasi-particle responsible for the charge transport, providing the fine details in the mode of electronic transport and the interactions among conjugated electrons (Watanabe et al. 2014). Polaronic quasi-particles (Kahol 2000; Chipara et al. 2003), which are essentially un-spin-paired ($s = 1/2$), are responsible for a single resonance line in an ESR spectrum located close to the g value of 2.003. The bipolaron, which forms as a result of the interaction of two half-spin polarons interacting to yield a zero-spin ($s = 0$) state, thereby presents no observable ESR spectrum (Lippe and Holze 1991; Yang and Li 1993).

The onset of doping and the associated increase of electrical conduction associated with charge carrier generations (increasing charge carrier density) can be easily identified with ESR signal (Kudo et al. 2013) (Figure 4.12). In recent work (Dennany et al. 2010, 2011), it has been shown that the loss of electron spin in a PAn system is a result of polarons interacting with each other to preferentially form spinless bipolarons. Significantly, this process occurs without the polymer composite itself undergoing an oxidation state change, which is usually associated with the change in electrical conductivity. This study also indicated that the chemical environment can have a strong influence over the inter- and intracharge transfer processes where the suppression of intrachain charge transfer to adjacent ICP chains can result in the preferential formation of the more stable spin-paired bipolaron.

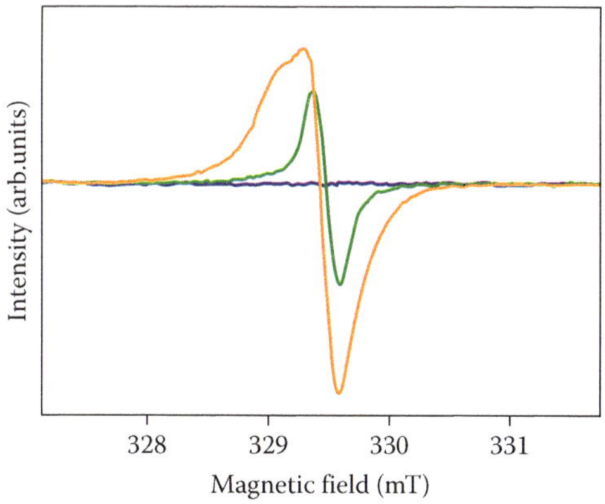

Figure 4.12 Electron spin resonance (ESR) spectroscopy results of in situ iodine doping of disubstituted polyacetylene. Blue line: 0 min of doping time; green line: 10 min; orange line: 60 min. (From Kudo, Y., et al., *J. Phys. Conf. Ser.*, 428, 012006/1–012006/6, 2013. With permission.)

4.5 REDOX CYCLE OF CONJUGATED POLYMERS

Switching a conducting polymer's redox state is accompanied by ion exchange between the polymer and the surrounding electrolyte. This is due to the oxidative (p-type) doping processes, resulting in the formation of the cationic polaron or dicationic bipolaron charge carriers distributed along the conjugated backbone, requiring the incorporation of a charge balancing anion (or dopant). Reduction of the conducting polymer subsequently removes the polaronic (or bipolaronic) charge carrier through injecting an electron to re-form a neutral conjugated backbone. As a consequence of this reduction, the dopant anion is typically expelled from the reduced polymer.

Gandhi et al. (1995) summarized that in general cases, there are four stages of the oxidation–reduction process that may occur during redox cycles:

1. Oxidation by cation ejection
2. Oxidation by anion incorporation
3. Reduction by anion ejection
4. Reduction by cation incorporation

These four processes are illustrated in Figure 4.13 (PPy is used as an example). While shown as two separate, reversible reactions, it is possible (perhaps likely) that both reactions occur simultaneously during oxidation and reduction of CPs.

Unlike other conjugated polymers, PAn only shows electrochemical activity under acid environments, which makes the redox process of PAn more complicated compared with other conjugated polymers (Figure 4.14). PAn is only electroactive at low pH ranges (<4), where it can undergo oxidation and reduction cycling between three states: leucoemeraldine (a neutral, fully reduced state, i.e., an insulator), emeraldine salt (a half-oxidized state and a conductor), and pernigraniline (a fully oxidized state and an insulator) (Smela et al. 2005).

4.5.1 ANION AND CATION EXCHANGE: SIZE RELATIONSHIP

Experimental evidence also supports the cation-dominated and anion-dominated ion exchange in CPs. CPs prepared with large dopant anions are usually cation dominated. Those large ions are immobile and trapped within the polymer during polymerization and cannot be readily released during subsequent reduction cycles. Similarly, if large anions are used in the electrolyte, they may be too large to penetrate into the polymer. In such cases, the redox cycling of a CP will be dominated by cation exchange. For instance, PPy doped with dodecylbenzenesulfonate (DBS) and PPy doped with *p*-toluene sulfonic acid sodium

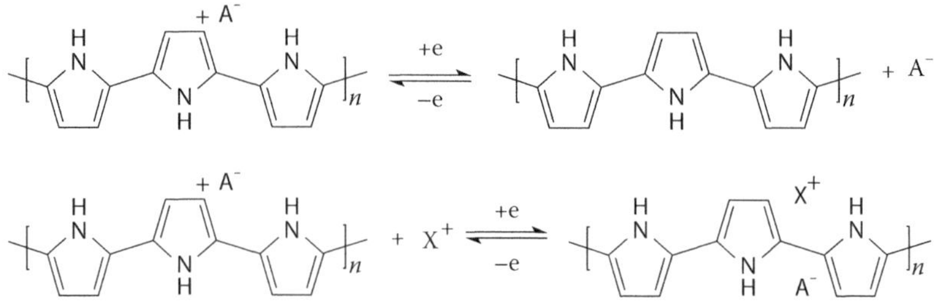

Figure 4.13 Switching property of polypyrrole where A^- represents anion incorporated into the polypyrrole and X^+ represents the cation from the electrolyte.

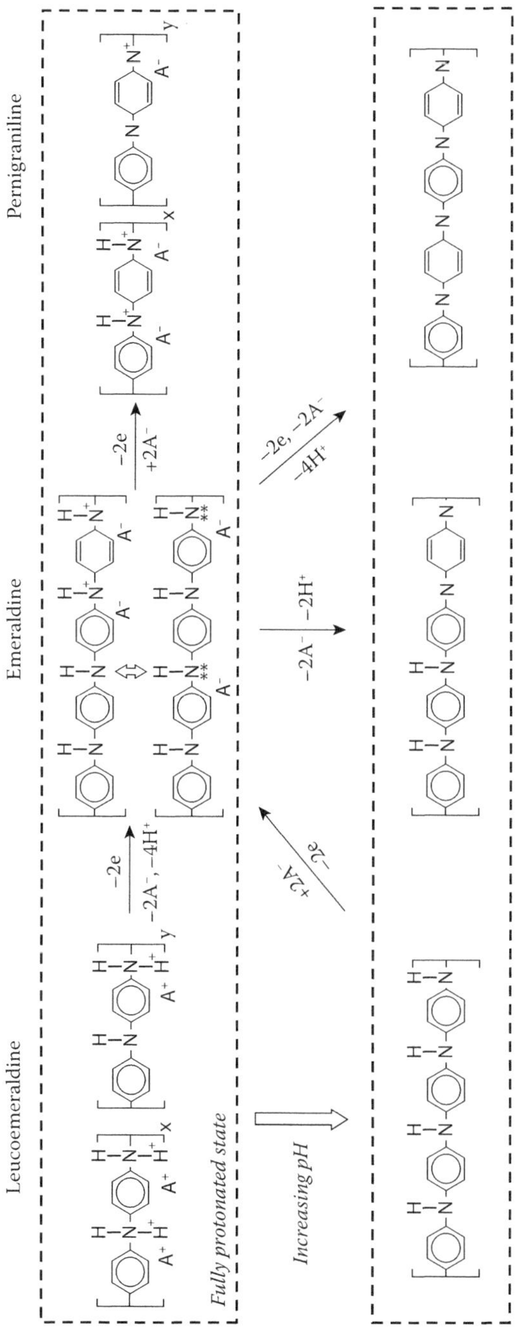

Figure 4.14 Electrochemical oxidation states of polyaniline.

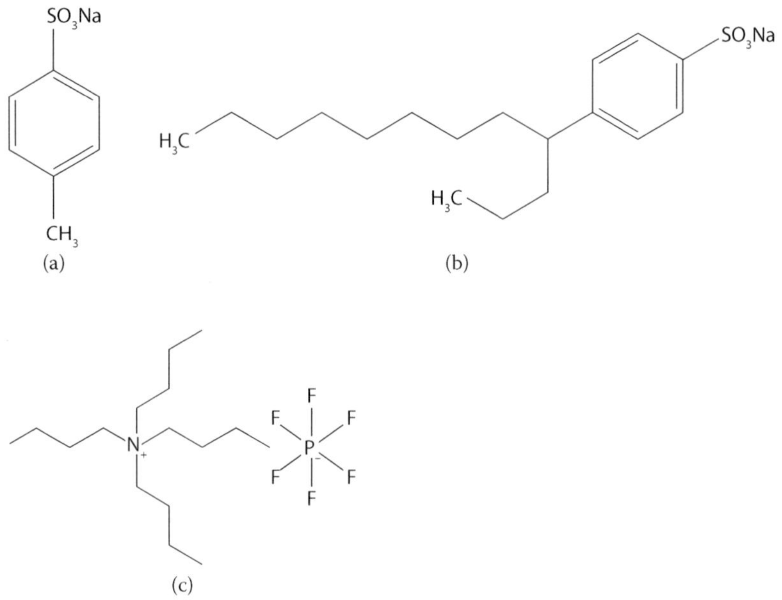

Figure 4.15 Chemical formulas of (a) NapTS, (b) NaDBS, and (c) $TBAPF_6$.

salt (*p*TS) are two typical cation exchange CPs, as shown in Figure 4.15, as only a small mobile cation like Na^+ can move in and out of the polymer due to *p*TS and DBS being entrapped within the polymer bulk (Pei and Inganaes 1993a, 1993b; Matencio et al. 1995). Conversely, if an electrolyte from which the CP was synthesized contained large or sterically bulky cations, but with a relatively small anion, the ion exchange of CPs would be dominated by anions. In the tetrabutylammonium hexafluorophosphate ($TBAPF_6$) system, due to the large size of TBA^+, PF_6^- dominates the ion-exchange process during redox cycling.

4.5.2 SWITCHING PROPERTIES OF CONDUCTING POLYMERS

The reversible oxidation and reduction of CPs are referred to in terms of their "switching" properties. While reducing or oxidizing, electrons flow in or out of the polymer to form charged or uncharged sites on the polymer chains, and ions must also then move in or out of the polymer to charge balance the polymer and achieve electrostatic neutrality. Such changes are accompanied by a number of observable physical changes in the CP, such as surface energy (hydrophobicity and hydrophilicity), optical (color or band gap), electrical (conductor and insulator transitions), mechanical (volume and shape), and magnetic (spin-state) changes.

4.5.3 SURFACE SWITCHING

Switching the redox state of a CP changes the surface energy, thereby changing the surface wetting ability (Figure 4.16) (Halldorsson et al. 2009, 2011) from hydrophobic to hydrophilic (Wang et al. 2008; Greco et al. 2013). By further extending this concept, Pei et al. (2012) demonstrated that the changes in ionic solution composition in the surface layer, resulting from oxidation and reduction of the CPs, trigger a switch in

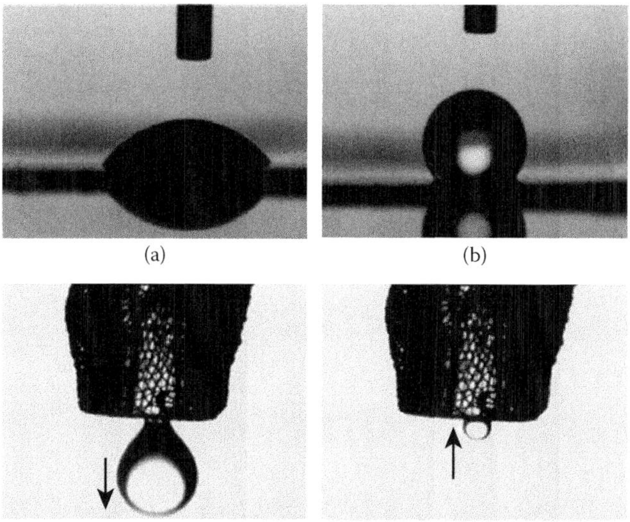

Figure 4.16 Water drop on surface of ClO_4-doped polyterthiophene (PTTh·ClO_4) sheet: (a) freshly deposited films and (b) electrochemically reduced films. Microfluidic control concepts of using PPy/DBS: (c) reduced state and (d) oxidized state. (From Halldorsson, J. A., et al., *Langmuir*, 25 [18], 11137–11141, 2009. With permission.)

conformation of the surface-bound polymer "brushes," which is capable of a change in wettability. By fine-tuning the surface structure of polymerization substrate, a thin layer of PPy with a honeycomb structure was electrochemically synthesized (Santos et al. 2013). Upon PPy oxidation–reduction, reversible switching of wettability is also obtained.

Generally, the surface morphology is found to be related to the doping level or oxidation state. The typical top surface morphology is characterized by a nodular structure. The top of the grains tends to be more oxidized than the surrounding area, and therefore more doped than the peripheral regions (Barisci et al. 2000). Significant thickness changes of up to 40% (Smela and Gadegaard 1999) have also been observed during electrochemical stimulation. Similarly, a significant volume change has been observed for the very top surface of the conducting polymer while experiencing the redox cycle (Higgins et al. 2009), resulting in morphology and roughness changes.

4.5.4 OPTICAL PROPERTIES

Coloration in conducting polymers depends on their band gap and band structure. Many conjugated polymers absorb light in the visible region as their optical band gap matches with the visible light wavelength. As a result of the reduction–oxidation processes, CPs undergo changes in their electronic structure that will alter the absorption of light (Pei et al. 1994). As a result, CPs have received special attention not only due to their ability to change color but also due to their ability to provide high light transmission contrast ratios, fast switching speed, and solution processability (Mortimer et al. 2009). Significantly, the color of electrochromic (EC) polymers can be tuned by modification of the monomer structure, which influences the band gap of the CP (Zou et al. 2007; Reeves et al. 2004). The Reynolds group has reported a method to fabricate dual-polymer complementary Electro Chromic Display(s) (ECDs) by assembling paired dark-state and charge-matched cathodically and anodically coloring EC polymers to increase color efficiency (Sapp et al. 1996, 1998).

Depending on the coloring type, EC devices utilizing conjugated polymers can be classified as anodically (oxidation) or cathodically (reduction) coloring (Gunbas and Toppare 2012). A key parameter to characterize EC performance is EC contrast (%*T*), which refers to the percent transmittance change. Switching time (response time) and coloration efficiency are also important to evaluate EC materials and devices. Classical conjugated polymers like PPy, PAn, and PT have shown to change their color when an electrical potential is applied (Camurlu 2014; Tong et al. 2014; Ferraris et al. 1997). PPy changes from yellow-green to blue grey when it is oxidized by the application of positive potentials. PAn displays a spectrum of color as different potentials are applied to it. PT can be switched from red to blue once it undergoes transition from the oxidized state to the reduced state. PEDOT and its derivatives are particularly interesting, given the diverse chemistries available to enable fine-tuning of the polymer band gap (Zhang et al. 2014). Recently, the polyselenophene family, together with poly(3,4-ethylenedioxyselenophene) (PEDOS) and its derivatives, attracted more attention as its band gap was lower (about 0.2 eV less) than that of PEDOT and its derivatives (Figure 4.17). Low band gap can absorb spectrum more efficiently, and thus the optoelectronic devices are expected to perform better (Patra et al. 2014).

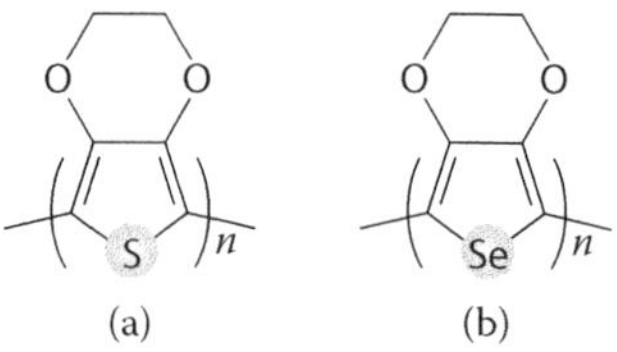

Figure 4.17 Chemical formulas of (a) PEDOT and (b) PEDOS.

An alternate approach for EC devices was successfully developed using complementary coloring Poly(3,4-propylenedioxythiophene) (PProDOT) and Poly(2-methoxyaniline-5-sulfonic acid) (PMAS) polymers from which tunable large-scale EC devices have been fabricated (Figure 4.18) (Weng et al. 2013). In this work, the possibility of using mixtures of PProDOT-$(Hx)_2$ and PProDOT-$(EtHx)_2$ to adjust the dark-state color, photopic contrast, and color saturation of the device was demonstrated without the need to modify the base polymer band gaps via monomer substitution. Device colorations from red-purple to dark blue were achieved by changing the ratios of the mixed PProDOT component (Table 4.1) (Weng et al. 2013). In addition, solvatochromic effects were noted whereby the device colors could be modified by selective use of different types of ionic electrolyte. In this instance, the solvent used changed the molecular conformation

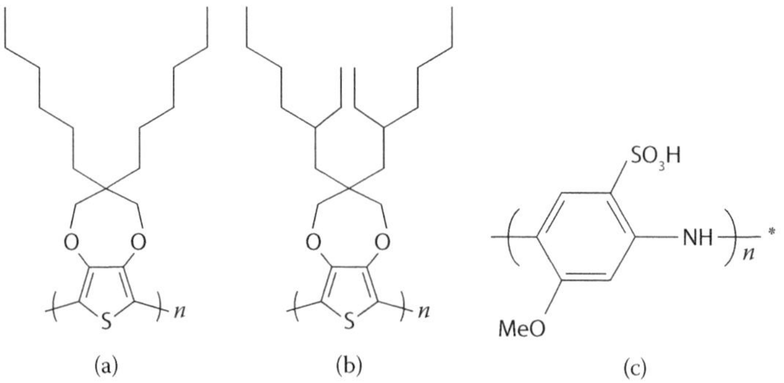

Figure 4.18 Chemical structure of (a) PProDOT-$(Hx)_2$, (b) PProDOT-$(EtHx)_2$, and (c) PMAS.

Table 4.1 Dark-state $L^*a^*b^*$ values and contrasts of PProDOT-$(Hx)_2$, PProDOT-$(EtHx)_2$, and mixed polymer electrochromics

ECDs	$L^*a^*b^*$ values	Transmittance of PProDOT slide (T%)	Transmittance of PAMS slide (T%)	Upper transmittance of ECD (T%)	Lower transmittance of ECD (T%)	Photopic contrast = Upper ECD transmittance – Lower ECD transmittance (%)
PProDOT$(EtHx)_2$	L^* = 50.43; a* = 13.4; b^* = −13.87; colour:	18.86	78.74	69.79	18.78	51.17
PProDOT$(Hx)_2$	L^* = 38.39; a* = 37.11; b^* = −28.19; colour:	18.06	78.25	64.08	10.89	57.19
Hx:Et = 2:1	L^* = 39; a* = 36.27; b^* = −28.48; colour:	18.2	78.6	69.05	12.63	56.72
Hx:Et = 1:1	L^* = 39.09; a* = 30.14; b^* = −31.62; colour:	17.88	77.41	68.85	10.14	58.71
Hx:Et = 1:2	L^* = 42.21; a* = 25.62; b^* = −29.38; colour:	18.09	78.46	67.38	10.66	56.42

Source: Weng, B., et al., *J. Mater. Chem. C.*, 1 (44), 7430–7439, 2013.

of the CP, which also influences the band gap of the polymer as a consequence of steric effects impacting the charge carriers along the CP backbone.

EC display devices, light-emitting diodes, and optical switches based on CPs have been reported (Matsushita et al. 2013; DeLongchamp and Hammond 2004; Burroughes et al. 1990; Chuang et al. 2014). A typical conjugated polymer light-emitting diode consists of very thin layers of polymer films sandwiched by two electrodes (Figure 4.19). There are two polymer layers, one of which functions as the hole-transporting layer and the other which functions as the light emission layer. An indium tin oxide (ITO) layer is generally used as the transparent anode. The metal layer provides conductance to the emitting polymer layer. While the potential applies, holes are injected from the anode to the hole-transporting layer, while electrons are injected from the cathode. The injected holes and electrons move toward opposite electrodes. When the recombination of holes and electrons occurs, the released energy of recombination can transfer to light.

Conjugated polymer–based light-emitting diodes have also developed due to their flexibility and the capability of mass production in a solution-assisted process (ink-jet printing or roll-to-roll coating). PPV can be conveniently made into high-quality films and show strong photoluminescence due to its large semiconductor band gaps (Burroughes et al. 1990). The band gap could also be altered electrochemically, so the photoluminescence will depend on operating voltage (Figure 4.20).

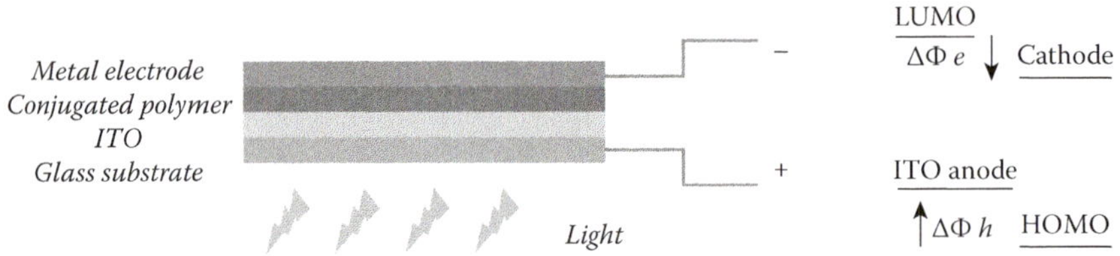

Figure 4.19 Classical structure of conjugated polymer–based-light-emitting diodes and its energy level.

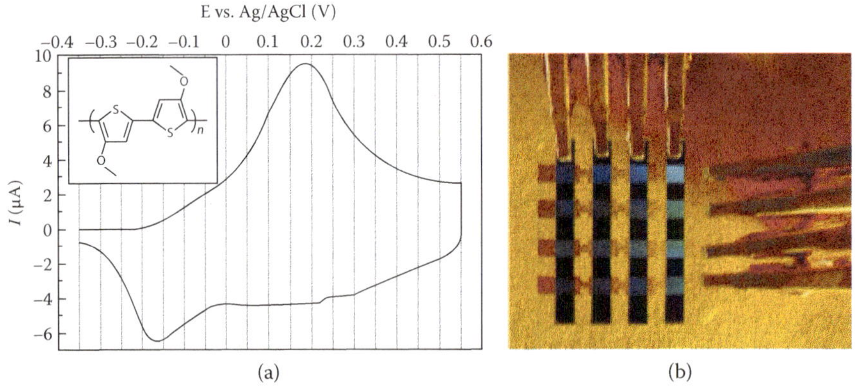

Figure 4.20 (a) Poly(4,4′-dimethoxy-bithiophene) (PDBT) cyclic voltammetry and (b) organic light-emitting diode (OLED) array. For panel (b), the electrochemical equilibrium potential increases from left to right, and the PDBT thickness increases from top to bottom. (Reprinted by permission from Macmillan Publishers Ltd., *Nature*, Gross, M., et al., Improving the performance of doped π-conjugated polymers for use in organic light-emitting diodes, 405 [6787], 661–665, 2000, copyright 2000.)

4.6 STABILITY OF CONJUGATED POLYMERS

In general, CPs are not very stable. Conductivity is known to decline upon aging in an ambient environment, especially for *n*-type conjugated polymers. Over time, the reduced polymer chain will undergo oxidation with oxygen in the air and lose conjugated length (Figure 4.21) (Yamaguchi and Uehara 2013).

In practical circumstances, the stability of conducting polymers is also commonly referred to as electrochemical stability, photochemical stability, thermal stability, and instability of the polymer, which is often due to the chemical reactions in those processes (Joergensen et al. 2008).

4.6.1 ELECTROCHEMICAL REASONS

Conformational change, overoxidation, and degradation of electrolyte are three main reasons for instability. Conformational change in CPs can sometimes result in irreversible structural change, especially during reduction (Otero et al. 1996). Otero et al. (1996) pointed out that during the cathodic potential step, which is accompanied by the neutralization of positive charges and the exclusion of the doping counterions, polymeric chains will diffuse into the free volume left by counterions. As a result, the "channel" for ions in and out will be closed and subsequently reopened, and these channel structures will thus require more energy and time. In normal electrochemical tests, CPs become compacted during reduction, which hinders the transfer of ions. Continued reduction from cycle to cycle can lead to a gradual compaction that diminishes the ability of the polymer to be oxidized. Later studies by the same author found that such conformational change is irreversible and responsible for the degradation of CPs (Otero et al. 2004).

Deterioration of the electroactivity of CPs has been reported when the cyclic voltammogram showed an obvious reduction in currents while cycling (Ding et al. 2003; Yamato et al. 2009). Overoxidation of CPs results in dedoping and formation of the carboxyl group and loss of conjugation (Li and Qian 2000). *In situ* spectroscopies (e.g., *in situ* FTIR and Raman) are efficient tools to track changes in chain conformation and deformation, as well as conjugation length. This effect is observed by FTIR by the shift of the ~1540 cm^{-1} peak toward higher frequencies, when cycled between 1.3 and –2.1 V. This shift indicates the diminution of the π-electron delocalization, which means a reduction in the chain conjugation length (Zerbi et al. 1994) (Figure 4.22b). Similarly, stretching vibration of C=C in the backbone of PPy shifts from a low wavenumber to a high wavenumber in Raman spectra, indicating a short conjugated length (Liu 2003) caused by overoxidation under high current density (Figure 4.22c).

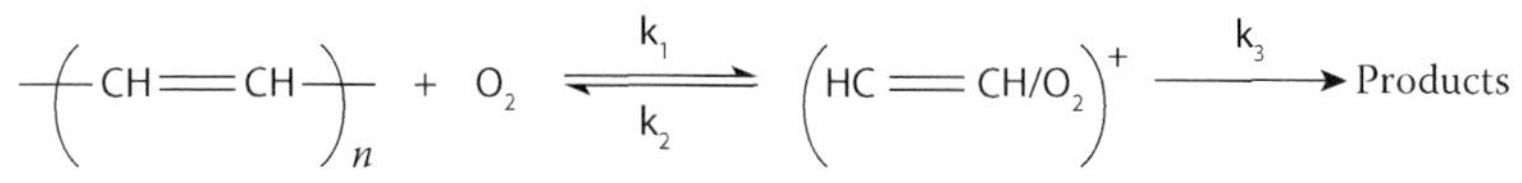

Figure 4.21 Reactions of polyacetylene with oxygen. The conductivity will increase in step 1 and dramatically decrease in step 2, which corresponds to the degradation of the conjugated chain. (From Pochan, J. M., et al., *Macromolecules*, 14 [1], 110–114, 1981. With permission.)

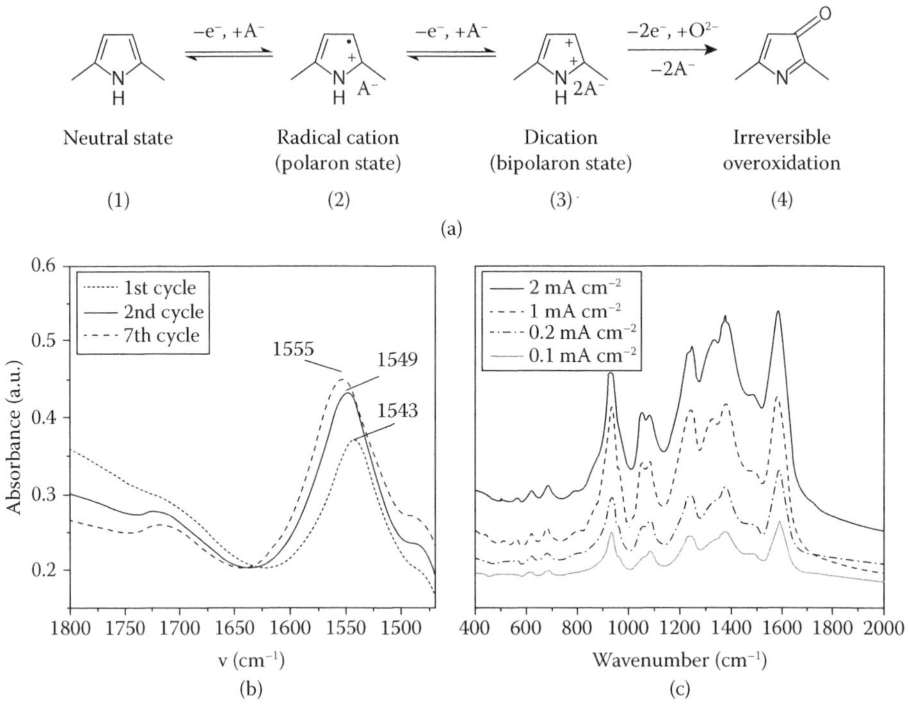

Figure 4.22 (a) Oxidation state of polypyrrole. (b) In situ FTIR spectra of PPy/PVS after having been electrochemically cycled from −2.1 to 0.8 V versus Ag/AgCl in 0.1 M $LiClO_4$ acetonitrile solution. (From Fernandez Romero, A. J., et al., *J. Phys. Chem. B*, 109 [44], 21078–21085, 2005. With permission.) (c) Raman spectra of PPy/pTS films polymerized under different current densities. (From Sui, J., et al., *J. Appl. Polym. Sci.*, 111 [2], 876–882, 2009. With permission.)

Degradation of the electrolyte will also lead to failure. For instance, anodic oxidation of propylene carbonate (PC) in water will produce H_2CO_3. Nucleophilic HCO_3^- and OH^- then attack polymer chains, interrupting conjugation and diminishing conjugation length (Novak et al. 1991). When using Cl^- as a dopant, some earlier studies illustrated that conjugated polymers could undergo excess chlorination in the presence of Cl^- ions above a potential of ~0.6 V (vs. Saturated Calomel Electrode [SCE]), which relates to the generation of ClO^- at the anode (Qi et al. 1996; Chen and Rajeshwar 1994). Obviously, carefully selecting the electrolyte is helpful for improving the stability. Ionic liquids are good choices, as they are environmentally stable and have high ionic conductivity, a high evaporating point, and most importantly, a wide potential window. Lu et al. (2002) described adopting ionic liquid as the electrolyte for conducting polymer devices that lead to long periods of stable device performance. It has also been found that CPs have a stable cyclic voltammetry (CV) in ionic liquid, indicating that the electroactivity is maintained in ionic liquids (Ding et al. 2003).

Spinks et al. (2004) pointed out another possible source of instability of CPs. During symmetrical CV, instead of returning to their original redox state, CPs shift toward a higher oxidized state from one cycle to another. Thus, a slow net oxidation causes a loss of redox capacity after several tens of cycles. In that study, current control provided a more stable electroactivity than voltage control. With current control, the amount of charge injected during oxidation could be matched to that occurring during reduction. After each cycle, the polymer returned to the same redox state.

4.6.2 PHOTOVOLTAIC AND THERMAL STABILITY

As mentioned before, conjugated polymers can easily lose conductivity in an ambient atmosphere environment. If being exposed to light for photovoltaic application (UV light, sunlight, etc.), such processes will be accelerated. Oxygen is readily active by UV illumination, especially when titanium oxide exists and is widely used in photovoltaic devices. The formation of superoxide will aggressively attack conjugated polymers. Poly-3-hexylthiophene (P3HT) is believed to be significantly stable in molecular structure but still susceptible to instability when under photovoltaic stimulation (Ratha et al. 2014). Thus, sealing of a photovoltaic device to avoid any oxygen and water in an ambient environment is a great challenge. Even when a complex sealing technique such as a sealed glass container or high-vacuum chamber is employed, oxygen and water are still capable of diffusing into devices, further causing degradation (Krebs 2006; Krebs et al. 2005; Krebs and Norrman 2007; Norrman et al. 2010).

The thermal stability of polymers always refers to the weight loss under high temperature. For conjugated polymers, the actual thermal degradation process is quite typical compared with the normal polymer (Figure 4.23). Typical Thermal Gravimetric Analysis (TGA) curves could contain four stages:

1. Volatile substances such as moisture and some additives
2. Polymer degradation
3. Carbonization
4. Inorganic compounds (dopant) and ash

The major concern of the thermal stability of a conjugated polymer is how stable the conjugated chain is under certain temperatures. Thermal stability is particularly important for photovoltaic devices that are usually operated under large temperature fluctuations. In general, the degradation speeds up while temperature increases (Tang et al. 2014). Fluorination on the backbone of conjugated polymer chains is believed to enhance the material crystallization and thermal stability (Subramanian et al. 2008). A higher degree

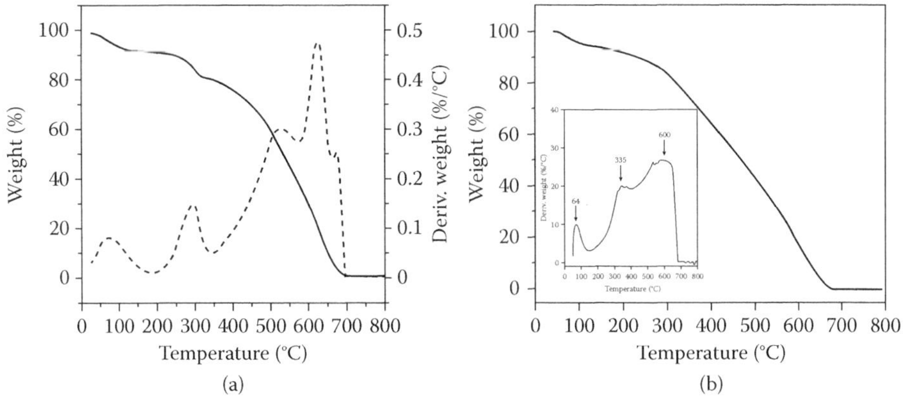

Figure 4.23 TGA analysis of (a) polyaniline (from Jeevananda, T., et al., *J. Appl. Polym. Sci.*, 109 [1], 200–210, 2008, with permission) and (b) chemically synthesized polypyrrole (from Cui, Z., et al., *Langmuir*, 30 [46], 14086–14094, 2014, with permission).

of fluorination leads to better stability (Liu et al. 2014). Other approaches to achieve high thermal stability include adding extra photo-cross-link groups to backbones (Chen et al. 2013), novel design of the molecular backbone structure (Wang et al. 2012), side chain engineering (Gu et al. 2015), and introducing carbon nanotubes or functionalized graphite oxide sheets (Lee et al. 2013).

REFERENCES

Baeriswyl, D., D. K. Campbell, and S. Mazumdar. An overview of the theory of π-conjugated polymers. *Springer Series in Solid-State Science* 102 (1992): 7–133.

Bakhshi, A. K. Theoretical designing of novel conducting polymers. *Indian Journal of Technology* 26 (10) (1988): 463–486.

Barisci, J. N., R. Stella, G. M. Spinks, and G. G. Wallace. Characterization of the topography and surface potential of electrodeposited conducting polymer films using atomic force and electric force microscopies. *Electrochimica Acta* 46 (4) (2000): 519–531.

Bernard, M. C., and G. A. Hugot-Le. Quantitative characterization of polyaniline films using Raman spectroscopy. I. Polaron lattice and bipolaron. *Electrochimica Acta* 52 (2) (2006a): 595–603.

Bernard, M. C., and G. A. Hugot-Le. Quantitative characterization of polyaniline films using Raman spectroscopy. II. Effects of self-doping in sulfonated polyaniline. *Electrochimica Acta* 52 (2) (2006b): 728–735.

Bertho, D., A. Laghdir, and C. Jouanin. Energetic study of polarons and bipolarons in polythiophene:Importance of Coulomb effects. *Physical Review B: Condensed Matter* 38 (17) (1988): 12531–12539.

Boyer, M. I., S. Quillard, G. Louarn, G. Froyer, and S. Lefrant. Vibrational study of the FeCl3-doped dimer of polyaniline; a good model compound of emeraldine salt. *Journal of Physical Chemistry B* 104 (38) (2000): 8952–8961.

Bredas, J. L., and G. B. Street. Polarons, bipolarons, and solitons in conducting polymers. *Accounts of Chemical Research* 18 (10) (1985): 309–315.

Burroughes, J. H., D. D. C. Bradley, A. R. Brown, R. N. Marks, K. Mackay, R. H. Friend, P. L. Burns, and A. B. Holmes. Light-emitting diodes based on conjugated polymers. *Nature (London)* 347 (6293) (1990): 539–541.

Camurlu, P. Polypyrrole derivatives for electrochromic applications. *RSC Advances* 4 (99) (2014): 55832–55845.

Chakrabarti, S., B. Das, P. Banerji, D. Banerjee, and R. Bhattacharya. Bipolaron saturation in polypyrrole. *Physical Review B: Condensed Matter and Materials Physics* 60 (11) (1999): 7691–7694.

Chen, C. C., and K. Rajeshwar. Chemical attack on polypyrrole by electrolytically generated solution species in aqueous chloride medium. *Journal of the Electrochemical Society* 141 (11) (1994): 2942–2946.

Chen, X., L. Chen, and Y. Chen. The effect of photocrosslinkable groups on thermal stability of bulk heterojunction solar cells based on donor-acceptor-conjugated polymers. *Journal of Polymer Science Part A: Polymer Chemistry* (19) (2013): 4156–4166.

Chipara, M., D. Hui, P. V. Notingher, M. D. Chipara, K. T. Lau, J. Sankar, and D. Panaitescu. On polyethylene-polyaniline composites. *Composites Part B* 34B (7) (2003): 637–645.

Chuang, C.-N., C.-Y. Chang, C.-L. Chang, Y.-X. Wang, Y.-S. Lin, and M.-K. Leung. Poly(9,9-dialkylfluorene-co-triphenylamine)s—A family of photo-curable hole-transport polymers for polymer light-emitting diodes applications. *European Polymer Journal* 56 (2014): 33–44.

Coskun, E., E. A. Zaragoza-Contreras, and H. J. Salavagione. Synthesis of sulfonated graphene/polyaniline composites with improved electroactivity. *Carbon* 50 (6) (2012): 2235–2243.

Cui, Z., C. Coletta, A. Dazzi, P. Lefrancois, M. Gervais, S. Neron, and S. Remita. Radiolytic method as a novel approach for the synthesis of nanostructured conducting polypyrrole. *Langmuir* 30 (46) (2014): 14086–14094.

DeLongchamp, D. M., and P. T. Hammond. Multiple-color electrochromism from layer-by-layer-assembled polyaniline/Prussian blue nanocomposite thin films. *Chemistry of Materials* 16 (23) (2004): 4799–4805.

Dennany, L., P. C. Innis, F. Masdarolomoor, and G. G. Wallace. ESR, Raman, and conductivity studies on fractionated poly(2-methoxyaniline-5-sulfonic acid). *Journal of Physical Chemistry B* 114 (7) (2010): 2337–2341.

Dennany, L., P. C. Innis, S. T. McGovern, G. G. Wallace, and R. J. Forster. Electronic interactions within composites of polyanilines formed under acidic and alkaline conditions. Conductivity, ESR, Raman, UV-vis and fluorescence studies. *Physical Chemistry Chemical Physics* 13 (8) (2011): 3303–3310.

Diaz, A. F., and J. F. Bargon. Electrochemical synthesis of conducting polymers. In *Handbook of Conducting Polymers*, ed. T. A. Skotheim. New York: Marcel Dekker, 1986, 81–115.

Diaz, A. F., K. K. Kanazawa, and G. P. Gardini. Electrochemical polymerization of pyrrole. *Journal of the Chemical Society: Chemical Communications* (14) (1979): 635–636.

Ding, J., D. Zhou, G. M. Spinks, G. G. Wallace, S. Forsyth, M. Forsyth, and D. MacFarlane. Use of ionic liquids as electrolytes in electromechanical actuator systems based on inherently conducting polymers. *Chemistry of Materials* 15 (12) (2003): 2392–2398.

Duluard, S., B. Ouvrard, A. Celik-Cochet, G. Campet, U. Posset, G. Schottner, and M. H. Delville. Comparison of PEDOT films obtained via three different routes through spectroelectrochemistry and the differential cyclic voltabsorptometry method (DCVA). *Journal of Physical Chemistry B* 114 (22) (2010): 7445–7451.

Fernandez Romero, A. J., J. J. Lopez Cascales, and T. F. Otero. In situ FTIR spectroscopy study of the break-in phenomenon observed for ppy/pvs films in acetonitrile. *Journal of Physical Chemistry B* 109 (44) (2005): 21078–21085.

Ferraris, J. P., M. S. K. Dhurjati, D. C. Loveday, N. N. Barashkov, M. Hmyene, and C. R. Henderson. Novel electrochromic materials based on polythiophene. *Proceedings of the Electrochemical Society* 96–24 (1997): 14–35.

Fukuda, T., H. Takezoe, K. Ishikawa, A. Fukuda, H. S. Woo, S. K. Jeong, E. J. Oh, and J. S. Suh. IR and Raman studies in three polyanilines with different oxidation levels. *Synthetic Metals* 69 (1–3) (1995): 175–176.

Furukawa, Y. Reexamination of the assignments of electronic absorption bands of polarons and bipolarons in conducting polymers. *Synthetic Metals* 69 (1–3) (1995): 629–32.

Furukawa, Y. Electronic absorption and vibrational spectroscopies of conjugated conducting polymers. *Journal of Physical Chemistry* 100 (39) (1996): 15644–15653.

Furukawa, Y., A. Sakamoto, H. Ohta, and M. Tasumi. Raman characterization of polarons, bipolarons and solitons in conducting polymers. *Synthetic Metals* 49 (1–3) (1992): 335–340.

Furukawa, Y., S. Tazawa, Y. Fujii, and I. Harada. Raman spectra of polypyrrole and its 2,5-13C-substituted and C-deuterated analogs in doped and undoped states. *Synthetic Metals* 24 (4) (1988): 329–341.

Galal, A., E. T. Lewis, O. Y. Ataman, H. Zimmer, and H. B. Mark Jr. Electrochemical synthesis of conducting polymers from oligomers containing thiophene and furan rings. *Journal of Polymer Science Part A: Polymer Chemistry* 27 (6) (1989): 1891–1896.

Gandhi, M. R., P. Murray, G. M. Spinks, and G. G. Wallace. Mechanism of electromechanical actuation in polypyrrole. *Synthetic Metals* 73 (3) (1995): 247–256.

Greco, F., A. Zucca, S. Taccola, B. Mazzolai, and V. Mattoli. Patterned free-standing conductive nanofilms for ultraconformable circuits and smart interfaces. *ACS Applied Materials & Interfaces* 5 (19) (2013): 9461–9469.

Gross, M., D. C. Muller, H.-G. Nothofer, U. Scherf, D. Neher, C. Brauchle, and K. Merrholz. Improving the performance of doped π-conjugated polymers for use in organic light-emitting diodes. *Nature (London)* 405 (6787) (2000): 661–665.

Gu, C., Z. Du, W. Shen, X. Bao, S. Wen, D. Zhu, T. Wang, N. Wang, and R. Yang. Optical, electrochemical, and photovoltaic properties of conjugated polymers with dithiafulvalene as side chains. *Journal of Applied Polymer Science* 132 (8) (2015): 41508/1–41508/6.

Gunbas, G., and L. Toppare. Electrochromic conjugated polyheterocycles and derivatives—Highlights from the last decade towards realization of long lived aspirations. *Chemical Communications (Cambridge, UK)* 48 (8) (2012): 1083–1101.

Halldorsson, J. A., S. J. Little, D. Diamond, G. M. Spinks, and G. G. Wallace. Controlled transport of droplets using conducting polymers. *Langmuir* 25 (18) (2009): 11137–11141.

Halldorsson, J. A., Y. Wu, H. R. Brown, G. M. Spinks, and G. G. Wallace. Surfactant-controlled shape change of organic droplets using polypyrrole. *Thin Solid Films* 519 (19) (2011): 6486–6491.

Higgins, M. J., S. T. McGovern, and G. G. Wallace. Visualizing dynamic actuation of ultrathin polypyrrole films. *Langmuir* 25 (6) (2009): 3627–3633.

Hillman, A. R., S. J. Daisley, and S. Bruckenstein. Ion and solvent transfers and trapping phenomena during n-doping of PEDOT films. *Electrochimica Acta* 53 (11) (2008): 3763–3771.

Hou, Y., L. Zhang, L. Y. Chen, P. Liu, A. Hirata, and M. W. Chen. Raman characterization of pseudocapacitive behavior of polypyrrole on nanoporous gold. *Physical Chemistry Chemical Physics* 16 (8) (2014): 3523–3528.

Hu, Z., T. Adachi, Y.-G. Lee, R. T. Haws, B. Hanson, R. J. Ono, C. W. Bielawski, V. Ganesan, P. J. Rossky, and B. D. A. Vanden. Effect of the side-chain-distribution density on the single-conjugated-polymer-chain conformation. *Chemphyschem* 14 (18) (2013): 4143–4148.

Jeevananda, T., Siddaramaiah, T. S. Lee, J. H. Lee, O. M. Samir, and R. Somashekar. Polyaniline-multiwalled carbon nanotube composites: Characterization by WAXS and TGA. *Journal of Applied Polymer Science* 109 (1) (2008): 200–210.

Joergensen, M., K. Norrman, and F. Krebs. Stability/degradation of polymer solar cells. *Solar Energy Materials and Solar Cells* 92 (7) (2008): 686–714.

Kahol, P. K. Magnetic susceptibility and electron spin resonance investigations of polyaniline and polyaniline-poly(methyl methacrylate) blend. *Solid State Communications* 117 (1) (2000): 37–39.

Krebs, F. C. Encapsulation of polymer photovoltaic prototypes. *Solar Energy Materials and Solar Cells* 90 (20) (2006): 3633–3643.

Krebs, F. C., J. E. Carle, N. Cruys-Bagger, M. Andersen, M. R. Lilliedal, M. A. Hammond, and S. Hvidt. Lifetimes of organic photovoltaics: Photochemistry, atmosphere effects and barrier layers in ITO-MEHPPV:PCBM-aluminium devices. *Solar Energy Materials and Solar Cells* 86 (4) (2005): 499–516.

Krebs, F. C., and K. Norrman. Analysis of the failure mechanism for a stable organic photovoltaic during 10 000 h of testing. *Progress in Photovoltaics* 15 (8) (2007): 697–712.

Kudo, Y., K. Kawabata, and H. Goto. A possibility for generation of two species of charge carriers along main-chain and side-chains for a π-conjugated polymer. *Journal of Physics: Conference Series* 428 (2013): 012006/1–012006/6.

Lee, R.-H., J.-L. Huang, and C.-H. Chi. Conjugated polymer-functionalized graphite oxide sheets thin films for enhanced photovoltaic properties of polymer solar cells. *Journal of Polymer Science Part B: Polymer Physics* 51 (2) (2013): 137–148.

Lei, T., J.-Y. Wang, and J. Pei. Roles of flexible chains in organic semiconducting materials. *Chemistry of Materials* 26 (1) (2014): 594–603.

Lemaire, M., W. Buchner, R. Garreau, A. H. Huynh, A. Guy, and J. Roncali. Synthesis of organic conductors using electrodesilylation. *Journal of Electroanalytical Chemistry and Interfacial Electrochemistry* 281 (1–2) (1990): 293–298.

Li, Y., and R. Qian. Electrochemical overoxidation of conducting polypyrrole nitrate film in aqueous solutions. *Electrochimica Acta* 45 (11) (2000): 1727–1731.

Lippe, J., and R. Holze. Electrochemical in-situ conductivity and polaron concentration measurements at selected conducting polymers. *Synthetic Metals* 43 (1–2) (1991): 2927–2930.

Liu, X., B. B. Y. Hsu, Y. Sun, C.-K. Mai, A. J. Heeger, and G. C. Bazan. High thermal stability solution-processable narrow-band gap molecular semiconductors. *Journal of the American Chemical Society* 136 (46) (2014): 16144–16147.

Liu Y. C., Evaluation of the conductivity of the polypyrrole film via its C=C bonds stretching on surface-enhanced Raman spectrum. *Electroanalysis* 15(13)(2003): 1134–1138.

Louarn, G., M. Lapkowski, S. Quillard, A. Pron, J. P. Buisson, and S. Lefrant. Vibrational properties of polyaniline-isotope effects. *Journal of Physical Chemistry* 100 (17) (1996): 6998–7006.

Lu, W., A. G. Fadeev, B. Qi, et al. Use of ionic liquids for π-conjugated polymer electrochemical devices. *Science (Washington, DC.)* 297 (5583) (2002): 983–987.

MacDiarmid, A. G. "Synthetic metals": A novel role for organic polymers (Nobel lecture). *Angewandte Chemie International Edition* 40 (14) (2001): 2581–2590.

Maddison, D. S., J. Unsworth, and R. B. Roberts. Electrical conductivity and thermoelectric power of polypyrrole with different doping levels. *Synthetic Metals* 26 (1) (1988): 99–108.

Mai, C.-K., R. A. Schlitz, G. M. Su, D. Spitzer, X. Wang, S. L. Fronk, D. G. Cahill, M. L. Chabinyc, and G. C. Bazan. Side-chain effects on the conductivity, morphology, and thermoelectric properties of self-doped narrow-band-gap conjugated polyelectrolytes. *Journal of the American Chemical Society* 136 (39) (2014): 13478–13481.

Matencio, T., M. A. De Paoli, R. C. D. Peres, R. M. Torresi, and S. I. Cordoba de Torresi. Ionic exchanges in dodecylbenzenesulfonate doped polypyrrole. Part 1. Optical beam deflection studies. *Synthetic Metals* 72 (1) (1995): 59–64.

Matsushita, S., Y. S. Jeong, and K. Akagi. Electrochromism-driven linearly and circularly polarised dichroism of poly(3,4-ethylenedioxythiophene) derivatives with chirality and liquid crystallinity. *Chemical Communications (Cambridge, UK)* 49 (19) (2013): 1883–1890.

Mizoguchi, K. Electronic states in conjugated polymers studied by electron spin resonance. *Synthetic Metals* 119 (1–3) (2001): 35–38.

Monk, P., R. Mortimer, and D. Rosseisky. *Electrochromism and Electrochromic Devices.* Cambridge: Cambridge University Press, 2007.

Mortimer, R. J., K. R. Graham, C. R. G. Grenier, and J. R. Reynolds. Influence of the film thickness and morphology on the colorimetric properties of spray-coated electrochromic disubstituted 3,4-propylenedioxythiophene polymers. *ACS Applied Materials & Interfaces* 1 (10) (2009): 2269–2276.

Mott, N. F., and E. A. Davis. *Electronic Processes in Non-Crystalline Materials.* Oxford: Oxford University Press, 2012.

Mu, S., J. Kan, J. Lu, and L. Zhuang. Interconversion of polarons and bipolarons of polyaniline during the electrochemical polymerization of aniline. *Journal of Electroanalytical Chemistry* 446 (1–2) (1998): 107–112.

Nalwa, H. S. Phase transitions in polypyrrole and polythiophene conducting polymers demonstrated by magnetic susceptibility measurements. *Physical Review B: Condensed Matter* 39 (9) (1989): 5964–5974.

Nguyen, T.-Q., V. Doan, and B. J. Schwartz. Conjugated polymer aggregates in solution: Control of interchain interactions. *Journal of Chemical Physics* 110 (8) (1999): 4068–4078.

Nguyen, T.-Q., I. B. Martini, J. Liu, and B. J. Schwartz. Controlling interchain interactions in conjugated polymers: The effects of chain morphology on exciton-exciton annihilation and aggregation in MEH-PPV films. *Journal of Physical Chemistry B* 104 (2) (2000): 237–255.

Niaura, G., R. Mazeikiene, and A. Malinauskas. Structural changes in conducting form of polyaniline upon ring sulfonation as deduced by near infrared resonance Raman spectroscopy. *Synthetic Metals* 145 (2–3) (2004): 105–112.

Norrman, K., M. V. Madsen, S. A. Gevorgyan, and F. C. Krebs. Degradation patterns in water and oxygen of an inverted polymer solar cell. *Journal of the American Chemical Society* 132 (47) (2010): 16883–16892.

Novak, P., B. Rasch, and W. Vielstich. Overoxidation of polypyrrole in propylene carbonate. An in situ FTIR study. *Journal of the Electrochemical Society* 138 (11) (1991): 3300–3304.

Otero, T. F., and I. Boyano. Characterization of polypyrrole degradation by the conformational relaxation model. *Electrochimica Acta* 51 (28) (2006): 6238–6242.

Otero, T. F., H. Grande, and J. Rodriguez. Conformational relaxation during polypyrrole oxidation: From experiment to theory. *Electrochimica Acta* 41 (11/12) (1996): 1863–1869.

Otero, T. F., M. Marquez, and I. J. Suarez. Polypyrrole: Diffusion coefficients and degradation by overoxidation. *Journal of Physical Chemistry B* 108 (39) (2004): 15429–15433.

Patra, A., M. Bendikov, and S. Chand. Poly(3,4-ethylenedioxyselenophene) and its derivatives: Novel organic electronic materials. *Accounts of Chemical Research* 47 (5) (2014): 1465–1474.

Pei, Q., and O. Inganaes. Electrochemical applications of the bending beam method. 2. Electroshrinking and slow relaxation in polypyrrole. *Journal of Physical Chemistry* 97 (22) (1993a): 6034–6041.

Pei, Q., and O. Inganaes. Electrochemical applications of the bending beam method; a novel way to study ion transport in electroactive polymers. *Solid State Ionics* 60 (1–3) (1993b): 161–166.

Pei, Q., G. Zuccarello, M. Ahlskog, and O. Inganaes. Electrochromic and highly stable poly(3,4-ethylenedioxythiophene) switches between opaque blue-black and transparent sky blue. *Polymer* 35 (7) (1994): 1347–1351.

Pei, Y., J. Travas-Sejdic, and D. E. Williams. Reversible electrochemical switching of polymer brushes grafted onto conducting polymer films. *Langmuir* 28 (21) (2012): 8072–8083.

Pereira Da Silva, J. E., M. L. A. Temperini, and S. I. Cordoba De Torresi. Secondary doping of polyaniline studied by resonance Raman spectroscopy. *Electrochimica Acta* 44 (12) (1999): 1887–1891.

Plocharski, J. Mechanisms of conductivity in conjugated polymers and relations to morphology. *Materials Science Forum* 42 (1989): 17–27.

Pochan, J. M., D. F. Pochan, H. Rommelmann, and H. W. Gibson. Kinetics of doping and degradation of polyacetylene by oxygen. *Macromolecules* 14 (1) (1981): 110–114.

Pratt, F. L., S. J. Blundell, W. Hayes, K. Nagamine, K. Ishida, and A. P. Monkman. Anisotropic polaron motion in polyaniline studied by muon spin relaxation. *Phys. Rev. Lett.* 79 (15) (1997): 2855–2858.

Qi, Z., N. G. Rees, and P. G. Pickup. Electrochemically induced substitution of polythiophenes and polypyrrole. *Chemistry of Materials* 8 (3) (1996): 701–707.

Ratha, R., P. J. Goutam, and P. K. Iyer. Photo stability enhancement of poly(3-hexylthiophene)-PCBM nanocomposites by addition of multi walled carbon nanotubes under ambient conditions. *Organic Electronics* 15 (7) (2014): 1650–1656.

Reeves, B. D., C. R. G. Grenier, A. A. Argun, A. Cirpan, T. D. McCarley, and J. R. Reynolds. Spray coatable electrochromic dioxythiophene polymers with high coloration efficiencies. *Macromolecules* 37 (20) (2004): 7559–7569.

Roncali, J. 1992. Conjugated poly(thiophenes): Synthesis, functionalization, and applications. *Chemical Reviews* 92 (4): 711–738.

Roncali, J., A. Guy, M. Lemaire, R. Garreau, and A. H. Huynh. Tetrathienylsilane as a precursor of highly conducting electrogenerated polythiophene. *Journal of Electroanalytical Chemistry and Interfacial Electrochemistry* 312 (1–2) (1991): 277–283.

Roth, S. Conducting polymers—Present state of physical understanding. *Materials Science Forum* 42 (1989): 1–16.

Sabouraud, G., S. Sadki, and N. Brodie. The mechanisms of pyrrole electropolymerization. *Chemical Society Reviews* 29 (5) (2000): 283–293.

Salafsky, J. S. Exciton dissociation, charge transport, and recombination in ultrathin, conjugated polymer-TiO2 nanocrystal intermixed composites. *Physical Review B: Condensed Matter and Materials Physics* 59 (16) (1999): 10885–10894.

Santos, L., A. G. Brolo, and E. M. Girotto. Study of polaron and bipolaron states in polypyrrole by in situ Raman spectroelectrochemistry. *Electrochimica Acta* 52 (20) (2007): 6141–6145.

Santos, L., P. Martin, J. Ghilane, P. C. Lacaze, and J.-C. Lacroix. Micro/nano-structured polypyrrole surfaces on oxidizable metals as smart electroswitchable coatings. *ACS Applied Materials & Interfaces* 5 (20) (2013): 10159–10164.

Sapp, S. A., G. A. Sotzing, J. L. Reddinger, and J. R. Reynolds. Rapid switching solid-state electrochromic devices based on complementary conducting polymer films. *Advance Materials (Weinheim, Ger.)* 8 (10) (1996): 808–811.

Sapp, S. A., G. A. Sotzing, and J. R. Reynolds. High contrast ratio and fast switching dual polymer electrochromic devices. *Chemistry of Materials* 10 (8) (1998): 2101–2108.

Schwartz, B. J. Conjugated polymers as molecular materials: How chain conformation and film morphology influence energy transfer and interchain interactions. *Annual Review of Physical Chemistry* 54 (2003): 141–72.

Scrosati, B. Electrochemical properties of conducting polymers. *Progress in Solid State Chemistry* 18 (1) (1988): 1–77.

Smela, E., and N. Gadegaard. Surprising volume change in PPy/DBS. An atomic force microscopy study. *Advanced Materials (Weinheim, Ger.)* 11 (11) (1999): 953–957.

Smela, E., W. Lu, and B. R. Mattes. Polyaniline actuators. *Synthetic Metals* 151 (1) (2005): 25–42.

Spinks, G. M., B. Xi, D. Zhou, V.-T. Truong, and G. G. Wallace. Enhanced control and stability of polypyrrole electromechanical actuators. *Synthetic Metals* 140 (2–3) (2004): 273–280.

Street, G. B. From powder to plastics. In *Handbook of Conducting Polymers*, ed. T. A. Skotheim. New York: Marcel Dekker, 1986, 265–291.

Subramanian, S., S. K. Park, S. R. Parkin, V. Podzorov, T. N. Jackson, and J. E. Anthony. Chromophore fluorination enhances crystallization and stability of soluble anthradithiophene semiconductors. *Journal of the American Chemical Society* 130 (9) (2008): 2706–2707.

Sui, J., J. Travas-Sejdic, S. Y. Chu, K. C. Li, and P. A. Kilmartin. The actuation behavior and stability of p-toluene sulfonate doped polypyrrole films formed at different deposition current densities. *Journal of Applied Polymer Science* 111 (2) (2009): 876–882.

Tang, H., Y. Ding, C. Zang, J. Gu, Q. Shen, and J. Kan. Effect of temperature on electrochemical degradation of polyaniline. *International Journal of Electrochemical Science* 9 (12) (2014): 7239–7252.

Tong, Z.-Q., H.-M. Lv, J.-P. Zhao, and Y. Li. Near-infrared and multicolor electrochromic device based on polyaniline derivative. *Chinese Journal of Polymer Science* 32 (8) (2014): 1040–1051.

Ullah, H., A.-u.-H. A. Shah, S. Bilal, and K. Ayub. Doping and dedoping processes of polypyrrole: DFT study with hybrid functionals. *Journal of Physical Chemistry C* 118 (31) (2014): 17819–17830.

Waller, A. M., and R. G. Compton. Simultaneous alternating current impedance/electron spin resonance study of electrochemical doping in polypyrrole. *Journal of the Chemical Society, Faraday Transactions 1* 85 (4) (1989): 977–990.

Wang, S., W. Hong, S. Ren, J. Li, M. Wang, X. Gao, and H. Li. New ladder-type conjugated polymer with broad absorption, high thermal stability, and low band gap. *Journal of Polymer Science Part A: Polymer Chemistry* 50 (20) (2012): 4272–4276.

Wang, X., M. Berggren, and O. Inganaes. Dynamic control of surface energy and topography of microstructured conducting polymer films. *Langmuir* 24 (11) (2008): 5942–5948.

Watanabe, S., K. Ando, K. Kang, S. Mooser, Y. Vaynzof, H. Kurebayashi, E. Saitoh, and H. Sirringhaus. Polaron spin current transport in organic semiconductors. *Nature Physics* 10 (4) (2014): 308–313.

Weng, B., S. Ashraf, P. C. Innis, and G. G. Wallace. Colour tunable electrochromic devices based on PProDOT-(Hx)2 and PProDOT-(EtHx)2 polymers. *Journal of Materials Chemistry C* 1 (44) (2013): 7430–7439.

Xu, Z., H. Tsai, H.-L. Wang, and M. Cotlet. Solvent polarity effect on chain conformation, film morphology, and optical properties of a water-soluble conjugated polymer. *Journal of Physical Chemistry B* 114 (36) (2010): 11746–11752.

Yamaguchi, I., and R. Uehara. Organometallic synthesis of n-type π-conjugated polymers with dopant cation trapping sites and stability of n-doping state against air. *Polymer International* 62 (5) (2013): 766–773.

Yamato, K., K. Tominaga, W. Takashima, and K. Kaneto. Stability of electrochemomechanical strains in polypyrrole films using ionic liquids. *Synthetic Metals* 159 (9–10) (2009): 839–842.

Yang, S. M., and C. P. Li. EPR studies of polyaniline. *Synthetic Metals* 55 (1) (1993): 636–641.

Yim, K.-H., G. L. Whiting, C. E. Murphy, J. J. M. Halls, J. H. Burroughes, R. H. Friend, and J.-S. Kim. Controlling electrical properties of conjugated polymers via a solution-based p-type doping. *Advanced Materials* 20 (17) (2008): 3319–3324.

Zerbi, G., M. Veronelli, S. Martina, A. D. Schlueter, and G. Wegner. π-Electron delocalization in conformationally distorted oligopyrroles and polypyrrole. *Advanced Materials (Weinheim, Ger.)* 6 (5) (1994): 385–388.

Zhang, S., J. Xu, B. Lu, L. Qin, L. Zhang, S. Zhen, and D. Mo. Electrochromic enhancement of poly(3,4-ethylenedioxythiophene) films functionalized with hydroxymethyl and ethylene oxide. *Journal of Polymer Science Part A: Polymer Chemistry* 52 (14) (2014): 1989–1999.

Zou, Y., W. Wu, G. Sang, Y. Yang, Y. Liu, and Y. Li. Polythiophene derivative with phenothiazine-vinylene conjugated side chaIn: Synthesis and its application in field-effect transistors. *Macromolecules (Washington, DC)* 40 (20) (2007): 7231–7237.

5 Industry-viable metal anticorrosion application of polyaniline

Yizhong Luo and Xianhong Wang
Key Laboratory of Polymer Ecomaterials, Chinese Academy of Sciences
Changchun, China

Contents

Metal corrosion is a natural phenomenon that we can see almost everywhere, since most metals are generally in favor of reacting with environmental factors to form oxides or other compounds different from their original states in nature, mainly due to the thermodynamical instability of pure metal. Metal corrosion causes serious damage to infrastructure, such as bridges, railways, and oil and gas pipelines, sometimes leading to a severe threat to public safety.

Fortunately, unlike sudden natural disasters such as earthquakes or severe weather, metal corrosion is a time-dependent damage that can be controlled and even prevented by various state-of-the-art techniques. One of such techniques is employing an organic coating containing corrosion-inhibiting pigments, by which the organic matrix offers the physical barrier, while corrosion like pinholes and small defects may be prevented by the inhibitory pigments. Chromate (Cr^{6+}) exhibits excellent protection by forming a passive layer (Zhao et al. 2001; Pommiers et al. 2014). However, because chromate is toxic and carcinogenic, its use has been strictly limited by the Waste Electrical and Electronic Equipment (WEEE) and Restriction of Hazardous Substances (ROHS) Directives since 2005. Another pigment is zinc powder, which is used as a sacrificial material to prevent metal corrosion. But as a heavy metal, zinc can spoil soil or marine environment. Therefore, novel environmentally friendly anticorrosive pigments are critically needed to replace chromate and zinc.

Polyaniline (PANI) is regarded as one of the most promising environmentally friendly corrosion-inhibiting pigments. In the past decades, PANI has been widely investigated and demonstrated efficient corrosion-inhibiting ability for various metals, leading to commercial products and industrial applications since the mid-1990s. Although the anticorrosion mechanism remains to be further clarified, the corrosion-protective capacity of PANI is believed to be definitely related to its unique redox property (Wessling 1996; Barisci et al. 1997). Due to the variable ratio of the phenylenediamine unit to the quinone diimine unit, PANI may exist in different states at certain media, and such change is reversible in many cases (Huang et al. 1986). Numerous studies show that upon immersion of PANI-coated metals, the interaction between PANI and metal occurs, where PANI acts as catalyst experiencing a reversible reduction and oxidation reaction, while metal is oxidized to form an oxide layer (Wessling 1996; Wang et al. 2003). The most possible mechanisms are anodic protection (i.e., ennobling metal surface and promoting a passive layer formation) and intelligent release of corrosion-inhibiting dopant anions; both seemingly depend on the interaction and redox properties of PANI (Spinks et al. 2002; Seegmiller et al. 2005). In this chapter, we discuss the redox properties of PANI and its interaction with various metals and alloys, such as iron, steel, copper, aluminum alloy, and magnesium alloy; the importance of the interface layer formed between the PANI coating and metal surface is explicated. Finally, the viable but tortuous developing process of a PANI anticorrosive coating in industry is commented on.

5.1 REVERSIBLE REDOX REACTION IN PANI

The corrosion protection performance of PANI for various metals is primarily attributed to its reversible redox property, as associated with its special structure containing reduced repeat units (phenylenediamine) and oxidized ones (quinone diimine), as shown in Figure 5.1. Based on the relative ratio of these two units, PANI may exist in different oxidation states. A value of $y = 1$ is assigned to the fully reduced state, the leucoemeraldine base (LB) form, while a value of $y = 0$ corresponds to the fully oxidized state, the pernigraniline base (PNB) form. And emeraldine base (EB), the most stable form in an ambient atmosphere, corresponds to the semioxidized state when $y = 0.5$ (MacDiarmid et al. 1987).

Figure 5.2 presents the interconversion of various redox states and protonated or deprotonated forms in PANI (Zhang 2007). In acidic media, PANI usually exists in its protonated form, emeraldine salt (ES), the electronically conductive form. ES can be produced by treatment of EB with Brønsted acid or directly through oxidation

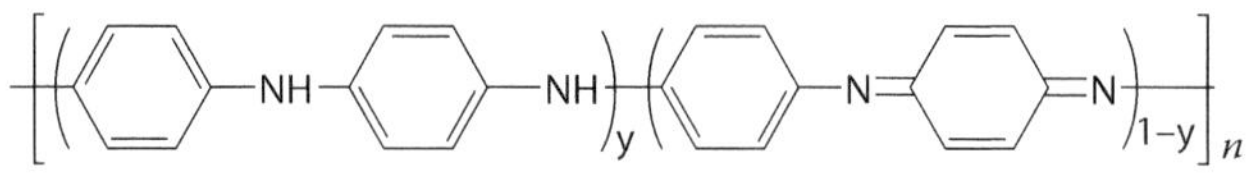

Figure 5.1 Chemical structure of PANI, showing phenylenediamine on the left and quinone diamine on the right.

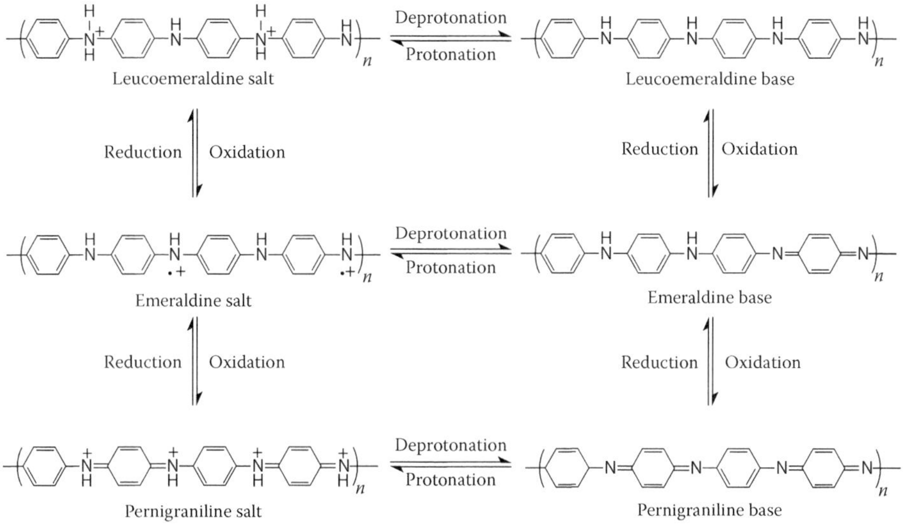

Figure 5.2 Interconversion of redox state and protonated or deprotonated form of PANI. (From Zhang, D. H., *Polym. Test.*, 26, 9–13, 2007. With permission.)

polymerization of aniline in acidic solution. Also, in acidic solution, ES can be reduced to leucoemeraldine salt (LES) (fully reduced state) or oxidized to pernigraniline salt (PNS) (fully oxidized state) (Huang et al. 1986).

Cyclic voltammetry has been widely employed to investigate the above interconverted processes. A typical cyclic voltammogram of HCl-doped ES film between −0.2 and 1.0 V versus standard hydrogen electrode (SHE) in 1 M HCl solution usually exhibits two anodic peaks and two cathodic peaks (Huang et al. 1986). These peaks are related to the redox processes of the different oxidation states in PANI: the first redox process, between 0 and 0.2 V, involves the oxidation of the fully reduced state to a semioxidized state; the second redox process, in the range of 0.6–0.8 V, corresponds to the change of the semioxidized state to a fully oxidized state (Huang et al. 1986). The small difference between the oxidation and reduction potentials indicates its good electrochemical reversibility (Wang et al. 2002).

Although EB is insulating, and quite weak redox peaks are observed in the cyclic voltammogram upon measuring in a neutral electrolyte (Cecchetto et al. 2007), it still has the capacity to be reduced to LB or oxidized to PNB by gaining or losing electrons. The reversible redox reaction in EB has been confirmed by ultraviolet–visible (UV–Vis) spectrophotometry and x-ray photoelectron spectroscopy (XPS).

5.2 INTERACTION OF PANI WITH METALS

The most common PANI forms are ES and EB; both are redox-active polymers with equilibrium potentials (0.4–1.0 V), which are more positive to the standard potential of common metals, such as iron (–0.44 V_{SHE}), aluminum (–1.66 V_{SHE}), and magnesium (–2.36 V_{SHE}) (Rohwerder 2009). Thus, the interaction of PANI with metals and the following alternation of corrosion behavior are anticipated when PANI is in contact with metal. The reaction between PANI and metal is depicted as Equations 5.1 and 5.2 (Kinlen et al. 1999):

$$(1/n)\ M + (1/m)\ PANI^{m+} + (y/n)\ H_2O \rightarrow (1/n)\ M(OH)_y^{(n-y)+} + (1/m)\ PANI^0 + (y/n)\ H^+ \tag{5.1}$$

$$(m/4)\ O_2 + (m/2)\ H_2O + PANI^0 \rightarrow PANI^{m+} + mOH^- \tag{5.2}$$

As can be seen in Equation 5.2, atmospheric or dissolved oxygen can reoxidize the reduced PANI and replenish the PANI charge consumed by metal dissolution.

5.2.1 REACTION BETWEEN POLYANILINE SALT AND IRON

Although the corrosion protection phenomenon of PANI on metal has been reported since 1985 (Deberry 1985), the interaction between PANI and metal was first verified by Wessling (1994, 1996) and Sathiyanarayanan (1994), with the evolution of the oxidation states of ES being observed using a UV–Vis spectrometer when ES–polypropylene (PP) composite films with and without iron powder were tested in neutral and acidic solutions. For the film with iron powder, reduction of ES appeared as a color change from green to pale yellow; also found was the subsequent reverse reaction to its original state, whereas there was no such change for the film without iron powder. In addition, when a pure PANI film was applied on iron, a significant shift of corrosion potential to a noble direction was observed upon immersing in 1 M NaCl solution (Wessling 1994), and a passive oxide layer formed on the iron surface consisting of a Fe_3O_4 inner layer and a γ-Fe_2O_3 outer layer (Lu et al. 1995). It was concluded that an interaction took place at the PANI–iron interface, where PANI was reduced, while iron was oxidized to form a passive oxide layer. With the help of oxygen, the reduced PANI was reoxidized to ES (in acid solution) or EB (in neutral or alkaline environment). The interaction between ES and iron was also confirmed *by ex situ* Raman spectrum (Pereira da Silva et al. 2005), and an increase of the relative intensity ratio of I_{1600} (the aromatic –C–C–) to I_{1336} (stretching mode of radical cation –C–N–) was found in the ES-coated iron after a 40-day immersion in 1 M H_2SO_4 solution, in comparison with that before immersion, indicating the reduction process of PANI upon interacting with iron.

5.2.2 REVERSE REACTION BETWEEN POLYANILINE EMERALDINE BASE AND IRON

Such an interaction also exists between EB and iron. In this group, the EB powder was stirred with reduced iron powder (1:10 in mole ratio of EB to Fe) in 3.5 wt% NaCl solution at room temperature, and the oxidation degree of EB was determined via UV–Vis spectra by dissolving a tiny amount of EB in *N*-methylpyrrolidone (NMP) solution (Lu et al. 2003). Figure 5.3 shows the time dependence of the relative intensity

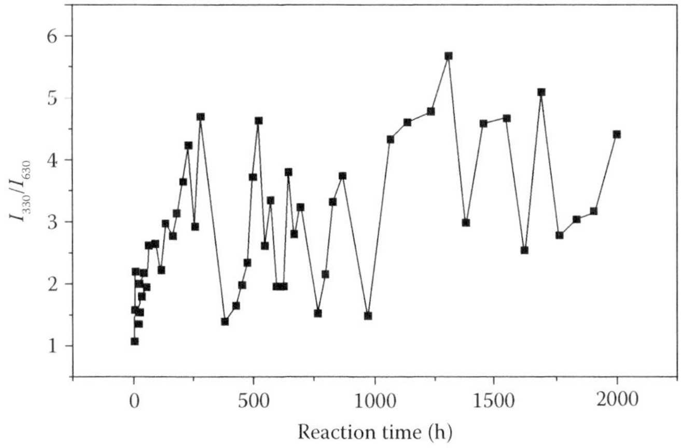

Figure 5.3 Time dependence of I_{330}/I_{630} for EB after interaction with Fe powder in 3.5 wt% NaCl solution at room temperature. (From Lu, J. L., et al., *Synth. Met.*, 135, 237–238, 2003. With permission.)

ratio of the absorption band of 330–630 nm (I_{330}/I_{630}). Usually, the absorption band at 630 nm is related to the quinoid unit of PANI, while that at 330 nm is associated with the benzoid content. The change in oxidation state of PANI can be reflected by the value of I_{330}/I_{630}.

Within the first 250 h, I_{330}/I_{630} increased from 1.1 to 4.8, indicating a significant reduction of EB at this stage. After that, I_{330}/I_{630} dropped down suddenly and became 1.3 at 400 h, indicating the reoxidation process of PANI. Interestingly, I_{330}/I_{630} increased again and approached 4.7 at 500 h, and then it decreased to 1.7 at 600 h. Subsequently, the quasi-reversible change in oxidation degree of EB continued until the end of the test.

5.2.3 REVERSE REACTION BETWEEN ES AND COPPER

The above method was also used to investigate the change of the oxidation states in ES upon interacting with copper (Chen 2007). It should be noted that dedoping of ES was inevitable when it was dissolved in NMP, but its oxidized state kept constant.

Figure 5.4a shows the time dependence of I_{330}/I_{630} of the dodecyl benzene sulfonic acid–doped PANI (DBSA-ES). Within the first 20 h, I_{330}/I_{630} increased from 1.12 to 1.58, indicating the reduction of DBSA-ES, while I_{330}/I_{630} decreased back to 1.16 after 40 h, reflecting the reoxidation. The redox reaction continued for the whole period of 260 h measurement. A similar phenomenon was observed between EB and copper, as shown in Figure 5.4b.

After a copper coated with 10 wt% EB–epoxy had been immersed in a 3.5 wt% NaCl solution for a certain time, the surface coating was removed to investigate the metal surface after immersion. The scratches on the copper surface disappeared, and a smooth, compact oxide layer was found with immersion time. Since no chlorine and sodium elements were detected on the substrate surface by x-ray energy-dispersive (EDX), the oxide layer was assumed to be induced by the interaction of EB and copper. Figure 5.5 presents a simple physical schematic of the proposed reactions in this corrosion-protective system (Chen 2007).

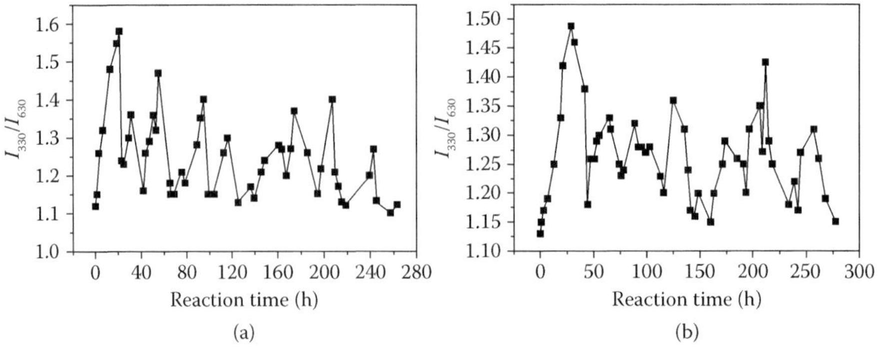

Figure 5.4 Evolution of I_{330}/I_{630} of (a) DBSA-ES and (b) EB as a function of time upon reaction with copper in 3.5 wt% NaCl solution at 60°C. (From Chen, Y., Study on anticorrosion mechanism of polyaniline on metals, PhD thesis, Changchun Institute of Applied Chemistry, Chinese Academy of Sciences, 2007. With permission.)

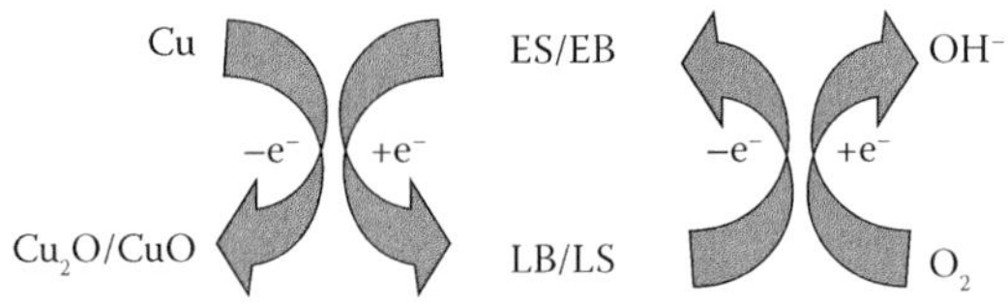

Figure 5.5 Passivation mechanism of PANI on copper. (From Chen, Y., Study on anticorrosion mechanism of polyaniline on metals, PhD thesis, Changchun Institute of Applied Chemistry, Chinese Academy of Sciences, 2007. With permission.)

5.2.4 REACTION BETWEEN PANI AND MG ALLOY

To follow the oxidation degree of PANI in the PANI–Mg alloy system upon immersion in deionized water, *in situ* Raman spectroscopy was recorded. As shown in Figure 5.6a, the peak at 1167 cm^{-1} was associated with two peaks that are difficult to separate, the C–H stretch of the quinoid ring at 1163 cm^{-1} and the C–H stretch of the benzoid ring at 1178 cm^{-1} (Gustavsson et al. 2009), while the two bands at 1220 and 1475 cm^{-1} were related to the C–N and C=N stretching vibrations (Izumi et al. 2006), respectively. The peak located at 1591 cm^{-1} corresponded to C=C stretching of the quinoid ring, while the peak at 1612 cm^{-1} was attributed to C–C stretching vibration in the benzoid ring (Gustavsson et al. 2009). Generally, the oxidation degree of PANI could be evaluated by the relative peak area ratio of the C–C band in the benzoid ring to the C=C band in the quinoid ring (I_{1612}/I_{1591}). The higher ratio of I_{1612}/I_{1591} represented the smaller amounts of semi-quinoid units in PANI, that is, reduction state (Gustavsson et al. 2009). The spectrum recorded after 4.5 h of immersion revealed an increase in the C–C band at 1612 cm^{-1}, indicating that EB was reduced after immersion. The highest intensity of the C–C band and the weakest C=C peak were observed after 9 h of immersion, and then the reduced PANI underwent reoxidation. After that, PANI was found to be reduced again after 22.5 h, as evidenced by the increase of I_{1612}/I_{1591} after 29.5 h. PANI subsequently returned to its original oxidation state (EB) after 37 h. For clearance,

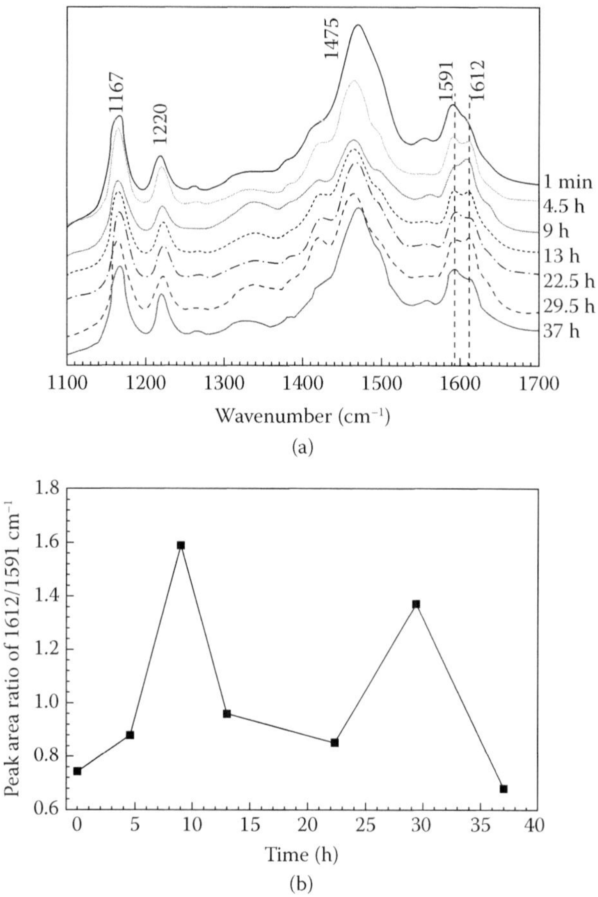

Figure 5.6 (a) *In situ* Raman spectra (λ = 632.8 nm) of EB film on AZ91D surface measured in deionized water and (b) the variation of I_{1612}/I_{1591} with immersion time.

the variation of I_{1612}/I_{1591} with immersion time was illustrated in Figure 5.6b, again displaying a quasi-reversible redox interaction between PANI and Mg alloy (EB-AZ91D).

5.3 ANTICORROSION PERFORMANCE OF POLYANILINE COATINGS

5.3.1 ANTICORROSION PROPERTY OF POLYANILINE AND ITS SALT ON MILD STEEL

The first report about the corrosion protection phenomenon of PANI dates back to 1985 (Deberry 1985), when Deberry showed that the electrochemically deposited PANI could significantly protect stainless steel from corrosion in dilute sulfuric acid solution. Later, Wessling first reported that various metals could also be passivated by a pure PANI film that was prepared in a chemical way and then applied on the metal surface (Wessling 1994). The corrosion potential of PANI-coated iron in 1 M NaCl solution increased by +800 mV compared with that of the bare iron. After removal of the PANI coating, the shift in the corrosion potential was still observable, which could increase with the thickness of PANI film. The passive oxide layer was also observed by scanning electron microscopy (SEM) and XPS, consisting of γ-Fe_2O_3 on the top layer and Fe_3O_4 in the inner layer (Lu et al. 1995). Therefore, PANI was considered to induce the passive oxide layer formation and repair the eventual chemical degradation of the metal oxide. This was the first time the role of the interactions between PANI and metal in corrosion protection was reported. Later, Hermas et al. (2005) pointed out that the stabilized oxide film underneath electrochemically deposited PANI film was different from that formed on steel by applying an external positive potential. The passive state of steel could remain for several days in acid solution after the PANI coating was completely removed, while the anodically formed passive layer was broken down immediately upon immersion in the same solution.

Although pure PANI film has shown a good corrosion protection effect, its corrosion protection capacity is not enough for long-term protection of metals due to its poor barrier property and weak adhesion with the metal surface. A promising strategy is to mix PANI with matrix polymers, such as epoxy resin (ER), polyurethane, polyvinyl butyral, and acrylic polymers (Riaz et al. 2014). It is noteworthy that different resin matrixes may sometimes determine whether the PANI loading can exert its corrosion protection ability (Diniz et al. 2013).

In addition to the conducting ES form, the nonconducting EB was also extensively studied for corrosion protection, although there is still considerable debate as to the efficiency and mechanism of the protection. McAndrew (1997) attributed the improved corrosion protection of EB-coated steel to the high diffusion resistance of EB to corrosive ions, while Talo et al. (1999) found that even after a hole was made to the coating, EB–epoxy-coated steel still offered an efficient protection compared with the reference sample without EB. These results indicated that EB offered more than a simple barrier protection. Chen et al. (2007b) found that the impedance of the EB-ER-coated steel obtained from Bode impedance plots increased gradually and was maintained at a high level after an initial decrease at the beginning of immersion; meanwhile, a passive film of Fe_2O_3/Fe_3O_4 was observed by SEM and XPS at the coating–metal interface. Therefore, the protection of EB on steel was mainly dominated by anodic protection, where the interaction between EB and steel promoted the oxidation of steel to form a passive oxide layer.

Except for the protective oxide layer on a steel surface induced by PANI due to anodic protection, another corrosion protection mechanism is ascribed to the controlled release of a corrosion-inhibiting dopant anion. As reported by many researchers (Kamaraj et al. 2012b; De Souza 2007; Pereira da Silva et al. 2007; Williams et al. 2006), PANI would be reduced and dopants would be released synchronously due to the interaction of PANI with a metal substrate. The corrosion reaction at defects would be prevented or slowed down, if dopants have anodic or cathodic inhibiting properties, or deposit as an insoluble dopant–metal salt with soluble metal ions.

Recently, nano-PANIs have also been of great interest in corrosion protection area due to their unique small size effect, as well as surface effect, which is believed to be beneficial for enhancing the dispersion of PANI in the matrix, increasing the interaction with matrixes as well as metal substrates, and improving the mechanical properties and barrier effect of coatings (Tian et al. 2014; Deshpande et al. 2014). Furthermore, the PANI composite containing other functional additives is popular to further enhance the barrier properties or offer other functions to coatings (Sababi et al. 2014; Zhang et al. 2013; Chang et al. 2012).

5.3.2 ENHANCED CORROSION RESISTANCE OF ALUMINUM ALLOY BY PANI

Aluminum alloys are important engineering structural metals, especially in the aerospace industry, due to their excellent strength-to-weight ratio. Corrosion of aluminum alloys mainly results from the galvanic coupling among alloy elements, which may cause enormous economic loss. Since PANI coating has been found to offer effective corrosion protection on iron and steel, the corrosion control of Al alloys using PANI is attractive. The earliest report was from Racicot et al. (1995, 1997), where a smooth surface free of any defect was observed even when a scratched PANI-coated aluminum sample was immersed in a 0.5 M NaCl solution for a week, whereas pits were formed on the uncoated aluminum sample within the same time period. The salt spray test and immersion experiments indicated that PANI–poly(methyl acrylate-co-acrylic acid)-coated Al alloy exhibited advanced protection compared with an Alodine-600 chromate conversion sample (Racicot et al. 1997), which was attributed to the formation of a protective oxide layer due to the reactive interaction between PANI and aluminum to prevent the penetration of the corrosive anions. This strong interaction was confirmed by Gustavsson et al. (2009) in a Raman study of the ES–copolymer composite coating on aluminum alloy 2024-T3, where an increase in the C–C/C=C ratio (1610/1590 cm^{-1}) and a decrease in the radical cation double band at 1314/1333 cm^{-1} were observed, indicating the reduction and dedoping of the ES. However, the authors claimed that this interaction resulted in a rapid oxidation of the exposed metal at coating defects, and no protective oxide was observed. This result was contrary to that from Seegmiller (2005), in which ES–polymethylmethacrylate (PMMA) was used as the coating and a combination of two mechanisms was proposed: growth of a protective oxide layer at the scratch and prevention of cathodic reactions via positive open-circuit potential (OCP).

Epstein et al. (1999) reported that the corrosion current of an EB film–coated 2024-T3 was 10-fold lower than that of the uncoated sample, and dramatically less corrosion products existed at the interface between the EB film and aluminum alloy. Lower copper content was observed on the substrate surface of the coated sample than on that of the bare Al 2024-T3, suggesting that EB could have facilitated the extraction of copper from the aluminum surface, hence weakening the galvanic couple of aluminum with copper, leading to less corrosion. Since EB shows a weak redox electrochemical activity in neutral solution, while the reduction of EB to LB upon interacting with Al alloy did happen, Cecchetto et al. (2007) suggested that the redox reaction between EB and an alloy substrate, as well as the low ion permeability of the EB film, is responsible for the improved corrosion resistance of Al alloy in a neutral environment.

Sometimes, multi-inhibitors were used in addition to PANI. Kamaraj et al. (2010) reported that electrochemical deposition of a ES film revealed a poor corrosion protection performance on aluminum alloy AA 7075 T6, but the protection property can be significantly increased by posttreatment in cerium salt solution to form a cerium oxide coating on the surface, along with the ES film. If this ES–cerium oxide film was further covered by an epoxy coating, the resistance (R_c) of the coated aluminum alloy remained above $10^6\ \Omega\ cm^2$ after 75 days of immersion in 3% NaCl solution, whereas the alloy coated with epoxy alone showed a R_c value of less than $10^4\ \Omega\ cm^2$ (Kamaraj et al. 2012a). Kartsonakis et al. (2012) reported a more complicated hybrid organic–inorganic multilayer coating, where two coating layers existed, with the inner layer consisting of a PANI–polypyrrole copolymer containing CeO_2 and 2-mercaptobenzothiazole, and the outer layer made of a sol–gel coating.

Dopant anion plays an important role in the corrosion control of aluminum. When an aluminum alloy substrate starts to dissolve as an anode, the anodic corrosion reaction will drive the reduction of the PANI film, causing the release of the inhibiting dopant anions, and thus stopping the corrosion. Kendig et al. (2003) found that the scratch in the ES region without inhibiting dopant showed white aluminum corrosion products, and the green color of ES turned into blue, while the scratch with inhibiting dopant only exhibited a slight tarnish and the ES film remained green in appearance. Apparently, the released inhibitors effectively prevented further corrosion, so the reduction of ES was stopped. Therefore, it is suggested that the dopant anions containing a phosphorous element with an inhibitor property be used, whereas inorganic protonic acids such as hydrochloric acid or sulfuric acid should not be recommended.

5.3.3 ANTICORROSION PERFORMANCE OF PANI ON COPPER

Although copper is not an active metal and is not as easily corroded as iron or aluminum, it does slowly react with atmospheric oxygen to form a green corrosion layer called verdigris, which is often seen on old copper statues. Wessling (1994) reported that a copper coupon coated with PANI dispersion exhibited a positive potential shift of +100 to +200 mV compared with a bare copper, and its corrosion current increased very slowly with an increasing extra voltage, with the PANI-coated copper recording only a one-third current density of the uncoated one, even at a voltage of 2.2 V. Later, Posdorfer and Wessling (2001) demonstrated that the interaction between ES and copper substrate could change the oxidation state of copper: a linear increase of Cu_2O and CuO was observed on an acid-etched Cu surface during the whole record time, while for the Cu immersed in a dispersion of PANI, the CuO growth was proportional to the square root of time and Cu_2O decreased exponentially with time until a constant thickness was formed. Based on the slope of the Arrhenius plot of rate against $1/T$, an activation energy of 35 kJ mol^{-1} was obtained for CuO formation, whereas it was 50 kJ mol^{-1} for that formed without PANI. SEM images displayed that the surface morphology was quite uniform for the oxide layer formed on the PANI-treated Cu, while the acid-etching Cu surface was rough.

Chen et al. (2007a) also claimed that EB–ER containing EB could provide a high protection efficiency to copper compared with ER coating alone, especially when EB content was around 3–10 wt%. Upon immersion, the impedance at 0.1 Hz of an epoxy coating dramatically decreased and reached a low value of less than $10^6\ \Omega\ cm^2$ after 30 days, while the impedance of an EB–epoxy coating, after a slight decrease on the first day, gradually grew to above $10^9\ \Omega\ cm^2$ and then became nearly constant. A similar evolution of OCP was also observed, again indicating the advanced protection and strong passivation effect of EB on Cu. By using the XPS etching technique, the passive copper oxide layer formed under the EB–epoxy coating was confirmed, mainly composed of a thin CuO outer layer, followed by a thick Cu_2O inner layer.

5.3.4 ANTICORROSION PERFORMANCE OF PANI COATING ON MAGNESIUM ALLOY

Magnesium is one of the lightest metals with a density of 1.7 g cm^{-3}; it has become the third most commonly used structural metal. Due to their low density and high specific strength, Mg alloys have drawn great attention in the automotive and aerospace industries. However, the poor corrosion resistance has limited their widespread application. To prevent Mg alloys from corrosion, numerous techniques have been explored, among which organic coating containing corrosion-inhibiting pigments is an economical method in industry. In the past 10 years, PANI including ES and EB has been employed as pigment for the corrosion control of Mg alloys. Sathiyanarayanan et al. (2008) reported that upon immersion in 0.1 wt% NaCl solution for 28 days, the resistance of an ES–epoxy coating was $3.32 \times 10^7\ \Omega\ cm^2$, 10 times higher than that of a chromate-containing coating and 100 times higher than the filler pigment-containing coatings, indicating the superior corrosion protection of the ES–epoxy coating on a Mg alloy. Additionally, a self-healing capability was found on a hybrid sol–gel/ES-coated Mg alloy upon immersion in Harrison's solution, where no pitting or filiform corrosion occurred around the scratched area, while a protective oxide layer was observed (Wang et al. 2010). This corrosion protection is primarily attributed to the redox potential of PANI, which facilitates the oxidation of the Mg alloy surface and promotes a protective oxide layer formation.

To improve the protection performance of PANI composite coatings, other filler materials, such as montmorillonite (Zhang et al. 2013), TiO_2 (Sathiyanarayanan et al. 2007), and SiO_2 (Chen et al. 2010), are added into coating along with PANI to enhance the barrier effect. In addition, concerning the dopant effect on corrosion protection, hydrofluoric acid (HF)-doped ES is used as a pigment in epoxy coating by Zhang et al. (2011), because MgF_2 is chemically inert and can work as a barrier when it forms on the Mg alloy surface. After 300 h of immersion, this HF–ES–epoxy coating showed an impedance of 30 $M\Omega\ cm^2$ at low frequency (0.01 Hz), higher than those of the ER coating (0.05 $M\Omega\ cm^2$) and EB–ER coating (1 $M\Omega\ cm^2$).

Luo et al. (2015) observed a steady improved protection on the EB–ER-coated Mg alloy when EB loading increased from 0 to 10 wt%, maintaining a complex impedance of 2 $G\Omega\ cm^2$ with 10 wt% EB loading even after an 80-day exposure in a 0.5 M NaCl solution, while that of the pure ER-coated analogue decreased to 0.17 $M\Omega\ cm^2$ after only 31 days. The surface structure and composition in the EB-induced interface oxide layer could be regulated by changing the EB loading in EB–ER to achieve different corrosion resistance of the Mg alloy.

5.4 UNIQUE CORROSION PROTECTION FEATURE OF PANI

5.4.1 SCRATCH RESISTANCE

A common organic coating has no scratch resistance capability. Therefore, once a scratch is created, the corrosion process will be initialized like the uncoated metals. Under this circumstance, coating delamination will readily occur due to the easy migration of electrolyte ions at the coating–metal interface, leading to the failure in corrosion protection of coatings. Extensive studies have shown that PANI exhibits a special "self-healing" capacity.

No matter whether pinholes or artificial defects occur on coatings, the exposed metal can still be protected by PANI coating from corrosion. Deberry (1985) observed that the potential can still be maintained at a high level in the last stage of immersion when a deep scratch was created in the PANI-coated steel, whereas the defect became active immediately in the same test solution for that without PANI coating. Later, Lu et al. (1995) also observed the corrosion inhibition phenomenon at the scratched mild steel coated with ES–epoxy, which displayed a corrosion rate 100 times less than that on an identically scratched sample coated with ER alone in 0.1 M HCl solution. According to the results of optical microscopy, electron spectroscopy for chemical analysis, and Auger spectroscopy, the corrosion protection of the exposed metal at the defect was assumed to be caused by the formation of a passive iron oxide layer (γ-Fe_2O_3 outer layer and Fe_3O_4 inner layer), which initially formed under the PANI coating and then gradually moved toward the center of the exposed area at scratch. The distance for protection of the exposed steel can extend to 2 mm, sometimes even 6 mm from the edge of the coating, largely depending on the corrosive environment.

However, there is an argument against this anodic protection mechanism (i.e., ennobling metal surface and maintaining it in a passive state). Some metals with a high passive potential or some active metals, such as aluminum and magnesium, cannot be passivated by the ES–LS or EB–LB redox couple of polarizing in an acidic or neutral chloride condition. Therefore, the corrosion protection of the metal exposed at the defect site is suggested to be primarily associated with a dopant inhibitory mechanism rather than the passivated effect, which is caused by the release of corrosion-inhibiting dopant ions upon reduction of PANI. Williams et al. (2006) reported that the oxide layer induced by ES on a pure iron surface had poor corrosion resistance and so thickened linearly with time in a humid atmosphere, but the delamination velocity of coatings close to the defect was significantly reduced when ES was added in a poly(vinyl butyral) (PVB) coating. The inhibitory effect was found to largely depend on the dopant ions in the ES backbone, for example, *p*-toluenesulfonic acid, camphorsulfonic acid, phosphoric acid, and phenylphosphonic acid. ES doped by phosphoric acid or phenylphosphonic acid gave the best positive result, which was attributed to the formation of $Fe_3(PO_4)_2$ and FePP salt film at the coating–substrate interface. Recently, Vimalanandan et al. (2013) synthesized a novel redox-sensitive PANI nanocapsule containing a self-healing inhibitor (3-nitrosali-cylic acid) in the core. When the PANI nanocapsule was added into the PVB coating, a distinct potential shift to the noble direction was observed in the defect area, providing an apparently slow coating delamination rate compared with the pure PVB coating.

Our recent work (Li et al. 2011b) shows that the EB film–coated mild steel (although without any inhibitor) can also protect the substrate exposed at a coating defect in a 3.5 wt% NaCl solution, where EB was regarded as an oxygen-scavenging agent to protect the steel exposed at a coating defect by reducing the oxygen concentration over the defect surface. A simple physical scheme of this protection mechanism is shown in Figure 5.7.

For the scratched surface, two regions exist: the defect region (A) and the EB-coated region (B). In region A, dissolved oxygen reacted with iron directly with a reduction rate v_1, while in region B, due to the interaction between EB and steel, EB was reduced to LB and reoxidation of LB to EB-consumed oxygen, leading to another reduction rate, v_2. If these two oxygen reduction rates are different, the oxygen concentration in these two regions will be different. Once v_2 was larger than v_1, the dissolved oxygen concentration in region B would be lower than that in region A. Then, the diffusion

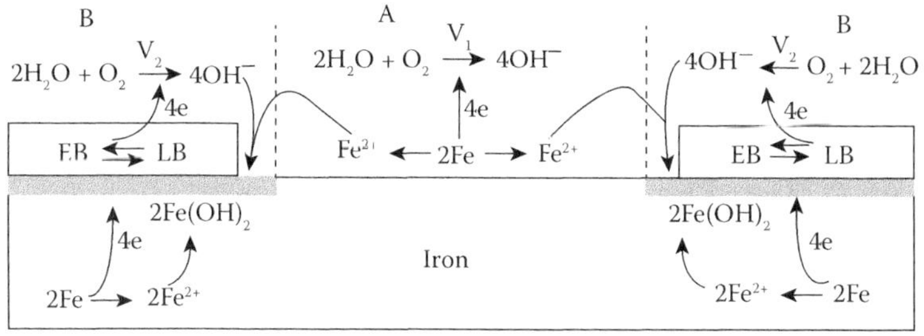

Figure 5.7 Oxygen-scavenging protection model of EB on the exposed steel at the EB coating defect. (From Li, Y. P., et al., *Synth. Met.*, 161, 2312–2317, 2011. With permission.)

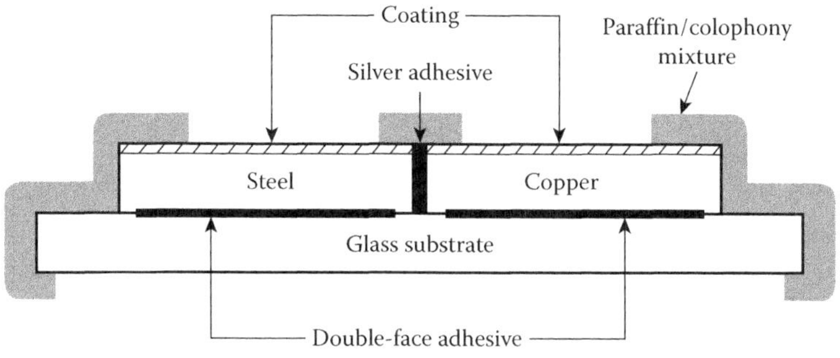

Figure 5.8 Schematic structure of a coated steel–copper couple. (From *Corros. Sci.*, 49, Chen, Y., et al., Long-term anticorrosion behaviour of polyaniline on mild steel, 3052–3063, Copyright 2007, with permission from Elsevier.)

of dissolved oxygen occurs from region A to region B, leading to a decrease in oxygen concentration in region A, and thus suppressing the corrosion of the exposed steel at the defect. The reduction rate of dissolved oxygen on the EB coating surface (v_2) was proved to be nearly three times higher than that on the steel surface (v_1) by using potentiodynamic polarization measurements, leading to suppressed corrosion in the scratched surface.

5.4.2 GALVANIC CORROSION PREVENTION FEATURE

When different metals are used together, a galvanic couple exists, and galvanic corrosion occurs easily in the corrosive electrolyte, where the less noble metal acts as an anode and the more noble metal acts as a cathode. Under this circumstance, corrosion of the anode member of the galvanic couple will be accelerated by the driving force, resulting from the potential difference between the dissimilar metals. Several years ago, we designed a copper–steel couple as shown in Figure 5.8 (Chen et al. 2007a). It was found that EB–ER coating could provide superior corrosion protection for mild steel coupled with copper. In salt spray tests, a ER coating with a thickness of 100 μm failed in the corrosion protection of the steel–copper couple after 1000 h, while an EB–ER coating exhibited effective protection for at least 2000 h. These observations were mainly attributed to the EB-induced passive oxide layers of Fe_2O_3/Fe_3O_4 on steel and CuO/Cu_2O on copper, thus suppressing both the anodic and cathodic reactions. In another report, Vera et al. (2010) showed that even when PANI coating was just applied on the cathodic Cu surface in Al–Cu galvanic couple, a reduction of Al damage was also observed in 0.1 M NaCl solution, compared with the Al connected with bare Cu, due to the decrease of an effective area of oxygen reduction in the cathode.

5.5 NATURE OF INTERFACE OXIDE LAYER

5.5.1 INTERFACE OXIDE LAYER BETWEEN PANI COATING AND MILD STEEL

The first study in corrosion protection of PANI already reported that a passive oxide layer was formed on the steel surface during electrochemical polymerization of the ES film (Deberry 1985). The passive oxide layer was also observed in the case of steel coated with ES film, which was deposited from pure ES dispersion without an electrochemical process. Wessling (1994) claimed that the passive oxide layer between the ES coating and metal occurred in two steps: first, a few microns of iron were dissolved due to the interaction of ES with substrate; and second, an oxide layer was formed on the iron surface. The passive oxide layer was mainly composed of γ-Fe_2O_3 on the outside and an Fe_3O_4 inner layer. Since then, similar findings have been reported, even for other metals without "active–passive" transition. In most studies, the positive shift of corrosion potential or complex impedance of the PANI-coated metals during immersion is usually regarded as evidence of the formation of this highly corrosion-resistant oxide layer.

For the corrosion protection of EB on steel, several research groups have demonstrated that EB, although it is nonconducting, could also promote a passive oxide film formation by withdrawing electrons from steel substrate because its oxidant capacity is similar to that of ES, capable of forming an EB–LB redox couple (Spinks et al. 2002; Fahlman et al. 1997). In the study of the corrosion behavior of the EB–epoxy-coated steel, we observed that the rough surface underneath the EB–epoxy coating became smoother and the polishing grooves were almost invisible with increasing immersion time from 0 to 20 days, but no chloride or sodium element was detected by energy-dispersive spectrometer (EDS) (Chen et al. 2007b), indicating that the growth oxides were not caused by the corrosive electrolyte. XPS results also revealed that metallic Fe peaks disappeared with immersion time, while two peaks at 710.7 and 724.1 eV corresponding to Fe 2p3/2 and 2p1/2 lines of trivalent iron in Fe_2O_3 became dominant. Upon etching, the Fe_2O_3 peaks at the outer layer vanished, whereas the Fe_3O_4 peaks appeared after 2 min of sputtering. These observations were found to be consistent with the corresponding evolution of impedance and the OCP against the immersion time of the EB–epoxy-coated steels, demonstrating the ability of EB to induce a protective oxide layer.

We evaluated the growth rate of the oxide layer formed at the EB–steel interface under different environments (pH, NaCl concentration, and temperature) using OCP measurement (Li et al. 2011a). During the period of oxide growth, the OCP of EB-coated electrode increased linearly with logarithmic time, indicating that the growth of the EB-induced iron oxide followed a direct logarithm law. It was found that the growth rate decreased with increasing NaCl concentration and temperature (T), while with increasing pH value, it decreased first and then increased. According to the linear relationship between the natural logarithm of the growth rate and $1/T$, an activation energy value of -39.8 kJ mol^{-1} was calculated for oxide layer growth, implying that the growth of the EB-induced oxide layer was under diffusion control.

5.5.2 INTERFACE OXIDE LAYER BETWEEN PANI AND MAGNESIUM ALLOY

In the past decade, PANI has received considerable interest as an eco-friendly inhibition pigment in organic coating for the protection of Mg alloys. Coatings containing ES (Sathiyanarayanan et al. 2007; Chen et al. 2010; Zhang et al. 2013) or EB (Shao et al. 2009) were found to exhibit efficient protection performance in comparison with the reference varnish coatings. This corrosion protection is primarily attributed to the interaction between PANI and Mg substrate, promoting the formation of a protective oxide layer at the interface (Sathiyanarayanan et al. 2007; Shao et al. 2009).

Our recent work (Luo et al. 2015) indicated that this interface oxide layer due to EB exhibited a morphology different from that formed under pure ER coating, which was further regulated by changing EB content in the coating so that the corrosion resistance of Mg alloy was altered correspondingly. The corrosion oxides underneath the epoxy coating were a porous layer consisting of large elongated crystals (Figure 5.9a), while with the increase of EB loading, this structure gradually disappeared with the pore space among the crystals filled with more homogeneous and dense corrosion products (Figure 5.9a–d), which finally became a quite compact oxide layer under the epoxy coating containing 10 wt% EB (Figure 5.9d). XPS results showed that the top surface of the oxide layer under either epoxy or EB–epoxy coating consisted of similar components: $Mg(OH)_2$, MgO, and a few $MgCO_3$. However, there existed a significant difference in proportion. With increasing EB loading, the characteristic peaks in the Mg2p and O1s spectra showed a tendency for moving to high energy, and the relative peak area ratio of MgO to $Mg(OH)_2$ ($I_{MgO}/I_{Mg(OH)_2}$) increased from 0.78 for ER

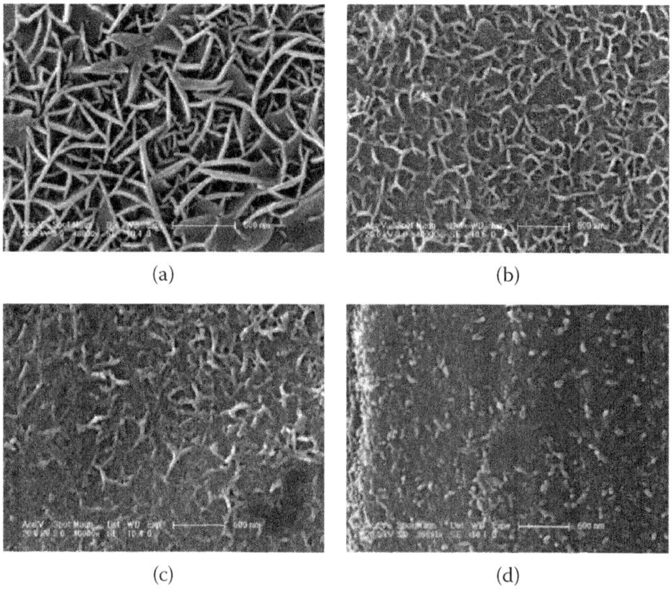

Figure 5.9 SEM images of interface oxide layers between Mg alloy (AZ91D) and EB–ER coating after an 80-day exposure in 0.5 M NaCl solution for coated samples of (a) ER, (b) 1% EB–ER, (c) 5% EB–ER, and (d) 10% EB–ER.

coating to 1.18 for 10% EB–ER coating. Similar results were also observed by x-ray diffraction (XRD) that the increase of EB loading in epoxy coating could significantly raise the MgO content in the interface oxide layer, which was expected to possess high corrosion resistance.

We considered that the interaction between EB and Mg alloy was the driving force to make the growth process of the interface layer underneath the EB–ER coating different from that beneath the pure epoxy coating, as shown in Figure 5.10, resulting in oxide layers with different morphologies, chemical structures, and corrosion resistance performance.

For the magnesium exposed to the corrosive electrolyte (Figure 5.10a), a thin permeable MgO layer is expected to form first with the production of Mg^{2+}, which then reacts with H_2O and partially changes to $Mg(OH)_2$ (Taheri et al. 2014). Due to the energetic hydration, the inner MgO layer undergoes a breakdown, producing a volume porosity; the egress of Mg cations to the surface oxide layer–solution interface is thus enhanced (Taheri and Kish 2013). Mg^{2+} ions diffuse to the solution and react with OH^-, resulting from the cathodic reaction to form $Mg(OH)_2$, which spontaneously precipitates on the metal surface (Taheri et al. 2012):

$$Mg \rightarrow Mg^{2+} + 2e^- \tag{5.3}$$

$$2H_2O + 2e^- \rightarrow H_2 + 2OH^- \tag{5.4}$$

$$Mg^{2+} + 2OH^- \rightarrow Mg(OH)_2 \tag{5.5}$$

As the dissolution–precipitation reaction is dominant during corrosion, a thick porous $Mg(OH)_2$ layer grows with immersion time on the Mg surface and a bilayer structure (MgO inner layer and $Mg(OH)_2$ outer layer) is obtained.

For the EB-coated system, EB can be regarded as a catalytic oxidant to metal substrates for corrosion protection; hence, the interaction between EB and Mg alloy may promote the oxidation of magnesium to form MgO directly and perhaps a few $Mg(OH)_2$ as well (Figure 5.10b). Various properties of oxides, such as crystallinity, granular size, roughness, or density, might have been altered during this process; thus, a compact and

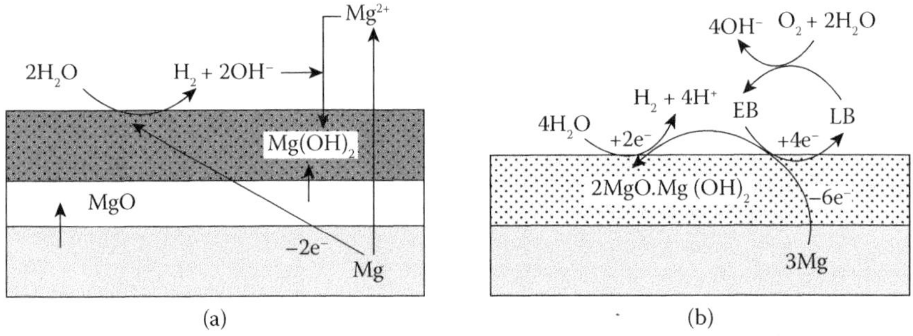

Figure 5.10 Schematic diagrams showing the growth behavior of the oxide layer naturally generated on bare Mg alloy (a) and that formed at the interface between an EB coating (arrow) and Mg alloy (b).

stable MgO-rich layer mixed with a few $Mg(OH)_2$ quickly grows on the alloy surface by a solid reaction, as shown in Equation 5.6:

$$(x + y/2)/n \, \mathrm{Mg} + 1/m \, \mathrm{EB}^{m+} + (x + y)/n \, \mathrm{H_2O} \rightarrow 1/n \, \mathrm{Mg}_{x+y/2}\mathrm{O}_x(\mathrm{OH})_y^{(n-2x-y)+} + 1/m \, \mathrm{LB}^0 + (2x + y)/n \, \mathrm{H}^+ \quad (5.6)$$

During this process, both the hydration of MgO and the dissolution–precipitation reaction occur less. Therefore, only a tiny amount of $Mg(OH)_2$ forms in the outer layer. This growth model is consistent with the results obtained by XPS sputtering in that MgO is dominant in the bulk oxide layer formed beneath EB film, and no significant bilayer structure exists.

5.6 PERSPECTIVES

5.6.1 PROS AND CONS OF PANI AS ANTICORROSION AGENT

As mentioned above, the unique redox feature in PANI and the interaction between PANI and metal substrate are extremely vital to exert the corrosion-protective ability of PANI, no matter whether in an anodic protection mechanism or in an intelligent dopant release mechanism. However, it is argued that these special properties may also cause the problems of anodic corrosion. Principally, the high redox potential of PANI should keep the stainless steel in the passive region upon interacting with the substrate surface; however, in a composite coating with a low content of PANI loading, there is only a small part of the metal surface having direct contact with PANI. If these PANI particles do not have enough charge to polarize the substrate surface covered by the resin matrix, the resulting weak polarization might lead to an accelerated corrosion of the substrate surface (Rohwerder 2009). This is in consistent with the finding of Zhu et al. (2004) that when the area ratio of the PANI film to 20A carbon steel was above 25:1, the steel in contact with PANI could maintain passivity, whereas for a ratio of less than 25:1, the PANI film would accelerate the corrosion of steel slightly. As shown by McMurray's group (Holness et al. 2005; Williams et al. 2006), the PANI-induced iron oxide layer exhibited a poor corrosion resistance, which could not stop further corrosion but thickened linearly with time when the PANI-coated iron was placed in a humid atmosphere.

As far as the release of a corrosion-inhibiting anion is concerned, Rohwerder et al. (Rohwerder and Michalik 2007; Rohwerder et al. 2009) pointed out that when ES was reduced once it obtained electrons from metal, dopant anions in the ES backbone would release or electrolyte cations would incorporate into ES to neutralize the charge of ES. Since the small cations in the electrolyte move faster in coating than the large anions in the dopant, the dopant anions remain in the coating, which act as fixed negative charges to raise the cation mobility, leading to a faster reduction of ES, and hence a faster coating delamination. Therefore, it is suggested that pure ES coating or composite coating with macroscopically extended percolation networks should be avoided to prevent the fast cation mobility in coating. PANI should be dispersed in a nonconducting resin matrix as isolated microscopic particles.

There is also a considerable debate on whether EB can offer corrosion protection for metals in a corrosive environment. Tiitu et al. (2005) found that for the EB–epoxy-coated steel, the complex resistance increased from 70 to 20 $G\Omega \, cm^2$, while the corrosion potential was shifted to the noble region from –700 to –200 mV during a 500-day immersion in a 3.5% NaCl solution. Akbarinezhad et al. (2009) also observed the increase of OCP as a function of immersion time for carbon steel coated with EB–epoxy, contributing to the passivation of EB. The superior protective performance of EB has been confirmed on various metals, such as copper (Chen et al. 2007a), aluminum alloy (Epstein et al. 1999), and magnesium alloy (Luo et al. 2015). However, Araujo et al. (2001) showed that a pure EB coating, as well as an EB–epoxy coating, did not protect steel in 0.01 M Na_2SO_4 due to the poor barrier property and adhesion to the substrate. Armelin et al. (2007) also found that the addition of EB could not improve the protective properties of ER coating and alkyd coating. In addition, the coating delamination studies of Williams's group (Williams et al. 2006, 2014) showed that no matter how high the amount of EB loading in PVB coating, it had basically no effect on the intact coating potential, as well as the coating delamination kinetics for iron and zinc.

Consequently, the corrosion protection of metals by PANI seems to be very complicated, which can be influenced by various factors. Even for the same PANI pigment, its corrosion-inhibiting capacity also depends on the type of resin matrix, use of metal substrate, amount of loading, pretreatment of metal surface, coating process, testing environment (such as pH, temperature, and ion concentration), and so forth.

5.6.2 PRACTICAL APPLICATIONS AND REMARKS

Currently, most experimental results definitely demonstrate the positive effect of PANI on corrosion protection for various metals, such as iron, steel, copper, zinc, aluminum alloys, and magnesium alloys, and also the feasibility of PANI anticorrosive coating in industry. However, the step from lab curiosity to practical technique is rather tortuous. In 1996, the first corrosion protection product, named as CORRPASSIV, was launched by Ormecon Chemie GmbH & Co. KG. However, there was only a few commercial orders for this unique corrosion protection product. In mid-1999, this company changed its original strategy to focus on other products. Similar results also seem to take place in other companies, although they claimed that their corrosion protection products possessed excellent protection performance in laboratory experiments, even under salt fog test. For example, Ancatt, a U.S. company, claimed that its nanodispersed PANI coating can guarantee steel panels to go through 13,000 h of salt fog exposure (ASTM B 117) tests, and aluminum panels more than 10,000 h, while 5,000 h is the maximum testing hours according to ASTM standards. Also in China, licensed by the authors' technology in 2005, China Hunan Ben'an Co. has commercialized a PANI-based anticorrosion coating on 1000 tons per year scale, leading to wide applications of this coating in many engineering projects, such as guardrails, bridges, and heavy machinery.

Although disputes exist regarding the anticorrosion mechanism, positive anticorrosion behavior of the PANI-based coatings has been confirmed, which may open a new era in anticorrosion techniques. PANI-based coating is special in that its anticorrosion behavior is driven by redox reactions between PANI and metals, which is totally different from sacrificial anode cathodic protection like the Zn slab. We believe that it may provide not only hexavalent chromium-free anticorrosion coating for aluminum and its alloys but also heavy-metal-free and scratch-tolerant coating for steel and even magnesium alloy. Of course, in addition to the bonus of an environmentally friendly feature, competitiveness against traditional anticorrosion systems is still the primary issue. Recent progress in industry application has witnessed the feasibility of the PANI coating in terms of cost, showing that a coating containing only 0.5–1.5 wt% PANI may provide comparable performance to that of the traditional zinc-rich coating. The low PANI loading in coating product ensures market competitiveness. Of course, there are still concerns about whether a real breakthrough is realized. Only a series of successful industry examples with long-term tests like zinc-rich coating or chromate coating can resolve such concerns. The confidence of the industry in the anticorrosion PANI coatings comes from its practical performance.

REFERENCES

Akbarinezhad, E., M. Ebrahimi, and H. R. Faridi. Corrosion inhibition of steel in sodium chloride solution by undoped polyaniline epoxy blend coating. *Progress in Organic Coatings* 64 (2009): 361–3664.

Ancatt. n.d. Anti-corrosion coating. http://www.ancatt.com (accessed February 25, 2016).

Araujo, W. S., I. C. P. Margarit, M. Ferreira, O. R. Mattos, and P. L. Neto. Undoped polyaniline anticorrosive properties. *Electrochimica Acta* 46 (2001): 1307–1312.

Armelin, E., C. Ocampo, F. Liesa, J. I. Iribarren, X. Ramis, and C. Aleman. Study of epoxy and alkyd coatings modified with emeraldine base form of polyaniline. *Progress in Organic Coatings* 58 (2007): 316–322.

Barisci, J. N., T. W. Lewis, G. M. Spinks, C. O. Too, and G. G. Wallace. Responsive systems based on conducting polymers. *Smart Materials, Structures, and Integrated Systems* 3241 (1997): 10–19.

Cecchetto, L., D. Delabouglise, and J. P. Petit. On the mechanism of the anodic protection of aluminium alloy AA5182 by emeraldine base coatings—Evidences of a galvanic coupling. *Electrochimica Acta* 52 (2007): 3485–3492.

Chang, C. H., T. C. Huang, and C. W. Peng, et al. Novel anticorrosion coatings prepared from polyaniline/graphene composites. *Carbon* 50 (2012): 5044–5051.

Chen, X., K. Shen, and J. Zhang. Preparation and anticorrosion properties of polyaniline-SiO_2-containing coating on Mg-Li alloy. *Pigment & Resin Technology* 39 (2010): 322–326.

Chen, Y. *Study on anticorrosion mechanism of polyaniline on metals.* PhD thesis, Changchun Institute of Applied Chemistry, Chinese Academy of Sciences, 2007.

Chen, Y., X. H. Wang, J. Li, J. L. Lu, and F. S. Wang. Long-term anticorrosion behaviour of polyaniline on mild steel. *Corrosion Science* 49 (2007a): 3052–3063.

Chen, Y., X. Wang, J. Li, J. Lu, and F. Wang. Polyaniline for corrosion prevention of mild steel coupled with copper. *Electrochimica Acta* 52 (2007b): 5392–5399.

Deberry, D. W. Modification of the electrochemical and corrosion behavior of stainless steels with an electroactive coating. *Journal of the Electrochemical Society* 132 (1985): 1022–1026.

Deshpande, P. P., N. G. Jadhav, V. J. Gelling, and D. Sazou. Conducting polymers for corrosion protection: A review. *Journal of Coatings Technology and Research* 100 (2014): 1331–1342.

De Souza, S. Smart coating based on polyaniline acrylic blend for corrosion protection of different metals. *Surface and Coatings Technology* 201 (2007): 7574–7581.

Diniz, F. B., G. F. De Andrade, C R. Martins, and W. M. De Azevedo. A comparative study of epoxy and polyurethane based coatings containing polyaniline-DBSA pigments for corrosion protection on mild steel. *Progress in Organic Coatings* 76 (2013): 912–916.

Epstein, A. J., J. A. O. Smallfield, H. Guan, and M. Fahlman. Corrosion protection of aluminum and aluminum alloys by polyanilines: A potentiodynamic and photoelectron spectroscopy study. *Synthetic Metals* 102 (1999): 1374–1376.

Fahlman, M., S. Jasty, and A. J. Epstein. Corrosion protection of iron/steel by emeraldine base polyaniline: An x-ray photoelectron spectroscopy study. *Synthetic Metals* 85 (1997): 1323–1326.

Gustavsson, J. M., P. C. Innis, J. He, G. G. Wallace, and D. E. Tallman. Processable polyaniline-HCSA/poly(vinyl acetate-co-butyl acrylate) corrosion protection coatings for aluminium alloy 2024-T3: A SVET and Raman study. *Electrochimica Acta* 54 (2009): 1483–1490.

Hermas, A. A., M. Nakayama, and K. Ogura. Enrichment of chromium-content in passive layers on stainless steel coated with polyaniline. *Electrochimica Acta* 50 (2005): 2001–2007.

Holness, R. J., G. Williams, D. A. Worsley, and H. N. McMurray. Polyaniline inhibition of corrosion-driven organic coating cathodic delamination on iron. *Journal of the Electrochemical Society* 152 (2005): B73–B81.

Huang, W. S., B. D. Humphrey, and A. G. MacDiarmid. Polyaniline, a novel conducting polymer—Morphology and chemistry of its oxidation and reduction in aqueous-electrolytes. *Journal of the Chemical Society, Faraday Transactions I* 82 (1986): 2385–2400.

Izumi, C. M. S., V. R. L. Constantino, A. M. C. Ferreira, and M. L. A. Temperini. Spectroscopic characterization of polyaniline doped with transition metal salts. *Synthetic Metals* 156 (2006): 654–663.

Kamaraj, K., V. Karpakam, S. S. Azim, and S. Sathiyanarayanan. Electropolymerised polyaniline films as effective replacement of carcinogenic chromate treatments for corrosion protection of aluminium alloys. *Synthetic Metals* 162 (2012a): 536–542.

Kamaraj, K., V. Karpakam, S. Sathiyanarayanan, and G. Venkatachari. Electrosynthesis of polyaniline film on AA 7075 alloy and its corrosion protection ability. *Journal of the Electrochemical Society* 157 (2010): C102–C109.

Kamaraj, K., T. Siva, S. Sathiyanarayanan, S. Muthukrishnan, and G. Venkatachari. Synthesis of oxalate doped polyaniline and its corrosion protection performance. *Journal of Solid State Electrochemistry* 16 (2012b): 465–471.

Kartsonakis, I. A., E. P. Koumoulos, A. C. Balaskas, G. S. Pappas, C. A. Charitidis, and G. Kordas. Hybrid organic-inorganic multilayer coatings including nanocontainers for corrosion protection of metal alloys. *Corrosion Science* 57 (2012): 56–66.

Kendig, M., M. Hon, and L. Warren. 'Smart' corrosion inhibiting coatings. *Progress in Organic Coatings* 47 (2003): 183–189.

Kinlen, P. J., V. Menon, and Y. W. Ding. A mechanistic investigation of polyaniline corrosion protection using the scanning reference electrode technique. *Journal of the Electrochemical Society* 146 (1999): 3690–3695.

Li, Y. P., H. M. Zhang, X. H. Wang, J. Li, and F. S. Wang. Growth kinetics of oxide films at the polyaniline/mild steel interface. *Corrosion Science* 53 (2011a): 4044–4049.

Li, Y. P., H. M. Zhang, X. H. Wang, J. Li, and F. S. Wang. Role of dissolved oxygen diffusion in coating defect protection by emeraldine base. *Synthetic Metals* 161 (2011b): 2312–2317.

Lu, J. L., N. J. Liu, X. H. Wang, J. Li, X. B. Jing, and F. S. Wang. Mechanism and life study on polyaniline anti-corrosion coating. *Synthetic Metals* 135 (2003): 237–238.

Lu, W. K., R. L. Elsenbaumer, and B. Wessling. Corrosion protection of mild-steel by coatings containing polyaniline. *Synthetic Metals* 71 (1995): 2163–2166.

Luo, Y. Z., Y. Sun, J. L. Lv, X. H. Wang, J. Li, and F. S. Wang. Transition of interface oxide layer from porous $Mg(OH)_2$ to dense MgO induced by polyaniline and corrosion resistance of Mg alloy therefrom. *Applied Surface Science* 328 (2015): 247–254.

MacDiarmid, A. G., J. C. Chiang, A. F. Richter, and A. J. Epstein. Polyaniline—A new concept in conducting polymers. *Synthetic Metals* 18 (1987): 285–90.

McAndrew, T. P. Corrosion prevention with electrically conductive polymers. *Trends in Polymer Science* 5 (1997): 7–12.

Pereira da Silva, J. E. P., S. I. Cordoba De Torresi, and R. M. Torresi. Polyaniline acrylic coatings for corrosion inhibition: The role played by counter-ions. *Corrosion Science* 47 (2005): 811–822.

Pereira da Silva, J. E., S. I. Cordoba De Torresi, and R. M. Torresi. Polyaniline/poly(methylmethacrylate) blends for corrosion protection: The effect of passivating dopants on different metals. *Progress in Organic Coatings* 58 (2007): 33–39.

Pommiers, S., J. Frayret, A. Castetbon, and M. Potin-Gautier. Alternative conversion coatings to chromate for the protection of magnesium alloys. *Corrosion Science* 84 (2014): 135–146.

Posdorfer, J., and B. Wessling. Oxidation of copper in the presence of the organic metal polyaniline. *Synthetic Metals* 119 (2001): 363–364.

Racicot, R., R. Brown, and S. C. Yang. Corrosion protection of aluminum alloys by double-strand polyaniline. *Synthetic Metals* 85 (1997): 1263–1264.

Racicot, R., R. L. Clark, H. B. Liu, S. C. Yang, M. N. Alias, and R. Brown. Thin film conductive polymers on aluminum surfaces: Interfacial charge-transfer and anti-corrosion aspects. *Optical and Photonic Applications of Electroactive and Conducting Polymers* 2528 (1995): 251–258.

Riaz, U., C. Nwaoha, and S. M. Ashraf. Recent advances in corrosion protective composite coatings based on conducting polymers and natural resource derived polymers. *Progress in Organic Coatings* 44 (2014): 743–756.

Rohwerder, M. Conducting polymers for corrosion protection: A review. *International Journal of Materials Research* 100 (2009): 1331–1342.

Rohwerder, M., L. M. Duc, and A. Michalik. In situ investigation of corrosion localised at the buried interface between metal and conducting polymer based composite coatings. *Electrochimica Acta* 54 (2009): 6075–6081.

Rohwerder, M., and A. Michalik. Conducting polymers for corrosion protection: What makes the difference between failure and success? *Electrochimica Acta* 53 (2007): 1300–1313.

Sababi, M., J. Pan, P.-E. Augustsson, P.-E. Sundell, and P. M. Claesson. Influences of polyaniline and ceria nanoparticle additives on corrosion protection of a UV-cure coating on carbon steel. *Corrosion Science* 84 (2014): 189–197.

Sathiyanarayanan, S., S. S. Azim, and G. Venkatachari. Corrosion protection of magnesium ZM 21 alloy with polyaniline–TiO_2 composite containing coatings. *Progress in Organic Coatings* 59 (2007): 291–296.

Sathiyanarayanan, S., S. S. Azim, and G. Venkatachari. Corrosion protection of magnesium alloy ZM21 by polyaniline-blended coatings. *Journal of Coatings Technology and Research* 5 (2008): 471–477.

Seegmiller, J. C., J. E. Pereira da Silva, D. A. Buttry, S. I. C. De Torresi, and R. M. Torresi. Mechanism of action of corrosion protection coating for AA2024-T3 based on poly(aniline)-poly(methylmethacrylate) blend. *Journal of the Electrochemical Society* 152 (2005): B45–B53.

Shao, Y. W., H. Huang, T. Zhang, G. Z. Meng, and F. H. Wang. Corrosion protection of Mg–5Li alloy with epoxy coatings containing polyaniline. *Corrosion Science* 51 (2009): 2906–2915.

Spinks, G. M., A. J. Dominis, G. G. Wallace, and D. E. Tallman. Electroactive conducting polymers for corrosion control—Part 2. Ferrous metals. *Journal of Solid State Electrochemistry* 6 (2002): 85–100.

Taheri, M., M. Danaie, and J. R. Kish. TEM examination of the film formed on corroding Mg prior to breakdown. *Journal of the Electrochemical Society* 161 (2014): C89–C94.

Taheri, M., and J. R. Kish. Nature of surface film formed on Mg exposed to 1 M NaOH. *Journal of the Electrochemical Society* 160 (2013): C36–C41.

Taheri, M., R. C. Phillips, J. R. Kish, and G. A. Botton. Analysis of the surface film formed on Mg by exposure to water using a FIB cross-section and STEM-EDS. *Corrosion Science* 59 (2012): 222–228.

Talo, A., O. Forsen, and S. Ylasaari. Corrosion protective polyaniline epoxy blend coatings on mild steel. *Synthetic Metals* 102 (1999): 1394–1395.

Tian, Z. F., H. J. Yu, L. Wang, M. Saleem, F. J. Ren, P. F. Ren, Y. S. Chen, R. L. Sun, Y. B. Sun, and L. Huang. Recent progress in the preparation of polyaniline nanostructures and their applications in anticorrosive coatings. *RSC Advances* 4 (2014): 28195–28108.

Tiitu, M., A. Talo, O. Forsen, and O. Ikkala. Aminic epoxy resin hardeners as reactive solvents for conjugated polymers: Polyaniline base/epoxy composites for anticorrosion coatings. *Polymer* 46 (2005): 6855–6861.

Vera, R., P. Verdugo, M. Orellana, and E. Munoz. Corrosion of aluminium in copper-aluminium couples under a marine environment: Influence of polyaniline deposited onto copper. *Corrosion Science* 52 (2010): 3803–3810.

Vimalanandan, A., L. P. Lv, T. H. Tran, K. Landfester, D. Crespy, and M. Rohwerder. Redox-responsive self-healing for corrosion protection. *Advanced Materials* 25 (2013): 6980–6984.

Wang, H. M., R. Akid, and M. Gobara. Scratch-resistant anticorrosion sol–gel coating for the protection of AZ31 magnesium alloy via a low temperature sol–gel route. *Corrosion Science* 52 (2010): 2565–2570.

Wang, X. H., J. L. Lu, J. Li, X. B. Jing, and F. S. Wang. Solvent-free polyaniline coating for corrosion prevention of metal. In *Electroactive Polymers for Corrosion Control*, ed. P. Zarras, J. D. Stenger-Smith, and Y. Wei. Washington, DC: American Chemical Society, 2003, 254–267.

Wang, Y. J., X. H. Wang, X. J. Zhao, J. Li, Z. S. Mo, X. B. Jing, and F. S. Wang. Conducting polyaniline confined in semi-interpenetrating networks. *Macromolecular Rapid Communications* 23 (2002): 118–121.

Wessling, B. Passivation of metals by coating with polyaniline: Corrosion potential shift and morphological changes. *Advanced Materials* 6 (1994): 226–228.

Wessling, B. Corrosion prevention with an organic metal (polyaniline): Surface ennobling, passivation, corrosion test results. *Materials and Corrosion* 47 (1996): 439–445.

Williams, G., A. Gabriel, A. Cook, and H. N. McMurray. Dopant effects in polyaniline inhibition of corrosion-driven organic coating cathodic delamination on iron. *Journal of the Electrochemical Society* 153 (2006): B425–B433.

Williams, G., H. N. McMurray, and A. Bennett. Inhibition of corrosion-driven organic coating delamination from a zinc surface using polyaniline pigments. *Materials and Corrosion* 65 (2014): 401–409.

Zhang, D. H. On the conductivity measurement of polyaniline pellets. *Polymer Testing* 26 (2007): 9–13.

Zhang, Y. J., Y. W. Shao, T. Zhang, G. Z. Meng, and F. H. Wang. The effect of epoxy coating containing emeraldine base and hydrofluoric acid doped polyaniline on the corrosion protection of AZ91D magnesium alloy. *Corrosion Science* 53 (2011): 3747–3755.

Zhang, Y. J., Y. W. Shao, T. Zhang, G. Z. Meng, and F. H. Wang. High corrosion protection of a polyaniline/organophilic montmorillonite coating for magnesium alloys. *Progress in Organic Coatings* 76 (2013): 804–811.

Zhao, J., L. Xia, A. Sehgal, D. Lu, R. L. McCreery, and G. S. Frankel. Effects of chromate and chromate conversion coatings on corrosion of aluminum alloy 2024-T3. *Surface and Coatings Technology* 140 (2001): 51–57.

Zhu, H., L. Zhong, S. H. Xiao, and F. X. Gan. Accelerating effect and mechanism of passivation of polyaniline on ferrous metals. *Electrochimica Acta* 49 (2004): 5161–5166.

6 Medical device implants for neuromodulation

Trenton A. Jerde
New York University
New York

Contents

6.1 INTRODUCTION

Electrical stimulation has a long history of modulating body function. In 15 AD, torpedo fish were used to treat pain after a man stepped on an electric fish and it relieved his pain from gout (Gildenberg 2005). In the late 1700s, Galvani evoked muscle contraction by applying an electric current to a dissected nerve. In 1870, Fritsch and Hitzig elicited limb movement by electric stimulation of the motor cortex. At present, deep brain stimulation (DBS) pacemakers are implanted in the brain or body as a therapy to alleviate the symptoms of otherwise treatment-resistant disorders, such as Parkinson's disease, tremor, dystonia, and chronic pain. To date, DBS has improved the lives of patients worldwide, with Medtronic (Minneapolis, Minnesota) having implanted its 100,000th DBS pacemaker in a patient in 2012.

The physiological properties of the normal and diseased brain depend on the electrical properties of brain cells. Just as a cardiac pacemaker rectifies heartbeat abnormalities, electrical stimulation from implantable devices modulates brain circuits to restore normal functioning. Although neuromodulation, at present, is primarily used for otherwise treatment-resistant disorders, it has advantages over options such as medications and permanent ablative procedures in surgery. For example, neuromodulation devices provide immediate delivery for therapeutic benefit; they are reversible in that treatment can be stopped by removing the device; they are programmable in that stimulation parameters can be adjusted; and they often present less risk of adverse effects than other therapies. Moreover, the safety and efficacy of neuromodulation techniques have been established in clinical trials.

This chapter presents an overview of the field of neuromodulation, with an emphasis on electronics in medicine. Neuromodulation involves contributions from basic science, applied science, technology, and medical practitioners. It includes academia, industry, and collaborations between them. It relies on funding from the government, foundations, philanthropies, and investment from private industry. It necessitates governmental oversight of device efficacy and patient safety. We here describe the translational continuum from basic research to the application of neural implants to benefit human lives.

6.2 DEFINITION OF A MEDICAL DEVICE

The term *medical device* has different meanings depending on context. Legally, medical devices are defined by the governing regulatory body. In the United States, medical devices are regulated by the Food and Drug Administration (FDA) Center for Devices and Radiological Health (CDRH). According to the FDA (http://www.fda.gov/MedicalDevices/DeviceRegulationandGuidance/Overview/ClassifyYourDevice/ucm051512.htm), a medical device is

> an instrument, apparatus, implement, machine, contrivance, implant, *in vitro* reagent, or other similar or related article, including a component part, or accessory which is:
>
> 1. recognized in the official National Formulary, or the United States Pharmacopoeia, or any supplement to them;
> 2. intended for use in the diagnosis of disease or other conditions, or in the cure, mitigation, treatment, or prevention of disease, in man or other animals; or
> 3. intended to affect the structure or any function of the body of man or other animals, and which does not achieve its primary intended purposes through chemical action within or on the body of man or other animals and which is not dependent upon being metabolized for the achievement of any of its primary intended purposes

The U.S. FDA classifies medical devices into Classes I–III based on potential risks to the patient and regulatory controls. Class I devices pose the lowest perceived risk, whereas Class III devices pose the highest perceived risk. Examples of Class III devices include cardiac and neural implantable pulse generators (IPGs), cochlear implants, and retinal prosthetics.

According to the European Union Medical Devices Directive, a medical device is defined as

> any instrument, apparatus, appliance, software, material or other article, whether used alone or in combination, including the software intended by its manufacturer to be used specifically for diagnostic and/or therapeutic purposes and necessary for its proper application, intended by the manufacturer to be used for human beings for the purpose of diagnosis, prevention, monitoring, treatment or alleviation of disease; diagnosis, monitoring, treatment, alleviation of or compensation for an injury or handicap; investigation, replacement or modification of the anatomy or of a physiological process; control of conception and which does not achieve its principal intended action in or on the human body by pharmacological, immunological or metabolic means, but which may be assisted in its function by such means. (p. 6)

The Medical Devices Bureau of Health Canada recognizes four classes of medical devices based on the level of control necessary to ensure the safety and effectiveness of the device. Class I devices present the lowest potential risk and do not require a license. Class II devices require the manufacturer's declaration of device safety and effectiveness. Class III and IV devices present a greater potential risk and are subject to in-depth scrutiny.

6.3 DEFINITION OF NEUROMODULATION

According to the North American Neuromodulation Society, neuromodulation is the "therapeutic alteration of (neural) activity either through stimulation or medication" delivered through implanted devices, such as an IPG for DBS or drug pumps. Neuromodulation may target the central or peripheral nervous systems, including the spinal cord, brain, or spinal nerves. Traditionally, therapeutic neuromodulation has treated chronic pain, movement disorders, epilepsy, gastrological disorders, and urological disorders (Lewis et al. 2016). Recently, neuromodulation has been studied for the treatment of traumatic spinal cord and brain injuries (Shin et al. 2014) and psychiatric disorders (Wichmann and Delong 2006). Neuromodulation may be applied or investigated by clinicians and researchers from a range of disciplines, including neurosurgery, neurology, neuroscience, psychiatry, physical therapy, and rehabilitation.

6.4 BRIEF HISTORICAL PERSPECTIVE

Therapeutic neuromodulation has a relatively short history. Clinical neuromodulation emerged from advances in cardiac pacing during the 1950s, when electrical stimulation for diagnostic and therapeutic applications became commonplace (Aquilina 2006). Prior to these developments, the main objective of neurosurgeons was to remove pathology by surgery. However, at about the time that cardiac pacing was introduced, neurosurgeons such as Penfield began to use electrical stimulation to distinguish the functions of specific brain areas prior to surgery for epilepsy (Jasper and Penfield 1954). The first modern neuromodulation procedure was DBS for chronic pain (Pool 1954; Heath et al. 1955); subsequent developments (Melzack and Wall 1965) led to the introduction of spinal cord stimulation (SCS) (Shealy et al. 1967) and peripheral nerve stimulation (Wall and Sweet 1967) as therapies for chronic pain.

6.4.1 SPINAL CORD SIGNALING MODULATION IN THE 1960s

Chronic pain is the greatest cause of disability in the world and the most economically costly of neurological and psychiatric conditions in Western societies. The gate control theory proposed by Melzack and Wall (1965) led to the clinical use of SCS by Shealy and colleagues (1967) for the treatment of chronic intractable neuropathic pain. In SCS, an implantable system delivers electric pulses via a lead to nerves in the dorsal aspect of the spinal cord. In this way, pain signals are inhibited before they reach the brain and replaced with a tingling sensation, called parasthesia. Importantly, SCS treats chronic pain produced by multiple indications, but it does not eliminate the source of pain or treat its underlying cause. SCS has gained FDA approval for the treatment of chronic pain conditions of the limbs and trunk, with the most common indications being failed back surgery syndrome and inoperable ischemic limb. As a historical note, Medtronic trademarked the term *deep brain stimulation* for the treatment of pain in the first commercially marketed devices introduced in the mid-1970s.

In SCS, an electrode with an array of contacts is implanted in the dorsal epidural space of the spinal cord. Large myelinated primary afferent dorsal column fibers are depolarized and excited along the spinal cord near the electrode, leading to the propagation of action potentials. The activation of the dorsal column axons is thought to be responsible for the paresthesia experienced by patients during stimulation. Unfortunately, about one-third of patients do not respond to SCS (Wolter 2014). The efficacy of SCS should improve, however, as technology develops and the biology of pain is better understood.

One challenge is that pain often accompanies body movements. To address this issue, position-adaptive SCS can be driven by acceleration sensors that assist with the proper stimulation levels during a shift in body position. A Medtronic technology, AdaptiveStim, determines the correlation between a change in body position and the level of required stimulation. Clinicians use these data to assess, evaluate, and optimize a patient's neurostimulation treatment. Other developments include rechargeable pulse generators, minimally invasive implantation of electrodes, development of multicolumn stimulation, and the capability of connecting multiple electrodes with up to 32 active contacts to a single generator.

In terms of pain management, a challenge is that the symptoms of pain are subjective and thus difficult to measure. An objective measure of pain would be useful as a biomarker to evaluate and treat patients. Recent research has addressed this issue by using machine-learning algorithms in functional magnetic resonance imaging (fMRI) to show that the pattern of activity across brain areas can act as a "pain signature" (Wager et al. 2013). An objective measure of pain of this type could be implemented in closed-loop systems (Section 6.9.2) to monitor and control pain by providing real-time feedback to neurostimulation or drug delivery devices.

6.4.2 COCHLEAR IMPLANT TECHNOLOGIES IN THE 1970s TO 1990s

The cochlear implant is the first example of a neural medical device that could substitute for a sensory body part, namely, the ear. A cochlear implant is an electronic device that provides the perception of sound to a person who is hard of hearing or deaf. In normal hearing, sound pressure waves travel down the ear canal, causing the eardrum to vibrate. The vibrations are transmitted to the cochlea by small bones of the middle ear, and these mechanical vibrations are transduced into action potentials that propagate to the brain, producing the perception of sound. The aim of a cochlear implant is to mimic the filtering normally performed by the bypassed portions of the auditory system. The cochlear implant dates to 1957, when Djourno and Eyriès evoked sound sensations in a deaf listener using an electrode implanted in the inner ear. Subsequently, the first commercialized multielectrode cochlear implant was implanted in 1978.

The implant has an external part worn behind the ear and an internal part surgically implanted under the skin. The external part consists of one or more microphones that pick up sound from the environment, a speech processor that selectively filters and arranges the sounds and directs electrical signals to the transmitter, and a transmitter that delivers power and processed sounds to the internal device by electromagnetic induction. The internal part consists of a receiver and stimulator that convert the signals into electric impulses. The impulses are sent to an array of electrodes surgically implanted along the cochlea, with the signal being delivered to regions of the auditory nerve. At present, an implant does not restore normal hearing but provides a useful representation of sounds in the environment and helps a person to understand speech. More than 300,000 deaf people in the world have been fitted with a cochlear implant.

6.4.3 BRAIN SIGNALING MODULATION IN THE 1980s TO 1990s

The origin of DBS is credited to Benabid et al. (1991), but the basic concepts date to the 1960s (e.g., Hassler et al. 1960; Albe Fessard et al. 1963). In 1987, modern DBS was introduced as a treatment for parkinsonian and essential tremor. Limousin et al. (1995) introduced stimulation of the subthalamic nucleus for the treatment of advanced Parkinson's disease, and Krauss et al. (1999) initiated stimulation of the pallidum for dystonia. These advances led to basic and clinical research that established DBS as a feasible method for the treatment of movement disorders.

6.4.4 MOVEMENT DISORDERS: DBS FOR PARKINSON'S DISEASE, DYSTONIA, AND TREMOR

Cooper and others applied electrical currents and chemicals to temporarily suppress movement disorder symptoms and identify brain sites for lesion therapy (Cooper 1973). These developments led to the application of therapeutic neuromodulation over the subsequent decades. In DBS, a surgically implanted medical device delivers electrical stimulation to targeted brain areas, thereby modulating pathological activity. DBS for the treatment of motor symptoms in Parkinson's disease was pioneered in the 1980s and 1990s with the discovery that electrical stimulation could mimic the effects of a lesion (Benabid et al. 1987; Pollak et al. 1993). The targeted brain areas for Parkinson's disease and dystonia are the subthalamic nucleus and globus pallidus interna. The targeted area for essential tremor is the ventral intermediate nucleus of the thalamus. In appropriately selected patients, DBS for movement disorders can provide greater than 50% improvement in clinical ratings of motor symptoms. The FDA approved DBS for essential tremor in 1997, Parkinson's disease in 2002, and primary dystonia in 2003.

A crucial factor in DBS for movement disorders is how long it takes for the beneficial effects to occur after the onset of stimulation. For tremor, the effects often occur soon after stimulation, thus providing the clinician with feedback on the effectiveness of DBS settings. For dystonia and certain nonmovement disorders, however, the therapeutic effects are not observed until later, which may be weeks or even months. The delayed therapeutic window necessitates the development of methods to optimize stimulation parameters at the time of implantation. Research is therefore underway to define biomarkers that can be used during the implantation phase to optimize DBS settings early in the clinical process.

6.4.5 BLINDNESS: RETINAL PROSTHETICS

Retinal degenerative diseases, such as age-related macular degeneration and retinitis pigmentosa, mainly affect photoreceptors, leaving the retina unable to sense light. However, neurostimulation can activate the remaining neurons in the retina. An electrode array can be implanted on the top surface of the retina (epiretinal), under the retina (subretinal), or between the sclera and choroid (suprachoroidal). Unlike deep brain stimulators, which utilize batteries and fixed settings, retinal implants are too small for batteries; additionally, retinal implants must be constantly updated with new data based on the input to the camera.

Two devices currently have regulatory approval for the treatment of retinitis pigmentosa. The Argus II Retinal Prosthesis (Second Sight Medical Products, Inc., Sylmar, California) has regulatory approval in Europe and the United States, and the Alpha IMS System (Retina Implant, Reutlingen, Germany) has regulatory approval in Europe. These devices provide patients with partial sight restoration and improvements in daily activities. Patients do not, however, perceive the world as in normal vision. For example, patients often have to move their head or make eye movements across objects to judge what they are viewing. Patients may also experience difficulty in recognizing the details of faces or objects. To improve the performance of retinal prosthetics, improvements are needed in the design of electrode arrays, electronic circuits, and hermetic packaging.

6.5 DEVICES USED TO DELIVER NEUROMODULATION

6.5.1 DIRECT ELECTRICAL STIMULATION (IPGs)

Electrical stimulation is the most common form of device-driven neuromodulation. In neurostimulation, the fundamental interaction is between the electric field generated by the stimulating electrode and the brain tissue surrounding it. For the successful delivery of electrical stimulation, three main issues arise. First, stimulation must be delivered to the targeted anatomical location. The neuronal population closest to the stimulating electrode will receive the greatest effect of electrical excitation; thus, a lower amount of current is needed to stimulate neurons close to the electrode. As the distance between the electrode and the targeted population of neurons increases, however, a larger amount of current is required. In such cases, neurons between the electrode and the targeted population are also activated, potentially yielding side effects or altering global brain function in unintended ways.

The second issue for the successful delivery of electrical stimulation is the selection of parameters for a targeted brain region in an individual patient for a specific disorder. Of note, the stimulation parameters of therapeutic value have been discovered largely by trial and error by observing the effects in patients during and after the implantation procedure. The key stimulation parameters are amplitude, frequency, and pulse width. The exact DBS parameters vary according to the disorder and the targeted brain region but are often between 1 and 9 V stimulus amplitude, 60 and 240 μs stimulus pulse duration, and low (5–50 Hz) and high (130–180 Hz) stimulus frequencies, and monopolar cathodic. For different disorders, the stimulation parameters may vary. For example, the typical therapeutic amplitudes range from 1.0 to 3.5 V for dystonia (Isaias et al. 2015) and from 2.4 to 7 V for epilepsy (Krishna et al. 2016).

The third issue for successful delivery of electrical stimulation is the interface between the electrode and the nerve tissue, which is critical for neuromodulation performance, especially over the long term. A challenge for implantable electrodes is to stimulate spatially restricted regions of the nervous system while inducing minimal tissue response. Simply decreasing the size of the electrode will not suffice, however, because the charge injection capability of the electrode becomes insufficient to deliver effective neuronal stimulation. One way to address this issue is to modify the electrodes with conductive polymers (CPs). CPs are porous in morphology and have a much larger electrochemically active surface area than metallic equivalents of the same dimensions. Advantages of CPs over metallic electrodes include improved biocompatibility, higher charge injection capacity, and lower electrode impedance (see Section 6.8.6). The long-term stability of CPs under physiological conditions is a concern for chronic applications, and research has sought to minimize the system impedance caused by glial scar tissue (Ouyang et al. 2014) and to improve the neural recording performance (Du et al. 2015).

6.5.2 DRUG INFUSION PUMPS

Implantable drug infusion pumps are electrical devices that contain and administer prescribed drugs or fluids to targeted sites in the body. Intrathecal drug delivery systems, for example, are implanted pumps that target pain relief or antispasm medication to a location on the spine that relays signals between an affected body area and the brain. Patients with chronic pain or muscle spasms may be considered for a drug infusion device if medications become ineffective or side effects become burdensome. Medications, in such systems, are released immediately into the fluid surrounding the spinal cord and reach the nerves responsible for the symptoms. By this means, lower amounts of the active ingredient are needed than when the medication is taken by tablet or intravenous infusion. One challenge in drug pumps is that the medication for chronic pain must be refilled every 6 weeks to 6 months at a doctor's office, depending on the drug concentration and the amount of medication needed by the patient.

6.5.3 EMERGING METHODS: TMS, tDCS, AND OPTOGENETICS

Electrical stimulation methods are *invasive* when they are implanted in the brain or body, and *noninvasive* when they are placed on the exterior of the brain or body. DBS is an example of an invasive method, in that a surgically implanted, battery-operated medical device delivers electrical stimulation to the targeted area. Examples of noninvasive neuromodulation are transcranial magnetic stimulation (TMS) and transcranial direct current stimulation (tDCS). In TMS, pulses of electric current are delivered to a coil placed on the patient's head to generate an electric field in the brain via electromagnetic induction. tDCS, on the other hand, influences the resting membrane potential of neurons to modulate their spontaneous firing rates. Two or more electrodes are positioned over the scalp and a weak direct current flows through the skull into the cerebral cortex. The current enters the brain through an anode, exits through a cathode, and depolarizes or hyperpolarizes neurons in the targeted region, depending on the polarity. Although there have been reports of therapeutic benefits for neurological and psychiatric disorders using noninvasive approaches (Fitzgerald and Daskalakis 2011; Fregni et al. 2006), the efficacy of the methods has not been established, and in some cases, the approaches are controversial (Horvath et al. 2014).

A problem with traditional neuromodulation methods is that stimulation produces undesirable effects at the targeted area and in other brain areas. This lack of precise control can yield side effects from stimulation, such as movement tremors. An optimal neurostimulation technique would produce excitation or inhibition of selected neuronal populations, without producing unwanted effects on other cell populations. Optogenetics is a relatively recent field in biotechnology that integrates genetic engineering, electrophysiology, and optical and electronic engineering (Deisseroth 2015). Optogenetics uses light to control cells, such as neurons, that have been genetically modified to express light-sensitive ion channels. Optogenetics provides millisecond-level control and the ability to selectively activate or inhibit particular genetically defined subpopulations of neurons within a larger neural circuit. At present, optogenetic neuromodulation is at an early stage of development. Technical advances and preclinical animal testing are needed before it can treat human disorders (Williams and Denison 2013).

6.6 OBJECTIVES FOR USING MEDICAL DEVICES FOR ELECTRICAL STIMULATION

6.6.1 MECHANISMS OF ACTION AND KNOWLEDGE GAPS

Neuromodulation reflects the interaction of basic and applied science to create technology that improves human lives. Although DBS and other neuromodulation methods are effective treatment options for many disorders, the mechanisms by which they work remain largely unknown. Indeed, DBS probably has multiple mechanisms of action, depending on several factors, including the particular disorder being considered, the distance of the targeted brain area from the electrode stimulation site, the spatial orientation of the electrodes, and the neural elements of interest, such as axons and cell bodies (Kringelbach et al. 2007). Axons, which are projections from a nerve cell that conduct electrical impulses, are more sensitive to DBS than cell bodies. DBS may alter the activity of axons projecting from a stimulated brain area (Vitek et al. 2012) and thereby block (Grill et al. 2004) or filter (Zimnik et al. 2015) the transmission of pathological signals to downstream brain areas. Research is investigating whether DBS inhibits or excites neurons and how it affects the information flow between brain areas (Chiken and Nambu 2016).

From a therapeutic perspective, some key questions are the following: What are the best anatomical areas to target with DBS for given disorders? What are the effects of DBS on nonmotor symptoms such as impulsivity? Should DBS be offered early in the course of a disorder or later, when the disorder has reached a critical threshold?

6.6.2 ENCOURAGE NORMAL PHYSIOLOGY

Much research has focused on the effects of DBS on brain physiology. One hypothesis is that DBS activates the output from the site of stimulation and adjacent fiber pathways. By this means, high-frequency DBS overrides pathological neuronal discharges and enacts a more orderly effect on downstream brain nuclei. In patients with Parkinson's disease, for example, DBS of the subthalamic nucleus may induce widespread normalization of activity by increasing activation of motor areas during movement execution, decreasing hyperactivity at rest, and modulating metabolic changes in other brain areas.

6.6.3 SUPPRESS ABNORMAL PHYSIOLOGY

Neuromodulation strategies may suppress abnormal physiology. Here, the therapeutic issue to be corrected or mitigated is an abnormal response, and the desired outcome is to suppress the response. Intraoperative recordings from patients with movement disorders suggest that DBS works by suppressing exaggerated low-frequency oscillations in the basal ganglia (Thompson et al. 2014). Indeed, the motor system improvement in Parkinson's disease that is mediated by DBS or medication therapy coincides with a decrease in abnormal beta-band (13–30 Hz) oscillations (Brown et al. 2001), while excessive beta activity can be induced in patients through administration of dopamine antagonists (Kühn et al. 2008). Essential tremor and dystonia are associated with abnormal oscillations in the theta-alpha range (4–12 Hz) (Silberstein et al. 2003), and the abnormal low-frequency oscillations in dystonia subside with DBS (Barow et al. 2014).

6.6.4 DIAGNOSTIC MONITORING AND ADJUSTMENT

An important function of neuromodulation is to monitor the symptoms, state, or progression of a disease. In open-loop neuromodulation, the stimulation parameters are adjusted manually and remain constant in the patient. Recent developments in closed-loop neuromodulation, however, optimize neuromodulation in real time by measuring signals that reflect changing biological conditions in the patient (Section 6.9.2). For example, the FDA-approved RNS System for epilepsy (NeuroPace, Inc., Mountain View, California) is a closed-loop system that detects epileptiform activity and delivers stimulation to alleviate it (Bergey et al. 2015). Closed-loop systems that track changes in patient physiology and modulate stimulation accordingly are a topic of current research and development in several applications of neuromodulation, including DBS for Parkinson's disease (Hariz 2014).

6.7 REGULATION AND OVERSIGHT OF MEDICAL DEVICES

6.7.1 REGULATORY CONSIDERATIONS

The design, manufacture, clinical testing, and marketing of neurostimulation systems were somewhat unregulated in the United States until the FDA became the regulatory body over implantable medical devices in the 1970s. The FDA employs two regulatory pathways for the marketing of medical devices. The most common pathway is the 510(k) process. A new medical device that can be demonstrated to be largely equivalent to a previously legally marketed device can be cleared by the FDA for marketing as long as general and special controls are met. Clinical trials are generally not required for the 510(k) pathway, and the majority of new medical devices enter the marketplace this way. The second regulatory pathway for new medical devices is the premarket approval process. This process is similar to the pathway for new drug approval, with clinical trials typically required for premarket approval.

Most neuromodulation devices would be considered Class III devices, ones for which insufficient information exists to ensure safety and effectiveness. To meet safety and effectiveness standards, such devices require scientific review and premarket approval. Class III devices are generally considered to support or sustain life, to prevent the serious impairment of health, or to present a risk for illness or injury. Examples of Class III devices that currently require a premarket notification include automated external defibrillators, implantable pacemakers, and pulse generators.

A distinction should be noted between *full approval* and *human device exemption* (HDE) approval. The FDA implemented the HDE application process for devices that are intended to treat or diagnose a condition or disease that affects, or is manifested in, fewer than 4000 individuals in the United States per year. A device with HDE status is exempt from the effectiveness requirements of premarket approval, as would be necessary for devices that treat diseases present in larger numbers of individuals. The HDE policy thus provides an incentive to offset research and development costs for the treatment of diseases or conditions that affect a small patient population. An example of a device in this category is the Argus II Retinal Prosthesis System used to treat advanced retinitis pigmentosa.

6.7.2 ROLE OF EARLY-PHASE CLINICAL RESEARCH

In early-phase clinical trials (Phase I), especially first-in-human trials, the primary objective is to evaluate safety. The evaluation of safety consists of assessing the nature and frequency of potential adverse reactions. Early-phase trials often assess not only the safety of specific modulation parameters but also the feasibility of delivery and the characteristics of the evoked neurophysiological activity. Small groups of people are tested first in early-phase research. Then, depending on the results, the treatment is given to a larger group of people in Phase II trials to assess whether it is effective and to further evaluate its safety.

6.7.3 LATER-PHASE CLINICAL TRIAL OBJECTIVES: SAFETY AND EFFICACY

In later-phase clinical trials (Phase III), the treatment is given to large groups of people. The objectives are to confirm the effectiveness of a treatment, monitor side effects, compare it to commonly used treatments, and acquire information that will allow it to be used safely. Phase IV trials are done after the treatment has been marketed to gather information on the effects of the treatment in different populations and to investigate side effects associated with long-term use.

A neuromodulation system must therefore be both safe and efficacious. Efficacy of stimulation refers to the ability to elicit the intended physiological response, which can include initiation, modulation, or suppression of brain activity in the targeted area. Safety has two primary characteristics. First, the stimulated tissue must not be damaged. And second, the stimulating electrode must not be compromised, for example, by corrosion. The safety considerations of implanting devices must be known by the neurosurgeon and communicated to the patient and caregivers (Grill 2015).

After the device is implanted, an important safety consideration is brain or body imaging of the patient. Imaging may be necessary to confirm the position of an electrode after implantation, or to detect movement or damage of the device if its efficacy is compromised. The patient may also develop a secondary medical condition that requires diagnostic or interventional studies. For the placement of DBS electrodes, postoperative MRI is helpful to determine the electrode location relative to the intended brain target, and for programming the stimulation parameters of the device (Vanegas-Arroyave et al. 2016). Ferromagnetic attraction, or the pulling on ferrous objects as they approach the magnet bore, is usually not a problem since most modern neuromodulation systems are made of nonferrous components. However, MRI can be problematic for implanted electrical devices because it may overheat the device or interfere with the functioning of the IPG.

6.8 DESIGN OF MEDICAL DEVICES

6.8.1 ANATOMICAL AND PHYSIOLOGICAL CONSIDERATIONS

The implantation of a medical device by surgery is an anatomical endeavor in that a specific site in the nervous system is the target of stimulation. Knowledge about anatomy, however, does not imply an understanding of physiology. That is, the targeting and implantation of an electrode in a particular brain site does not indicate how stimulation will impact the dynamic properties of the brain in an individual patient. Behavior, especially aberrant behavior caused by neurological or psychiatric disorders, emerges from the functioning of the nervous system as a whole. Neuromodulation approaches will benefit greatly as more basic research is conducted on brain physiology at all levels, from cells to neural networks.

The interactions among several variables must therefore be considered, including the properties of the electrode, the stimulation parameters (amplitude, frequency, pulse width, and duration), the intrinsic physiological properties of the brain region being stimulated, the geometric configuration of the surrounding neural tissue, and the effects of stimulation on other brain regions and brain networks.

6.8.2 ROLE OF COMPUTER MODELING

Computer modeling of DBS reduces the number of animal experiments, creates a virtual testing environment for new stimulation approaches, and enhances the understanding of DBS mechanisms (McIntyre and Foutz 2013). In particular, computer modeling has led to improved electrode placement and selection of stimulation parameters in patients undergoing DBS. Traditionally, surgical targeting in DBS focused on electrode placement in subcortical gray matter nuclei, the assumption being that stimulation of the cell bodies of neurons in the targeted population was responsible for the therapeutic effect. However, experimental and theoretical studies showed that the primary effect of DBS is to stimulate axons that surround the electrode. Such axons may project from the implanted nucleus, or they may project to, or nearby, the nucleus.

In the case of subthalamic DBS for Parkinson's disease, patient-specific computational models have demonstrated that activation of the white matter dorsal to the subthalamic nucleus is most associated with therapeutic benefit (Maks et al. 2009). Furthermore, when this dorsal region is targeted for electrode placement (Plaha et al. 2006), the outcomes are improved relative to stimulation concentrated on the subthalamic nucleus itself (Butson et al. 2011). These findings illustrate how computational models can provide clinically relevant input on surgical targeting and the selection of stimulation parameters.

6.8.3 ROLE OF PRECLINICAL ANIMAL MODELING

Although human studies are important for neuroscience research, their scientific impact is limited by practical challenges and the ethical boundaries of clinical research. Preclinical animal models are thus essential for the development of safe and effective commercialized central nervous system therapies. Animal models can deliver proof-of-concept data and provide critical inputs for later-phase clinical studies. In medical device development, a large animal with an intact nervous system is preferred because it allows for the use of human-scaled devices and controlled testing, which cannot be replicated by computer modeling or bench testing. To be clinically relevant, such models must reproduce the patterns of tissue damage observed in humans.

The gold standard large-animal model for Parkinson's disease has been the nonhuman primate (NHP) treated with the neurotoxin 1-methyl-4-phenyl-1,2,3,6-tetrahydropyridine (MPTP), which causes SNpc dopaminergic neurodegeneration and Parkinson's disease-like motor symptoms (Blesa and Przedborski 2014). Major advances in our understanding of both Parkinson's disease and DBS can be attributed directly to research using the MPTP NHP model (reviewed in Capitanio and Emborg 2008). This model helped to identify potential targets such as the subthalamic nucleus, which was subsequently shown to be an effective DBS target in monkeys and then in humans. Indeed, the discovery of the MPTP primate model and the subsequent elucidation of the physiology of the basal ganglia led to the revival of movement disorder surgery in the 1990s.

6.8.4 MEDICAL IMPLANT DESIGN AND MANUFACTURING

The limitations of current DBS technology are being addressed by collaborations among engineers, neuroscientists, neurosurgeons, physicists, and other specialists. Implantable neural devices need to be as small as possible, making them less invasive in the body. The batteries in neural implants present several problems, such as limited lifetime, chemical side effects, and large size. Challenges in design and manufacturing include the development of devices with low-power consumption, efficient energy transfer, high data rate, effective power amplifiers, small size, and low cost.

New neurostimulator designs aim to make the following improvements: stimulation of different contacts on the same electrode with different electrical parameters, simultaneous stimulation of multiple brain targets at different frequencies, and stimulation of several electrodes in one target.

6.8.5 BIOCOMPATIBILITY CONSIDERATIONS

Inflammation is the early response to material implantation. Even for biocompatible materials, chronic implantation of a device in the nervous system provokes a biological response. The response is characterized by hypertrophy of the surrounding astrocytes, infiltration of microglia and foreign body giant cells, and thickening of the neighboring tissue in the form of a capsule around the device. Although biocompatibility of the entire neural implant is important, the electrodes in particular are critical because they produce the intended effect. The noble metals, such as platinum, gold, iridium, palladium, and rhodium, have been commonly used for electrical stimulation mainly due to their relative resistance to corrosion.

6.8.6 CONDUCTIVE POLYMERS TO INCREASE ELECTRODE BIOCOMPATIBILITY

Different material coatings for neural electrode surfaces are being developed to improve the integration of electrodes in the brain. CPs represent one potential way to increase neural electrode biocompatibility. CPs are organic materials that can be used as electrode coatings to improve performance and stability. They do so by lowering the impedance of electrodes, increasing the charge density, and improving the mechanical interface between the electrode and brain tissue. Polypyrrole (PPy) is a CP coating that has great biocompatibility and conductive properties, making it a potential option for neural electrodes. Poly(3,4-ethylenedioxythiophene) (PEDOT) is another conducting polymer for coating electrode surfaces. Recent research has shown that biomimetic PEDOT materials allow for stable binding and electrical communication over long periods of time (Zhu et al. 2014). Ongoing research is investigating the applications of CPs for nerve regeneration and neuroprosthetic devices and sensors.

6.8.7 ENERGY CONSIDERATIONS

Current DBS neurostimulators utilize nonrechargeable battery technology, giving the devices a finite service life that varies according to the stimulation parameters, the number and polarity of electrode contacts, and the time period of stimulation during the day. For DBS of Parkinson's disease, the typical stimulation parameters yield an estimated battery life of approximately 3.5 years. Surgical replacement is thus required every few years due to the battery life of the stimulator. Surgery has clear disadvantages, including the risk of infection, scarring, damage to connections, and cost of treatment. Externally rechargeable batteries are being developed, along with ways to increase the energy efficiency of the individual stimulus pulses generated by the IPG.

6.9 EMERGING INDICATIONS AND CONCEPTS

6.9.1 HUNT FOR TRANSLATIONAL BIOMARKERS

A biomarker is an objectively measured indicator of normal biological processes, pathogenic processes, or modulatory responses to a therapeutic intervention. To be translational, a biomarker needs to be evaluated in experimental settings, for example, in the measurement of normal or pathological activity in animals or humans. Five examples of current or previous biomarkers for DBS include the following: (1) beta-band (13–30 Hz) power in the local field potential recorded from the subthalamic region of the basal ganglia in Parkinson's disease (Yang et al. 2014), (2) increased positron emission tomography (PET)–derived blood flow in the subgenual cingulate cortex in treatment-resistant depression (Mayberg et al. 1997), (3) pathological high-frequency oscillations in focal epilepsy (Bragin et al. 2004), (4) theta-band (4–8 Hz) oscillations in the field potential recorded from the medial pallidum in dystonia (Liu et al. 2002), and (5) idiopathic rapid eye movement (REM) sleep behavior disorder in Parkinson's disease (Postuma et al. 2010). In these examples, the objective of DBS therapy is to modulate the electrophysiological biomarkers of disease states.

Another approach is to use functional neuroimaging to examine whole-brain responses evoked by DBS. Neuroimaging techniques such as PET and fMRI measure changes in neural activity indirectly by recording alterations in blood flow, blood oxygenation, and glucose consumption. During DBS implantation, for example, researchers may measure the fMRI signal in motor areas of the brain (e.g., motor cortex and cerebellum) while systematically varying the stimulation settings, such as amplitude and frequency. This procedure can be fine-tuned for a given patient to optimize the electrode placement and stimulation parameters, rather than relying on a suggested checklist of parameters (Jerde et al. 2016). Such a biomarker-based approach could be effective for disorders in which the therapeutic benefits do not occur until a period after implantation, for example, dystonia.

The clinical symptoms of many brain disorders probably reflect abnormalities in brain networks and not just a disruption in a focal brain region. As the effects of neuromodulation on brain network physiology become better known, biomarkers may expand to include measurements of brain dynamics. For example, a recent method called *dynamic network biomarkers* is designed to detect the clinically silent predisease state that occurs prior to the first clinical onset of a disorder (Chen et al. 2012).

6.9.2 CLOSED-LOOP STIMULATION

Traditional DBS devices are open-loop systems that apply continuous stimulation regardless of changes in the brain or patient symptoms. Although some stimulation parameters can be altered after surgery, open-loop systems function without a physiologically recorded control signal. Open-loop DBS thus yields suboptimal performance in therapeutic efficacy and in technical considerations, such as wasted battery power. The lack of focal and controlled stimulation also increases the likelihood of stimulation-induced side effects.

Recently developed closed-loop DBS systems enable the sensing and recording of brain activity while simultaneously providing targeted DBS therapy (Stypulkowski et al. 2013). Closed-loop DBS systems utilize a biomarker to monitor a patient's disease state, and then modulate the delivery of stimulus pulses based on abnormal brain activity (Stypulkowski et al. 2014). Implantable closed-loop neurostimulation systems have been used to treat several disorders, including pain, epilepsy, and Parkinson's disease. For example, a closed-loop SCS system for pain, the RestoreSensor system (Medtronic), has been granted FDA approval to provide online adjustments in stimulation according to a patient's body position. A second example is the FDA-approved RNS System for epilepsy. In this system, a neurostimulator is implanted in the cranium and connected to one or two recording and stimulating leads that are surgically positioned at the seizure foci in the brain. When abnormal brain activity is detected, electrical stimulation is applied to the foci of the seizure.

Several scientific and technical challenges remain for closed-loop systems, such as identifying the biomarkers for specific symptoms, isolating the anatomical targets in which to sense abnormalities and trigger stimulation for a given disorder, and discovering algorithms to map the current physiological state to the stimulation parameters.

6.9.3 EPILEPSY: BRAIN AND VAGUS NERVE STIMULATION

Epilepsy afflicts approximately 1% of the world population. Effective treatment options include pharmacological therapy and resective surgery to remove the part of the brain that causes the seizures. But millions of patients still have recurrent seizures or cannot undergo resective surgery. For such patients, neurostimulation is a therapeutic option. The history of electrical stimulation for epilepsy has included a variety of anatomical targets (caudate nucleus, thalamic and subthalamic areas, cerebellum, and hippocampus), a variety of stimulation parameters, and several outcome measures for evaluation. Originally limited to vagus nerve stimulation (VNS), DBS has become a treatment option due to a better understanding of the physiology of epilepsy, precise stereotactic techniques, and controlled clinical studies. Two DBS therapies have obtained regulatory approval for more widespread use in patients: anterior nucleus of thalamus (ANT) DBS (Fisher et al. 2010) outside of the United States but not within it, and the RNS Stimulator within the United States but not outside of it.

6.9.4 PSYCHIATRIC DISORDERS: DBS FOR DEPRESSION AND OBSESSIVE–COMPULSIVE DISORDER

The long-term efficacy of DBS for the treatment of movement disorders such as Parkinson's disease, tremor, and dystonia, as well as for chronic pain, encouraged the expansion of neuromodulation methods to include psychiatric disorders such as depression and obsessive–compulsive disorder (OCD). The use, or potential use, of DBS for psychiatric disorders is based on several factors (Wichman and Delong 2006). First, the efficacy of DBS for movement disorders occurred over decades of clinical and experimental research, suggesting that DBS can be used for psychiatric disorders as we better understand their underlying neurobiology. Second, the drug treatments for psychiatric conditions are not effective in a substantial number of patients, so new therapeutic methods are needed. Third, the National Institute of Mental Health and other agencies have sought to classify psychiatric disorders in terms of behavior and neurobiological mechanisms (Cuthbert and Insel 2013). And fourth, the public is more aware of the neuroscientific basis of psychiatric disorders and therefore more willing to attempt to alleviate them by neuromodulation.

Major depressive disorder is a leading public health problem that affects approximately 14 million Americans each year. Approximately, one-third of patients suffering from major depression are not helped by conventional drug treatments. These patients experience a reduced quality of life, high risk of suicide, and little hope of recovery. For such patients, neuromodulation provides a potential therapeutic option. Thus far, however, clinical studies using tDCS and TMS for depression have produced mixed results. DBS studies have shown promise, but large randomized controlled trials have not yielded positive results (Deng et al. 2015).

For OCD, several neuromodulatory techniques have been investigated (Bais et al. 2014), including electroconvulsive therapy, tDCS, TMS, and DBS. Early reports using tDCS and TMS have not been impressive. Recently, the FDA approved DBS for OCD within the HDE program. DBS of subcortical targets in the frontostriatal network has shown a response rate of 60%, suggesting that DBS may be effective in treating OCD. Despite these preliminary results, however, even the pioneers of DBS have criticized the FDA as being premature in approving DBS for OCD (Fins et al. 2011).

There are several reasons to believe that DBS for psychiatric disorders will not be as successful as DBS for movement disorders. First, the symptoms in psychiatric disorders are heterogeneous, complex, and not well delineated in terms of the underlying neurobiological mechanisms (Cuthbert and Insel 2013). Second, there are many brain circuits involved in these disorders, which partially explains why so many brain targets have been proposed and evaluated. For example, DBS has been tried in at least 10 brain areas for depression and 8 areas for OCD. And third, animal models are lacking for research on the biological mechanisms of psychiatric disorders, because animals do not develop psychiatric or neurological disorders as humans do. For these reasons, there are substantial gaps in the translational continuum from basic science to the clinic in psychiatric disorders.

6.9.5 OTHER DISORDERS: SCHIZOPHRENIA, ADDICTION, EATING DISORDERS, AND TRAUMATIC BRAIN INJURY

Neuromodulation techniques have been investigated in other psychiatric and neurological disorders, with mixed results. It should be noted that serendipity has played a part in several "discoveries" of treatment therapies. For example, some patients who underwent DBS of the nucleus accumbens for OCD subsequently lost their dependence on alcohol, nicotine, or heroin. Should further clinical studies be conducted by applying DBS to this brain area in patients with drug addiction? Or would this treatment be a premature application of neuromodulation? These questions will be germane as neuromodulation is considered for the treatment of complex disorders for which the underlying physiology is largely unknown.

Recently, noninvasive neuromodulation approaches, such as TMS and tDCS, have been investigated for a number of disorders. For example, several studies have reported that TMS can reduce persistent auditory hallucinations in schizophrenia (Aleman et al. 2007). But the effects are generally small, and a key problem with noninvasive techniques is that they cannot penetrate to the depths of the brain required to stimulate subcortical areas, which are important biological substrates of the disorders.

Traumatic brain injury (TBI) may be amenable to treatment with DBS. TBI is a significant public health problem and a leading cause of death and disability in many countries. There has yet to be a successful Phase III clinical trial investigating a pharmacological intervention for TBI, and there are currently no FDA-approved therapeutic modalities for mitigating the consequences of TBI. However, because some of the common sequelae of TBI are movement disorders, including tremor, dystonia, and parkinsonism, recent studies have reported positive results using implanted brain stimulation devices to treat patients suffering from the motor and cognitive consequences of TBI (Shin et al. 2014).

6.9.6 NEUROPROSTHETICS

The neuromodulation approaches described thus far (DBS, SCS, VNS, etc.) use implantable devices to stimulate a region of the brain and thereby relieve the symptoms of a disorder. The goal in neuroprosthetics, on the other hand, is to record signals from the brain, decode them, and use the decoded information to control an external device, such as a robotic limb or a computer cursor. Thus far, neuroprosthetics have mainly been used for movement disorders, such as spinal cord injury, paralysis, and loss of limb.

A neuroprosthetic technology for movement consists of three main components: a sensor to record brain signals, a signal processor to decode the intended movement from the neural signals, and an effector that can be physical (e.g., a robotic limb) or virtual (e.g., a cursor) to implement the intended movement. Neuroprosthetic devices present similar technical and biological challenges as neuromodulation, including electrode technology, signal longevity, the inflammatory response of brain tissue to electrodes or sensors, patient safety, miniaturization, and durability, and the reader is encouraged to consult the literature on this topic (e.g., Lee et al. 2013). In the future, as technology develops and our understanding of brain physiology advances, neuromodulation and neuroprosthetics will likely be combined to improve patient lives.

DISCLAIMER

The author was a paid consultant for Medtronic during the writing of this chapter.

ACKNOWLEDGMENTS

The author thanks Dr. Robert S. Raike of Medtronic for helpful suggestions on this chapter.

REFERENCES

Albe Fessard, D., Arfel, G., Guiot, G., et al. Characteristic electric activities of some cerebral structures in man. *Ann. Chir* 17 (1963): 1185–1214.

Aleman, A., Sommer, I. E., Kahn, R. S. Efficacy of slow repetitive transcranial magnetic stimulation in the treatment of resistant auditory hallucinations in schizophrenia: a meta-analysis *J Clin Psychiatry* 68 (2007): 416–421.

Aquilina, O. A brief history of cardiac pacing. *Images Paediatr Cardiol* 8 (2006): 17–81.

Bais, M., Figee, M., and Denys, D. Neuromodulation in obsessive compulsive disorder. *Psychiatr Clin North Am* 37 (2014): 393–413.

Barow, E., Neumann, W. J., Brücke, C., et al. Deep brain stimulation suppresses pallidal low frequency activity in patients with phasic dystonic movements. *Brain* 137 (2014): 3012–3024.

Benabid, A. L., Pollak, P., Gervason, C., et al. Long-term suppression of tremor by chronic stimulation of the ventral intermediate thalamic nucleus. *Lancet* 337 (1991): 403–406.

Benabid, A. L., Pollak, P., Louveau, A., Henry, S., and de Rougemont, J. Combined (thalamotomy and stimulation) stereotactic surgery of the VIM thalamic nucleus for bilateral Parkinson disease. *Appl Neurophysiol* 50 (1987): 344–346.

Blesa, J., and Przedborski, S. Parkinson's disease: Animal models and dopaminergic cell vulnerability. *Front Neuroanat* 8 (2014): 155.

Bragin, A., Wilson, C. L., Almajano, J., Mody, I., and Engel, J., Jr. High-frequency oscillations after status epilepticus: Epileptogenesis and seizure genesis. *Epilepsia* 45 (2004): 1017–1023.

Brown, P., Oliviero, A., Mazzone, P., et al. Dopamine dependency of oscillations between subthalamic nucleus and pallidum in Parkinson's disease. *J Neurosci* 21 (2001): 1033–1038.

Butson, C. R., Cooper, S. E., Henderson, J. M., Wolgamuth, B., and McIntyre, C. C. Probabilistic analysis of activation volumes generated during deep brain stimulation. *Neuroimage* 54 (2011): 2096–2104.

Capitanio, J. P., and Emborg, M. E. Contributions of non-human primates to neuroscience research. *Lancet* 371 (2008): 1126–1135.

Chen, L., Liu, R., Liu, Z. P., Li, M., and Aihara, K. Detecting early-warning signals for sudden deterioration of complex diseases by dynamical network biomarkers. *Sci Rep* 2 (2012): 342.

Chiken, S., and Nambu, A. Mechanism of deep brain stimulation: Inhibition, excitation, or disruption? *Neuroscientist* 22 (2016): 313–322.

Cooper, I. S. Effect of chronic stimulation of anterior cerebellum on neurological disease. *Lancet* 1 (1973): 206.

Cuthbert, B. N., and Insel, T. R. Toward the future of psychiatric diagnosis: The seven pillars of RDoC. *BMC Med* 11 (2013): 126.

Deisseroth, K. Optogenetics: 10 years of microbial opsins in neuroscience. *Nat Neurosci* 18 (2015): 1213–1225.

Deng, Z., McClintock, S. M., Oey, N. E., Luber, B., and Lisanby, S. H. Neuromodulation for mood and memory: From the engineering bench to the patient bedside. *Curr Opin Neurobiol* 30C (2015): 38–43.

Djourno, A., and Eyriès, C. Auditory prosthesis by means of a distant electrical stimulation of the sensory nerve with the use of an indwelt coiling. *Presse Med* 65 (1957): 1417.

Du, Z. J., Luo, X., Weaver, C., and Cui, X. T. Poly (3,4-ethylenedioxythiophene)-ionic liquid coating improves neural recording and stimulation functionality of MEAs. *J Mater Chem C Mater Opt Electron Devices* 3 (2015): 6515–6524.

Fins, J. J., Mayberg, H. S., Nuttin, B., et al. Misuse of the FDA's humanitarian device exemption in deep brain stimulation for obsessive-compulsive disorder. *Health Aff* 30 (2011): 302–311.

Fisher, R., Salanova, V., Witt, T., et al. Electrical stimulation of the anterior nucleus of thalamus for treatment of refractory epilepsy. *Epilepsia* 51 (2010): 899–908.

Fitzgerald, P. B., and Daskalakis, Z. J. The effects of repetitive transcranial magnetic stimulation in the treatment of depression. *Expert Rev Med Devices* 8 (2011): 85–95.

Fregni, F., Boggio, P. S., Nitsche, M. A., et al. Treatment of major depression with transcranial direct current stimulation. *Bipolar Disord* 8 (2006): 203–204.

Fritsch, G., and Hitzig, E. Über die elektrische Erregbarkeit des Grosshirns. *Arch Anat Physiol Wissen* 37 (1870): 300–332.

Gildenberg, P. L. Evolution of neuromodulation. *Stereotact Funct Neurosurg* 83 (2005): 71–79.

Grill, S. Implementing deep brain stimulation in practice: Models of patient care. In *Deep Brain Stimulation Management*, ed. W. J. Marks Jr. 2nd ed. Cambridge: Cambridge University Press, 2015, 182–191.

Grill, W. M., Snyder, A. N., and Miocinovic, S. Deep brain stimulation creates an informational lesion of the stimulated nucleus. *Neuroreport* 15 (2004): 1137–1140.

Hariz, M. I. Deep brain stimulation: New techniques. *Parkinsonism Relat Disord* 20 (Suppl. 1) (2014): S192–S196.

Hassler, R., Riechert, T., Mundinger, F., et al. Physiological observations in stereotaxic operations in extrapyramidal motor disturbances. *Brain* 83 (1960): 337–350.

Horvath, J. C., Carter, O., and Forte, J. D. Transcranial direct current stimulation: Five important issues we aren't discussing (but probably should be). *Front Syst Neurosci* 8 (2014): 2.

Isaias, I. U., Fadil, H., and Tagliati, M. Managing dystonia patients treated with deep brain stimulation. In *Deep Brain Stimulation Management*, ed. W. J. Marks Jr. 2nd ed. Cambridge: Cambridge University Press, 2015, 108–117.

Jasper, H., and Penfield, W. *Epilepsy and the Functional Anatomy of the Human Brain*. 2nd ed. Boston, MA: W. Little, Brown and Co., 1954.

Jerde, T. A., Kelly, M., Reinking, N., Billstrom, T., Lentz, L., and Raike, R. S. Feasibility of 1.5T fMRI BOLD activation patterns for guiding deep brain stimulation targeting and parameter selection demonstrated in a large-animal model. Presented at Society for Neuroscience Conference Annual Meeting, San Diego, CA, 2016.

Krauss, J. K., Pohle, T., Weber S., Ozdoba, C., and Burgunder, J. M. Bilateral stimulation of globus pallidus internus for treatment of cervical dystonia. *Lancet* 354 (1999): 837–838.

Kringelbach, M. L., Jenkinson, N., Owen, S. L., and Aziz, T. Z. Translational principles of deep brain stimulation. *Nat Rev Neurosci* 8 (2007): 623–635.
Krishna, V., King, N. K., Sammartino, F., Strauss, I., Andrade, D. M., Wennberg, R. A., and Lozano, A. M. Anterior nucleus deep brain stimulation for refractory epilepsy: Insights into patterns of seizure control and efficacious target. *Neurosurgery* 78 (2016): 802–811.
Kühn, A. A., Brücke, C., Schneider, G. H., et al. Increased beta activity in dystonia patients after drug-induced dopamine deficiency. *Exp Neurol* 214 (2008): 140–143.
Lee, B., Liu, C. Y., and Apuzzo, M. L. A primer on brain-machine interfaces, concepts, and technology: A key element in the future of functional neurorestoration. *World Neurosurg* 79 (2013): 457–471.
Lewis, P. M., Thomson, R. H., Rosenfeld, J. V., and Fitzgerald, P. B. Brain neuromodulation techniques: A review. *Neuroscientist* 22 (2016).
Limousin, P., Pollak, P., Benazzouz, A., et al. Effect of parkinsonian signs and symptoms of bilateral subthalamic nucleus stimulation. *Lancet* 345 (1995): 91–95.
Liu, X., Griffin, I. C., Parkin, S. G., et al. Involvement of the medial pallidum in focal myoclonic dystonia: A clinical and neurophysiological case study. *Mov Disord* 17 (2002): 346–353.
Maks, C. B., Butson, C. R., Walter, B. L., Vitek, J. L., and McIntyre, C. C. Deep brain stimulation activation volumes and their association with neurophysiological mapping and therapeutic outcomes. *J Neurol Neurosurg Psychiatry* 80 (2009): 659–666.
Mayberg, H. S., Lozano, A. M., Voon, V., et al. Deep brain stimulation for treatment-resistant depression. *Neuron* 45 (2005): 651–60.
McIntyre, C. C., and Foutz, T. J. Computational modeling of deep brain stimulation. *Handb Clin Neurol* 116 (2013): 55–61.
Melzack, R., and Wall, P. D. Pain mechanisms: A new theory. *Science* 150 (1965): 971–979.
Ouyang, L., Shaw, C. L., Kuo, C. C., Griffin, A. L., and Martin, D. C. In vivo polymerization of poly(3,4-ethylenedioxythiophene) in the living rat hippocampus does not cause a significant loss of performance in a delayed alternation task. *J Neural Eng* 11 (2014): 026005.
Plaha, P., Ben-Shlomo, Y., Patel, N. K., and Gill, S. S. Stimulation of the caudal zona incerta is superior to stimulation of the subthalamic nucleus in improving contralateral parkinsonism. *Brain* 129 (2006): 1732–1747.
Pollak, P., Benabid, A. L., Gross, C., et al. Effects of the stimulation of the subthalamic nucleus in Parkinson disease. *Rev Neurol (Paris)* 149 (1993): 175–176.
Pool, J. L. Psychosurgery in older people. *J Am Geriatr Soc* 2 (1954): 456–466.
Postuma, R. B., Gagnon, J. F., Rompré, S., and Montplaisir, J. Y. Severity of REM atonia loss in idiopathic REM sleep behavior disorder predicts Parkinson disease. *Neurology* 74 (2010): 239–244.
Shealy, C. N., Mortimer, J. T., and Reswick, J. B. Electrical inhibition of pain by stimulation of the dorsal columns: Preliminary clinical report. *Anesth Analg* 46 (1967): 489–491.
Shin, S. S., Dixon, C. E., Okonkwo, D. O., and Richardson, R. M. Neurostimulation for traumatic brain injury. *J Neurosurg* 121 (2014): 1219–1231.
Silberstein, P., Kühn, A. A., Kupsch, A., et al. Patterning of globus pallidus local field potentials differs between Parkinson's disease and dystonia. *Brain* 126 (2003): 2597–2608.
Stypulkowski, P. H., Stanslaski, S. R., Denison, T. J., and Giftakis, J. E. Chronic evaluation of a clinical system for deep brain stimulation and recording of neural network activity. *Stereotact Funct Neurosurg* 91 (2013): 220–232.
Stypulkowski, P. H., Stanslaski, S. R., Jensen, R. M., Denison, T. J., and Giftakis, J. E. Brain stimulation for epilepsy—Local and remote modulation of network excitability. *Brain Stimul* 7 (2014): 350–358.
Thompson, J. A., Lanctin, D., Ince, N. F., and Abosch, A. Clinical implications of local field potentials for understanding and treating movement disorders. *Stereotact Funct Neurosurg* 92 (2014): 251–263.
Vanegas-Arroyave, N., Lauro, P. M., Huang, L., Hallett, M., Horovitz, S. G., Zaghloul, K. A., and Lungu, C. Tractography patterns of subthalamic nucleus deep brain stimulation. *Brain* 139 (2016): 1200–1210.

Vitek, J. L., Zhang, J., Hashimoto, T., Russo, G. S., and Baker, K. B. External pallidal stimulation improves parkinsonian motor signs and modulates neuronal activity throughout the basal ganglia thalamic network. *Exp Neurol* 233 (2012): 581–586.

Wager, T. D., Atlas, L. Y., Lindquist, M. A., et al. An fMRI-based neurologic signature of physical pain. *N Engl J Med* 368 (2013): 1388–1397.

Wall, P. D., and Sweet, W. H. Temporary abolition of pain in man. *Science* 155 (1967): 108–109.

Werginz, P., Benav, H., Zrenner, E., and Rattay, F. Modeling the response of ON and OFF retinal bipolar cells during electric stimulation. *Vision Res* 111 (2015): 170–181.

Wichmann, T., and Delong, M. R. Deep brain stimulation for neurologic and neuropsychiatric disorders. *Neuron* 52 (2006): 197–204.

Williams, J. C., and Denison, T. From optogenetic technologies to neuromodulation therapies. *Sci Transl Med* 5 (2013): 177ps6.

Wolter, T. Spinal cord stimulation for neuropathic paln: Current perspectives. *J Pain Res* 7 (2014): 651–663.

Yang, A. I., Vanegas, N., Lungu, C., and Zaghloul, K. A. Beta-coupled high-frequency activity and beta-locked neuronal spiking in the subthalamic nucleus of Parkinson's disease. *J Neurosci* 34 (2014): 12816–12827.

Zhu, B., Luo, S. C., Zhao, H., Lin, H. A., Sekine, J., Nakao, A., Chen, C., Yamashita, Y., and Yu, H. H. Large enhancement in neurite outgrowth on a cell membrane-mimicking conducting polymer. *Nat Commun* 5 (2014): 4523.

Zimnik, A. J., Nora, G. J., Desmurget, M., and Turner, R. S. Movement-related discharge in the macaque globus pallidus during high-frequency stimulation of the subthalamic nucleus. *J Neurosci* 35 (2015): 3978–3989.

7 The electromagnetic nature of protein–protein interactions

Anna Katharina Hildebrandt
Max Planck Institute for Informatics
Saarbrücken, Germany

Thomas Kemmer and Andreas Hildebrandt
Johannes Gutenberg University Mainz
Mainz, Germany

Contents

7.1 IMPORTANCE OF PROTEIN–PROTEIN INTERACTIONS

In 2000, while the Human Genome Project and its competitor Celera Genomics were rapidly closing in on sequencing the human genome, the British bioinformatician Ewan Birney started a betting pool—called Genesweep—among leading genomics researchers. The goal was to predict, as accurately as possible, the number of genes encoded for in the human genome. As larger contiguous stretches of the genome became available, the bet became simpler, and hence the cost of a bet rose from initially \$1 to \$20 in 2002. In total, the pool rose to \$1140.

According to Crick's central dogma (Crick 1970), genes encode for proteins through an almost universal mapping (Nirenberg 1972) between triplets of nucleotides in the genomic sequence and amino acids in the protein. Hence, the number of genes is strongly connected with the number of proteins.* Proteins are the "workhorses" of the cell—they transport matter or energy (Hsia 1998), catalyze reactions (Garcia-Viloca et al. 2004), provide structuring elements (McGhee and Felsenfeld 1980), cut other proteins (Glickman and Ciechanover 2002) or glue them together (Deshaies and Joazeiro 2009), and do whatever else is required for the functioning of the cell at the molecular level. Since humans are a seemingly very complex biological system with many biological functions at the molecular level, it seems obvious that the number of different proteins encoded for by the human genome should be very large—larger than for simpler organisms, at least. Some genomes had already been sequenced and queried for the genes they contained. For the model plant *Arabidopsis thaliana*, for example, a small flowering plant with a relatively small genome that was sequenced in 2000, gene number estimates† range from 25,000 to 32,000 (Arabidopsis Genome Initiative 2000; Parinov and Sundaresan 2000). Hence, entries in the Genesweep contest went up to as large as 300,000 genes. When the prize was finally awarded in 2003, the winning entry, submitted by the American bioinformatician Lee Rowen, turned out to be the lowest entry of the whole contest—a bet of 25,947 genes. This number, which most experts would have regarded as much too small just a few months before, was still too large. Today, we estimate that there are about 20,500 genes in the human genome, with a recent study suggesting that the number might in truth be as low as 19,000 (Ezkurdia et al. 2014).

The number of different proteins available to the human cell turns out to be larger due to alternative splicing during transcription. Also, proteins can be modified posttranslationally (i.e., after they have been synthesized by the cell) by adding certain chemical groups. But even with these modifications, the number of human proteins seems nonsensically low at first glance. Is it truly possible that humans have fewer genes than, say, the water flea, which has about 31,000 of them (Colbourne et al. 2011)? How do we arrive at the complexity of the human organism?

The answer to this conundrum seems to be that it is not so much the number of proteins that counts, but rather their *interactions*, which give rise to the variety of biological phenomena. In many cases, biological function seems to be "implemented" through biomolecular interactions. For this to work, nature needs a way to evolve such interactions—creating new ones to gain new functions and adapting existing ones to improve or adapt existing functions to new circumstances. Proteins are ideally suited for this task: through mutations of the genome, nature can change the sequence of amino acids of a protein. As we will see in Section 7.3, this can in turn lead to changes in the interactions the modified protein can participate in.

7.2 PROTEIN STRUCTURE

To understand protein–protein interactions, it is crucial to first have an understanding of protein structure. This can easily be seen as follows: the force that binds a protein to its interaction partner is composed of small interactions between atoms in both of

* They are not necessarily equal, due to alternative splicing.

† Finding genes in a genome is by no means trivial and requires probabilistic approaches.

the interacting molecules. Since the individual contributions are mostly weak, many of them are required to have a strong total effect. Since, in addition, the individual specific interactions are relatively short-ranged, many pairs of interacting atoms need to come into close contact with each other. This is only possible if not only the physicochemical properties of the molecules are compatible, but also their geometric shapes.

7.2.1 THE PROTEIN STRUCTURE HIERARCHY

Proteins are biopolymers that are composed of amino acid monomers, connected in a chain through peptide bonds between the C-terminus of one and the N-terminus of another amino acid in the chain. While all amino acids share the same backbone,* they differ in the chemical composition of the so-called side chain or residue. Through the different side chains, each amino acid has characteristic physicochemical properties, such as its size, its overall charge in a particular environment, or the ability to form hydrophobic interactions or hydrogen bonds through their side chains. While in principle, a large number of amino acids exist, only 21 of these are typically found in eukaryotic proteins. These are the amino acids encoded for by the genetic code.

Each amino acid is conventionally assigned a single letter from the Latin alphabet. The letter *A*, for instance, denotes the amino acid alanine. At the most abstract level of description, a protein is hence a string of text, where each letter corresponds to an amino acid, and the sequence is understood to be read from the N- to the C-terminal end. This text is known as the *primary structure* of the protein.

The primary structure alone already contains a wealth of information about the protein. In fact, since the primary structure is (nearly†) all that is encoded in the genome, it can be argued that it should contain (nearly) all there is to know about the protein, including complete knowledge of its potential interactions. Extracting and using this information, on the other hand, often turns out to be much too difficult. Instead, a more detailed description of proteins that uses atomic resolution is often employed. Considering that each possible geometric arrangement of a given protein brings different amino acids in contact with each other, and considering that these feature different chemical properties, each geometric arrangement is associated with a certain internal energy. Hence, the protein will try to assume the structure leading to the smallest free energy of folding.

Since all amino acids share the same backbone, atomic arrangements that lead to low-energy conformations of these backbone atoms occur very frequently in many proteins. For instance, if the backbone atoms are arranged in a spiral of roughly 3.6 residues per turn and a translation along the spiral's axis of 1.5 Å, the resulting geometry is just right for the formation of hydrogen bonds between the C=O group of one amino acid and the N–H group of one amino acid and the N–H group of the amino acid four positions later in the primary structure. This so-called *α-helix* is a common motif in proteins (other kinds of helices exist, but are much less frequent). Similarly, a characteristic zigzag pattern of the backbone atoms of consecutive amino acids—called a *β-strand*—allows the formation of hydrogen bonds between backbone atoms of two strands that are close in space, but might be remote in the primary structure. The α-helices and β-strands are often connected by short *turns* or longer *loops* that are rather flexible. In addition, *random coil* elements denote parts of the protein without such a regular structure. These different

* The amino acid proline is a special case, as the side chain connects back to the backbone to form a ring.

† Subtle effects of codon usage, for example, complicate matters.

motifs are known as *secondary structure elements*, and the sequence of secondary structure elements of a protein is known as its *secondary structure*. Knowledge of the secondary structure is highly useful and might be sufficient to answer many questions of scientific interest. But a detailed understanding of protein function or protein–protein interactions typically requires knowledge of the complete three-dimensional configuration of the protein, that is, the coordinates of all of its atoms. This is known as the protein's *tertiary structure*. Finally, biological function is often not carried by a single protein alone, but rather by a complex composed of multiple protein subunits. The arrangement of these subunits with respect to each other is known as the *quaternary structure* of the complex.

7.2.2 OBTAINING PROTEIN STRUCTURES

Obtaining information about the different levels of protein structure is an important problem in both the experimental and the theoretical sciences. While obtaining the primary structure is often comparatively simple, and rough information about the secondary structure can be gained through spectroscopic means, measuring the tertiary and quaternary structures is typically a much more difficult problem. In practice, a protein structure is measured (resolved) by either x-ray crystallography, nuclear magnetic resonance, or electron microscopy, but all of these have characteristic drawbacks and challenges. In fact, today, many proteins of importance have not yet been resolved due to various difficulties. Of those that have been resolved (nearly 100,000 at the time of writing [Worldwide Protein Data Bank 2015]), many required chemical or sequence modification (e.g., to facilitate crystal formation), potentially altering their structure. Even then, it is often only possible to resolve parts of the protein structure, leaving gaps where the atomic arrangement is unclear or not well defined. Finally, it is crucial to realize that all protein structures are, in essence, models that have been derived through an error-prone, indirect process and have been observed under experimental conditions that often differ strongly from those encountered in the protein's natural environment. Even worse, the whole notion of a protein's tertiary structure as the collection of its atomic positions is only an approximation, even when ignoring quantum mechanics: the protein's partition function is not necessarily dominated by a single configuration. Instead, the free energy landscape of many proteins contains a huge number of local minima, which sometimes differ significantly in structure. Depending on the system conditions, many of these system states might be occupied, and the protein will transition between different configurations over time. These problems also render the prediction of the protein structure from a sequence a formidable task.

While it is important to keep these limitations in mind, protein structure models are still of tremendous importance. For reasons we discuss later in this chapter, specificity of biomolecular interactions often imposes restrictions on the variability of the protein's structure, at least in the regions of interest. In addition, the internal motion of a protein over time can often be approximately simulated, using force field–based techniques as described in Section 7.3.

7.3 PHYSICS OF PROTEIN–PROTEIN INTERACTIONS

The amount of biological functions that a cell or a multicellular organism requires at the molecular level for its survival is huge and diverse, on many different levels, not only in spatial dimensions and time spans. The digestion of ingested proteins, for instance, is necessarily a very unspecific process (all ingested proteins should be digested, independent of their sequence or structure), while catalyzing a certain reaction needs to be highly specific. Consequently, nature needs powerful mechanisms for regulating such interactions, which can be tailored—through evolutionary means—to the needs of a specific biological function. Evolutionary changes to the primary structure of a protein can result in a different three-dimensional structure, as well as in a different chemical composition. It is this mechanism that allows proteins to fill such a large variety of different roles for so many different purposes. But the resulting diversity of protein structure and interaction modes poses serious challenges for modeling protein interactions: for most intents and purposes, the description of proteins and their interactions cannot be strongly coarse-grained without significant loss of precision. Instead, most application scenarios require a description at the atomic level. In the following sections, we discuss the necessary ingredients and present different models for protein interactions.

7.3.1 THERMODYNAMIC CONTROL VERSUS KINETIC CONTROL

From elementary thermodynamics, it is obvious that the association of two proteins into a protein complex occurs spontaneously (or, more generally, the association of n molecules into a molecular complex) if its associated Gibbs free energy ΔG is negative. This free energy can be decomposed into an enthalpic and an entropic term as follows:

$$\Delta G = \Delta H - T\Delta S$$

where ΔH denotes the change in enthalpy, T denotes temperature, and ΔS is the change in entropy. Considering, for example, the aggregation of proteins A and B to the complex C, we have

$$A + B \underset{k_{\text{off}}}{\overset{k_{\text{on}}}{\rightleftharpoons}} C$$

with on-rate constant k_{on} and off-rate constant k_{off}. In equilibrium, the reaction is described by the association constant

$$K_a = \frac{k_{\text{on}}}{k_{\text{off}}} = \frac{[C]}{[A][B]}$$

or, equivalently, the dissociation constant $K_{\text{d}} = 1/K_{\text{a}}$. From thermodynamics, we know that

$$\Delta G = -RT\ln(K_a) = RT\ln(K_d)$$

Hence, changes in Gibbs free energy lead to exponential changes in the concentration of bound versus unbound molecules.

The enthalpic contributions to binding all arise in one way or another from electromagnetism, but modeling them as such would require a very detailed picture and a quantum-level description. Instead, different kinds of forces are typically used as ad hoc phenomenological models for these effects. A prime example of this strategy is the van der Waals interaction, which is typically treated as a rapidly decaying interaction between uncharged particles without resorting to induced dipoles and the corresponding electrostatic interactions. We will later discuss the phenomenological models that are typically used to describe such interactions (so-called *force field models*). At this point, it is sufficient to see that the enthalpic contribution of a protein–protein interaction can be fine-tuned* through evolutionary processes by mutating the binding partners to improve their interactions, for example, by putting a hydrogen bond donor on one side of the complex and an acceptor on the other.

Each instance of these interactions, for example, each individual van der Waals interaction between two atoms of the complex, is relatively weak and short-ranged. To stabilize a complex against random thermal movement through enthalpy will thus typically require several favorable interactions between pairs of atoms in both binding partners. This, in turn, means bringing many atoms in one protein in close contact with many atoms of the other. Hence, protein binding implies not only a compatible physicochemistry (favorable interactions), but also a compatible geometry—large surface contact areas allow for many favorable interactions. These geometries may be achieved through large, flat parts of the protein, and often, this is indeed the case. On the other hand, building a concavity on one protein and a corresponding, closely fitting convex part on the other not only allows for multiple interactions, but also aids specificity.

The entropic contribution to the changes in Gibbs free energy can be quite complex and challenging to quantify. An important example of a binding-related effect on entropy is the loss or freezing of translational and rotational degrees of freedom: before binding, both binding partners could translate along three axes and rotate along three angles, for a total of 12 rigid degrees of freedom. After binding, the complex moves as a whole, with six degrees of freedom of its own. This loss of six degrees of freedom is obviously associated with an entropic cost. Similarly, internal degrees of freedom of the proteins (flexible degrees of freedom that allow for a certain shape distortion) can also be restricted upon binding. But more subtle effects have to be taken into account as well, such as solvent–solute interactions. Binding reduces the solvent-exposed protein surface. Around hydrophobic patches of the surface, water forms relatively rigid cages or solvation shells, restricting the water's degrees of freedom. Reducing the surface area can reduce the size of the hydrophobic cages, mobilize water molecules, and hence increase entropy in the system (Pratt and Chandler 1977; Hummer et al. 1996; Pratt and Pohorille 2002).† In other instances, individual water molecules are bound strongly somewhere inside the binding area of the binding partners in their unbound form and have to be released upon binding, again increasing entropy. These entropic contributions of solvent release are sometimes the main driving force of the binding.

Typically, a particular binding is either entropy or enthalpy dominated, with the other term being comparatively small. This phenomenon is known as entropy–enthalpy compensation (Gallicchio et al. 1998).

Inherent to the considerations of this section so far is a somewhat static model of the thermodynamics of binding. Enthalpic and entropic contributions are compared between the bound and the unbound configuration of the system, leading to a change in Gibbs free energy and a resulting concentration of bound versus unbound protein. The dynamics or kinetics of the binding process, on the other hand, are more or less ignored. But in reality, the kinetics of the interaction might be just as important as the thermodynamic picture discussed so far, and might in fact even dominate.

* Obviously, this is not a directed process; instead, mutations to the protein's gene are accepted with a higher probability if the resulting change in its interactions is favorable for the organism under given environmental conditions.

† This "hydrophobic effect" is often modeled as an enthalpic contribution in practice.

Since the specific interactions between proteins are short-ranged, they only become relevant after the binding partners have already encountered one another, that is, after the partners have come sufficiently close to one another that the specific interactions are no longer dominated by the thermal energy, and have oriented themselves with respect to the other such that the respective binding interfaces point toward another. In fact, spatial separations as small as a few angstroms and angular deviations of just a few degrees can render such interactions nearly invisible (Alsallaq and Zhou 2007). The necessary arrangements can take place purely by chance, through random diffusion processes, but this will take considerable time on average. Instead, longer-ranged but less specific electrostatic interactions can be used to guide this association phase. In this way, the kinetics of protein association is as important to the understanding of protein interactions as the thermodynamics. Or, to be more precise, owing to the huge diversity in protein structure, the association rate of any given protein complex can be dominated either by kinetics, by thermodynamics, or by a mixture of the two.

To understand the distinction further, it is useful to split up the reaction of the two binding partners to the complex into a two-step process: first, the partners must meet and orient themselves to form a near-native intermediate state called the *encounter complex* (Gabdoulline and Wade 1997) or *transient complex* (Schreiber et al. 2009). This encounter complex then reacts through flexible rearrangements—and potential further changes such as covalent bond formation—to form the native complex C, yielding the scheme

$$A+B \underset{k_{-AB}}{\overset{k_{AB}}{\rightleftharpoons}} A*B \xrightarrow{k_C} C$$

where the first part denotes the encounter phase, and the second the reaction. This scheme clarifies the two main ways electromagnetic interactions can influence protein binding: in the first step, guiding the encounter away from a purely random process toward a steered attraction can speed up the encounter phase, leading to greatly improved rate constants. In fact, when considering the whole spectrum of observed protein association rates—which has an enormous variety, ranging from less than 10^3 M^{-1} s^{-1} to more than 10^6 M^{-1} s^{-1}—long-ranged attraction is clearly needed to overcome the limit of purely diffusive association rates, which is estimated to lie somewhere between 10^5 M^{-1} s^{-1} and 10^6 M^{-1} s^{-1} (Schlosshauer and Baker 2004; Gabdoulline and Wade 2001). In biomolecular systems, this long-ranged attraction can only be provided by electrostatics. One well-studied example of this behavior is the ligand binding to acetylcholinesterase (AChE), a hydrolase for the neurotransmitter acetylcholine (ACh) that is responsible for the termination of signal transmissions to certain synapses in the central nervous system. AChE is surrounded by a strong electrostatic field leading positively charged ligands right to its active site, located within a deep gorge in the protein surface (Tan et al. 1993; Ripoll et al. 1993). The importance of electrostatics in the second stage is a little less pronounced, as it is not the long range, but often rather the specificity that becomes important.

7.3.2 MODELS OF PROTEIN INTERACTIONS

The interactions occurring between proteins, or between proteins and other biomolecules, such as DNA, RNA, or small ligands, all arise in one way or another from electromagnetism, as weak and strong interactions, as well as gravity, play virtually no role in the length scales involved. On the other hand, the behavior at molecular length scales clearly follows the laws of quantum mechanics rather than classical physics. While quantum descriptions of proteins and protein complexes exist—albeit with their own simplifications—they are far too costly and complex for most application scenarios.

A reasonable alternative is the use of semiclassical models, which condense expert knowledge about quantum chemical behavior into virtual interactions that can be described classically. A prime example is the treatment of chemical bonds as (potentially anharmonic) oscillators. Since these interaction models are heuristics rather than *ab initio* descriptions of the underlying physics, they typically contain several parameters that can be used as tuning knobs to fit the predicted interactions directly or indirectly to experimental data, or to data derived from more expensive models of the protein chemistry, such as quantum chemical calculations. A collection of interaction models, together with values for all the parameters, is known in molecular modeling as a *force field*.*

Many different force fields have been proposed in the literature, often with different characteristics. For instance, a force field aimed at accurately treating small molecules will take more care than a force field mostly aimed at proteins to handle anharmonicities in chemical bonds. Important force fields for protein simulation include Amber (Case et al. 2014), GROMOS (Schmid et al. 2011), and CHARMM (Brooks et al. 2009). The Amber force field, for example, has the following general functional form:

$$V(r) = \sum_{\substack{l \,\in\, \text{covalent} \\ \text{bonds}}} k_b (l(r) - l_0)^2 + \sum_{\substack{\alpha \,\in\, \text{covalent} \\ \text{bond angles}}} k_a (\alpha(r) - \alpha_0)^2$$

$$+ \sum_{\substack{\omega \,\in\, \text{torison} \\ \text{angles}}} \sum_{n} \frac{V_n}{2} (1 + \cos[n\omega(r) - \gamma]) + \sum_{i \in \text{atoms}} \sum_{j \in \text{atoms}\, j<i} \left[\frac{A_{ij}}{r_{ij}^{12}} - \frac{B_{ij}}{r_{ij}^{6}} \right]$$

$$+ \sum_{i \in \text{atoms}} \sum_{j \in \text{atoms}\; j<i} \frac{q_i q_j}{4\pi \varepsilon \varepsilon_0}$$

The first term, often called the stretch term, treats each covalent bond in the system as a harmonic spring with preferred length l_0 and spring constant k_b. These two parameters depend on the atoms involved in the spring, that is, on their chemical element and potentially on their chemical environment, which is subsumed in their so-called *atom type*. Similarly, the second term, known as the bend term, treats the angle between three atoms that are connected in a chain as a harmonic spring, too, to treat the bending of bond angles, where α_0 denotes the preferred bond angle and k_a the spring constant.

The third term, known as the (proper) torsion or dihedral term, models the rotatability of certain single bonds in molecules, where the groups connected to the respective ends of the bond can rotate against each other around the axis formed by the bond. In practice, a few distinct angle configurations will be preferable to the system, while some others will be highly unfavorable. A simple example is the case of ethane ($H_3C–CH_3$), where the methyl groups at both ends can rotate around the axis formed by the C–C bond. Owing to the electrostatic repulsion of hydrogens on both ends of the bond,

* Despite the name, a force field does not, typically, prescribe forces, but rather potential energies. Forces are then derived through analytical or numerical differentiation.

the three distinct configurations that stagger the hydrogens are energetically much more favorable than those eclipsing them. To treat such ensembles of distinct minima and maxima, a finite-order Fourier series is used (typically truncated at $n = 3$) with several parameters that again depend on the types of atoms involved: the Fourier coefficients V_n, the frequencies n, and the phases γ.

The fourth term, known as the van der Waals term, uses a Lennard–Jones potential to model interactions between pairs of neutral atoms. It consists of a strongly repulsive term that models Pauli exclusion between the electrons of the interacting atoms and a less pronounced attractive term that models interactions between induced dipoles. The parameters A_{ij} and B_{ij} tune the position and height of the resulting minimum, which occurs when the involved atoms "touch."* Depending on the force field and the user's choice, not all pairs of potential van der Waals interactions are actually included. Interactions between covalently bonded atoms (1–2 interactions) are usually excluded, just as those between pairs of atoms sharing a bonding partner (1–3 interactions); 1–4 interactions might or might not be included or weighted differently.

Finally, the fifth term, known as the Coulomb or polar term, uses a very simplistic model of molecular electrostatics: each atom is treated individually as a point charge, either in vacuum or in a space completely filled with a homogeneous medium. The parameters q_i and q_j denote the charges of the atoms i and j. These might be "full" charges, corresponding to the lack or excess of electrons (e.g., after deprotonation of an acidic amino acid), or, more commonly, partial charges due to different electronegativities of the atoms involved in a chemical bond. The Coulomb term is the only long-ranged term in the force field, and hence the only one that can substantially influence the encounter stage of protein interactions.

Force fields can be used in many different ways to study protein structure and protein interactions. Among the most important application areas are molecular dynamics (MD) simulations, where Newton's or Hamilton's equations of motion are solved numerically using the given force field equation and a set of initial conditions to approximate the trajectory of the atoms in the system. Accurate simulations require very small time steps for the numerical integration of the equations of motion, leading to long simulation times even for moderately sized proteins. But given sufficient computational resources—such as D. E. Shaw Research's Anton MD-optimized supercomputer—MD simulations can be used to study protein–protein, protein–ligand, or protein–DNA interactions in unparalleled detail (Shaw et al. 2008, 2014).

In situations where the computational cost of MD is too high for the application scenario at hand, force fields can be used in a more static fashion, such as estimators of the enthalpy difference between a bound and an unbound conformation of a protein complex in what is known as protein docking. Neglecting the dynamics of the system in this way typically leads to a drastic loss in accuracy, and indeed, force fields do not, typically, predict such energy differences well if used in such a simple fashion (Merz 2010). One main factor for this deficiency is the simple Coulombic electrostatics used in most force fields. Atoms in proteins simply are not point charges in vacuum, and even if they were, the partial charge model is often too simplistic, as the partial charges might react to changes in the environment, such as the polarization of covalent bonds. To further complicate matters, proteins do not live in

* Considering that the radius of an atom is ill-defined, as the probability density of its constituents is not compact, the position of the van der Waals minimum of two atoms of the same type is often used to define a radius for the atoms, assuming that this is the distance at which the atoms touch.

a vacuum, but are surrounded by a complex conglomerate of molecules. In the simplest case, the protein is at least embedded in water and typically also surrounded by mobile ions.

7.3.3 CONTINUUM ELECTROSTATICS

The treatment of the solvent is a major challenge in force field–based methods. An accurate representation requires the explicit inclusion of water molecules—and usually counterions—in the simulation. The amount of water atoms can easily surpass the amount of protein atoms, and most of the time spent in the computation of potentials and forces is spent inside the solvent. In addition, such an explicit treatment does not really work in the more static application described above: applying the force field to a snapshot of the system will not produce a very meaningful result, as the dynamics cannot be easily neglected. Instead, the solvent is sometimes treated as a structureless continuum—basically, a polarizable vacuum—that permeates even the protein. Then, Coulomb interactions are just scaled by the inverse dielectric constant of this medium. While this approach is computationally very efficient, as water molecules are no longer represented explicitly, it is also very inaccurate, as the assumption of a medium permeating the solute is not applicable in the case of a solvent as polar as water.

A more accurate implicit representation of solvent effects on the electrostatic interaction is provided by what is known as the *cavity model.* Here, the protein is modeled as a cavity of low dielectric constant—often taken in the range of 2–4 due to polarizable covalent bonds (Gilson and Honig 1986)—embedded in a medium of high dielectric constant, typically taken to be ~78 in the case of water (Malmberg and Maryott 1956; Uematsu and Frank 1980). James Clerk Maxwell's laws of electrodynamics (Maxwell 1865; Jackson 1962) are then used—in their macroscopic formulation to treat solvent and solute polarizability—to set up a system of differential equations and corresponding interface conditions for the potentials of interest. Here, we want to quickly recapitulate the most important aspects of this theory for the case of biomolecular systems.

In their full generality, Maxwell's laws allow us to determine the electric and magnetic fields—or, respectively, their potentials—for a given system of charges and currents. On the microscopic level, that is, treating all charges and currents in the system explicitly, they take the form

$$\begin{aligned} \nabla \cdot E &= \frac{\rho}{\varepsilon_0} & \nabla \cdot B &= 0 \\ \nabla \times E &= -\frac{\partial B}{\partial t} & \nabla \times B &= \mu_0 j + \varepsilon_0 \mu_0 \frac{\partial E}{\partial t} \end{aligned} \qquad (7.1)$$

Maxwell's laws also allow us to derive certain *interface conditions* that hold everywhere in space:

$$(E_2 - E_1) \cdot \hat{n} = \sigma \qquad (E_2 - E_1) \times \hat{n} = 0$$

When working with proteins, currents and magnetic fields are typically neglected, and only those parts of Maxwell's equations are used that pertain to static electric charges, that is, the laws of electrostatics. The charge distribution in Equation 7.1 is typically taken to be a collection of point charges, that is, a sum of Dirac δ-distributions, weighted by the charge's value. This so-called *partial charge model*, however, is not trivial to set up. In a first approximation, point charges are centered at the locations of the nuclei in the system, where the charge value is composed of the positive nuclear charge and a fraction of the electronic negative charges in the system, depending on the electronic probability distribution around that atom. Representing the quite localized nuclear charges in this way is typically unproblematic, but the discrete representation of the continuous negative charges to model the more complex electronic wave functions is more challenging. For small molecules, such as the solvent, partial charge values are often derived by computing multipole moments or electrostatic potentials at select points in space and fitting the partial charge values on the molecule to reproduce them as accurately as possible. For water, for instance, the dipole moment is very dominant and can be reproduced by suitable placement of partial charges on the oxygen and the hydrogens. For larger systems, such as a whole protein, the necessary computations are typically infeasible—otherwise, we would not need to resort to classical approximations of the molecular physics anyhow. Here, the polymeric character of proteins (and of DNA and RNA) helps by computing such values for the monomers individually, leading to a set of, for example, 20 partial charge assignments for the 20 different amino acids.

But this simple model neglects that the charge distributions in a real system will react to their environment. Some of these effects, such as the (de-)protonation of acidic and basic residues according to the environmental pH value, can be predicted relatively accurately and taken into account explicitly. Other, more subtle effects, such as the rearrangement of electronic wave functions to react to their environment, will render the predicted partial charges inaccurate.

In the biomolecular setting, we need to account for two different forms of such *polarization* effects: *orientational* polarization, which arises from molecular rotations, and *electronic* polarization, which arises from internal electronic rearrangements. The relative importance of these two effects differs dramatically between solute and solvent. In the protein, steric constraints and internal stabilizing interactions severely restrict the ability of charged groups to rotate to accommodate an applied electric field. In the solvent, on the other hand, the situation is quite different: the natural solvent of proteins is water, which is a small, and hence mobile, molecule with a pronounced dipole moment of ~1.85 D (Gregory et al. 1997).

In the bulk, that is, without further external influences, water dipoles will arrange in a way that, on average, the resulting electrostatic field vanishes. But the presence of an external electric field, such as the one generated by a protein embedded in water, will induce the dipole moments of the surrounding water molecules to reorient parallel to it (Böttcher 1973). This reorientation will not be perfect, due to random thermal movement and solvent–solvent interactions (more details will be given below). Still, it will lead to an additional field that is superimposed onto the solute's, screening all electrostatic interactions as compared with a biomolecule in vacuum.* As a consequence, a dipolar medium such as water has a crucial influence on how an immersed biomolecule is visible to its environment.

Treating this effect in a microscopic fashion using Maxwell's equations would be impossibly complex in practice. Instead, protein electrostatics typically relies on a

* In addition to this bulk response, water can also play a site-specific ligand-like role in proximity to the solute (Smolin et al. 2005; Damjanović et al. 2007).

continuum description, Maxwell's well-known macroscopic electrostatic theory, which extends the classic microscopic theory by macroscopic material equations. These typically treat the medium as a homogenous featureless continuum—known as the assumption of *locality*—and average its screening effect. More specifically, the material equations capture the net changes of the medium's dipole fields owing to the introduction of the solute's charge distribution. Neglecting molecular interdependencies and free ions, we can expect the medium's dipole moments to align with the solute's field, implying that neighboring molecules within a small patch of space are oriented alike. This then yields the *orientational polarization* of the medium and allows averaging of the dipole moments over macroscopically small volumes, leading to the *macroscopic electric polarization vector* $\boldsymbol{P}(\boldsymbol{r})$, which represents the average dipole moment per unit volume at position $\boldsymbol{r}$. Using the polarization vector, the electric component $\boldsymbol{E}(\boldsymbol{r})$ of the electromagnetic field can be expressed in terms of a new macroscopic quantity, the *dielectric displacement field* $\boldsymbol{D}(\boldsymbol{r})$:

$$\boldsymbol{D}(\boldsymbol{r}) = \varepsilon_0 \boldsymbol{E}(\boldsymbol{r}) + \boldsymbol{P}(\boldsymbol{r})$$

with the so-called *vacuum permittivity* ε_0. The displacement field replaces the electric component of the electromagnetic field in the microscopic Maxwell equations to form its macroscopic counterpart.

Obviously, the quality of the macroscopic approximation vitally depends on the material equations chosen. Two particular assumptions are usually being made, the first one concerning the relationship between the electromagnetic and the dielectric displacement field (linear or nonlinear), and the second one concerning the interdependencies of medium molecules (local or nonlocal). In the simplest case, the molecules are assumed to be completely independent of one another and the fields are coupled through a linear operator ε:

$$\boldsymbol{D}(\boldsymbol{r}) = \varepsilon\,(\boldsymbol{E}(\boldsymbol{r}))$$

In this linear local continuum description, the medium's response to the solute's electrostatic field can be described by a function that only depends on position, the so-called *dielectric function* $\varepsilon(\boldsymbol{r})$:

$$\boldsymbol{D}(\boldsymbol{r}) = \varepsilon_0 \varepsilon(\boldsymbol{r})\,\boldsymbol{E}(\boldsymbol{r})$$

Here, the spatial dependency can be used to model continuously varying dielectric responses, for example, due to changes in solvent density around a solute, but also to differentiate between spatial regions filled with different media. In the case of proteins, this would typically imply one region, the cavity, filled with protein and another one filled with water. $\varepsilon(\boldsymbol{r})$ would then take different values inside and outside the protein, where the value inside would mainly be dominated by electronic polarization, and the one outside mainly by orientational polarization. For proteins embedded in water, typical values are 2–4 inside and 78 outside the protein. However, this jump means that the dielectric function, and hence the displacement field, becomes nondifferentiable along the two-dimensional manifold describing the boundary of the molecule. Instead of interpreting, say, $\nabla \boldsymbol{D}(\boldsymbol{r})$ in the weak sense, it is customary to set up an interface problem, that is, to solve Maxwell's equations on the inside and on the outside and to couple them through the interface conditions on the boundary. We discuss numerical solution schemes to this problem later in this chapter.

The linearity assumption holds for moderate field strengths (Böttcher 1973). While this is sufficient for typical biomolecular applications, neglecting complex internal solvent interactions has a considerable impact on the quality of the macroscopic approximation. The significance of such interactions, which will lead to nontrivial correlations between the solvent's dipole moments, becomes clear when considering the fact that water, despite being electrically neutral as a whole, is highly structured and tends to interact with surrounding water molecules in order to build a dense network of hydrogen bonds. With all three of its atoms being able to form these bonds, there exist numerous different water complexes of comparable energies (Moore Plummer and Chen 1987; Lenz and Ojamäe 2005). Under the influence of an external electric field, such as the one provided by the protein, the water molecules will now have to decide to follow the guidance of that field, or to maintain their network of hydrogen bonds. Only in exceedingly rare circumstances will both be possible simultaneously, and hence the system will be frustrated, leading to a complex interplay of solvent–solvent correlations and reactions to external fields.

To model the effects of solvent–solvent correlations, *nonlocal electrostatics* assumes that the dielectric function depends not only on a single point $\boldsymbol{r}$ but also on a reference point $\boldsymbol{r}'$, representing the location of a possible interaction partner in the solvent. Obviously, the water at each point might interact with the water at any other point inside the solvent, and all of these interactions have to be accounted for. In practice, the interactions at each point propagate up to a certain length scale λ, which is usually on the order of several angstroms. Consequently, the relation between $\boldsymbol{D}(\boldsymbol{r})$ and $\boldsymbol{E}(\boldsymbol{r})$ becomes considerably more complex:

$$\boldsymbol{D}(\boldsymbol{r}) = \varepsilon_0 \int_{\boldsymbol{r}' \in \Sigma} \varepsilon(\boldsymbol{r}, \boldsymbol{r}', \lambda)\, d\boldsymbol{r}'\, \boldsymbol{E}(\boldsymbol{r})$$

where $\varepsilon(\boldsymbol{r}, \boldsymbol{r}', \lambda)$ is known as the nonlocal dielectric function of water and Σ denotes the solvent-filled exterior of the protein. While nonlocal electrostatics has been shown to lead to accurate models of molecular solvation (Hildebrandt 2005), it is also computationally and conceptually more demanding. In the following, we will hence focus on local electrostatics. For more information on nonlocal electrostatics, the interested reader is referred to Hildebrandt et al. (2004), Bardhan (2011), Kornyshev et al. (1978), Vorotyntsev (1978), and Kornyshev and Vorotyntsev (1979).

Up to now, we have neglected the influence of mobile ions inside the solvent, but in biological systems, such ions are of great importance. In principle, ions are included as a further source term in the Maxwell equations with a charge distribution ρ_{ion}. Using a discrete charge model leads to similar problems as an explicit treatment of the solvent molecules: an abundance of source terms that complicate the equations and a fundamental need to represent dynamics, as the ions are highly mobile. Instead, mobile ions are usually modeled through a potential of mean force (PMF) ansatz, where the influence exerted on a single ion by all the others in the system is treated in a mean-field manner. This naturally leads to a Boltzmann distribution for the charge distribution of the ions, resulting in the so-called Poisson–Boltzmann equation:

$$-\varepsilon_0 \nabla \varepsilon(r) \nabla \varphi(r) = \rho(r) + \sum_i e z_i n_i^0 \exp\left(-\frac{e z_i \varphi(r)}{k_B T}\right)$$

for the electrostatic potential $\varphi(\boldsymbol{r})$, where i iterates over the ion species, e is the elementary charge, z_i denotes the integral charge of species i, and n_i^0 its number density. This partial differential equation is obviously highly nonlinear, leading to difficult challenges when attempting to solve it. For small ion concentrations ($ez_i\,\varphi \ll k_BT$ everywhere), we can instead replace the Boltzmann source term by the linear term or its Taylor series, leading to the so-called Debye–Hückel (DH) equation, which, for an overall neutral ionic solution $\left(\sum_i ez_in_i^0 = 0\right)$, reads

$$-\varepsilon_0\nabla\varepsilon(r)\nabla\varphi(r) = \rho(r) - \varphi(r)\sum_i \frac{e^2 z_i^2 n_i^0}{k_BT}$$

This DH theory forms the basis for most applications of continuum electrostatics to protein systems. The fully nonlinear Poisson–Boltzmann (PB) equation, or even theories that go beyond the mean-field approach, are rarely required for biomolecular systems in natural conditions, but become very important in certain nonstandard environments (Zhang et al. 2008, 2010; Jordan et al. 2014).

Given the equations discussed in this section, protein potentials in water can be predicted in several ways (cf. Section 7.4). In the following, we consider how knowledge of the electrostatic potential allows us to study kinetic and thermodynamic aspects of protein interactions.

7.3.4 SIMULATING THE KINETICS

To simulate the kinetics of protein association, we essentially need to simulate the Brownian trajectories of the interaction partners toward the encounter complex, where the motion is biased by the electrostatic interaction. A fully atomistic simulation would again lead to an MD approach with its associated computational cost. To simplify matters, it is a common practice to ignore intramolecular degrees of freedom (i.e., molecular flexibility) and to use macroscopic continuum electrostatics instead of an explicit solvent representation. The resulting techniques are known as *Brownian dynamics* (BD) simulations. Since BD would essentially require a recomputation of the electrostatic potentials in each step—at least when the proteins are close, as influence of the low-dielectric cavities of the binding partners on their potentials can then not be ignored—practical applications typically rely on cost-effective approximations, such as the Coulombic treatment of one of the molecules.

A classical algorithm for BD simulations was proposed by McCammon and coworkers in the 1980s (Northrup et al. 1984): First, one of the binding partners is chosen as the reference, which remains fixed during the simulation. Starting conformations for the second protein are then generated by drawing a random orientation, as well as a random translation, where the translation is drawn uniformly from a spherical shell centered in the center of mass of the immobile protein. This shell is supposed to be sufficiently far away from the immobile protein that, ideally, its electrostatic potential is completely dominated by the monopole moment. From this starting point, a biased Brownian motion is then simulated for the mobile protein's translational and rotational degrees of freedom, biased by the electrostatic potential. A trajectory is terminated, if the mobile protein "escapes," that is, reaches a second, far larger, spherical shell centered around the immobile protein, or if it "finds the encounter complex." This condition can be tested in several different ways, such as atomic distances between the different proteins.

Finally, the association rate is approximated from the fraction of trajectories that led to an encounter complex versus those that escaped.

For more information about BD simulations and for alternative algorithms, the interested reader is referred to Schreiber et al. (2009).

7.3.5 SIMULATING THE THERMODYNAMICS

The quantity governing the thermodynamics of protein interactions is the *free energy of binding* ΔG_{bind}, which is composed of polar and nonpolar contributions. The polar component, that is, the electrostatic contribution to the free energy of binding, can again be decomposed into three different contributions (Jackson and Sternberg 1995): the electrostatic interaction between the two binding partners A and B, $\Delta G_{\text{int}}^{A-B}$, and the changes in polar interactions of both A and B with the solvent, $\Delta\Delta G_{\text{sol}}^{A}$ and $\Delta\Delta G_{\text{sol}}^{B}$. These latter two terms describe the change in the electrostatic contributions to the *solvation free energies* of both partners upon binding. Predicting $\Delta G_{\text{int}}^{A-B}$ can be done as follows: Choose the protein A as a reference and compute its electrostatic potential $\varphi_A(\boldsymbol{r})$ in the presence of an uncharged molecule B, modeled by a low-dielectric cavity in the water. Then, add the charges in B to the system to compute their electrostatic energy in the field of A, that is,

$$\Delta G_{\text{int}}^{A-B} = \sum_{i \in \text{Charges of B}} q_i \varphi_B(r_i)$$

where q_i denotes the value of the ith charge of B and $\boldsymbol{r}_i$ its position. The changes in solvation free energy represent the energetic effects of solvent polarization and ionic redistributions due to the presence of the binding partner. They can be similarly computed by uncharging molecule A (B), computing the reaction field potential for the resulting system, and evaluating it at the locations of the charges of A (B). The reaction field potential can be computed either directly, if the numerical solver allows it, or by computing the difference in electrostatic potentials in water and vacuum.

The nonpolar contributions to ΔG_{bind} can also be decomposed into those arising from the partial desolvation and those resulting from interactions between A and B. The solvation terms are associated with the formation of a molecule-sized cavity in the water, and are typically approximated to show a simple dependence on the volume or surface area of the cavity. The contribution to the interactions contains an enthalpic term, typically taken from force field computations, and an entropic term that is notoriously hard to predict and is often neglected, or taken to be proportional to the number of degrees of freedom frozen by the complexation.

7.4 NUMERICAL SOLUTION SCHEMES FOR PROTEIN ELECTROSTATICS

The complex geometries and large number of charges involved render analytical solutions to protein electrostatics infeasible. Several different numerical strategies are typically employed instead, each with its own advantages and shortcomings. The simplest, most efficient, but least accurate, solution technique uses a simple Coulombic ansatz of point charge potentials in vacuum or an unstructured medium, as described in our discussion of force fields.

A similarly simple, but much more accurate, approach can be used if the goal is not the prediction of the electrostatic potential per se, but rather the electrostatic contribution to the free energy of solvation. As Max Born showed in 1920, for a spherically symmetric ion with charge q and radius a (including the first *solvation shell*, i.e., the first shell of water molecules around the ion), this quantity can be solved for exactly to yield

$$\Delta G_{\mathrm{sol}} = \frac{1}{4\pi\varepsilon_0}\frac{q^2}{2a}\left(\frac{1}{\varepsilon_\Omega} - \frac{1}{\varepsilon_\Sigma}\right)$$

with the dielectric constant of the solvent, ε_Σ, and of the ion, ε_Ω (typically taken to be 1). In the *generalized Born* approach (Still et al. 1990) for the computation of ΔG_{sol}, the total free energy of the molecule is decomposed into interactions between all pairs of charges in the molecule as follows:

$$\Delta G_{\mathrm{sol}} = \frac{1}{4\pi\varepsilon_0}\frac{1}{2}\left(\frac{1}{\varepsilon_\Omega} - \frac{1}{\varepsilon_\Sigma}\right)\sum_{i,j}\frac{q_i q_j}{\sqrt{r_{ij}^2 + R_i R_j \exp\left(-\frac{g r_{ij}^2}{R_i R_j}\right)}}$$

with a positive constant g for which different values have been proposed in the literature, and the atomic "radii" R_i, R_j, which are computed by first estimating the solvation free energy of charge i (j) only in an otherwise uncharged system and then inverting Born's equation to yield a corresponding radius. Computing this energy directly would lead to the same effort as computing the solvation free energy directly without the use of the generalized Born approach. Instead, it is typically approximated in practice. For a discussion of such approximation schemes, the interested reader is referred to Wei (2013).

If the potential itself is required, or if the previous approximations are deemed insufficient, numerical solvers of different kinds can be used for the PB or DH equations. The simplest of these are *finite difference* (FD) strategies, where the differential equation is replaced by a difference equation on a finely resolved spatial grid. The first difficulty arises from the need to define the inside and outside of the protein, for which there is no canonical definition, even if radii have already been assigned to its atoms. In the simplest case, the surface of the protein is defined to be the surface of the collection of atomic spheres with these radii, known as the van der Waals surface. However, since the outside of the protein should denote the region of space where water molecules can be found, this definition is typically deemed inappropriate, as it will contain numerous small clefts inaccessible to water molecules. Instead, the solvent accessible surface (SAS) is defined by rolling a sphere with a radius compatible with that of water over the protein's atoms and tracing out the manifold traversed by its center. A much smoother surface, and the standard definition of molecular surfaces in this field, is instead given by the path traced out of the shell of this sphere, that is, the surface where a water molecule would touch the protein. This is known as the solvent excluded surface (SES).

A second problem results from the finite size of the simulated volume. In the FD method, values for the potential and suitable derivatives at the boundary of the volume need to be known to bootstrap the computations. These values are typically approximated through monopole or dipole fields of the protein, or simply set to be zero.

In principle, the differential operators could then be discretized on the grid, and the PB or DH equation could then be solved, using a vanishing ion strength on the inside and different values for the dielectric constant in the protein and in water. However, the jump in this quantity leads to numerical challenges, which can be resolved in different ways, such as through the use of immersed interface methods (IIMs) (Weggler et al. 2010; Wang et al. 2009).

Alternatively, the problem can be decomposed into two individual sets of partial differential equations on the inside and outside, which are coupled through the interface conditions. In the *finite element method* (FEM), a region of space encompassing the molecule is represented by simple geometric objects, typically tetrahedra, of finite size, where the surface of the molecule is represented by a collection of triangles, which form the bases of some of these tetrahedra. The differential equation is then cast into a weak form and solved using a Galerkin approach, using an expansion in simple basis functions defined over the tetrahedra.

Alternatively, if a fundamental solution (or Green's function) to the differential equation is known, a similar Galerkin ansatz can be formulated, using basis functions only on the surface of the molecule. In these *boundary element methods* (BEMs), Gauss's and Stokes' theorems are used to convert between differential equations in the volume and integral equations on the surface. Solving the equations yields the full Cauchy data on the surface, from which all desired potentials (including the reaction field) everywhere in space can be computed through numeric integration over the surface. In contrast to FD and FEM approaches, BEM does not need to specify values at an outer boundary.

FEM and BEM approaches typically yield more accurate solutions than FD approaches, but often come with additional computational costs, and indeed, all three of these are used in practice (e.g., Bardhan and Hildebrandt 2011; Honig and Nicholls 1995; Holst et al. 2000).

7.5 SOFTWARE PACKAGES AND DATABASES FOR THE STUDY OF PROTEIN INTERACTIONS

Experimental information about protein interaction is provided by several important databases, such as STRING (Franceschini et al. 2013) and KEGG (Ogata et al. 1999). The protein structures themselves are deposited to the Protein Data Bank (PDB) (Kouranov et al. 2006), from where they can be easily downloaded for further studies. Many different software packages are available, both free and commercially, for the study of protein–protein interactions in general, and their electromagnetic nature in particular, based on the information contained in these databases. Due to the enormous complexity of the problem, fully automated solutions are often only part of a pipeline that requires manual intervention and expertise in different stages. This is supported by molecular viewers, such as BALLView (Moll et al. 2005, 2006) Yasara (Krieger and Vriend 2014), and Maestro (Schrödinger 2014), with built-in modeling functionality, such as force field computations or PB solvers, allowing the simultaneous visual study of geometric and energetic aspects, even using modern human-computer interaction (HCI) concepts such as real-time ray tracing (Marsalek et al. 2010), augmented reality (Nickels et al. 2012), and web-based user interfaces

(Nickels et al. 2013). More specialized software packages focus on MD methods and force field applications (Case et al. 2014; Brooks et al. 2009; Bowers et al. 2006), electrostatic potentials (Holst et al. 2000; Baker et al. 2001), or the prediction of protonation states (Anandakrishnan et al. 2012). New software can be written with the help of software libraries such as BALL (Hildebrandt et al. 2010), and tools can be combined using powerful workflow packages (Berthold et al. 2007; Hildebrandt et al. 2014). Protein docking software (Kramer et al. 1999; Cosconati et al. 2010; Lyskov and Gray 2008) attempts to predict the conformation and free energy of binding for a protein–protein, protein–DNA, or protein–ligand complex, with varying degrees of success, depending on the molecules under consideration (Janin 2005). Going beyond single instances of protein interactions, whole interaction graphs or networks can be defined, and conveniently studied in software such as Cytoscape (Shannon et al. 2003).

7.6 CONCLUSION

Protein interactions are an essential aspect of life at the molecular level. Through them, reactions can be catalyzed, matter or energy transported, and a multitude of further biological functions realized. Enabling a better understanding of protein interactions is thus a crucial task, not only for basic science, but also for more practical aspects. Drugs, for example, typically work by having a small molecule exert an influence on a protein that modifies its interactions with other biomolecules. Only if we understand the involved protein–protein, protein–DNA, and protein–ligand interactions sufficiently well can we hope to design new drugs in a guided and systematic manner.

To implement the enormous variety in biological functions required for life, a similarly large diversity in the properties of protein interactions is needed, such as widely varying degrees of efficiency or specificity. These are fine-tuned through the structure and physicochemical composition of the protein: mutations are more likely to be accepted if they modify the kinetics and thermodynamics of the interactions with other binding partners in a way that is biologically advantageous.

On the physical level, the interactions between the binding partners are electrodynamic in origin. To avoid a full quantum chemical description, many aspects of these interactions are typically described through heuristic approximations by virtual classical interactions, but the electrostatic component of protein interactions remains a major component even in these force field–level descriptions. Here, electrostatics not only provides an important enthalpic contribution toward the binding free energies, but also greatly impacts the kinetics of the interaction by biasing the Brownian motion of the interaction partners toward the encounter complex.

Interestingly, the same versatility of proteins—that is, the wide diversity of structure and interactions—that renders them so crucial for life poses severe challenges for the scientist trying to model them. Physics has a long history of successes in abstracting away from the unnecessary detail toward elegant theories on, say, simplified geometries. But the complexity of protein interactions has so far defied attempts to describe them in a simple and general manner for most application scenarios. Hence, understanding the detailed electromagnetic nature of protein interactions, and thus understanding protein interactions themselves, will with all probability remain a difficult challenge for years to come.

REFERENCES

Alsallaq, R., and Zhou, H.-X. 2007. Energy landscape and transition state of protein-protein association. *Biophysical Journal* 92 (5), 1486–1502.

Anandakrishnan, R., Aguilar, B., and Onufriev, A. V. 2012. H++ 3.0: Automating pK prediction and the preparation of biomolecular structures for atomistic molecular modeling and simulations. *Nucleic Acids Research* 40: W537–W541.

Arabidopsis Genome Initiative. 2000. Analysis of the genome sequence of the flowering plant *Arabidopsis thaliana*. *Nature* 408, 796–815.

Baker, N. A., Sept, D., Joseph, S., Holst, M. J., and McCammon, J. A. 2001. Electrostatics of nanosystems: Application to microtubules and the ribosome. *Proceedings of the National Academy of Sciences of the United States of America* 98, 10037–10041.

Bardhan, J. P. 2011. Nonlocal continuum electrostatic theory predicts surprisingly small energetic penalties for charge burial in proteins. *Journal of Chemical Physics* 135 (10), 104113.

Bardhan, J. P., and Hildebrandt, A. 2011. A fast solver for nonlocal electrostatic theory in biomolecular science and engineering. Presented at Design Automation Conference (DAC), San Diego, CA, 2011 48th ACM/EDAC/IEEE.

Berthold, M. R., Cebron, N., Dill, F., et al. 2007. Knime. *Web* 1–8.

Böttcher, C. J. F. 1973. *Theory of Electrostatic Polarization: Dielectrics in Static Fields*. 2nd ed. Amsterdam: Elsevier.

Bowers, K. J., Chow, E., Xu, H. X. H., et al. 2006. Scalable algorithms for molecular dynamics simulations on commodity clusters. Presented at ACM/IEEE SC 2006 Conference (SC'06), Tampa, FL.

Brooks, B. R., Brooks, C. L., Mackerell, A. D., et al. 2009. CHARMM: The biomolecular simulation program. *Journal of Computational Chemistry* 30 (10), 1545–1614.

Case, D. A., Babin, V., Berryman, J. T., et al. 2014. Amber 14, University of California, San Francisco.

Colbourne, J. K., Pfrender, M. E., Gilbert, D., et al. 2011. The ecoresponsive genome of *Daphnia pulex*. *Science* 331 (6017), 555–561.

Cosconati, S., Forli, S., Perryman, A. L., Harris, R., Goodsell, D. S. and Olson, A. J. 2010. Virtual screening with AutoDock: Theory and practice. *Expert Opinion on Drug Discovery* 5, 597–607.

Crick, F. 1970. Central dogma of molecular biology. *Nature* 227 (5258), 561–563.

Damjanović, A., Schlessman, J. L., Fitch, C. A., García, A. E., and García-Moreno, E. B. 2007. Role of flexibility and polarity as determinants of the hydration of internal cavities and pockets in proteins. *Biophysical Journal* 93 (8), 2791–2804.

Deshaies, R. J., and Joazeiro, C. A. P. 2009. RING domain E3 ubiquitin ligases. *Annual Review of Biochemistry* 78, 399–434.

Ezkurdia, I., Juan, D., Rodriguez, J. M., et al. 2014. Multiple evidence strands suggest that there may be as few as 19,000 human protein-coding genes. *Human Molecular Genetics* 23 (22), 5866–5878.

Franceschini, A., Szklarczyk, D., Frankild, S., et al. 2013. STRING v9.1: Protein-protein interaction networks, with increased coverage and integration. *Nucleic Acids Research* 41: D808–D815.

Gabdoulline, R. R., and Wade, R. C. 1997. Simulation of the diffusional association of barnase and barstar. *Biophysical Journal* 72 (5), 1917–1929.

Gabdoulline, R. R., and Wade, R. C. 2001. Protein-protein association: Investigation of factors influencing association rates by Brownian dynamics simulations. *Journal of Molecular Biology* 306 (5), 1139–1155.

Gallicchio, E., Kubo, M. M., and Levy, R. M. 1998. Entropy-enthalpy compensation in solvation and ligand binding revisited. *Journal of the American Chemical Society* 120 (20), 4526–4527.

Garcia-Viloca, M., Gao, J., Karplus, M., and Truhlar, D. G. 2004. How enzymes work: Analysis by modern rate theory and computer simulations. *Science* 303 (5655), 186–195.

Gilson, M. K., and Honig, B. H. 1986. The dielectric constant of a folded protein. *Biopolymers* 25, 2097–2119.

Glickman, M. H., and Ciechanover, A. 2002. The ubiquitin-proteasome proteolytic pathway: Destruction for the sake of construction. *Physiological Reviews* 82 (2), 373–428.

Gregory, J., Clary, D., Liu, K., Brown, M., and Saykally, R. 1997. The water dipole moment in water clusters. *Science* 275 (5301), 814–817.

Hildebrandt, A. 2005. Biomolecules in a structured solvent: A novel formulation of nonlocal electrostatics and its numerical solution. Dissertation, Universität des Saarlandes, Saarbrücken, Germany.

Hildebrandt, A., Blossey, R., Rjasanow, S., Kohlbacher, O., and Lenhof, H.-P. 2004. Novel formulation of nonlocal electrostatics. *Physical Review Letters* 93 (10), 108104.

Hildebrandt, A., Dehof, A. K., Rurainski, A., et al. 2010. BALL—Biochemical Algorithms Library 1.3. *BMC Bioinformatics* 11 (1), 531.

Hildebrandt, A. K., Stöckel, D., Fischer, N. M., et al. 2014. Ballaxy: Web services for structural bioinformatics. *Bioinformatics* 31 (1), 121–122.

Holst, M., Baker, N., and Wang, F. 2000. Adaptive multilevel finite element solution of the Poisson-Boltzmann equation. I. Algorithms and examples. *Journal of Computational Chemistry* 21 (15), 1319–1342.

Honig, B., and Nicholls, A. 1995. Classical electrostatics in biology and chemistry. *Science* 268 (5214), 1144–1149.

Hsia, C. C. 1998. Respiratory function of hemoglobin. *New England Journal of Medicine* 338 (4), 239–247.

Hummer, G., Garde, S., Garcia, A. E., Pohorillet, A., and Pratr, L. R. 1996. An information theory model of hydrophobic interactions. *Proceedings of the National Academy of Sciences of the United States of America* 93 (17), 8951–8955.

Jackson, J. D. 1962. *Classical Electrodynamics*. New York: John Wiley.

Jackson, R. M., and Sternberg, M. J. 1995. A continuum model for protein-protein interactions: Application to the docking problem. *Journal of Molecular Biology* 250 (2), 258–75.

Janin, J. 2005. Assessing predictions of protein-protein interaction: The CAPRI experiment. *Protein Science: A Publication of the Protein Society* 14 (2), 278–283.

Jordan, E., Roosen-Runge, F., Leibfarth, S., et al. 2014. Competing salt effects on phase behavior of protein solutions: Tailoring of protein interaction by the binding of multivalent ions and charge screening. *Journal of Physical Chemistry B* 118 (38), 11365–11374.

Kornyshev, A. A., Rubinshtein, A. I., and Vorotyntsev, M. A. 1978. Model nonlocal electrostatics. I. *Journal of Physics C: Solid State Physics* 11 (15), 3307–3322.

Kornyshev, A. A., and Vorotyntsev, M. A. 1979. Model non-local electrostatics. III. Cylindrical interface. *Journal of Physics C: Solid State Physics* 12 (22), 4939–4946.

Kouranov, A., Xie, L., de la Cruz, J., et al. 2006. The RCSB PDB information portal for structural genomics. *Nucleic Acids Research* 34, D302–D305.

Kramer, B., Metz, G., Rarey, M., and Lengauer, T. 1999. Ligand docking and screening with FLEXX. *Medicinal Chemistry Research* 9, 463–478.

Krieger, E., and Vriend, G. 2014. YASARA View—Molecular graphics for all devices—From smartphones to workstations. *Bioinformatics* 30: 2981–2982.

Lenz, A., and Ojamäe, L. 2005. A theoretical study of water clusters: The relation between hydrogen-bond topology and interaction energy from quantum-chemical computations for clusters with up to 22 molecules. *Physical Chemistry Chemical Physics* 2005 (7), 1905–1911.

Lyskov, S., and Gray, J. J. 2008. The RosettaDock server for local protein-protein docking. *Nucleic Acids Research* 36: W233–W238.

Malmberg, C. G., and Maryott, A. A. 1956. Dielectric constant of water from 0 to 100 C. *Journal of Research of the National Bureau of Standards* 56 (I), 1–8.

Marsalek, L., Dehof, A. K., Georgiev, I., Lenhof, H.-P., Slusallek, P., and Hildebrandt, A. 2010. Real-time ray tracing of complex molecular scenes. In *2010 14th International Conference on Information Visualisation (IV)*, London, UK, 239–245.

Maxwell, J. C. 1865. A dynamical theory of the electromagnetic field. *Philosophical Transactions of the Royal Society of London* 155, 459–512.

McGhee, J. D., and Felsenfeld, G. 1980. Nucleosome structure. *Annual Review of Biochemistry* 49, 1115–1156.

Merz, K. M. 2010. Limits of free energy computation for protein-ligand interactions. *Journal of Chemical Theory and Computation* 6 (4), 1018–1027.

Moll, A., Hildebrandt, A., Lenhof, H.-P., and Kohlbacher, O. 2005. BALLView: An object-oriented molecular visualization and modeling framework. *Journal of Computer-Aided Molecular Design* 19 (11), 791–800.

Moll, A., Hildebrandt, A., Lenhof, H.-P., and Kohlbacher, O. 2006. BALLView: A tool for research and education in molecular modeling. *Bioinformatics* 22 (3), 365–366.

Moore Plummer, P. L., and Chen, T. S. 1987. Investigation of structure and stability of small clusters: Molecular dynamics studies of water pentamers. *Journal of Chemical Physics* 86 (12), 7149–7155.

Nickels, S., Sminia, H., Mueller, S. C., et al. 2012. ProteinScanAR—An augmented reality web application for high school education in biomolecular life sciences. In *2012 16th International Conference on Information Visualisation (IV)*, Montpellier, France, 578–583.

Nickels, S., Stockel, D., Mueller, S. C., Lenhof, H.-P., Hildebrandt, A., and Dehof, A. K. 2013. PresentaBALL—A powerful package for presentations and lessons in structural biology. In *2013 IEEE Symposium on Biological Data Visualization (BioVis)*, 33–40.

Nirenberg, M. 1972. The genetic code. In *Nobel Lectures, Physiology or Medicine 1963–1970.* Amsterdam: Elsevier, 372–395.

Northrup, S. H., Allison, S. A., and McCammon, J. A. 1984. Brownian dynamics simulation of diffusion-influenced bimolecular reactions. *Journal of Chemical Physics* 80, 1517–1524.

Ogata, H., Goto, S., Sato, K., Fujibuchi, W., Bono, H., and Kanehisa, M. 1999. KEGG: Kyoto encyclopedia of genes and genomes. *Nucleic Acids Research* 27, 29–34.

Parinov, S., and Sundaresan, V. 2000. Functional genomics in *Arabidopsis*: Large-scale insertional mutagenesis complements the genome sequencing project. *Current Opinion in Biotechnology* 11 (2), 157–161.

Pratt, L. R., and Chandler, D. 1977. Theory of the hydrophobic effect. *Journal of Chemical Physics* 67 (8), 3683.

Pratt, L. R., and Pohorille, A. 2002. Hydrophobic effects and modeling of biophysical aqueous solution interfaces. *Chemical Reviews* 102 (8), 2671–2692.

Ripoll, D. R., Faerman, C. H., Axelsen, P. H., Silman, I., and Sussman, J. L. 1993. An electrostatic mechanism for substrate guidance down the aromatic gorge of acetylcholinesterase. *Proceedings of the National Academy of Sciences of the United States of America* 90 (11), 5128–5132.

Schlosshauer, M., and Baker, D. 2004. Realistic protein-protein association rates from a simple diffusional model neglecting long-range interactions, free energy barriers, and landscape ruggedness. *Protein Science: A Publication of the Protein Society* 13, 1660–1669.

Schmid, N., Eichenberger, A. P., Choutko, A., et al. 2011. Definition and testing of the GROMOS force-field versions 54A7 and 54B7. *European Biophysics Journal* 40, 843–856.

Schreiber, G., Haran, G., and Zhou, H.-X. 2009. Fundamental aspects of protein-protein association kinetics. *Chemical Reviews* 109 (3), 839–860.

Schrödinger. 2014. *Schrödinger release 2014-4: Maestro. Version 10.0.* New York: Schrödinger.

Shannon, P., Markiel, A., Ozier, O., et al. 2003. Cytoscape: A software environment for integrated models of biomolecular interaction networks. *Genome Research* 13, 2498–2504.

Shaw, D. E., Chao, J. C., Eastwood, M. P., et al. 2008. Anton, a special-purpose machine for molecular dynamics simulation. *Communications of the ACM* 51 (7), 91–97.

Shaw, D. E., Grossman, J. P., Bank, J. A., et al. 2014. Anton 2: Raising the bar for performance and programmability in a special-purpose molecular dynamics supercomputer. In Damkroger T., Dongorra J., and Kellenberger P. (eds.), *SC14: IEEE International Conference for High Performance Computing, Networking, Storage and Analysis*, New Orleans, LA, 41–53.

Smolin, N., Oleinikova, A., Brovchenko, I., Geiger, A., and Winter, R. 2005. Properties of spanning water networks at protein surfaces. *Journal of Physical Chemistry B* 109 (21), 10995–11005.

Still, W. C., Tempczyk, A., Hawley, R. C., and Hendrickson, T. 1990. Semianalytical treatment of solvation for molecular mechanics and dynamics. *Journal of the American Chemical Society* 112, 6127–6129.

Tan, R. C., Truong, T. N., McCammon, J. A., and Sussman, J. L. 1993. Acetylcholinesterase: Electrostatic steering increases the rate of ligand binding. *Biochemistry* 32 (2), 401–403.

Uematsu, M., and Frank, E. U. 1980. Static dielectric constant of water and steam. *Journal of Physical and Chemical Reference Data* 9, 1291–1306.

Vorotyntsev, M. A. 1978. Model nonlocal electrostatics. II. Spherical interface. *Journal of Physics C: Solid State Physics* 11(15), 3323–3331.

Wang, J., Cai, Q., Li, Z. L., Zhao, H. K., and Luo, R. 2009. Achieving energy conservation in Poisson-Boltzmann molecular dynamics: Accuracy and precision with finite-difference algorithms. *Chemical Physics Letters* 468, 112–118.

Weggler, S., Rutka, V., and Hildebrandt, A. 2010. A new numerical method for nonlocal electrostatics in biomolecular simulations. *Journal of Computational Physics* 229 (11), 4059–4074.

Wei, C. 2013. Computational methods for electromagnetic phenomena: Electrostatics in solvation, scattering and electron transport. *Contemporary Physics* 54 (5), 263–264.

Worldwide Protein Data Bank. 2015. RCSB PDB— Holdings report.

Zhang, F., Skoda, M. W. A., Jacobs, R. M. J., et al. 2008. Reentrant condensation of proteins in solution induced by multivalent counterions. *Physical Review Letters* 101 (14), 3–6.

Zhang, F., Weggler, S., Ziller, M. J., et al. 2010. Universality of protein reentrant condensation in solution induced by multivalent metal ions. *Proteins: Structure, Function, and Bioinformatics* 78 (16), 3450–3457.

8 The impact of electric fields on cell processes, membrane proteins, and intracellular signaling cascades

Trisha M. Pfluger
Juno Biomedical, Inc.
Mountain View, California

Siwei Zhao
Tufts University
Medford, Massachusetts

Contents

8.1 INTRODUCTION

Bioelectricity was first discovered in human tissues more than 150 years ago. Since then, researchers across the globe have sought to understand *how* and *why* electric signals are generated in the human body. For many years after the initial discovery of bioelectricity in the 1800s, embryogenesis, wound healing, and regenerative biology were studied from a chemotactic perspective. Research fields have evolved, however, as we begin to understand the complex spatially and temporally controlled regulatory effects of bioelectric signals in three-dimensional tissue patterning (Levin 2012). Accumulating evidence suggests bioelectric cues are predominant to chemical cues in tissue development and healing (Zhao et al. 2006; Wu and Lin 2011). Subsequently, a strong understanding of the role bioelectricity plays throughout the human life span is fundamental to understanding human physiology. Most importantly, it is imperative to the development of cutting-edge therapies for disorders affecting millions of people worldwide, including stroke, traumatic brain injury, Alzheimer's disease, diabetes, and cancer.

In this chapter, we provide an overview of the molecular mechanisms underlying both the cause of naturally occurring electric signals in human tissues and the effects bioelectric signals have on stem, progenitor, and fully differentiated cells. Our understanding of bioelectricity and electrophysiology has advanced a great deal in the past few decades. Therefore, a comprehensive review is beyond the scope of a single book chapter. As this body of knowledge continues to grow, the development of novel therapies for clinical applications advances in parallel. In this chapter, we also discuss the use of conductive polymers to convey electrical cues to cells for applications in tissue engineering and regenerative medicine.

8.2 ENDOGENOUS ELECTRIC CURRENTS IN THE HUMAN BODY

8.2.1 HISTORY OF THE DISCOVERY OF ENDOGENOUS ELECTRIC CURRENTS

An Italian physician named Luigi Galvani first discovered "animal electricity" in the late eighteenth century. Galvani performed a series of experiments on frogs. In one of these experiments, he applied an electric field (EF) to the legs of dead frogs and observed the electricity had the ability to induce twitches in their muscle tissue. Interestingly, he also found the muscle twitches continued even in the absence of applied electricity. Galvani interpreted this to mean the frog had a way to generate electricity internally.

Galvani's findings mark the beginning of the exciting and medically relevant research field known as bioelectricity (Sabbatini 1998).

Technology in Galvani's time was not advanced enough to measure small bioelectric currents. Consequently, research in the century following made very few advances toward understanding this new phenomenon. Then in the mid-nineteenth century, a German physiologist by the name of Emil Heinrich Du Bois-Reymond created his own sensitive apparatus, which he used in combination with the galvanometer created two decades earlier. Using his device, Du Bois-Reymond measured the propagation of a bioelectric current through nerves in frog legs, finding it moved at a fixed speed. Shortly thereafter, he published his findings, calling his discovery an "action current." Du Bois-Reymond's findings marked the beginning of the field of electrophysiology and this action current became known as the action potential (Sabbatini 1998).

8.2.2 ELECTRIC CURRENTS ARE GENERATED BY ACTIVE TRANSPORT OF IONS

Technology and research tools have advanced significantly since the time of Galvani and Du Bois-Reymond, leading to many new discoveries in the fields of bioelectricity and electrophysiology. The invention of the highly sensitive and noninvasive self-referencing vibrating probe (described in Reid and Zhao 2011) and other highly sensitive instruments, such as the four-point probe system, ion-selective probes, and patch clamps, has allowed scientists to detect and measure naturally occurring direct current (DC) EFs in developing embryos, in a variety of biological systems, and at sites of tissue injury (Borgens 1982; Mittal and Bereiter-Hahn 1985; Tai et al. 2009).

Bioelectricity is generated in many tissues, including skin, cornea, and brain, with an EF strength ranging from 3 to 500 mV/mm^2—levels referred to as physiological strength or physiological magnitude (Jahanshahi et al. 2014). Endogenous EFs have additionally been studied in brain development during embryogenesis (Keller 2005), where they are found to exist in the wall of the primary neural tube of vertebrate embryos. Researchers propose that these electric signals act as dominant neuronal guidance cues during craniocaudal development (Jahanshahi et al. 2014).

In a recent study by Cao and colleagues (2013), researchers used the vibrating probe and four-point probe system to measure electric currents in whole brain tissue of adult mice. The group discovered the presence of an endogenous EF along the rostral migratory stream (RMS)—a specialized migratory route for neuronal precursors that runs from the subventricular zone (SVZ) to the olfactory bulb (OB) in some mammals. Using the vibrating probe, Cao et al. (2013) measured an average inward current of -1.6 ± 0.4 μA/cm^2 located along the lateral ventricle wall and an outward current of 1.5 ± 0.6 μA/cm^2 at the surface of the OB (Figure 8.1a and b), resulting in an average field strength of 3 mV/mm along the RMS. Based on their findings, the authors concluded this endogenous EF might act as a guidance cue for neuroblast migration from the neurogenic region in the SVZ to their final location in the OB. The olfactory system is one of the few mammalian tissues known to possess the natural ability to regenerate throughout the animal's life span (Ma et al. 2014).

Cao and colleagues (2013) then sought to further investigate how this voltage across the RMS is generated. To do so, they used antibodies to label the beta 1 subunit of Na^+/K^+-ATPases and fluorescence imaging to observe the expression of these ion pumps. They found a significantly high expression located exclusively on the apical surface of the epithelial cell layer of the OB (Figure 8.1c). These data led them to

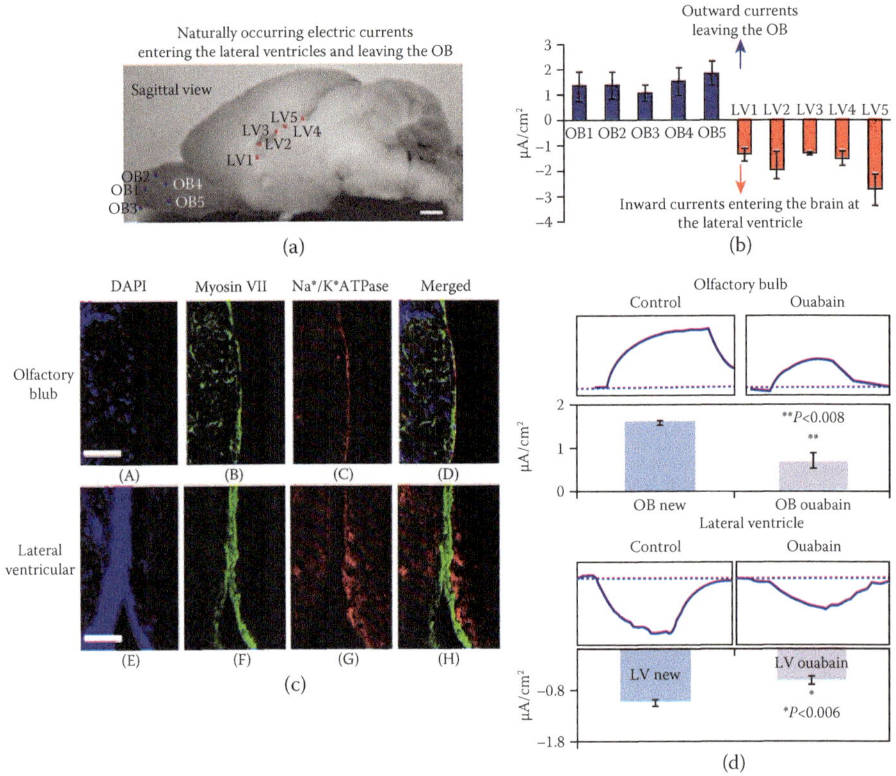

Figure 8.1 Naturally occurring electric currents flow from the SVZ to the OB. (a,b) An inward electric current of –1.6 ± 0.4 mA/cm² and an outward electric current of 1.5 ± 0.6 mA/cm² were detected at the lateral ventricular wall (LV1–5) and the surface of olfactory blub (OB1–5), respectively, with the vibrating probe system. Scale bar, 1mm. (c) Naþ/Kþ-ATPases distribute at the apical side of epithelial layer of OB (c,d), whereas those electrogenic molecules distribute at the basal side of the lateral ventricular wall (g,h). Myosin was stained in green (b,f). Nuclei blue (a,e). Scale bars, 100 mm. (d) Treatment with Ouabain significantly decreased both the inward currents at the LV and outward currents at the OB. Data were from three or more independent experiments and are presented as mean±s.e.m. DAPI, 4,6-diamidino-2-phenylindole; OB, olfactory bulb; SVZ, subventricular zone. (From Cao, L., et al., *EMBO Rep* 14: 184–190, 2013. With permission.)

conclude that the voltage drop found in the OB could be a result of Na^+/K^+-ATPases driving the outward flow of positive charges across the tissue. To confirm this, investigators treated the tissues with a Na^+/K^+-ATPase inhibitor, ouabain, to determine the effect, if any, on the generation of the bioelectric signals. Notably, they found Na^+/K^+-ATPase inhibition significantly reduced the electric current at both the OB and the ventricular wall of the SVZ (Figure 8.1d). These data led them to conclude that the measured EFs of 3–5 mV/mm found along the RMS are generated by the orchestration of the outward flow of current (cathode) at the OB and the inward flow of current (anode) at the SVZ (Cao et al. 2013). These and other data supporting a directed movement of cells along the RMS from the SVZ to the OB coincide with the data showing that the majority of cell types tested, when exposed to an EF, migrate away from the anode and toward the cathode.

Studies have begun to unravel the intricate mechanisms underlying bioelectric currents in an attempt to understand their temporal and spatial regulation. A complete understanding is especially important for medical applications, so scientists and clinicians can treat disorders using an integrative approach: understanding the physical, chemical, and electrical underpinnings to developmental disorders and disease states. As our knowledge of natural regeneration in mammalian systems expands, researchers and clinicians seek to utilize these inborn processes to enhance regeneration of central nervous tissue in patients suffering from stroke, traumatic brain injury, and neurodegenerative diseases such as dementia and Alzheimer's. For example, it is possible to enhance the body's natural ability to repair damaged tissue by applying electric current directly to areas of damage. Electrical stimulation of this nature has the potential to guide the patient's endogenous stem cells to the location of damaged tissue, or to work in combination with stem cell transplantations. Section 8.3 highlights the specific effects electric currents of physiological magnitude have on various fundamental cellular processes.

8.3 CELLULAR RESPONSES IN AN ELECTRIC FIELD

Endogenous bioelectric signals play a key role in cell function. To study this, researchers have applied an exogenous EF of physiological magnitude to multiple cell types, allowing the researchers to carefully observe and quantify the effects EFs have on cellular processes. We now know that applied EFs have profound effects on cell morphology, including shape and cytoskeletal changes, and on basic cellular processes such as survival, proliferation, polarity, migration, and differentiation (Robinson 1985; Özkucur et al. 2009; Borys 2012; Jahanshahi et al. 2013; Chang and Minc 2014; Cortese et al. 2014). Different cell types respond to external stimuli in different ways, but data from decades of research support the same general conclusion: cells have mechanisms by which to *sense* and *respond* to an EF (Li et al. 2008; Meng et al. 2012; Cao et al. 2013; Jahanshahi et al. 2013; Chang and Minc 2014; Cortese et al. 2014; Li et al. 2014). These data have been gathered from a combination of *in vitro*, *ex vivo*, and *in vivo* model systems.

Taken together, these data and data from multiple additional *in vivo* studies over the last 10 years strongly support that it is the presence and magnitude of the electric current at the location of wounding or tissue damage that is responsible for proper wound healing. Naturally following, there is also much interest in how these fundamental bioelectric processes can be harnessed and used to enhance tissue repair after cornea damage, broken bones, chronic impaired wound healing, and even neurological damage due to stroke, traumatic brain injury, or neurodegenerative disease. Multiple therapies are now being developed around the world that harness this fundamental biological process by applying exogenous currents directly to damaged tissue to increase the body's ability to repair itself after tissue damage caused by these diseases.

8.3.1 CELL ELECTROTAXIS

Small electric signals have a powerful effect on the outcome of embryogenesis, tissue maintenance, and tissue repair. These microcurrents accomplish this by carefully regulating critical cell processes seen in stem, progenitor, and differentiated cell lineages. To study this, researchers apply exogenous EFs of varying magnitudes to cell cultures or tissue slices in a spatially and temporally controlled manner using a specialized piece of equipment known as the *electrotaxis chamber* (Meng et al. 2012). Electrotaxis chambers are often used in combination with fluorescence time-lapse imaging and MATLAB® or other quantitative tracking methods to analyze cell movements during electrical stimulation (Djamgoz et al. 2001; Meng et al. 2012). Although EFs are applied externally, the same fields are found in native tissues, and thus this method is a way to recapitulate natural cell processes (Zhao 2009) and gleam insights into the specific regulatory mechanisms taking place during tissue growth and repair.

Using this technique, researchers have found that EFs of physiological strength induce directed migration toward the source of the EF in more than 15 different cell types (Nuccitelli 2003a, 2003b), with the vast majority of cell types exhibiting directional movement toward the cathode (Djamgoz et al. 2001; Meng et al. 2012; Cao et al. 2013; Chang and Minc 2014). Figure 8.2 shows an analysis of the movements of neural precursor cells (NPCs) during exposure to EFs of physiological magnitude, displaying a strong trajectory and direction of cell movement toward the cathode in an applied EF (Meng et al. 2012) (Figure 8.2). Cell electrotaxis has been extensively studied in the discovery period for more than three decades now, and thus researchers and companies around the globe are now turning their attention to investigating the existence of electrotactic movements *in vivo*.

Our ability to track the movement of cells in humans is limited, but the effects of cell electrotaxis can be measured through standard clinical assessments. Therefore, clinical studies must be carefully designed to assess the safety and efficacy of electrotaxis-based therapies. Because of this, we must look to rodent and other animal models to gleam insights about fundamental mammalian biological processes that take place in all mammals, including humans. One interesting study by Jahanshahi and colleagues (2013) (described in greater detail in Section 8.3) used bromodeoxyuridine (BrdU) to quantify neurogenesis and directed cell migration after brain electrical stimulation in adult rat models. In total, their data support not only the existence of *in vivo* electrotaxis, but also the enhancement of this process when microcurrents are applied to damaged regions of brain tissue (Jahanshahi et al. 2013). A more recent study that came out of the University of California–Davis Institute for Regenerative Cures strongly supported that the magnitude of electric currents is significantly lower

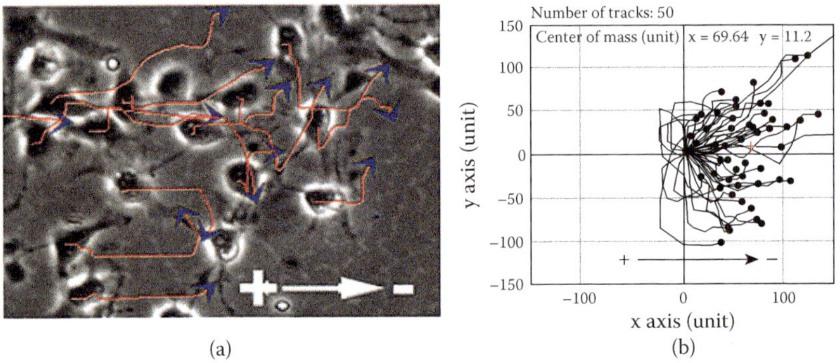

Figure 8.2 (a) NPCs show directed migration in EFs. PCs showed highly directed migration towards the cathode when exposed to EFs, red lines and blue arrows represent trajectories and direction of cell movement. (b) Migration paths of NPCs. Bar: 50 μm. (From Meng, X. et al., *J Vis Exp*. 60: 3453 (2012). With permission.)

at the edge of cornea wounds in diabetic rat and mouse, which was also significantly correlated with impaired wound healing (Shen et al. 2016). Further, researchers found that local stem and transient amplifying cells isolated from the diabetic cornea were not significantly impaired in their ability to electrotax when exposed to an exogenous electric current.

8.3.2 EFFECTS OF AN ELECTRIC FIELD ON CELL MORPHOLOGY

8.3.2.1 Cell shape and cytoskeletal changes

Cells exposed to DC EFs undergo morphological changes such as reorientation, elongation, and perpendicular alignment to the DC EF source (Hronik-Tupaj and Kaplan 2012). Intracellular calcium signaling is known to regulate signal transduction, cytoskeletal reorganization, and cell orientation and migration (Cortese et al. 2014). Thus, it is no surprise that researchers seek to understand the role of calcium and calcium-binding messenger proteins, such as calmodulin, in field-induced directional movement. Many early studies attempted to identify the role of calcium by treating cells with calcium channel blockers and depleting extracellular calcium levels. In one of these early studies, investigators stimulated mouse embryo fibroblasts with EFs up to 10 V/cm and found a field-induced increase in intracellular free calcium levels taking place concurrently with a disruption in cytoskeletal stress fiber organization, cell shape changes, and preferential cell movement toward the cathode (Onuma and Hui 1988). Furthermore, cell shape changes and cytoskeletal rearrangements were inhibited when cells were then treated with a calcium channel blocker or a calmodulin antagonist. These data led investigators to conclude that the field-induced cell shape changes and directional cell movement are calcium dependent, caused by an influx of calcium ions across the cell membrane (Onuma and Hui 1988).

Although the exact role of calcium in cell shape changes and cytoskeletal rearrangements remains unclear, it is generally accepted that EF-induced membrane polarization and ion channel activation contribute to the cytoskeletal disassembly through ion efflux (Mycielska and Djamgoz 2004; Borys 2012). Multiple studies have reported that cells migrating toward the cathode have a high calcium concentration at the leading (cathodal) edge of the cell (Bray 2001; Guo et al. 2010; Borys 2013). Additionally, studies have reported that following calcium level increases, the leading edge contracts to propel the cell toward the cathode, followed by eventual depolarization (opening) of the voltage-gated calcium channels (VGCCs) and influx of calcium near the cathodal side of the cell (Mycielska and Djamgoz 2004; Gao et al. 2011).

In a more recent study, researchers exposed both calvarial osteoblasts and SaOS-2 cells to DC EFs ranging from 1 to 14 V/cm for 5 mins and found that cells elongated and oriented perpendicular to the EF vector in a time- and voltage-dependent manner (Özkucur et al. 2009). Both cell types displayed electrotactic ability, but the calvarial osteoblasts migrated toward the cathode and the SaOS-2 cells migrated toward the anode. Investigators proposed this could be due to the fact that initiation of the local rise in intracellular calcium levels existed on opposite sides of the two cell types. Furthermore, they found that weaker EFs induced a steady-state level of intracellular calcium, whereas stronger EFs triggered a rapid increase in intracellular calcium levels. To confirm their conclusion that the cytoskeletal rearrangements were coupled with ion movement, investigators then blocked VGCCs and observed the effects. They found inhibited cells displayed a significant reduction in the peak magnitude of calcium levels in both cell types and an absence of cytoskeletal rearrangements that occur in the osteoblasts during EF-induced morphological changes (Özkucur et al. 2009).

Additional studies should be conducted to fully elucidate the involvement of calcium ion movement in the cytoskeletal rearrangements and cell shape changes seen during cell electrotaxis. However, taken together, the data above and data from additional studies in this field strongly support that the movement and changes in cellular concentration of calcium ions play a key role during cell orientation, elongation, and alignment when exposed to an EF. Further, the data indicate VGCCs could additionally play a key role during cell electrotaxis.

8.3.3 EFFECTS OF AN ELECTRIC FIELD ON OTHER CELLULAR PROCESSES

8.3.3.1 Neurogenesis and proliferation

It is also well known that endogenous EFs are critical for normal nervous system development (McCaig et al. 2005; Li et al. 2014), with disruption of these bioelectric signals resulting in failure of a normal nervous system to form (McCaig et al. 2005). Consequently, a strong understanding of cell neurogenesis and proliferation is critical to the understanding of nervous system development and regeneration. For years, scientists adhered to the dogma that the mammalian central nervous system was incapable of undergoing neurogenesis and cell proliferation. However, many data now point to the contrary, and thus we now know this is not true. Studies have shown that both neurogenesis and proliferation, in addition to widespread neural migration, including to the cerebral cortex, take place in the adult mammalian brain (Hatten 2002). These findings over the last couple decades have many implications for modern-day medicine, and because of this, therapies and technologies are being explored that can utilize the stem cells already present in the human brain to enhance repair, either in lieu of or in combination with the transplantation of exogenous stem cells.

Jahanshahi and colleagues are especially interested in how this translates to the adult mammalian brain's ability to regenerate throughout the life span. In an attempt to understand *in vivo* electrotaxis, Jahanshahi and colleagues (2013) electrically stimulated the motor cortex (MC) of adult rats, with and without induced traumatic brain injury, and using small alternating current (AC) EFs. Investigators divided subjects into three groups: low-frequency stimulation (LFS) (100 μA/electrode, 30 Hz), high-frequency stimulation (HFS) (100 μA/electrode, 330 Hz), and sham (no electrical stimulation). They performed a craniotomy on all subjects, implanted the electrodes, and stimulated the right hemisphere of the MC of the experimental groups for 4 h a day, over a period of 31 days. Using BrdU labeling and immunohistochemical staining, investigators found a significantly enhanced progenitor cell migration to the site of damaged motor cortex tissue in the groups receiving electrical stimulation, with no significant difference between the LFS and HFS groups. Investigators double-stained the tissues with BrdU, a marker of cell proliferation, and with NeuN, a marker of differentiation into neuronal cell subtypes, and found a significantly higher number of BrdU/NeuN cells in the right hemisphere of the LFS and HFS groups than in the sham group and compared with the number of BrdU-labeled cells in the left hemisphere (no stimulation) (Figure 8.3). Based on their findings, investigators concluded that electrical signals have the potential to regulate neurogenesis in the adult mammalian central nervous system

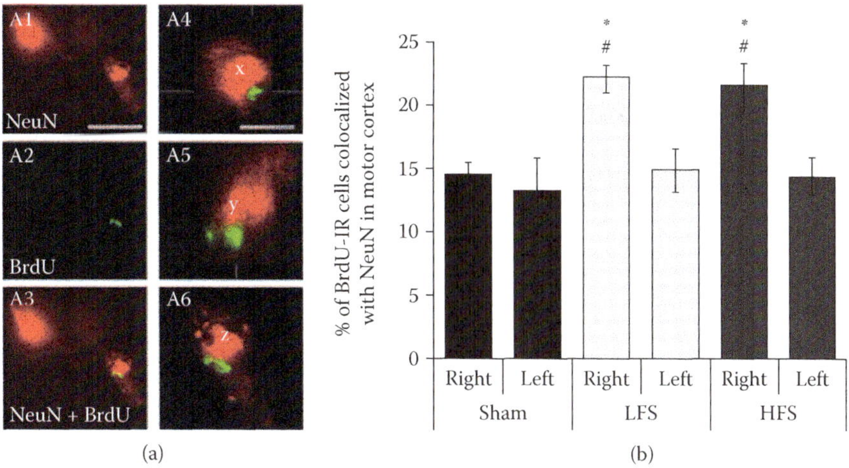

Figure 8.3 Representative high-power photomicrograph taken from the motor cortex section of a rat with high frequency stimulation of the motor cortex (MC) using a confocal microscope. (a) A double-labeled cell with BrdU and NeuN (a1–a3) (*scale bar* = 20 μm), in a three-dimensional constructed image in three anatomical planes (a4–a6) (*scale bar* = 10 μm). (b) The cumulative data on the effect of epidural stimulation of the MC at low and high frequency on neuronal differentiation in the MC. Please note the significant increase in the percentage of BrdU-containing cells, which are double labeled with NeuN in animals stimulated at both low and high frequency (*LFS* and *HFS*) in comparison with controls. In addition, there is a significant difference between stimulated (*right*) and non-stimulated sides (*left*) of the motor cortex, irrespective of the frequency of stimulation. Data are presented as mean ± standard error of the mean (SEM). $^*p < 0.05$; $\#p < 0.05$ compared to the sham and non-stimulated side, respectively. (From Jahanshahi, A., et al., *Exp Brain Res* 231: 165–177, 2013. With permission.)

(Jahanshahi et al. 2013). Additional studies report similar findings—that electrical stimulation using small, applied electric currents significantly affects cell proliferation *in vivo* (Becker et al. 2010; Li et al. 2010), a common indicator of regeneration in animal tissue.

8.3.3.2 Migration rate and directedness

It is also well documented that DC EFs induce directional cell migration. After discovering the existence of an endogenous EF along the RMS in mammalian brain tissue, Cao and colleagues (2013) additionally sought to identify if biophysical cues could be responsible for the migration of neuroblasts away from the SVZ in mice. To investigate this, researchers isolated and cultured neuroblasts from the mouse SVZ and exposed the cultured cells to EFs of 3.5 and 5 mV/mm magnitudes. Interestingly, investigators found that the EFs induced significantly increased migration directedness, with neuroblast migration further enhanced as EF strength increases. Another study in 2014 further confirmed EFs have the ability to directionally guide migration of neural precursors derived from embryonic stem cells. Their data showed that cells migrated toward the cathode, with migration directedness and displacement increasing with increased field strength (from 50 to 100 mV/mm) (Li et al. 2014). Many additional studies have demonstrated this same phenomenon (Li et al. 2008; Zhao 2009; Feng et al. 2012).

Following their *in vitro* studies, Cao and colleagues (2013) then sought to further assess the EF-guided migration of neuroblasts *ex vivo*. They injected fluorescence-labeled neuroblasts into the SVZ of mouse brain tissue, and then harvested and sliced the regions of interest. Next, they applied a 50 mV/mm EF to the tissues of interest and analyzed cell migration. Not surprisingly, their results showed that neuroblasts migrated toward the cathode in whole brain tissue slices. In order to determine if their results were artifact, they then reversed the polarity of the EF and found that the labeled injected cells migrated in the opposite direction—toward the new cathode.

8.3.3.3 Additional effects on neurons

Accumulating evidence additionally support a role of EFs in multiple neural cell processes such as growth cone alignment (McCaig et al. 2005), neurite outgrowth (Hinkle et al. 1981; Koppes et al. 2011), and axonal sprouting (Brus-Ramer et al. 2007; Sharma et al. 2009), with the extent of the effects varying based on species and cell type (Cortese et al. 2014). For example, neurites from *Xenopus* neurons grow toward the cathode in fields smaller than 10 mV/mm (Hinkle et al. 1981). By contrast, neurons from zebrafish show no response to fields up to 100 mV/mm (Cormie and Robinson 2007). In cell types that do respond to an EF, data suggest that nerve growth cone growth is voltage dependent and occurs toward the cathode (Cortese et al. 2014). Furthermore, some studies link applied DC EFs with increased production of messenger RNA expression and protein synthesis, followed by cell differentiation (Hronik-Tupaj and Kaplan 2012). EF stimulation has also been found to promote the production of molecules critical for neuronal growth and maintenance, such as brain-derived neurotrophic factor (BDNF) and other necessary growth factors (Huang et al. 2010). Further, researchers in a study by Yao and colleagues (2008) found that hippocampal neurons polarize with the leading process (growth cone) facing toward the cathode in an applied EF of 300 mV/mm.

The application of these data is critical to the development of innovative therapies to treat persons with neurological trauma, disorders, and disease from dementia, Alzheimer's, and so forth. The beneficial effects of electrical stimulation of neurons for the purpose of rebuilding brain tissue after damage continue to pile up. It is because of this that human clinical trials must be conducted in order to show not only the efficacy but also safety of these techniques. If shown safe and effective, cell electrotaxis–based therapies could open a whole new division of medical science that is currently limited to surgery and pharmaceutical use—both of which continue to show many side effects, while often only masking the symptoms of the disease. The use of electric currents in medicine and the use of regenerative and neurorestorative therapies in general offer the potential to treat the root cause of disease and thus give the patient back function lost by the disease. We have now established that cells respond to an electric current, with the current having effects on multiple cellular processes. Section 8.4 attempts to shed some light on how the cells sense electrical cues.

8.4 IMPACT OF AN ELECTRIC FIELD ON CELL MEMBRANES

We know that cells respond to electric currents in definable and even predictable ways, but we still do not understand exactly *how* this process takes place. Which part of the cell is responsible for sensing the electric current, and how is this current translated into a "language" cells understand so well? The answer to these questions appears to lie in the cell membrane.

All cells exhibit an electric potential difference across their plasma membrane. This electrical membrane potential is a result of a charge difference between the intracellular and extracellular spaces, which is dynamically regulated by ion pumps and ion channels. Membrane potentials change as cell surface proteins and ion channels are activated or inhibited. It is well documented that the immediate target of an exogenous EF is likely the plasma membrane. Researchers propose that the electrical state of cell membranes is regulated by EF-induced changes in extracellular ion distribution and ionic currents through careful regulation of cell surface proteins and ion channels (Huang et al. 2010; Cortese et al. 2014).

Although the exact mechanisms underlying EF activation and inhibition of membrane proteins are not known, researchers have elucidated this mechanism in part. Puc and colleagues (2004) report that ion flux through membrane channels may result in membrane permeability, followed by changes in cytoskeletal arrangement (Kojima et al. 1992), leading to the changes in cellular processes mentioned in Section 8.3. Furthermore, researchers hypothesize that EFs could be disruptive to cell membrane potential, and that in order to minimize these potentially disruptive effects, cells align perpendicularly to the source of the EF (Cooper and Keller 1984). As electrotaxis mechanisms have been observed to fail when the extracellular pH falls below 6, membrane charge is likely critical to the electrotactic ability of cells (Allen et al. 2013). Researchers have made great headway in elucidating the mechanisms underlying how a cell senses an EF. The above data outline a few seminal studies that strongly support the involvement of cell surface proteins and ion channels. The entire picture is likely an intricate involvement of many cell surface proteins, secondary messengers, and signaling pathways.

8.4.1 INDUCTION OF ASYMMETRICAL PROTEIN DISTRIBUTION

Membrane receptor reorganization is likely an initial and essential step in EF-guided cell migration (Cortese et al. 2014). Studies from more than 30 years ago showed EFs have profound effects on a variety of membrane receptors (Robinson 1985). Decades later, we have a better understanding of protein redistribution during EF-guided migration and the downstream effects it has on cell processes such as proliferation, migration, and differentiation.

As membrane proteins redistribute in response to exogenous EFs, negatively charged membrane proteins accumulate on the cell surface facing the cathode and positively charged proteins accumulate on the cell surface facing the anode (Cortese et al. 2014). This EF-induced redistribution of membrane proteins is hypothesized to be the primary physical mechanism involved in a cell's ability to sense an EF. For example, a study by Zhang and Peng (2011) reports that acetylcholine receptors (AChRs)—a group of membrane-bound receptors—redistribute in response to applied EFs. It is further hypothesized by some that membrane polarization could initiate downstream intracellular signaling pathways canonical to those seen during chemotaxis (Allen et al. 2013), although this is somewhat controversial.

8.4.2 ACTIVATION OF PROTEIN RECEPTORS

A few of the specific signaling cascades involved in cell electrotaxis have been pinpointed and studied in depth. Among these are epidermal growth factor receptor (EGFR), vascular endothelial growth factor (VEGF), P2Y1, and integrins. EF-induced directed cell migration requires that cells (1) receive a local extracellular cue and

(2) activate intracellular signaling pathways in an asymmetrical fashion. According to Cortese et al. (2014), EGFR is a critical component in the latter. Interestingly, evidence also supports that EGFR is expressed in high levels in highly invasive lung adenocarcinoma cells (Tsai et al. 2013) and fibroblast-like cells and human keratinocytes (Wolf-Goldberg et al. 2013). Some data are conflicting regarding the exact role of EGFR, but many studies suggest this membrane protein plays a fundamental role in EF-induced cellular responses. In addition to EGFRs, signaling molecules implicated in EF-guided cell migration include integrins, purinergic receptors, and VEGF, among others (Cao et al. 2013; Zhao et al. 2006; Jaffe and Nuticelli 1977; Ramadan et al. 2008). Furthermore, it is well documented that signaling molecules are generally activated in the cathode-facing side of the cell (Nishimura et al. 1996), which is hypothesized to be critical to the cell's guidance toward a wound center during wound healing (Zhao et al. 1999).

8.4.2.1 P2Y1

In the study by Cao et al. (2013), investigators identified that the purinergic cell surface receptor P2Y1—a receptor that is activated by adenoside nucleotides—might also be a candidate molecule in directional responses seen during EF-induced neuroblast migration. They used three cell treatments to determine the involvement of P2Y1 purinergic receptor: (1) nonspecific inhibition of purinergic receptors, (2) P2Y1-specific inhibition, and (3) P2Y1-specific knockdown using siRNA. Investigators exposed each treatment group to an EF comparable to that seen endogenously in ectodermally derived tissues, and quantified the directedness of each cell population using fluorescence time-lapse imaging and MATLAB software and found that cells in the first treatment group displayed a markedly decreased electrotactic ability. Possibly their most intriguing finding was that the electrotaxis phenotype of the cells incubated with a general purinergic receptor was virtually identical to that of the cells incubated with a P2Y1-specific inhibitor and to that of cells with siRNA knocked-down P2Y1. Taken together, these data strongly suggest that the P2Y1 cell surface receptor somehow acts as a "receiver" of the EF signal during EF-guided neuroblast migration (Cao et al. 2013) (Figure 8.4).

8.4.2.2 Integrins

It is not likely that electric signals are sensed by only one cell surface receptor. Studies from many labs interested in elucidating the EF-sensing mechanisms of cells show that multiple receptors are involved, providing complex tissues with an ability to carefully regulate EF-induced responses in a spatially and temporally controlled manner. Additionally, responses to EFs are found to be cell-type specific. For example, Pullar and colleagues (2006) identified a potential role of α6β4 integrin in the EF-guided directional migration of keratinocytes. Investigators demonstrated that by knocking out β4 integrin, keratinocytes were unable to electrotax in the absence of EGF. To confirm their findings that β4 integrin plays a role in cell electrotaxis, researchers then transfected cells to express β4 integrin and, by doing so, were able to recover the electrotaxis phenotype. Furthermore, they demonstrated a cooperative interaction of EGF and β4 integrin during EF-guided cell migration. A study a few years prior demonstrated a role of Rac1 in EGF- and β4 integrin–mediated directional migration in keratinocytes in response to an applied EF in a manner similar to that found in EGF-driven chemotaxis (Russell et al. 2003). These findings added great value to the understanding of underlying electrotactic

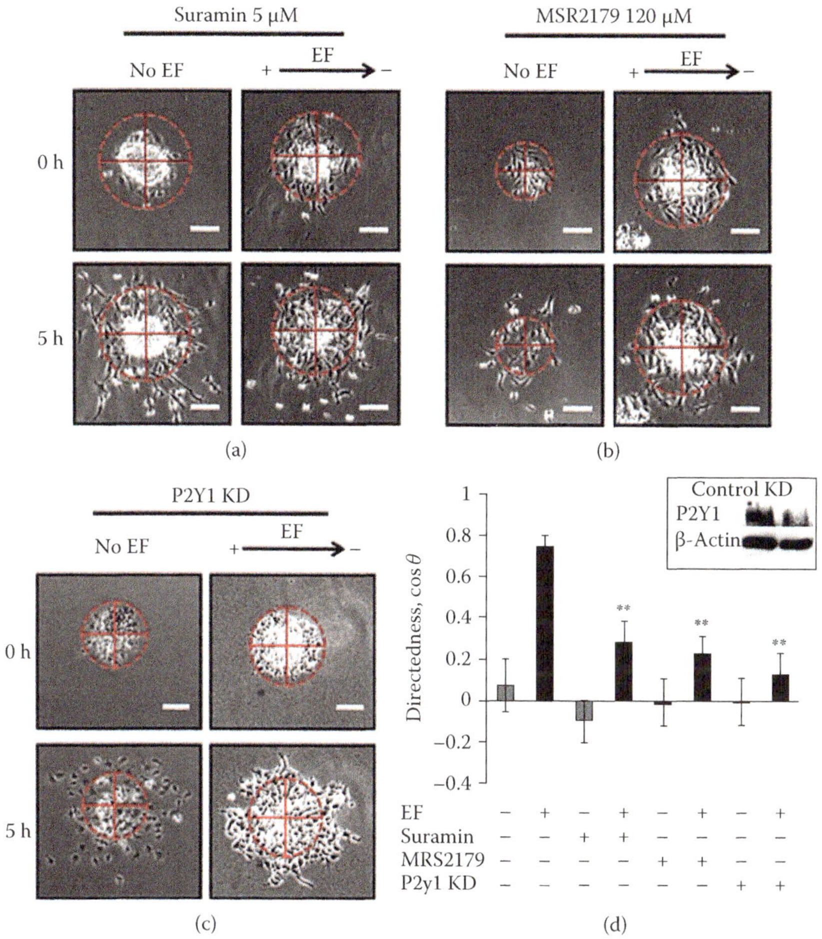

Figure 8.4 P2Y1 receptor mediated the directional migration of SVZ neuroblasts in an EF. (a) Suramin (general purinergic receptor inhibitor, 5 mM) did not affect cell motility rate. However, the direction of EF induced migration was significantly inhibited when cells were treated with 5 mM suramin. (b) A specific inhibitor of P2Y1 (MSR2179, 120 mM) significantly inhibited the directedness of neuroblasts in an EF without affecting migration rate significantly. (c) Knockdown of P2Y1 with siRNA (P2Y1 KD) significantly reduced the directedness of neuroblast migration. (d) Suramin, MSR2179 and $siRNA_{P2Y1}$ significantly reduced the directedness (Cosy) of neuroblasts migration in an applied EF. The western blot inset shows downregulation of P2Y1 expression by $siRNA_{P2Y1}$. Trajectories of cell migration are shown for 5 h. The directedness value (Cosy) is mean±s.e.m. Total number of cells n=45–55. **P<0.01 (compared with that exposed to an EF without drug or siRNA treatment with unpaired t-test). Scale bars, 100 mm. EF, electric field; siRNA, small interfering RNA; SVZ, subventricular zone. (From Cao, L., et al., *EMBO Rep* 14: 184–190, 2013. With permission.)

mechanisms, but the exact involvement of β4 integrin continues to remain elusive. Pullar and colleagues (2006) further speculate that the role of β4 integrin in electrotaxis might involve regulation of cell turning or lamellipodium stabilization during directed cell migration and suggest additional studies should be conducted to further elucidate the underlying mechanisms.

8.4.2.3 VEGF

In addition to studying the nervous system, researchers in the field of regenerative medicine are also interested in understanding how bioelectric cues can guide angiogenesis, or the growth of new blood vessels. In one of their studies, Zhao and colleagues (2004) found that VEGF plays a key role in this process. Specifically, they found that applied EFs of physiological strength induce enhanced VEGF production in muscle cells, suggesting angiogenesis might be controlled electrically through the VEGF receptor signaling pathway (Zhao et al. 2004). Zhao and colleagues are especially interested in elucidating the underlying mechanisms of angiogenesis for the purpose of developing novel therapies to treat ischemia and vascular disorders.

8.4.3 ACTIVATION OF VOLTAGE-GATED ION CHANNELS

Since the early discoveries that ion channels are involved in electrotaxis and EF-induced cell shape and cytoskeletal changes, researchers have sought to understand the potential role of voltage-gated ion channels. To investigate this, Djamgoz and colleagues (2001) treated highly metastatic MAT-LyLu cells with (1) tetrodotoxin, a voltage-gated sodium channel (VGSC) inhibitor, and (2) veratridine, a VGSC activator, and found the treatments suppressed and enhanced the electrotactic response, respectively (Djamgoz et al. 2001). Taken together, their data suggest a role of VGSCs in cell motility (Djamgoz et al. 2001). Many studies support the hypothesis that cell membrane proteins are responsible for sensing electrical cues during embryogenesis, tissue maintenance, and tissue repair. Section 8.5 takes a look at what changes within the cell can be seen after sensing an EF.

8.5 IMPACT OF AN ELECTRIC FIELD ON INTRACELLULAR SIGNALING EVENTS

Another exciting area of study in bioelectricity and electrophysiology attempts to answer the question, what happens to cells internally after they sense electrical cues? Intuitively, the answer to this question involves carefully controlled regulation of a multitude of intracellular signaling events downstream of the changes taking place at the cell membrane (Hronik-Tupaj and Kaplan 2012)—with a strong role of various phosphorylation events. In 2006, a groundbreaking study by Zhao et al. identified, for the first time, genes that modulate EF-guided keratinocyte movements during wound healing. Investigators found that EFs guide cell migration through phosphatidylinositol-3-OH kinase-g (PI(3)Kg) and phosphatase and tensin homolog (PTEN). Specifically, they found that loss of PTEN expression enhanced EF-induced Akt and ERK expression in keratinocytes (Figure 8.5a and b), which leads to increased cell migration (Figure 8.5c and d). Furthermore, they observed the effects of PTEN deficiency during EF-guided monolayer wound healing and found it significantly increased the migration of keratinocytes into the wound (Figure 8.5e and g) and away from the wound (Figure 8.5f and g) when fields were directed into the wound center and away from the wound center, respectively.

Additional studies have identified a role of PI3K/Akt activation in the proliferation and subsequent migration of NPCs (McCaig et al. 2005) and a redistribution of PIP3—a downstream effector of PI3K—in response to an applied EF (Meng et al. 2011). Furthermore, signaling cascades downstream of *N*-methyl-D-aspartate (NMDA) receptors, including the Rac1/Tiam1/Pak-1 pathways, are also reported to be involved

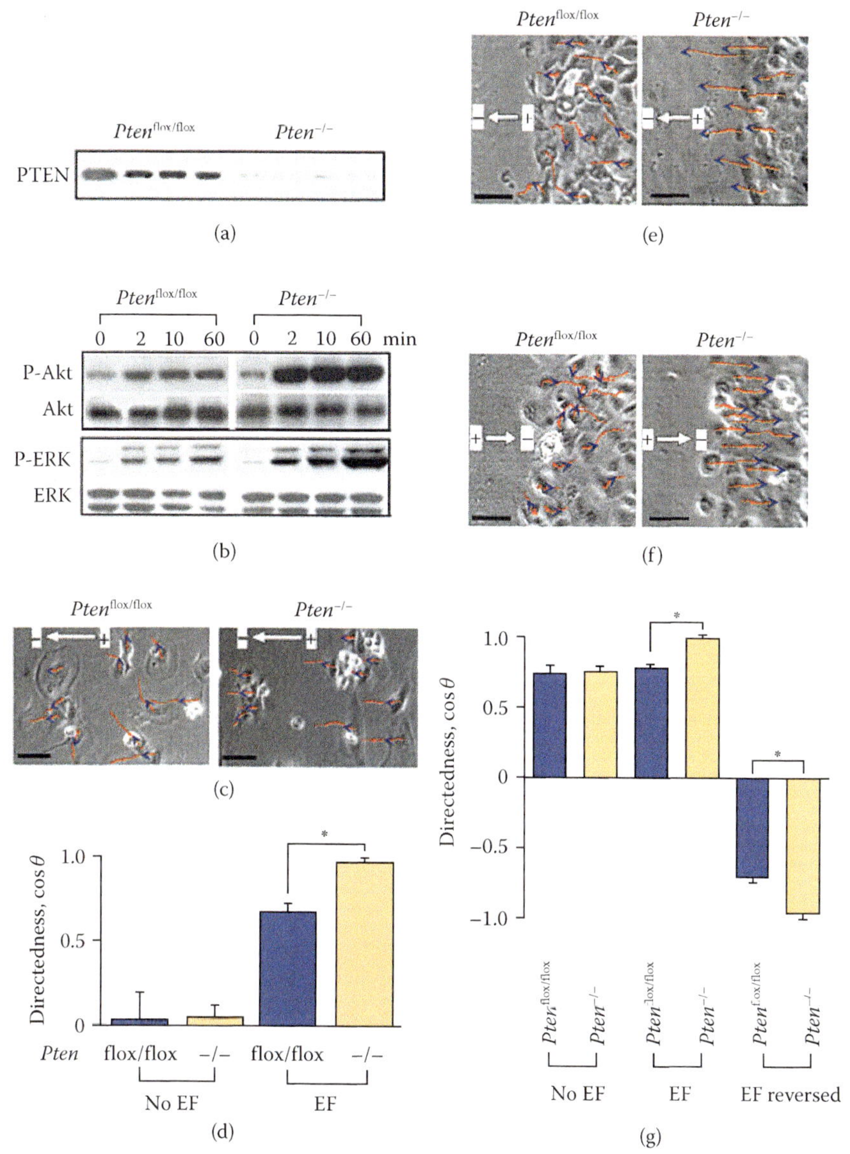

Figure 8.5 The tumour suppressor PTEN negatively regulates electrotaxis. (a) PTEN protein expression in keratinocytes. Four different cultures for each genotype are shown. (b) Loss of PTEN expression in keratinocytes results in enhanced electric field (EF)-induced activation of Akt and ERK. (c,d) Increased electrotactic migration of nonconfluent Pten-deficient keratinocytes. (e-g) Loss of PTEN increases migration of keratinocytes in monolayer wound healing experiments in response to electric fields directed into (e,g) or away from (f,g) the wound. Red lines and blue arrows represent trajectories and direction of cell movement, respectively. Data are representative of at least four independent experiments with similar results. Quantification data are the mean ± s.e.m. (d,g). Scale bars, 50 mm, EF 1/4 200 mV mm21. *P < 0.05, Student's t-test. See also Supplementary Movies 12 (for c,d) and 13 (for e–g). (Reprinted by permission from Macmillan Publishers Ltd., *Nature*, Zhao, M., et al., Electrical signals control wound healing through phosphatidylinositol-3-OH kinase-gamma and PTEN, 442 [6787], 457–460, copyright 2006.)

in EF-induced calcium-dependent cell movements (Li et al. 2008). The mitogen-activated protein kinase (MAPK) superfamily signaling cascades have also been implicated in some studies to have direct involvement in electrotaxis (Wang et al. 2003). In 2005, Pullar and Isseroff reported they demonstrated for the first time that cyclic AMP is capable of mediating electrotaxis in keratinocytes. There is no doubt that intracellular signaling cascades are involved in the electrotactic response of multiple cell types. In this chapter, we could not report the findings of every study involving the intracellular signaling events that take place during electrotaxis. Furthermore, some of the results are conflicting and the activation and inhibition of intracellular signaling cascades should be studied further to fully elucidate the responses of cells to an EF. However, intracellular signaling events are a critical piece to the larger puzzle and will undoubtedly continue to be studied for many years to come. Excitingly, as we are led further into the cell to continue to understand the intracellular events that take place during electrical stimulation, we begin to embark on other fields of study that also have direct relevance in today's medicine, such as genetic engineering and gene editing. There is no doubt we live in exciting times for medicine, especially as we begin to better understand our ability to harness and refine the use of a fundamental process as powerful as bioelectricity. Could this field be the third division of medicine, after pharmaceutical treatment and surgery? It is absolutely possible that we are embarking on the golden age of not only regenerative medicine, but also the use of electric currents to treat a range of disorders and disease, while maximizing safety and efficacy for the patient. Section 8.6 talks about one of these exciting developments in the field of medicine: the use of conductive polymers, together with electrical stimulation, to further enhance cellular processes, cell growth, and even the building of entirely new tissues after disease or damage.

8.6 USE OF ELECTRICAL STIMULATION IN COMBINATION WITH CONDUCTIVE POLYMERS IN BIOLOGICAL SYSTEMS

Electrical stimulation shows great potential for use in various medical fields. Li et al. (2008) state that applied electrical stimulation has been shown to be a safe physical approach in clinical applications and has potential to repair brain damage. At the same time, the use of conductive polymers in translational and medical research has boomed over the past decade (Mawad et al. 2012; Abidian et al. 2012; Guarino et al. 2013). Conductive polymers are a type of electroactive biomaterial that, when carefully tuned, allow for tremendous control of externally applied electrical stimulus. Furthermore, the electrical, chemical, and physical properties of conductive polymers can be tailored for specific applications. Balint et al. (2014, p. 2341) claim that "considering the vast amount of new possibilities conductive polymers offer, we believe they will revolutionize the world of tissue engineering."

Conductive polymers are often used to coat metal electrodes in order to improve the tissue–electrode interface with long-term stability while also reducing cytotoxicity and potential immune responses (Green et al. 2008; Zhou et al. 2010). A conductive polymer–coated electrode can be either a stimulating electrode or a sensing electrode. In the first case, the electrode is used as a means to apply electrical stimulation to living cells or tissues to impact their cellular activities. In the second case, the electrode is designed to monitor the status of the cell or tissue samples. In both scenarios, an interface between conductive polymer electrode and the biological sample has to be established. Furthermore, this interface has to be properly designed and conditioned for proper function of the electrode. In this section, we are particularly interested in the use of conductive polymers to stimulate biological tissues and systems. We discuss how the electrical signal is transmitted to the biological sample, what happens at the interface, and how biological samples respond to polymer-coated electrode stimulation. We also provide a practical guide to the design of the stimulation system, namely, the equivalent circuit method.

8.6.1 PHYSICAL EFFECTS INVOLVED IN THE USE OF CONDUCTIVE POLYMERS FOR STIMULATION

8.6.1.1 Charge delivery at electrode interface

In order to stimulate a biological sample, the external circuit needs to deliver charges to the biological sample through the electrodes. This can be achieved by either interfacial electrochemical reactions or the charging and discharging of the interfacial capacitor (Zhou et al. 2010; Martinsen and Grimnes 2011).

In the case of interfacial electrochemical reaction (or faradaic reaction), the conversion between electron current and ion current happens at the interface between the stimulating electrode and the biological solution through the redox reactions of the electrode materials or electrolytes. For example, if we apply a DC stimulation to normal saline (0.9% NaCl solution) through a pair of Ag/AgCl electrodes, we will find that the AgCl coating on the cathode gradually decomposes. Furthermore, if the stimulation is applied over a prolonged period of time, the AgCl coating will be eventually depleted. On the anode side, however, more AgCl is generated at and coated on the electrode surface. In this example, the AgCl decomposition and generation are the major chemical reactions happening at the electrode–electrolyte interface, which allow the current to pass with little resistance. In another example, the Ag/AgCl electrodes are replaced with Pt electrodes and a DC stimulation is applied again. When using this setup, gas bubbles are found to form on both electrodes. This is caused by the water decomposition, which generates H_2 gas on the cathode and O_2 gas on the anode. Furthermore, this process is accompanied by a change in pH around both electrodes: with a more basic pH around the cathode and a more acidic pH around the anode.

The second mode of charge delivery is the charging and discharging of the interfacial capacitor. As an electrode is immersed in the electrolytic solution, a thin layer of charged ions is absorbed from the solution to the surface of the electrode via chemical interactions. Additionally, a loose layer of counterions is attracted to the interface electrically by the first thin layer. This double-layer structure forms an interfacial capacitor, called a double-layer capacitor, as shown in Figure 8.6. It can be charged and discharged by an AC stimulation, which further induces an AC field inside of the electrolytic solution. Since this mechanism only involves charge oscillation with no electrochemical reaction at the electrode, the adverse effects associated with DC stimulation, such as pH change and gas bubble generation, do not exist.

8.6.1.2 Electrode polarization

Electrode materials can be divided into two categories: polarizable electrodes and nonpolarizable electrodes. Ideal polarizable electrodes do not permit charge transfer at the electrode–electrolyte interface, so they can be considered capacitors. On the

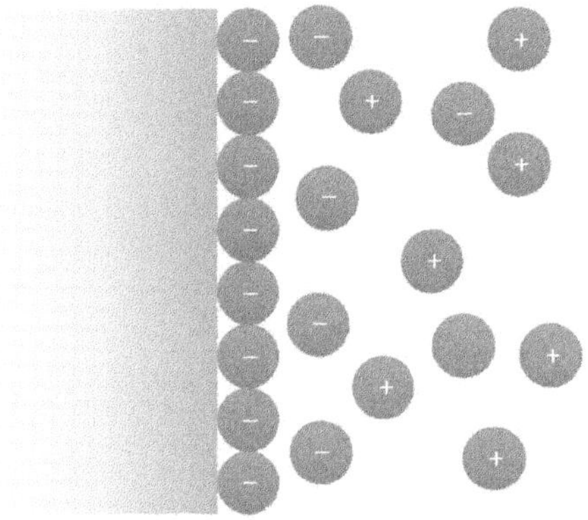

Figure 8.6 Electric double layer model consisting a thin layer of absorbed charges on electrode surface and a loose layer of counterions (diffusion layer) attracted electrically to the absorbed layer.

contrary, ideal nonpolarizable electrodes allow free charge transfer at the interface, and thus they are normally modeled as resistors. According to the definition, ideal polarizable electrodes can only be used for charge delivery through a charging and discharging mechanism, while the ideal nonpolarizable electrodes only allow charge transfer through faradaic reaction. Gold is a good example of polarizable electrode, while Ag/AgCl is a typical nonpolarizable electrode. For certain electrode materials, the polarizability depends not only on the material itself, but also on the medium with which the electrode is interfacing. For example, noble metal Pt is a polarizable electrode when it interfaces with normal saline solution. However, it acts as a nonpolarizable electrode when immersed in hydrogen gas and hydrogen ion–containing medium (Paunovic and Schlesinger 2006). Conductive polymer electrodes are normally less polarizable than their metal counterparts when interfacing with aqueous medium (Sasso et al. 2010).

8.6.1.3 Selection of proper electrodes for electrical field application

For electrical stimulation in biological environments, selection of the electrode material becomes critically important. Ideally, electrical energy should be delivered to the biological sample with minimal attenuation, time lag, side effects, and toxicity. The signal attenuation is highly dependent on the electrical conductivity of the electrode, which can be further divided into two parts: bulk conductivity of the electrode materials (i.e., conductivity of the metal and the conductive polymer) and interfacial conductivity of the electrode–electrolyte interface. On average, the bulk conductivity of conductive polymers is lower than that of metal by two to three orders of magnitude. It is also strongly dependent on the selection of dopant molecules. For example, it is reported that polypyrrole (PPy) has a bulk conductivity of 100 S/cm when doped with sodium *para*-toluenesulfonate (pTS) (Khalkhali 2005). However, the conductivity drops to 18 S/cm if the dopant is switched to polystyrene sulfonate (PSS) (Green et al. 2007).

Interfacial conductivity is largely controlled by the interfacial reaction and the surface conditions. As aforementioned, the surface of a polarizable electrode can be modeled as a capacitor, which means a larger surface always leads to higher interfacial conductance. Therefore, creating surface roughness can effectively increase the current injection. Furthermore, the interfacial conductivity depends on signal frequency, which is related to the capacitive impedance in the AC range.

In addition to electrical properties, biocompatibility and long-term stability are also important factors to consider when designing electrodes that interface with biological environments. Noble metals, such as platinum, are highly biocompatible and stabile when used for long-term stimulation. Recently, the biocompatibility of conductive polymers, especially PPy, has been extensively studied. Results show that conductive polymer electrodes support *in vitro* cell survival for up to 6 months (Wang et al. 2004). Furthermore, the *in vivo* tests show that PPy electrodes do not induce significant inflammatory response (Wang et al. 2004). When used for long-term stimulation, it is important to ensure strong adhesion between the electrode and the underlying substrate. A small adhesive molecule (such as thiol) can be used to provide better bonding between conductive polymers and the underlying metal seed layer (Smela 1998). Roughening the metal seed surface can also improve the adhesion strength of the polymer layer (Liu et al. 2007).

8.6.2 RESPONSE OF BIOLOGICAL SYSTEMS TO ELECTRICAL STIMULATION

8.6.2.1 A physical view

Biological systems are water-based ionically conductive systems due to electrolytes. Electrolytes, such as Na^+, K^+, Ca^{++}, Cl^-, and ionized biomolecules, are found in abundance in biological systems. The conductivity of biological systems is on the order of 1 S/m, which remains constant when the external stimulating frequency is below 10 MHz (Cooper 1946). At low frequency or DC stimulation, the resistance of an electrolytic solution can be easily determined by Ohm's law: $R = \rho L/A$, where R is the resistance, ρ is the resistivity of the solution, A is the cross-sectional area of the solution, and L is the length of the solution. The resistivity is dependent largely on electrolyte concentration, but is also related to other factors, such as temperature. At higher frequency, the biological solution tends to store energy instead of dissipating the energy, and thus its response to electrical stimulation becomes more capacitive than resistive. For example, simple saline solution (without any polar biomolecules) is considered a good conductor at low frequency but starts exhibiting capacitive properties when the stimulating frequency exceeds 250 MHz (Martinsen and Grimnes 2011). Large biomolecules, such as proteins, behave more like a dipole. They are not able to migrate fast enough to respond to the changing EF vector. So at HFS, they orient themselves with the EF (polarization), instead of migration. Lipids are the major building block of the membrane structure of biological systems. Since they are amphiphilic small molecules, they can form monolayer or bilayer membranes in an aqueous biological system. The lipid membrane can be modeled as a capacitor because it does not normally allow current to pass through. However, a leaking resistance might need to be considered when modeling the activities of the ion and water channels on the cell membrane. Lipid membranes are extremely thin (several nanometers in thickness), and thus the unit area capacitance of the lipid membrane can be very high. Although a DC signal is not allowed to pass the membrane, a high-frequency AC signal may be able to pass. The ability of AC stimulation to penetrate the membrane structure is determined by the capacitive impedance equation, which can be expressed as $Z = 1/(j\omega C)$, where Z is the capacitive impedance, ω is the angular frequency of the signal, C is the capacitance of the lipid membrane, and j is the imaginary unit. Under normal conditions, cells carry negative charges due to the carbohydrates on the cell membrane and on the proteins. This is the reason that free-floating cells repel each other and flow toward the anode in a DC electrical field. The electrical properties of tissue and organs are more complicated. Tissues and organs are highly heterogeneous, consisting of extracellular fluid, membranes, and intracellular fluid. The interfacial phenomenon also plays an important role in the electrical behavior except for the bulk properties. The specific tissue electrical properties have been extensively studied. For further details, please refer to the reviews by Foster and Schwan (1989) and Gabriel et al. (1996).

8.6.2.2 Equivalent circuit model

Although biological systems are complex, continuous, and largely nonlinear in their electrical properties, it is quite helpful to model them using equivalent electric circuits, even though these models only consist of linear, discrete, and passive elements. The use of equivalent circuit model simplifies the electrical stimulation distribution calculations in biological systems, and helps to obtain an approximation of the biological response and guide the system design.

In the equivalent circuit method, we consider electrolyte solutions—such as the interstitial fluid and the intracellular fluid—a resistor. The interface that does not allow charge to pass is considered a capacitor, such as the nonpolarizable electrode and the cell membrane. As a simple example, a uniform cell suspension in culture medium is modeled using an equivalent circuit (Figure 8.7). C_{mem} is the capacitance of the cell membrane, R_{in} is the resistance of the intracellular fluid, and R_{ex} is the resistance of the extracellular fluid. Human blood closely resembles this example. The distribution of the electrical signal can thus be calculated using the impedance information and the voltage division rule.

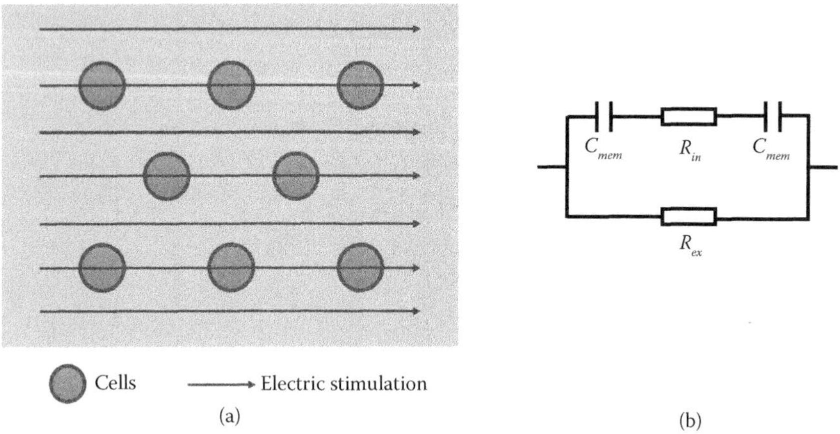

Figure 8.7 (a) Uniform cell suspension in culture medium and (b) its equivalent circuit model.

It should be noted that the electrical properties of the tissue are very specific to the tissue type and also can be time dependent. Different tissue type, from the multi-layer human skin to the anisotropic muscle, from the gas-filled lung to the mineralized bone tissue, can have vastly different electrical properties. A pulsing heart has a different volume during systole and diastole, which can also affect its electrical properties. Therefore, these factors have to be carefully considered when developing equivalent circuit models for tissues.

8.6.3 USE OF CONDUCTIVE POLYMERS FOR NEURAL STIMULATION

The use of conductive polymers as the electrical interface for neural stimulation has recently attracted broad attention (Green et al. 2008; Zhou et al. 2010). Conductive polymers are normally coated on metal microelectrodes (also called a seed layer) as a thin layer. This is accomplished through electrochemical polymerization. The commonly used conductive polymer materials for neural stimulation include PPy and poly(3,4-ethylene-dioxythiophene)/PSS (PEDOT/PSS).

The use of conductive polymer–coated electrodes has two major benefits. First, it introduces rough surface morphology, which increases the surface area of the electrode and thus decreases the interfacial impedance. For example, D. C. Martin et al. coated the metal microelectrode with a PEDOT/PSS conductive polymer layer and found that the rough surface morphology lowers the electrode impedance by two orders of magnitude (Cui and Martin 2003). Second, the conductive polymer can be decorated with functional biomolecules, which renders the electrode surface more bio-friendly and integratable with the biological sample. In research report by Cui and Martin (2003), two laminin fragments were used to dope the electropolymerized PPy electrodes. The peptide-modified PPy neural interface not only showed reduced interfacial impedance compared with gold electrode, but also supported better neuron cell culture with significantly longer primary neuritis (Stauffer and Cui 2006).

Conductive polymers have shown promise as stimulating electrode materials by improving the tissue–electrode interface. A better understanding of the electron transfer process at the electrode surface will be critical for the future enhancement of electrode performance. Novel fabrication and modification methods will also allow better control of the electrical properties and long-term stability of conductive polymer electrodes.

8.7 SUMMARY

Bioelectricity is generated in living systems and plays a substantial, if not predominant, role in embryogenesis, wound healing, and regeneration. In this chapter, we provided an overview of how stem, progenitor, and terminally differentiated cells can sense and respond to electrical cues. Specifically, we discussed how electric signals have the ability to alter membrane potentials, activate or inhibit cell surface receptors, and rearrange membrane proteins. We additionally highlighted data showing that once sensed by the cell, electrical fields induce changes in cytoskeletal arrangements, alterations in intracellular signaling cascades, and ultimately changes in cell processes such as polarization, migration, proliferation, and differentiation. Work by Levin (2012) takes this one step further, proposing that the regulation of cell membrane potential at the *tissue* level has a direct influence on morphogenetic processes such as embryonic patterning, tissue architecture, and organ regeneration. We are seeing in many medical fields today that electrical stimulation has great therapeutic potential in treating a range of disorders, including neurological trauma and disease. Excitingly, we are now seeing clinical studies demonstrating that electrical stimulation can improve both cognitive and motor outcomes in patients. Further, due to their electrical properties, biocompatibility, and ability to be tailored to specific applications, conductive polymers are also of great interest to those seeking to develop novel clinical therapies. When used in combination, electrical stimulation and conductive polymers have the potential to revolutionize the field of regenerative medicine by offering safe and effective ways for patients to rebuild tissue after disease or trauma.

REFERENCES

Abidian, M. R., E. D. Daneshvar, B. M. Egeland, D. R. Kipke, P. S. Cederna, and M. G. Urbanchek. Hybrid conducting polymer-hydrogel conduits for axonal growth and neural tissue engineering. *Adv Healthcare Mater* 1 (2012): 762–767.

Allen, G. M., A. Mogilner, and J. A. Theriot. Electrophoresis of cellular membrane components creates the directional cue guiding keratocyte galvanotaxis. *Curr Biol* 23 (2013): 560–568.

Balint, R., N. J. Cassidy, and S. H. Cartmell. Conductive polymers: Towards a smart biomaterial for tissue engineering. *Acta Biomater* 10 (2014): 2341–2353.

Becker, D., D. S. Gary, E. S. Rosenzweig, W. M. Grill, and J. W. McDonald. Functional electrical stimulation helps replenish progenitor cells in the injured spinal cord of adult rats. *Exp Neurol* 222 (2010): 211–218.

Borgens, R. B. What is the role of naturally produced electric current in vertebrate regeneration and healing. *Int Rev Cytol* 76 (1982): 245–298.

Borys, P. On the biophysics of cathodal galvanotaxis in rat prostate cancer cells: Poisson-Nernst-Planck equation approach. *Eur Biophys J* 41 (2012): 527–534.

Borys, P. The role of passive calcium influx through the cell membrane in galvanotaxis. *Cell Mol Biol Lett* 18 (2013): 187–199.

Bray, D. *Cell Movements: From Molecules to Motility.* New York: Garland Publishing, 2001.

Brus-Ramer, M., J. B. Carmel, S. Chakrabarty, and J. H. Martin. Electrical stimulation of spared corticospinal axons augments connections with ipsilateral spinal motor circuits after injury. *J Neurosci* 27 (2007): 13793–13801.

Cao, L., D. Wei, B. Reid, S. Zhao, J. Pu, T. Pan, E. Yamoah, and M. Zhao. Endogenous electric currents might guide rostral migration of neuroblasts. *EMBO Rep* 14 (2013): 184–190.

Chang, F., and N. Minc. Electrochemical control of cell and tissue polarity. *Annu Rev Cell Dev Biol* 30 (2014): 317–336.

Cooper, M. S., and R. E. Keller. Perpendicular orientation and directional migration of amphibian neural crest cells in dc electrical fields. *Proc Natl Acad Sci USA* 81 (1984): 160–164.

Cooper, R. The electrical properties of salt-water solutions over the frequency range 1–4000 Mc/s. *J Inst Electr Eng* 93 (1946): 69–75.

Cormie, P., and K. R. Robinson. Embryonic zebrafish neuronal growth is not affected by an applied electric field in vitro. *Neurosci Lett* 411 (2007): 128–132.

Cortese, B., I. E. Palama, S. D'Amone, and G. Gigli. Influence of electrotaxis on cell behaviour. *Integr Biol (Camb)* 6 (2014): 817–830.

Cui, X., and D. C. Martin. Electrochemical deposition and characterization of poly(3,4-ethylenedioxythiophene) on neural microelectrode arrays. *Sens Actuators B Chem* 89 (2003): 92–102.

Djamgoz, M. B. A., M. Mycielska, Z. Madeja, S. P. Fraser, and W. Korohoda. Directional movement of rat prostate cancer cells in direct-current electric field: Involvement of voltage-gated Na^+ channel activity. *J Cell Sci* 114 (2001): 2697–2705.

Feng, J. F., J. Liu, X. Z. Zhang, L. Zhang, J. Y. Jiang, J. Nolta, and M. Zhao. Guided migration of neural stem cells derived from human embryonic stem cells by an electric field. *Stem Cells* 30 (2012): 349–355.

Foster, K. R., and H. P. Schwan. Dielectric properties of tissues and biological materials: A critical review. *Crit Rev Biomed Eng* 17 (1989): 25–104.

Gabriel, C., S. Gabriel, and E. Corthout. The dielectric properties of biological tissues. I. Literature survey. *Phys Med Biol* 41 (1996): 2231–2249.

Gao, R. C., X. D. Zhang, Y. H. Sun, Y. Kamimura, A. Mogilner, P. N. Devreotes, and M. Zhao. Different roles of membrane potentials in electrotaxis and chemotaxis of dictyostelium cells. *Eukaryot Cell* 10 (2011): 1251–1256.

Green, R. A., N. H. Lovell, G. G. Wallace, and L. A. Poole-Warren. Conducting polymers for neural interfaces: Challenges in developing an effective long-term implant. *Biomaterials* 29 (2008): 3393–3399.

Green, R. A., L. A. Poole-Warren, and N. H. Lovell. 2007. Novel neural interface for vision prosthesis electrodes: Improving electrical and mechanical properties through layering. In Third International IEEE EMBS Conference on Neural Engineering, Kohala Coast, HI, vol. (2007): 97–100.

Guarino, V., M. A. Alvarez-Perez, A. Borriello, T. Napolitano, and L. Ambrosio. Conductive PANi/PEGDA macroporous hydrogels for nerve regeneration. *Adv Healthcare Mater* 2 (2013): 218–227.

Guo, A., B. Song, B. Reid, Y. Gu, J. V. Forrester, C. A. Jahoda, and M. Zhao. Effects of physiological electric fields on migration of human dermal fibroblasts. *J Invest Dermatol* 130 (2010): 2320–2327.

Hatten, M. E. New directions in neuronal migration. *Science* 297 (2002): 1660–1663.

Hinkle, L., C. D. McCaig, and K. R. Robinson. The direction of growth of differentiating neurones and myoblasts from frog embryos in an applied electric field. *J Physiol* 314 (1981): 121–135.

Hronik-Tupaj, M., and D. L. Kaplan. A review of the responses of two- and three-dimensional engineered tissues to electric fields. *Tissue Eng Part B Rev* 18 (2012): 167–180.

Huang, J., X. Hu, L. Lu, Z. Ye, Q. Zhang, and Z. Luo. Electrical regulation of Schwann cells using conductive polypyrrole/chitosan polymers. *J Biomed Mater Res A* 93 (2010): 164–174.

Jaffe, L. F., and R. Nuccitelli. Electrical controls of development. *Annu Rev Biophys Bioeng* 6 (1977): 445–476.

Jahanshahi, A., L. Schonfeld, M. L. Janssen, S. Hescham, E. Kocabicak, H. W. Steinbusch, J. J. van Overbeeke, and Y. Temel. Electrical stimulation of the motor cortex enhances progenitor cell migration in the adult rat brain. *Exp Brain Res* 231 (2013): 165–177.

Jahanshahi, A., L. M. Schonfeld, E. Lemmens, S. Hendrix, and Y. Temel. In vitro and in vivo neuronal electrotaxis: A potential mechanism for restoration? *Mol Neurobiol* 49 (2014): 1005–1016.

Keller, R. Cell migration during gastrulation. *Curr Opin Cell Biol* 17 (2005): 533–541.

Khalkhali, R. A. Electrochemical synthesis and characterization of electroactive conducting polypyrrole polymers. *Russ J Electrochem* 41 (2005): 950–955.

Kojima, J., H. Shinohara, Y. Ikariyama, M. Aizawa, K. Nagaike, and S. Morioka. Electrically promoted protein production by mammalian cells cultured on the electrode surface. *Biotechnol Bioeng* 39 (1992): 27–32.

Koppes, A. N., A. M. Seggio, and D. M. Thompson. Neurite outgrowth is significantly increased by the simultaneous presentation of Schwann cells and moderate exogenous electric fields. *J Neural Eng* 8 (2011): 046023.

Levin, M. Morphogenetic fields in embryogenesis, regeneration, and cancer: Non-local control of complex patterning. *Biosystems* 109 (2012): 243–261.

Li, L., Y. H. El-Hayek, B. Liu, et al. Direct-current electrical field guides neuronal stem/progenitor cell migration. *Stem Cells* 26 (2008): 2193–2200.

Li, Q., M. Brus-Ramer, J. H. Martin, and J. W. McDonald. Electrical stimulation of the medullary pyramid promotes proliferation and differentiation of oligodendrocyte progenitor cells in the corticospinal tract of the adult rat. *Neurosci Lett* 479 (2010): 128–133.

Li, Y., M. Weiss, and L. Yao. Directed migration of embryonic stem cell-derived neural cells in an applied electric field. *Stem Cell Rev* 10 (2014): 653–662.

Liu, Y., Q. Gan, S. Baig, and E. Smela. Improving PPy adhesion by surface roughening. *J Phys Chem C* 111 (2007): 11329–11338.

Ma, L., Y. Wu, Q. Qiu, H. Scheerer, A. Moran, and C. R. Yu. A developmental switch of axon targeting in the continuously regenerating mouse olfactory system. *Science* 344 (2014): 194–197.

Martinsen, O. G., and S. Grimnes. *Bioimpedance and Bioelectricity Basics*. Burlington, MA: Academic Press, 2011.

Mawad, D., E. Stewart, D. L. Officer, T. Romeo, P. Wagner, K. Wagner, and G. G. Wallace. A single component conducting polymer hydrogel as a scaffold for tissue engineering. *Advanced Functional Materials* 22 (2012): 2692–2699.

McCaig, C. D., A. M. Rajnicek, B. Song, and M. Zhao. Controlling cell behavior electrically: Current views and future potential. *Physiol Rev* 85 (2005): 943–978.

Meng, X., M. Arocena, J. Penninger, F. H. Gage, M. Zhao, and B. Song. PI3K mediated electrotaxis of embryonic and adult neural progenitor cells in the presence of growth factors. *Exp Neurol* 227 (2011): 210–217.

Meng, X., W. Li, F. Young, R. Gao, L. Chalmers, M. Zhao, and B. Song. Electric field-controlled directed migration of neural progenitor cells in 2D and 3D environments. *J Vis Exp* 60 (2012): 3453.

Mittal, A. K., and J. Bereiter-Hahn. Ionic control of locomotion and shape of epithelial cells. I. Role of calcium influx. *Cell Motil* 5 (1985): 123–136.

Mycielska, M. E., and M. B. Djamgoz. Cellular mechanisms of direct-current electric field effects: Galvanotaxis and metastatic disease. *J Cell Sci* 117 (2004): 1631–1639.

Nishimura, K. Y., R. R. Isseroff, and R. Nuccitelli. Human keratinocytes migrate to the negative pole in direct current electric fields comparable to those measured in mammalian wounds. *J Cell Sci* 109 (Pt. 1) (1996): 199–207.

Nuccitelli, R. Endogenous electric fields in embryos during development, regeneration and wound healing. *Radiat Prot Dosimetry* 106 (2003a): 375–383.

Nuccitelli, R. A role for endogenous electric fields in wound healing. *Curr Top Dev Biol* 58 (2003b): 1–26.

Onuma, E. K., and S. W. Hui. Electric field-directed cell shape changes, displacement, and cytoskeletal reorganization are calcium dependent. *J Cell Biol* 106 (1988): 2067–2075.

Özkucur, N., T. K. Monsees, S. Perike, H. Q. Do, and R. H. Funk. Local calcium elevation and cell elongation initiate guided motility in electrically stimulated osteoblast-like cells. *PLoS One* 4 (2009): e6131.

Paunovic, M., and M. Schlesinger. *Fundamentals of Electrochemical Deposition*. 2nd ed. Indianapolis, IN: Wiley, 2006.

Puc, M., S. Corovicm, K. Flisarm, M. Petkovsekm, J. Nastranm, and D. Miklavcicm. Techniques of signal generation required for electropermeabilization: Survey of electropermeabilization devices. *Bioelectrochemistry* 64 (2004): 113–124.

Pullar, C. E., B. S. Baier, Y. Kariya, A. J. Russell, B. A. Horst, M. P. Marinkovich, and R. R. Isseroff. beta4 integrin and epidermal growth factor coordinately regulate electric field-mediated directional migration via Rac1. *Mol Biol Cell* 17 (2006): 4925–4935.

Pullar, C. E., and R. R. Isseroff. Cyclic AMP mediates keratinocyte directional migration in an electric field. *J Cell Sci* 118 (2005): 2023–2034.

Ramadan, A., M. Elsaidy, and R. Zyada. Effect of low-intensity direct current on the healing of chronic wounds: A literature review. *J Wound Care* 17 (2008): 292–296.

Reid, B., and M. Zhao. Measurement of bioelectric current with a vibrating probe. *J Vis Exp* (2011): 1–6.

Robinson, K. R. The responses of cells to electrical fields: A review. *J Cell Biol* 101 (1985): 2023–2027.

Russell, A. J., E. F. Fincher, L. Millman, R. Smith, V. Vela, E. A. Waterman, C. N. Dey, S. Guide, V. M. Weaver, and M. P. Marinkovich. Alpha 6 beta 4 integrin regulates keratinocyte chemotaxis through differential GTPase activation and antagonism of alpha 3 beta 1 integrin. *J Cell Sci* 116 (2003): 3543–3556.

Sabbatini, R. M. E. The discovery of bioelectricity. *Brain Mind Mag* 2 (1998) Available at: http://www.cerebromente.org.br/n06/historia/bioelectr_i.htm.

Sasso, L., P. Vazquez, I. Vedarethinam, J. Castillo-Leon, J. Emneus, and W. E. Svendsen. Conducting polymer 3D microelectrodes. *Sensors (Basel)* 10 (2010): 10986–11000.

Sharma, N., L. Coughlin, R. G. Porter, L. Tanzer, R. D. Wurster, S. J. Marzo, K. J. Jones, and E. M. Foecking. Effects of electrical stimulation and gonadal steroids on rat facial nerve regenerative properties. *Restor Neurol Neurosci* 27 (2009): 633–644.

Shen, Y., T. Pfluger, F. Ferriera, J. Liang, M. F. Navedo, Q. Zeng, B. Reid, and M. Zhao. Diabetic cornea wounds produce significantly weaker electric signals that may contribute to impaired healing. *Nat Sci Rep* 6 (2016): 26525.

Smela, E. Thiol-modified pyrrole monomers. 4. Electrochemical deposition of polypyrrole over 1-(2-thioethyl)pyrrole. *Langmuir* 14 (1998): 2996–3002.

Stauffer, W. R., and X. T. Cui. Polypyrrole doped with 2 peptide sequences from laminin. *Biomaterials* 27 (2006): 2405–2413.

Tai, G., B. Reid, L. Cao, and M. Zhao. Electrotaxis and wound healing: Experimental methods to study electric fields as a directional signal for cell migration. *Methods Mol Biol* 571 (2009): 77–97.

Tsai, H. F., C. W. Huang, H. F. Chang, J. J. Chen, C. H. Lee, and J. Y. Cheng. Evaluation of EGFR and RTK signaling in the electrotaxis of lung adenocarcinoma cells under direct-current electric field stimulation. *PLoS One* 8 (2013): e73418.

Wang, E., M. Zhao, J. V. Forrester, and C. D. McCaig. Electric fields and MAP kinase signaling can regulate early wound healing in lens epithelium. *Invest Ophthalmol Vis Sci* 44 (2003): 244–249.

Wang, X., X. Gu, C. Yuan, S. Chen, P. Zhang, T. Zhang, J. Yao, F. Chen, and G. Chen. Evaluation of biocompatibility of polypyrrole in vitro and in vivo. *J Biomed Mater Res A* 68 (2004): 411–422.

Wolf-Goldberg, T., A. Barbul, N. Ben-Dov, and R. Korenstein. Low electric fields induce ligand-independent activation of EGF receptor and ERK via electrochemical elevation of H(+) and ROS concentrations. *Biochim Biophys Acta* 1833 (2013): 1396–1408.

Wu, D., and F. Lin. A receptor-electromigration-based model for cellular electrotactic sensing and migration. *Biochem Biophys Res Commun* 411 (2011): 695–701.

Yao, L., L. Shanley, C. McCaig, and M. Zhao. Small applied electric fields guide migration of hippocampal neurons. *J Cell Physiol* 216 (2008): 527–535.

Zhang, H. L., and H. B. Peng. Mechanism of acetylcholine receptor cluster formation induced by DC electric field. *PLoS One* 6 (2011): e26805.

Zhao, M. Electrical fields in wound healing—An overriding signal that directs cell migration. *Semin Cell Dev Biol* 20 (2009): 674–682.

Zhao, M., H. Bai, E. Wang, J. V. Forrester, and C. D. McCaig. Electrical stimulation directly induces pre-angiogenic responses in vascular endothelial cells by signaling through VEGF receptors. *J Cell Sci* 117 (2004): 397–405.

Zhao, M., A. Dick, J. V. Forrester, and C. D. McCaig. Electric field-directed cell motility involves up-regulated expression and asymmetric redistribution of the epidermal growth factor receptors and is enhanced by fibronectin and laminin. *Mol Biol Cell* 10 (1999): 1259–1276.

Zhao, M., B. Song, J. Pu, et al. Electrical signals control wound healing through phosphatidylinositol-3-OH kinase-gamma and PTEN. *Nature* 442 (2006): 457–460.

Zhou, D. D., X. T. Cui, A. Hines, and R. J. Greenberg. 2010. Conducting polymers in neural stimulation applications. In *Implantable Neural Prostheses 2: Techniques and Engineering Approaches*, ed. D. D. Zhou and E. Greenbaum. Berlin: Springer Science & Business Media, 217–250.

9 Lipid–protein electrostatic interactions in the regulation of membrane–protein activities

Natalia Wilke, María B. Decca, and Guillermo G. Montich
Universidad Nacional de Córdoba
Ciudad Universitaria, Córdoba, Argentina

Contents

9.1 INTRODUCTION

Electrostatic interactions are inherently related to cell membrane processes. Strong electric fields are generated by gradients of ions between compartments separated by cell membranes, fixed net charges in the membranes, and oriented dipoles. If we consider a protein molecule as a collection of dipoles and partial and net charges, it is easy to imagine that within the membrane milieu, the interactions with electric fields constitute an excellent tool to modulate their activity, conformation, and functionality (Tsong 1990; Astumian 1994; Teissie 2007; Marsh 2008). As examples of the relevance of electrostatics in lipid membranes, we can consider that two fundamental processes that support life on earth, photosynthesis and respiration, rely on the generation of electrochemical gradients between aqueous compartments separated by a biological membrane. The development of a nervous system in complex organisms was also completely dependent on mechanisms that include electrochemical gradients and electrostatic control of protein activities to translate information.

The components of lipid membranes in a living cell have a well-defined orientation and long-range organization. In water, lipids assemble spontaneously in a large variety of structures, among them the bilayer (Gennis 1989; Jakli and Saupe 2006). Within the same lipid molecule, a water-soluble portion, the polar head group, coexists with a chemical group, the hydrocarbon tail, with low solubility in water. Lipids in the bilayer are held together not by covalent bonds, but because the water-insoluble portion is segregated from the aqueous solvent (Tanford 1980). The order and orientation are imposed by the thermodynamic preference of the head groups for being in contact with water and the hydrocarbon chains to segregate from the aqueous solvent. Van der Waals, hydration, hydrogen bonding, and screened electrostatic interactions also participate in defining the membrane structure (Israelachvili 2011).

Proteins also acquire their structure as a consequence, to a large extent, of a balance between opposing solubilities (Privalov 1992; Richards 1992). Side chains with remarkable different solubilities in water are present in a given sequence of amino acids. In the aqueous medium, the best solution to this thermodynamic pressure is to fold, hiding nonpolar residues in the inner core and exposing the polar, water-soluble residues to the solvent. The fate for some sequences is to be soluble or membrane-associated peripheral extrinsic proteins. For other sequences, the best solution is to fold and insert into the lipid membrane, optimizing the interactions of nonpolar residues with the hydrophobic core of the membrane, yielding the transmembrane intrinsic proteins. A static snapshot of a cell membrane is shown in Figure 9.1. To highlight the dynamical nature of the ensemble, the bidimensional and rotational diffusion coefficients of the membrane components are indicated in Figure 9.1.

Chemical reactions are intrinsically vectorial: for a collision to be effective, reactants must be in a defined orientation relative to each other and the products leave the transition state in a defined direction. The anisotropic organization of the components in a biological membrane introduces geometrical and energetic restrictions that make evident the intrinsic anisotropy of chemical reactions.

We find basically two different approaches in the literature, or a mixture of them, to evaluate the contribution and role of electrostatics in lipid–protein interactions. In some cases, the lipid membrane, protein molecule, and electrolyte solution are considered a continuous medium. In others, the discrete nature of charges and the microscopic arrangement

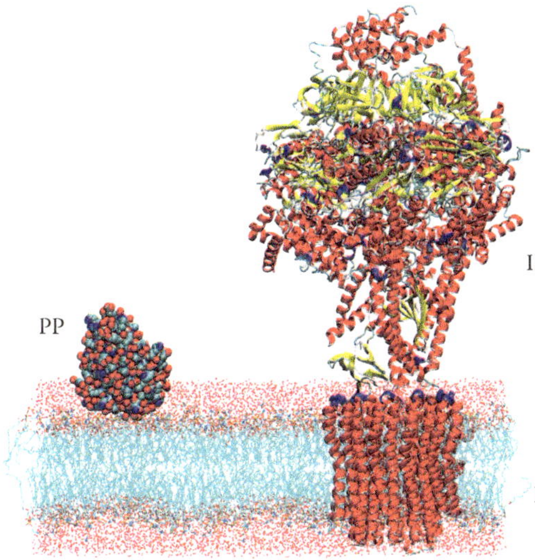

Figure 9.1 Elementary representation of a lipid bilayer with associated peripheral and integral proteins. Lipid bilayer (LB) is pure POPC; both lipids and water molecules are shown. (Data from http://www.ucalgary.ca/tieleman/) The peripheral protein (PP) is L-BABP, a protein of 14 kDa. (From Nichesola, D., et al. Biochemistry 43: 14072–14079, 2004. With permission.) The integral protein (IP) is the ATP synthase from *Paracoccus denitrificans* [PDB 5DN6]. (From Morales-Rios, E., et al. *Proc Natl Acad Sci USA* 112: 13231–13236, 2015. With permission.) The ensemble is very dynamic: bidimensional diffusion coefficients are $D_{lipids} = 10^{-10}$–10^{-7} cm^2 s^{-1} in pure lipid bilayer, the corresponding average displacements are $l = 2 \times 10^2$ and 6×10^3 nm in 1 s. D_{lipids} in natural cell membranes are in the order of 10^{-8} cm^2 s^{-1}. Dperipheral proteins = 10^{-7} cm^2 s^{-1}. Dintrinsic proteins = 10^{-9} cm^2 s^{-1} in fluid membranes, experimental systems without interactions with cytosolic components (average displacement 6×10^2 nm s^{-1}). In natural membranes, interacting with citosolic components, diffusion coefficients of integral membrane proteins are reduced to 10^{-12}–10^{-11} cm^2 s^{-1}, with an average displacement of 1–6 nm s^{-1}.

of atoms and molecules are taken into account. The electric field E generated by a charge q at a distance r in a medium considered as continuous is $E = q/\varepsilon_o\, \varepsilon\, r$, where ε_o is the permittivity of vacuum and ε is the dielectric constant. The dielectric constant depends on the nature of the medium, particularly its dielectric polarizability. The dielectric constant is a parameter that takes into account the shielding of the electric field produced by a continuous medium. But at a molecular level, the dielectric constant loses its meaning. Let us suppose we want to evaluate the energy difference between two configurations; in one of them, two charges are separated by a distance r_1, and in the other by r_2. In a continuous medium, the energy difference is $\Delta E = (qq'/\varepsilon_o\, \varepsilon)/(1/r_2 - 1/r_1)$. The dielectric constant is used to take into account the decrease of the electrostatic energy due to the material medium compared with the vacuum. The situation is quite different if the discrete nature of the material, the charges, and their fixed locations are explicitly considered. In this case, it is the particular arrangement of chemical groups between charges that defines the change in energy between both configurations (Figure 9.2). The dynamic characteristics of the system also define how suitably it can be described by a continuous model: if the timescale for the charge separation is slow compared with the fluctuation of charges in the medium, the electrostatic screening can be averaged to a dielectric constant. A rigorous and extensive analysis of this problem is addressed by Warshel and Aqvist (1991).

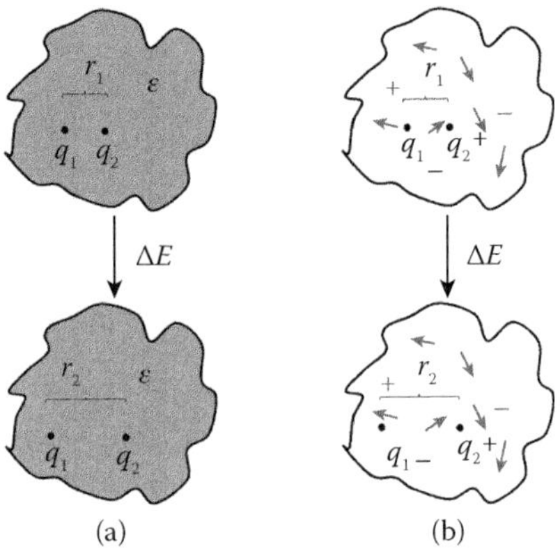

Figure 9.2 Continuous versus microscopic evaluation of electrostatic interactions. Both systems represent a change in the electrostatic energy due to an increase in the distance between charges q_1 and q_2. (a) represents the change in a continuous medium of dielectric constant ε. Arrows in (b) represent dipoles, and + and – symbols represent fixed charges. In this case, energy change results from evaluating the interactions of q_1 and q_2 with each dipole and charge.

9.2 PROFILE OF ELECTRIC FIELD ACROSS THE LIPID MEMBRANE

An electric test charge translocating between two cell compartments separated by a lipid membrane detects a complex profile of electrostatic potentials (Figure 9.3). Several components can be described separately, and to some extent, they have a separate and specific influence on membrane processes. A point charge, deprived of any chemical characteristic, describes only the electric contribution to the potentials. Because of the chemical interactions, the energy profiles would be quantitatively different if different ions were used as test charges.

9.2.1 TRANSMEMBRANE POTENTIAL (Ψ_{TM})

An electrostatic potential difference arises between two aqueous compartments separated by a lipid membrane when soluble ions with different permeabilities are at different concentrations in the two compartments. Let us take a membrane that separates two aqueous compartments, *a* and *b*, with different concentrations of KCl and NaCl. The Ψ_{TM} potential for a system in steady state is described by the Goldman–Hodgkin–Katz equation:

$$\Delta\psi_{TM} = -\frac{RT}{F}\ln\left(\frac{P_{Na}\left[Na^+\right]_b + P_K\left[K^+\right]_b + P_{Cl}\left[Cl^-\right]_a}{P_{Na}\left[Na^+\right]_a + P_K\left[K^+\right]_a + P_{Cl}\left[Cl^-\right]_b}\right) \tag{9.1}$$

P_{Na+}, P_{K+}, and P_{Cl^-} are the membrane permeabilities for the corresponding ions. Terms in brackets are the concentrations, or activities, of the ion in compartments *a* and *b*. In the living cell, the concentration gradients of ionic species are built up and supported by pumping machineries with the expenses of metabolic energy. A thorough explanation of the origin of Ψ_{TM} potential and a detailed derivation of Equation 9.1 can be found in Jackson (2006).

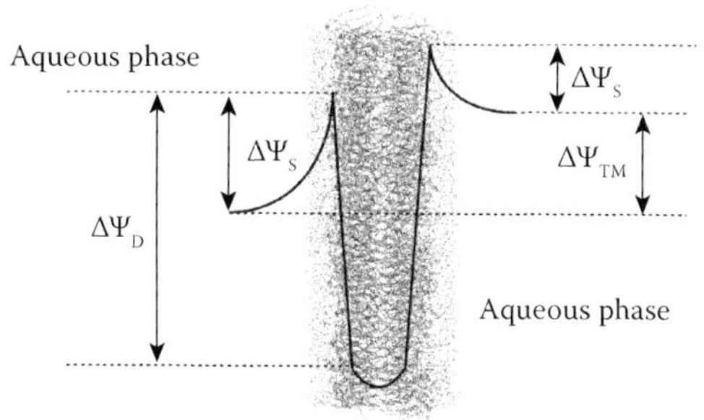

Figure 9.3 Electrostatic potential across a lipid membrane. $\Delta\Psi_{TM}$, transmembrane potential; $\Delta\Psi_S$, surface potential; $\Delta\Psi_D$, dipole potential. The inner potential, $\Delta\Psi_I$, is superimposed on the dipole potential; in the absence of other potentials, it would be the potential difference between the inner hydrocarbon core of the membrane and the aqueous phase. The profile corresponds to a negatively charged lipid membrane. Then, the upward direction in the trace represents negative values of potential. (Courtesy of Peter Tieleman's Biocomputing Group, University of Calgary.)

9.2.2 SURFACE POTENTIAL (Ψ_S)

Lipid polar head groups can be zwitterionic or support a net charge, negative in the case of natural cell membranes. These charges generate an electric field that, on the side of the aqueous solution, produces a redistribution of ionic species. Considering the charges on the membrane as uniformly distributed on a plane and that the ions in solution distribute along the axis normal to the interface according to the Boltzmann distribution, the Poisson equation can be solved, leading to the Gouy–Chapman equation for the surface potential (Aveyard and Haydon 1973).

$$\Psi_{S(0)} = \frac{\sigma}{\epsilon \epsilon_0 \kappa}, \kappa = \sqrt{\frac{2(zeF)^2 C}{\epsilon \epsilon_0 RT}} \tag{9.2a}$$

$$\Psi_{S(x)} = \Psi_{S(0)} e^{-\kappa x} \tag{9.2b}$$

$$C_{i(0)} = \Psi C_{i(\text{solution})} e^{\frac{-Z_i \Psi_{S(0)}}{\kappa T}} \tag{9.2c}$$

Subindex 0 stands for the distance normal to the plane of charges $x = 0$. Ψ_S decreases with the distance according to Equation 9.2b. σ is the surface concentration of fixed charges in C m^{-2}. According to Equation 9.2b, $1/\kappa$ is the distance at which Ψ_S decreases to $1/e$ of the value of $\Psi_{S(0)}$, it is known as the Debye length, and it depends on the bulk concentration of water-soluble ions. $C_{i(0)}$ is the concentration in the interface of a soluble ion whose concentration in the bulk solution is $C_{i(\text{solution})}$. Equations 9.2 are valid for small values of $\Psi_{S(0)}$, about 0.1 V, and monovalent soluble salts in the order of 0.1 mole L^{-1}.

9.2.3 INNER POTENTIAL (Ψ_I)

The inner core of the membrane is populated by hydrocarbon chains with little possibility for charge separation or redistribution of charges, a property that for a bulk material is associated with a low dielectric constant. Consequently, an electric charge inserted in this core generates a larger electric field than the same charge in the aqueous, electrolytic solution. Then, transferring a charge from the aqueous phase to the inner phase of the lipid membrane requires work. This is the origin of the electrostatic inner potential, and its related work can be estimated using the Born equation:

$$\Delta W_I = \frac{q^2}{2r}\left(\frac{1}{\varepsilon_1} - \frac{1}{\varepsilon_2}\right) \tag{9.3}$$

q is the charge on the ion, r is its radius, and ε_1 and ε_2 are the dielectric constant of membrane and water, respectively. To some extent, this model considers the chemical nature of the translocating ion because it takes into account its size: for the same charge, smaller ions will monitor a larger inner potential. Inner potential accounts for the low solubility of ions in the hydrocarbon membrane phase and, consequently, the low intrinsic permeability of the lipid membrane for ionic or polar molecules.

9.2.4 DIPOLE POTENTIAL (Ψ_D)

Because of the anisotropic organization of the lipid membrane, the dipoles associated with the polar head groups, the hydrocarbon chains, and the water bound to the polar groups have a preferential orientation. As a consequence, an electrostatic potential difference of 300–800 mV, positive inside, arises within a slab of about 1 nm between the polar group–water interface and the hydrocarbon core (Brockman 1994; Clarke 2001; Peterson et al. 2002; Starke-Peterkovic and Clarke 2009; Wang 2012; Lairion and Disalvo 2009). The Ψ_D can be evidenced and measured in lipid monolayers at the air–water or oil–water interface by measuring the electrostatic potential difference between the two bulk phases. In lipid bilayers, the Ψ_D can be calculated by measuring the ionic conductivity of the lipid membrane (Wang 2012).

We have introduced a universally accepted profile of electric potential along the z axis, perpendicular to the lipid bilayer. We will describe several basic studies dedicated to understanding the contribution of electrostatics in protein–membrane interactions in which the lateral distribution of charged lipids is assumed to be uniform. This is assuming that the profile remains the same at any point of the bidimensional interface. Nevertheless, the lateral distribution of lipids is far from homogeneous even in simple experimental systems, and the local values of the profile in Figure 9.3 should change from point to point in the plane of the membrane. This has a deep influence on the binding of proteins and the modulation of their biological activity.

To introduce how electrostatic interactions participate in the functioning of a lipid–protein system, we comment on how peripheral proteins interact with lipid membranes, how lipid organization influences these interactions, and how the function of some membrane integral proteins is controlled by electrostatic fields.

9.3 ELECTROSTATIC FORCES CONTRIBUTE TO THE MEMBRANE BINDING OF PERIPHERAL PROTEINS

Several proteins develop their physiological function within the lipid membrane interface; this is the region where lipid polar head groups are in contact with water. Some are constitutive peripheral proteins, permanently associated with the lipid membrane, while others translocate between a soluble state and a membrane-bound state and are designated amphitropic proteins (Johnson and Cornell 1999). Several interactions of different physical natures determine the binding of peripheral proteins, like the insertion into the membrane of aliphatic chains covalently attached to the protein, hydrophobic interactions due to partial penetration into the hydrocarbon core of the membrane, and electrostatic interactions between protein and charged lipids.

We can basically identify three approaches that have been used to evaluate the energy of electrostatic membrane binding: (1) concentration of peptides within the Ψ_S according to the Boltzmann distribution, (2) evaluation of electrostatic energy using the finite differential Poisson–Boltzmann (FDPB) equation, and (3) explicit consideration of atoms and evaluation of energy by molecular dynamics (MD) or Monte Carlo simulations.

9.4 CONCENTRATION OF PEPTIDES WITHIN THE Ψ_S ACCORDING TO THE BOLTZMANN DISTRIBUTION

The membrane binding of peptides can be described by a partition constant according to

$$P_{solution} \leftrightarrow P_{membrane}; K = [P_{membrane}]/[P_{Solution}] \quad (9.4)$$

$P_{solution}$ is the peptide free in solution, and $P_{membrane}$ is the membrane-bound peptide. The terms in brackets are the concentrations of the free and bound peptide. The meaning of K and the process for which the binding free energy is evaluated according to $\Delta G_{binding} = RT \ln K$ depends on the phase volume used to define the concentration (volume of the lipid membrane, volume of the aqueous phase, or total volume of the sample). For a discussion about the meaning and suitable concentration units to evaluate the binding to lipid membranes, see White and Wimley (1999).

Electrostatics participates in the equation because the concentration of the free peptide we have to consider is not the analytical concentration of the peptide in the bulk solution, but the actual concentration of peptide in the membrane interface, which is modified by electrostatic interactions. For a charged membrane and a charged peptide, this concentration can be evaluated according to Equation 9.2c. Then, the surface concentration of peptide is $[P_{surface}] = P_{bulk\ solution} \times \exp(-zF\Psi_S/RT)$, and the effective binding constant is $K = K_{intrinsic} \times \exp(zF\Psi_S/RT)$, where z is the effective valence of the peptide. $K_{intrinsic}$ is a binding constant that does not depend on the accumulation of peptide in the interface due to electrostatic attraction. A peptide has a finite size and asymmetrical shape, and the charges are spread along its structure; it can hardly be considered a point charge. The whole molecule cannot "feel" the full value of surface potential; the reduction of the analytical charge to an effective value allows treating the problem according to

Equations 9.2. Kim et al. (1991) studied the binding of anionic polylysine peptides and found that the relationship between the strength of binding and the amount of charges in the peptide, the surface density of anionic lipids, and the ionic strength of the medium can successfully be described by the physical background implicit in Equations 9.2.

This approach is essential to discriminate the Boltzmann accumulation from the intrinsic interactions and then understand the forces driving the binding and conformational changes of peptides and proteins, including water displacement from the interface, hydrogen bonding, and lipid reorganization (Seelig 2004).

9.4.1 EVALUATION OF ELECTROSTATIC ENERGY USING FINITE DIFFERENTIAL POISSON–BOLTZMANN EQUATION

In this approach, the system is divided into a three-dimensional grid. For each cell in the grid, the Poisson–Boltzmann equation is solved. The result obtained is the electrostatic potential for each point in the grid (Sharp and Honig 1990). Valuable conclusions were obtained from the studies reviewed by Mulgrew-Nesbitt et al. (2006) comparing the results from several proteins: even when patches of basic, cationic residues in a protein constitute evident binding sites for anionic lipid membranes, they are not necessarily required. It is rather the global electrostatic potential profile in the protein that contains the information for membrane binding. It is the asymmetric *distribution* of charges, rather than the net charge, that determines binding and orientation. Actually, anionic proteins do bind to anionic lipid membranes (Mulgrew-Nesbitt 2006; Galassi at al. 2014). Besides the contribution of nonpolar protein–membrane interactions, this occurs because Ψ_S decreases to zero within a length scale comparable to the dimension of a protein. For a defined orientation of the protein in the interface, the domains with positive electrostatic potential will "feel" the whole value of Ψ_S, while the negative domains can be located where Ψ_S already decreased to zero. Similar conclusions were further obtained by computer simulations considering explicitly all the atoms in a MD simulation (Galassi et al. 2014).

9.4.2 MOLECULAR DYNAMICS IN THE STUDY OF PERIPHERAL PROTEINS

The increased capacity of MD simulations to cope with systems containing a large amount of atoms opened the possibility to studying the membrane–protein system. To construct the system, all atoms, including the solvent, are considered explicitly. A force field that represents the energy of the system as a function of the coordinate reaction, that is, the geometric location of all atoms, is constructed from parameters that represent the energy of the torsion angles, Van der Waals interactions, and electrostatics of every atom. The behavior of the system is the result of evaluating explicitly the interactions between all the component atoms.

Several studies addressed the interactions of peripheral proteins (Kalli and Sansom 2014; Rogaski and Klauda 2012). Convenient examples are the studies by Zamarreño et al. (2012) and our own results about the interactions of fatty acid binding proteins (FABPs). This is a family of low-molecular-weight proteins that bind nonpolar ligands. It is proposed that their multiple functions in the cell include the transport of the ligands between cell compartments. Some members of the family deliver the ligand after an effective collision with the acceptor lipid membrane, while others do not require an effective binding to the interface, and the nonpolar ligand reaches the membrane after a release and diffusion step. Zamarreño et al. (2012) evaluated the electrostatic energy of the system, protein plus lipid membrane, using MD simulation and FDPB for several FABPs in zwitterionic and anionic lipid membranes. For FABPs that follow the collisional mechanism, they found a well-defined minimum, in coordinates of distance and orientation, of electrostatic energy. For those members of the family expected not to collide with the membrane, the energy landscape was flat, indicating no preferential orientation.

In free-run MD simulations, we observed that two different FABPs spontaneously acquire an orientation dictated by the interaction of the protein macrodipole and the surface potential electric field. FABPs have a well-defined asymmetric structure; on one side of the molecule there is a domain composed of two α-helix segments, and on the other side, we can identify the bottom of the β-barrel. In the liver bile acid binding protein, the electric macrodipole points the positive end toward the bottom of the barrel (Villarreal et al. 2008). Independently of the starting orientation in the simulation, this protein rotates while approaching the anionic lipid membrane and interacts with the membrane through the bottom of the barrel. For the structurally similar ReP1-NCXSQ, the macrodipole is oriented right in the opposite direction. Consistently, ReP1 orients in the simulations with the α-helix domain pointing to the membrane (Galassi et al. 2014).

The observations regarding the orientation within the electric field are consistent with an elementary evaluation of the interaction energy between the protein dipole and membrane electric field. The energy of a dipole in an electric field is $U = p.\ E \cdot \cos(\alpha)$. U is the electrostatic energy, p is the magnitude of the dipole moment, E is the magnitude of the electric field, and α is the angle between vectors **p** and **E**. Boltzmann distribution indicates that the probability of finding a dipole in an angle between α and $\alpha + d\alpha$ is

$$P(\alpha) = \frac{e^{\frac{-U(\alpha)}{kT}}}{\int_0^{\pi} e^{\frac{-U(\alpha)}{kT}}\, d\alpha} \tag{9.5}$$

Protein macrodipoles have values ranging between 150 and 400 Debye (Porschke 1997), and the electric field associated with the surface potential can take values of 10^7 and 10^8 V m^{-1}. Figure 9.4 shows that proteins with a macrodipole of 200 D, which

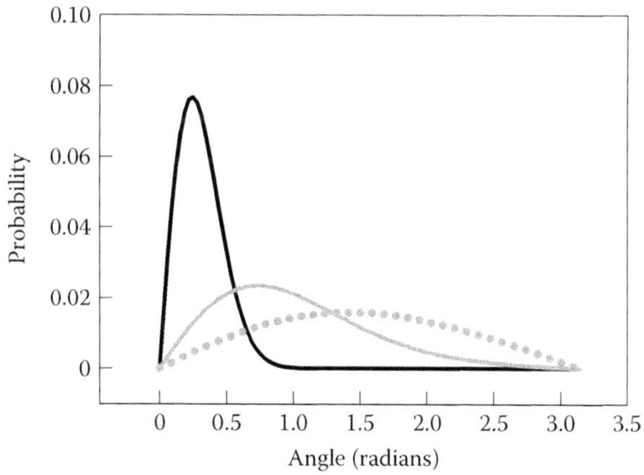

Figure 9.4 Probability distribution for the orientation of a dipole in an electric field. The ordinate is the probability of finding the angle between the dipole and the electric field within a value α and $\alpha + d\alpha$. The combinations of dipoles and electric fields are black line, 200 D and 108 V m^{-1}; grey line, 200 D and 10^7 V m^{-1}; and dotted grey line, 1.85 D and 10^8 V m^{-1}. The angle of 0 radians corresponds to the orientation of lower energy and 3.14 radians to the higher energy.

is about the value for L-BABP and ReP1-NCXSQ, have a defined bias to be oriented against the electric field at 25°C. For comparison, water molecules, with a dipole of 1.85 D, are almost randomly oriented within this field.

9.5 CONTROL OF THE INTERACTIONS

According to the previous presentation, electrostatic binding of proteins to lipid membrane could seem to be a static process: given a membrane charge density, an ionic strength, and a number of charges in the protein or peptide, the binding constant and, consequently, the amount of bound protein or peptide should be fixed. This is a landscape poor in alternatives to account for the large capacity of regulation processes in the living cell. Lipid–protein interactions in the biological membranes are highly dynamic. Local concentration of charges in the membrane, in the protein, or in the peptide can change; proteins can undergo conformational changes, and bind and detach from the membrane populating active and inactive states. Section 8.5.1 gives a brief insight into how the complexity of lateral distribution of lipids is modulated by electrostatics and how it can modulate electrostatic protein binding, charge, and function.

9.5.1 CHANGES IN THE PROTEIN ELECTROSTATIC CHARGE DUE TO INTERACTIONS WITH THE MEMBRANE

The global, net charge in the protein is mainly given by the balance between basic, positively charged side chains (lysine, arginine, and histidine) and acidic, negatively charged side chains (glutamic and aspartic) at pH = 7, and charged groups like phosphate attached by posttransductional modifications. The pKa of these groups, that is, the pH at which they change from the protonated to the unprotonated state, is modified because of the interaction with the lipid membrane. In the interface, due to the lower dielectric constant, any charged group is in a higher energy level than when it is in bulk water. Then, in the interface, the protonated (neutral) state of acidic groups and the unprotonated (neutral) state of basic groups are favored. The low dielectric constant of the interface has the effect of increasing the pK of acidic groups and decreasing the pK of basic groups. The negative surface potential determines that both acidic and basic groups prefer the protonated state: acidic groups yield a neutral group (otherwise, it would be a negative group within a negative potential, which is energetically unfavorable), and basic groups yield a positively charged group (more favorable than a neutral group within a negative potential). Equivalently, it can be considered that the negative Ψ_S produces an increased concentration of H^+ ions in the interface according to Equation 9.2c. Then, if we take as reference the measured pH in the bulk solution, all protonation processes in the negatively charged membrane will occur at an apparently higher pK (Tocanne and Teissié 1990; see also a brief, interesting discussion under the title "Thermodynamic Cycle for Membrane Binding" in Mihajlovic and Lazaridis 2006).

The control of interfacial pH has important consequences. The regulation of many processes depends on the protonation of specific amino acids. Changes in the local Ψ_S due to changes in the local surface concentration of anionic lipids, for example, could modulate the protonation state of proteins in the interface, and consequently their activities. As an example, we consider the work by Bustad et al. (2012). A large group of proteins involved in different cell processes have in common a "zinc finger" domain called FYVE (Stenmark et al. 1996). For one member of this group, it was demonstrated that binding to phosphatidylinositol 4,5-bisphosphate (PIP2) in the cell lipid membrane through this domain is determined by the protonation state of a His residue (Bustad et al. 2012). It is immediate to envisage how the control of the local pH through changes in surface potential should have influence on the membrane binding of this protein. Caldcrón and Cerbón (1992) demonstrated that in a membrane protein, the pH-dependent activity has apparent pK values that depend on the composition anionic lipids in the cell membrane.

For intrinsic transmembrane proteins, the protonation of specific residues, through modulation of the interfacial pH, can be an effective control mechanism of protein function (Asandei et al. 2008; Nielsen and McCammon 2003).

9.5.2 LATERAL ORGANIZATION OF LIPIDS

Lipid aggregates acquire phase states defined by the lipid composition, the temperature, and the aqueous phase composition. The balance of intermolecular interactions defines the phase state: strong attractive interactions lead to membranes in a densely packed state (gel or solid phase), while weaker intermolecular attraction results in more fluid, loosely packed membranes with increased degrees of freedom of the hydrocarbon chain (liquid crystalline, liquid ordered, or liquid disordered phase) (Marsh 1991). In multicomponent systems, lipids can mix to form a homogeneous phase or segregate into phases containing a pure segregated lipid or a mixture enriched (depleted) in one of the components.

Electrostatics plays a major role in defining the phase state, temperature of the phase transition, and size and shape of lipid domains. The phase boundaries in the diagram are defined, within other factors, by the electrostatic interaction between lipids, and thus can be regulated by the ionic strength of the aqueous milieu, by ions that interact specifically with the lipids and by the local pH.

The effect of pH and ionic strength on the phase state of a membrane composed of a pure ionizable surfactant can be modeled considering that lipid melting implies an increase in the area per lipid, and therefore the phase transitions of charged molecules are accompanied by a decrease in the electrostatic free energy due to the bilayer expansion. As a result, the melting temperature of a charged lipid membrane is lower than that of an equivalent neutral membrane. For a uniform charge distribution, the Gouy–Chapman theory of the electrical double layer predicts a decrease of the transition temperature by about 10 K at low ionic strength, which is also the observed experimental value (Heimburg 2008).

In mixed membranes containing a charged component, electrostatic repulsion usually prevents the formation of large clusters of charged molecules (Huang et al. 1993). Therefore, the biphasic region of the phase diagram for lipid mixtures is highly reduced when a neutral lipid is replaced by a charged one (Vequi-Suplicy et al. 2010), and the thermal stability of the two-phase region of mixtures also decreases. As expected, the miscibility depends strongly on pH, due to its influence on the degree of ionization of the charged molecule (Garidel et al. 1997; Garidel and Blume 1998). This effect is greatly modulated by the ionic strength of the solution through its influence on the energy accumulated in the double layer (Equations 9.2) (Vega Mercado et al. 2012).

9.5.3 LIPID DOMAINS

When two phases are present, lipid domains are generated with different molecular densities and thus different properties, and their presence affects the membrane properties. The modulation of the global and local membrane properties by phase coexistence depends on the domain size, amount, and shape, all these parameters defining the membrane texture, which has been an issue of intensive research (Wilke 2015). Two general cases can be defined for the membrane texture: equilibrium and out-of-equilibrium domain shapes. In both cases, rounded or elongated domains are observed, but the driving force that determines their shapes and stabilities is different.

The shape of a domain at equilibrium may be rounded or flower-like, depending on the size of the domain compared with a critical radius, which is determined by three major forces (McConnell 1989; Iwamoto and Ou-Yang 2004): line tension at the domain boundary, dipolar repulsion inside the domains, and domain–domain dipolar interactions. The dipolar repulsion refers to the interaction between the molecular dipoles of the surfactants that are aligned in the membrane (Brockman 1994; Smaby and Brockman 1990; Taylor 2000). Line tension is defined as the change of free energy upon an elongation of the phase boundary, and it arises from the difference in the intermolecular interactions within each phase and the thickness mismatch of the phases. This parameter has been determined experimentally and evaluated theoretically in monolayers of lipid at the air–water interface (Sriram at al. 2012; Palmieri et al. 2014). Its magnitude is usually in the piconewton range in monolayers and also in bilayers.

In the case of bilayers, a clear and accepted model similar to that of McConnell in monolayers has not been postulated yet. However, it is proposed that the electrostatic repulsion would be screened by the presence of the aqueous media (Iwamoto et al. 2008; Konyakhina et al. 2011) and the curvature of the membrane would appear as another important factor (Parthasarathy et al. 2006; Semrau et al. 2009).

Out-of-equilibrium domains are generated when the kinetics for the phase transition is faster than the migration of the molecules to the domain. This case has been described using the theory for diffusion-limited aggregation (Blanchette et al. 2007), which predicts that domains will grow faster in the more curved regions, and thus flower-like shapes are expected. The shape of these domains depends on the relative positions of the other domains (Bernchou et al. 2009; Vega Mercado et al. 2012), which in turn depends on the nucleation process.

A broad distribution of domain sizes is attained. Nanometer-sized domains have been observed (Tokumasu et al. 2003), but micron-sized ones are more frequently reported since optical microscopy is the more widely employed technique to study lipid domains.

The nonhomogeneous lateral distribution of molecules bearing a dipole or a net charge generates a complex distribution of local electrostatic fields in the plane of the membrane. Then, it can be expected that the organization of bound molecules will also follow a heterogeneous distribution. A striking proof was observed using microspheres of about 1–3 μm. These particles spontaneously acquire a dipole in the interface of lipid membranes, and are attracted to the border of domains composed of neutral molecules in a liquid-condensed phase (Nassoy et al. 1996), and depending on the interaction strength, the diffusion of the particle suffers a transition from two-dimensional to one-dimensional (Forstner et al. 2008). The behavior of the particles is very sensitive to the interaction strength; small differences in the potential result in orders of magnitude changes of the diffusion coefficient. This is a very interesting result since the interaction strength can be easily altered by changing the domain size, regardless of domain composition (Caruso et al. 2014). Additionally, two diffusing species with only a small difference in their interactions with domains will have significantly different propagations within the same environment. Thus, the presence of domains can selectively regulate the diffusion of the dipolar or charged species inserted in the membrane, according to the electrostatic properties of the species and the size of the domain. Little is known about how proteins can diffuse along the borders of lipid domains (Raine and Norris 2007; Lafontaine et al. 2007; Baneyx and Vogel 1999), but it can be easily envisaged in this process to be a factor for regulating interactions and activities.

9.5.4 LIPID DEMIXING INDUCED BY PROTEIN BINDING

We should also consider the lipid mobility within the same phase. Protein binding can produce an increase in the surface concentration of anionic lipids in the vicinity of the cationic binding domain. Opposing driving forces define whether the local, bidimensional concentration of lipids increases in the proximity of cationic domains of peripheral proteins. The decrease in mixing entropy and the increase in electrostatic repulsion play against recruitment of lipids. On the other hand, the local increase in surface charge density increases the binding constant of cationic domains, which helps to decrease the free energy of the system (Heimburg et al. 1999). Evaluation of the electrostatic contribution to the binding free energy of model cationic peptides demonstrated both theoretically and experimentally that there is an optimum total peptide concentration that produces the formation of lipid domains enriched in anionic lipid (Denisov et al. 1998). A delicate balance between these driving forces was also evidenced in the work by Kiselev et al. (2011): PIP2 is an anionic lipid with a net charge of −4, while PS has a charge of −1. The recruitment of phosphatidylserine (PS) by a basic domain of the enzyme PKC was enhanced by the presence of PIP2. In light of the work by Kiselev et al., it can be explained by the increased binding of the protein due to the presence of PIP2. The process seems to be quite general, and it is also present in the binding of cytochrome *c* (Gorbenko et al. 2009).

The experimental observation and theoretical evaluation of lipid demixing do not tell us much about the microscopic dynamics of the process. Does the protein or peptide attach to a lipid domain or slip over the surface while the domains are formed and disassembled? This question was addressed by Kiselev et al. (2011; see also Khelashvili et al. 2008). They used Monte Carlo simulation of a system consisting of a pentalysine interacting with a lipid membrane containing PC, PS, and PIP2, allowing for diffusion of peptides and lipids. As a streaking result, they found that a lateral gradient of the anionic lipid produced a net displacement or drift of the peptide, which was largely increased when a very small amount of PIP2 was present in the lipid domain.

Recruitment of anionic lipids coupled to the binding of proteins can be considered a specific case of a more general problem. The lateral organization of fluid membranes in response to naturally occurring electric fields in the plane of the membrane (e.g., those caused by a buried charge) has been explored theoretically and experimentally (Lee et al. 1994; Groves at al. 1998). It was shown that membrane reorganization occurs in a range of several nanometers, on the microsecond timescale. Two distinctively different attributes of the reorganization have been described: electrostatic reorganization (a bidimensional rearrangement similar to that predicted by the Gouy–Chapman model) and critical demixing effects (in the case of lipid mixtures close to a phase coexistence border). Both effects lead to enhancement or depletion of certain lipid species around the charged object, and thus may correspond to a sorting mechanism.

9.6 INFLUENCE OF DIPOLE POTENTIAL ON THE ACTIVITY OF MEMBRANE PROTEINS

Several membrane processes and the activity of membrane proteins are modulated by the insertion of molecules that modify the dipole potential or by direct modification of the potential by external fields; examples are the water conductivity of lipid membranes, the rate of redox reactions occurring in the lipid interface (Pilotelle-Bunner et al. 2009; Alakoskela and Kinnunen 2001), the assembling of conduction channels (Efimova et al. 2014), the binding of peptides (Voglino et al. 1998; Zhan and Lazaridis 2011), and changes in the activity of phospholytic enzymes in the interface (Maggio 1999).

The Ψ_D modifiers are molecules of a varied chemical nature, including sterols, polyphenols, and saccharides that bind and insert into the lipid membrane. The way they change the value of Ψ_D is varied and not always completely understood. They change Ψ_D by contributing directly with their own dipole or changing the global dipolar organization of the interface (Przybylo et al. 2014; Lairion and Disalvo 2007).

We selected one particular example to show how Ψ_D can determine protein function. Alamethicin is a small (20 amino acid) α-helix peptide. In the absence of a transmembrane potential, it binds as a monomer, with the major axis parallel to the plane of the lipid membrane. When a Ψ_{TM} potential is established, alamethicin inserts into the membrane with the major axis perpendicular to the plane of the membrane and self-associates to yield oligomers that allow for ionic conductance proportional to the Ψ_{TM}. Using the black lipid membrane system, Mereuta et al. (2011) prepared a lipid bilayer containing alamethicin, established a Ψ_{TM}, and obtained a conducting bilayer. Then, they incorporated modifiers of the Ψ_D in *one hemilayer*. Phlorizin, a modifier that produces a decrease in the positive inner value of Ψ_D, produced an increase in the conductivity. The styrylpyridinium dye RH421, a modifier that increases Ψ_D, produced a decrease in the conductivity. The explanation proposed for this effect can be observed in Figure 9.5. When phlorisin is added in one

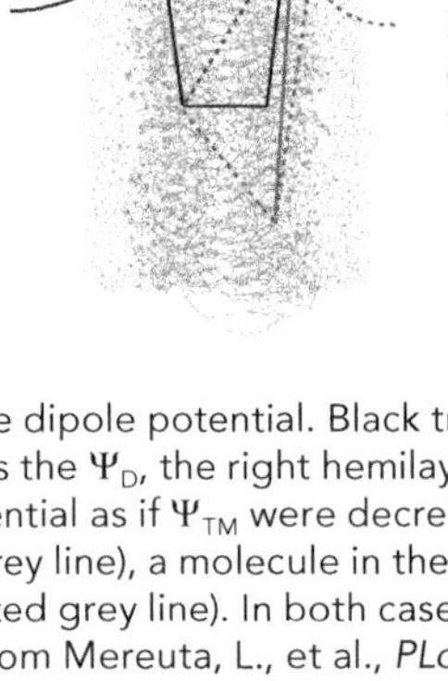

Figure 9.5 Changes in the dipole potential. Black trace is the normal electrostatic profile. When a modifier increases the Ψ_D, the right hemilayer (solid grey line), a molecule in the membrane core detects a potential as if Ψ_{TM} were decreased (dotted grey line). For a modifier that decreases the Ψ_D (solid grey line), a molecule in the membrane core detects a potential as if Ψ_{TM} were decreased (dotted grey line). In both cases, modifiers do not change the value of the surface potential. (Data from Mereuta, L., et al., *PLoS One*, 6, e25276, 2011. Courtesy of Peter Tieleman's Biocomputing Group, University of Calgary.)

hemilayer, alamethicin "feels" as if a larger Ψ_{TM} were applied, consequently increasing the conductance. Symmetrically, when Ψ_D is increased by RH421, the conductivity decreases. The key conceptual background, as explained in the paper by Mereuta et al. (2011), is that membrane electrostatic potentials are additive and the profile must be continuous.

9.7 VOLTAGE-DEPENDENT CHANNELS

Electrostatic interactions regulate the activity of peripheral and amphitropic proteins mainly by changing the membrane–protein affinity, switching the protein between bound and unbound states, placing the protein in defined orientations, and helping enzymes to interact with substrates in the membrane. The location and orientation of integral membrane proteins is instead almost completely defined by the arrangement of hydrophobic transmembrane domains and hydrophilic anchor domains, with large energy barriers for changing global orientations. The control of functions of transmembrane proteins by electrostatic tools in the cell relies on stabilizing particular conformations that are coupled to changes in charge separation and movements of charges within the protein.

A remarkable example is the electrostatic control of voltage-gated channels. In a hypothetical equilibrium state, if the cytoplasm membrane of a neuron cell were permeable only to Na^+ ions (only $P_{Na^+} > 0$ in Equation 9.1, others equal zero), the Ψ_{TM} should be +55 mV. If it were permeable only to K^+ (only $P_{K^+} > 0$, others equal zero), Ψ_{TM} should be –75 mV. Transmission of the nerve impulse is a complex mechanism where a local, steady-state Ψ_{TM} alternates between values approaching these equilibrium potentials. The key control relies in how the ionic channels are closed or opened, and consequently regulating the ionic permeability. Voltage-dependent ionic channels are integral proteins. What make them suitable for transmitting information is that, in turn, the probability of being in an open or closed state depends, in a first kinetic step, on the Ψ_{TM} potential. It is remarkable, for a mechanism that translates information, that the Ψ_{TM} potential can be switched on and off, positive inside to positive outside, by turning on and off the conductivity of specific ion channels. The paradigmatic Na^+ and K^+ channels, related to the transductions of nerve impulse, are proteins of four subunits (e.g., see Mathews et al. 2012). Each subunit is made of six transmembrane α-helices. A large amount of functional and structural information shows that the detector of changes in the Ψ_{TM} potential and the channel that allows ion permeation are in separate domains. The first four helices constitute a domain that changes its conformation as a response to Ψ_{TM} changes. This conformational change is coupled, or transduced, to the opening of the channel.

The conformational changes in sensor are related to electric charge displacement within the membrane (Bezanilla 2008). These charges are due to arginine, lysine, glutamic, and aspartic residues in the helices. The whole protein domain acquires the lowest-energy conformation according to the externally imposed Ψ_{TM} potential. This system also serves as an example of considering continuous, macroscopic electrostatic models versus a microscopic evaluation of the electrostatics. In an experimental system where electrodes are located in the aqueous media at both sides of the membrane, the displacement of fixed charges within the membrane must have as a consequence an ionic current, due to relocation of counterions in the aqueous phase, which can be measured. By relating this measured current with a simple model that considers the membrane a slab of low dielectric constant, basic and valuable information about the structure of the sensor can be obtained. A microscopic model, including coarse grain simulation, together with crystallographic information, provided an accurate picture about the distance traveled by the fixed charges and the extent to which the aqueous media penetrates within the protein domains (Kim et al. 2014).

9.8 MACROSCOPIC EFFECT OF ELECTRIC FIELD ON PROTEIN STRUCTURE

In Section 9.7, we described rather specific conformational changes as a consequence of interactions with electric fields. We may ask if global, unspecific changes in the protein structure can arise by the action of electric fields, like those present in the lipid membrane. The group of atoms involved in the peptide bond, Cα—(C=O)—NH, are fixed within a plane. Because there is (almost) free rotation along the bonds connecting the planes, a protein can be considered a polymer of planes and its structure defined by the list of angles between the planes. Because these planes contain a dipole, it is straightforward to expect that an electric field would influence the conformation. Some experimental systems have shown protein conformational changes under electric fields, but due to indirect thermal (Zhao and Yang 2010) and viscous drag (Bekard and Dunstan 2014) effects. Computer calculations instead allowed us to establish a direct relationship between the electric field and dipoles to yield a conformational change. States with an α-helix structure were neatly populated at the expense of unfolded and β-strands under an electric field (Ojeda-May and Garcia 2010). In other systems, strands were favored over the α-helix (Toschi et al. 2009). We observed a small unfolding process in a FABP, L-BABP, upon binding to a charged membrane in a MD simulation. The same change was observed when an external field was applied in the absence of a membrane (Villarreal et al. 2008). Consistently, the conformational changes predicted by MD in a defined domain of L-BABP were actually observed by Fourier transform infrared (FTIR) spectroscopy (Nolan et al. 2003) and correlated with changes in the electrostatic membrane surface potential (Decca et al. 2007, 2010).

9.9 FINAL STATEMENT

The electrostatic interactions are of paramount importance for the regulation of biological processes. Although they all derive from a coulombic potential, a particular distribution of charges leads to molecular interactions with high affinity and specificity. The scenario of electrostatic interactions is dynamic: when local changes in the ionization state of the molecules, ionic strength, and dielectric constant of the media, among others, are taken into account, complex and multiple modulation pathways emerge. Therefore, despite these kinds of interactions having been studied for centuries, new levels of complexity continue to come to light, as more is known about the dynamics and structure of proteins and lipid membranes.

ACKNOWLEDGMENTS

Our work is supported by grants from CONICET, MINCyT, and SECyT UNC. NW, MBD, and GGM are career members of CONICET. We appreciate the help of Julia Montich in making the figures.

REFERENCES

Alakoskela, J. I., Kinnunen, P. K. Control of a redox reaction on lipid bilayer surfaces by membrane dipole potential. *Biophys. J.* 80 (2001): 294–304.

Asandei, A., Mereuta, L., Luchian, T. Influence of membrane potentials upon reversible protonation of acidic residues from the OmpF eyelet. *Biophys. Chem.* 135 (2008): 32–40.

Astumian, R. D. Electroconformational coupling of membrane proteins. *Ann. N. Y. Acad. Sci.* 31 (1994): 136–140.

Aveyard, R., Haydon, D. A. *An Introduction to the Principles of Surface Chemistry.* Cambridge: Cambridge University Press, 1973.

Baneyx, G., Vogel, V. Self-assembly of fibronectin into fibrillar networks underneath dipalmitoyl phosphatidylcholine monolayers: Role of lipid matrix and tensile forces. *Proc. Natl. Acad. Sci. U S A.* 96 (1999): 1251–1252.

Bekard, I., Dunstan, D. E. Electric field induced changes in protein conformation. *Soft Matter* 10 (2014): 431–437.

Bernchou, U., Ipsen, J. H., Simonsen, A. C. Growth of solid domains in model membranes: Quantitative image analysis reveals a strong correlation between domain shape and spatial position. *J. Phys. Chem. B* 113 (2009): 7170–7177.

Bezanilla, F. How membrane proteins sense voltage. *Nat. Rev. Mol. Cell. Biol.* 9 (2008): 323–332.

Blanchette, C. D., Lin, W. C., Orme, C. A., Ratto, T. V, Longo, M. L. Using nucleation rates to determine the interfacial line tension of symmetric and asymmetric lipid bilayer domains. *Langmuir* 23 (2007): 5875–5877.

Brockman, H. Dipole potential of lipid membranes. *Chem. Phys. Lipids.* 73 (1994): 57–79.

Bustad, H. J., Skjaerven, L., Ying, M., et al. The peripheral binding of 14-3 3γ to membranes involves isoform-specific histidine residues. *PLoS One* 7 (2012): e49671.

Calderón, V., Cerbón, J. Interfacial pH modulation of membrane protein function in vivo. Effect of anionic phospholipids. *Biochim. Biophys. Acta* 1106 (1992): 251–256.

Caruso, B., Villarreal, M., Reinaudi, L., Wilke, N. Inter-domain interactions in charged lipid monolayers. *J. Phys. Chem. B* 118 (2014): 519–529.

Clarke, R. J. The dipole potential of phospholipid membranes and methods for its detection. *Adv. Coll. Int. Sci* 89–90 (2001): 263–281.

Decca, M. B., Galassi, V. V., Perduca, M., Monaco, H. L., Montich, G. G. Influence of the lipid phase state and electrostatic surface potential on the conformations of a peripherally bound membrane protein. *J. Phys. Chem. B* 46 (2010): 15141–15150.

Decca, M. B., Perduca, M., Monaco, H. L., Montich, G. G. Conformational changes of chicken liver bile acid-binding protein bound to anionic lipid membrane are coupled to the lipid phase transitions. *Biochim. Biophys. Acta* 1768 (2007): 1583–1591.

Denisov, G., Wanaski, S., Luan, P., Glaser, M., McLaughlin, S. Binding of basic peptides to membranes produces lateral domains enriched in the acidic lipids phosphatidylserine and phosphatidylinositol 4,5-bisphosphate: An electrostatic model and experimental results. *Biophys. J.* 74 (1998): 731–744.

Efimova, S. S., Schagina, L. V., Ostroumova, O. S. Channel-forming activity of cecropins in lipid bilayers: Effect of agents modifying the membrane dipole potential. *Langmuir* 30 (2014): 7884–7892.

Forstner, M. B., Martin, D. S., Rückerl, F., Käs, J. A., Selle, C. Attractive membrane domains control lateral diffusion. *Phys. Rev. E Stat. Nonlin. Soft Matter Phys.* 77 (2008): 051906.

Galassi, V. V., Villarreal, M. A., Posada, V., Montich, G. G. Interactions of the fatty acid-binding protein ReP1-NCXSQ with lipid membranes. Influence of the membrane electric field on binding and orientation. *Biochim. Biophys. Acta* 1838 (2014): 910–920.

Garidel, P., Blume, A. Miscibility of phospholipids with identical headgroups and acyl chain lengths differing by two methylene units: Effects of headgroup structure and headgroup charge. *Biochim. Biophys. Acta* 1371 (1998): 83–95.

Garidel, P., Johann, C., Blume, A. Nonideal mixing and phase separation in phosphatidylcholine-phosphatidic acid mixtures as a function of acyl chain length and pH. *Biophys. J.* 72 (1997): 2196–2210.
Gennis, R. B. *Biomembranes:Molecular Structure and Function.* Berlin: Springer-Verlag, 1989.
Gorbenko, G. P., Trusova, V. M., Molotkovsky, J. G., Kinnunen, P. K. Cytochrome c induces lipid demixing in weakly charged phosphatidylcholine/phosphatidylglycerol model membranes as evidenced by resonance energy transfer. *Biochim. Biophys. Acta* 1788 (2009): 1358–1365.
Groves, J. T., Boxer, S. G., McConnell, H. M. Electric field-induced critical demixing in lipid bilayer membranes. *Proc. Natl. Acad. Sci. U.S.A.* 95 (1998): 935–938.
Heimburg, T. *Thermal Biophysics of Membranes.*Hoboken, NJ: Wiley, 2008.
Heimburg, T., Angerstein, B., Marsh, D. Binding of peripheral proteins to mixed lipid membranes: Effect of lipid demixing upon binding. *Biophys. J.* 76 (1999): 2575–2586.
Huang, C., Li, S., Wang, Z. Q., Lin, H. N. Dependence of the bilayer phase transition temperatures on the structural parameters of phosphatidylcholines. *Lipids* 28 (1993): 365–370.
Israelachvili, J. N. *Intermolecular and Surface Forces.* 3rd ed. London: Academic Press, 2011.
Iwamoto, M., Liu, F., Ou-Yang, Z. C. Shapes of lipid monolayer domains: Solutions using elliptic functions. *Eur. Phys. J. E Soft Matter* 27 (2008): 81–86.
Iwamoto, M., Ou-Yang, Z. C. Shape deformation and circle instability in two-dimensional lipid domains by dipolar force: A shape- and size-dependent line tension model. *Phys. Rev. Lett.* 93 (2004): 206101.
Jackson, M. B. *Molecular and Cellular Biophysics.* Cambridge: Cambridge University Press, 2006.
Jakli, A., Saupe, A. *One- and Two-Dimensional Fluids: Properties of Smectic, Lamellar and Columnar Liquid Crystals.* Boca Raton, FL: CRC Press, 2006.
Johnson, J. E., Cornell, R. B. Amphitropic proteins: Regulation by reversible membrane interactions. *Mol. Membr. Biol.* 16 (1999): 217–235.
Kalli, A. C., Sansom, M. S. Interactions of peripheral proteins with model membranes as viewed by molecular dynamics simulations. *Biochem. Soc. Trans.* 42 (2014): 1418–1424.
Khelashvili, G., Weinstein, H., Harries, D. Protein diffusion on charged membranes: A dynamic mean-field model describes time evolution and lipid reorganization. *Biophys. J.* 94 (2008): 2580–2597.
Kim, I., Chakrabarty, S., Brzezinski, P., Warshel, A. Modeling gating charge and voltage changes in response to charge separation in membrane proteins. *Proc. Natl. Acad. Sci. U.S.A.* 111 (2014): 11353–11358.
Kim, J., Mosior, M., Chung, L. A., Wu, H., McLaughlin, S. Binding of peptides with basic residues to membranes containing acidic phospholipids. *Biophys. J.* 60 (1991): 135–148.
Kiselev, V. Y., Marenduzzo, D., Goryachev, A. B. Lateral dynamics of proteins with polybasic domain on anionic membranes: A dynamic Monte-Carlo study. *Biophys. J.* 100 (2011): 1261–1270.
Konyakhina, T. M., Goh, S. L., Amazon, J., et al. Control of a nanoscopic-to-macroscopic transition: Modulated phases in four-component DSPC/DOPC/POPC/Chol giant unilamellar vesicles. *Biophys. J.* 101 (2011): L08–L10.
Lafontaine, C., Valleton, J. M., Orange, N., Norris. V., Mileykovskaya, E., AlexandreS. Behaviour of bacterial division protein FtsZ under a monolayer with phospholipid domains. *Biochim. Biophys. Acta* 1768 (2007): 2812–2821.
Lairion, F., Disalvo, E. A. Effect of trehalose on the contributions to the dipole potential of lipid monolayers. *Chem. Phys. Lipids* 150 (2007): 117–124.
Lairion, F., Disalvo, E. A. Effect of dipole potential variations on the surface charge potential of lipid membranes. *J. Phys. Chem. B* 113 (2009): 1607–1614.
Lee, K. Y., Klingler, J. F., McConnell, H. M. Electric field-induced concentration gradients in lipid monolayers. *Science* 263 (1994): 655–658.
Maggio, B. Modulation of phospholipase A2 by electrostatic fields and dipole potential of glycosphingolipids in monolayers. *J. Lipid Res.* 40 (1999): 930–939.
Marsh, D. General features of phospholipid phase transitions. *Chem. Phys. Lipids* 57 (1991): 109–120.

Marsh, D. Protein modulation of lipids, and vice-versa, in membranes. *Biochim. Biophys. Acta* 1778 (2008): 1545–1575.

Mathews, C. K. van Holde, K. E., Appling, D. R., Anthony-Cahill, S. J. *Biochemistry*. 4th ed. Upper Saddle River, NJ: Pearson-Prentice Hall, 2012.

McConnell, H. M. Theory of hexagonal and stripe phases in monolayers. *Proc. Natl. Acad. Sci. U.S.A.* 86 (1989): 3452–3455.

Mereuta, L., Asandei, A., Luchian, T. Meet me on the other side: Trans-bilayer modulation of a model voltage-gated ion channel activity by membrane electrostatics asymmetry. *PLoS One* 6 (2011): e25276.

Mihajlovic, M., Lazaridis, T. Calculations of pH-dependent binding of proteins to biological membranes. *J. Phys. Chem. B* 110 (2006): 3375–3384.

Morales-Rios, E., Montgomery, M. G., Leslie, A. G., Walker, J. E. Structure of ATP synthase from Paracoccus denitrificans determined by X-ray crystallography at 4.0 Å resolution. *Proc Natl Acad Sci USA*. 112 (2015): 13231–13236.

Mulgrew-Nesbitt, A., Diraviyam, K., Wang, J., et al. The role of electrostatics in protein–membrane interactions. *Biochim. Biophys. Acta* 1761 (2006): 812–826.

Nassoy, P., Birch, W. R., Andelman, D., Rondelez, F. Hydrodynamic mapping of two-dimensional electric fields in monolayers. *Phys. Rev. Lett.* 76 (1996): 455–458.

Nichesola, D., Perduca, M., Capaldi, S., Carrizo, M. E., Righetti, P. G., Monaco, H. L. Crystal structure of chicken liver basic fatty acid-binding protein complexed with cholic acid. *Biochemistry* 43 (2004): 14072–14079.

Nielsen, J. E., McCammon, J. A. Calculating pKa values in enzyme active sites. *Protein Sci.* 12 (2003): 1894–1901.

Nolan, V., Parduca, M., Monaco, H. L., Maggio, B., Montich, G. G. Interactions of chicken liver basic fatty acid binding-protein with lipid membranes. *Biochim. Biophys. Acta* 1611 (2003): 98–106.

Ojeda-May, P., Garcia, M. E. Electric field-driven disruption of a native beta-sheet protein conformation and generation of a helix-structure. *Biophys. J.* 99 (2010): 595–599.

Palmieri, B., Yamamoto, T., Brewster, R. C., Safran, S. A. Line active molecules promote inhomogeneous structures in membranes: Theory, simulations and experiments. *Adv. Coll. Int. Sci.* 208 (2014): 58–65.

Parthasarathy, R., Yu, C. H., Groves, J. T. Curvature-modulated phase separation in lipid bilayer membranes. *Langmuir* 22 (2006): 5095–5099.

Peterson, U., Mannock, D. A., Lewis, R. N., Pohl, P., McElhaney, R. N., Pohl, E. E. Origin of membrane dipole potential: Contribution of the phospholipid fatty acid chains. *Chem. Phys. Lipids* 117 (2002): 19–27.

Pilotelle-Bunner, A., Beaunier, P., Tandori, J., Maroti, P., Clarke, R. J., Sebban, P. The local electric field within phospholipid membranes modulates the charge transfer reactions in reaction centres. *Biochim. Biophys. Acta* 1787 (2009): 1039–1049.

Porschke, D. Macrodipoles. Unusual electric properties of biological macromolecules. *Biophys. Chem.* 66 (1997): 241–257.

Privalov, P. L. Physical basis of the stability of the folded conformations of proteins. In *Protein Folding*, ed. T. E. Creighton. New York: Freeman, pp. 83–126, 1992.

Przybylo, M., Procek, J., Hof, M., Langner, M. The alteration of lipid bilayer dynamics by phloretin and 6-ketocholestanol. *Chem. Phys. Lipids* 178 (2014): 38–44.

Raine, D. J., Norris, V. Lipid domain boundaries as prebiotic catalysts of peptide bond formation. *J. Theor. Biol.* 246 (2007): 176–185.

Richards, F. M. Folded and unfolded proteins: An introduction. In *Protein Folding*, ed. T. E. Creighton. New York: Freeman, pp. 1–58, 1992.

Rogaski, B., Klauda, J. B. Membrane-binding mechanism of a peripheral membrane protein through microsecond molecular dynamics simulations. *J. Mol. Biol.* 423 (2012): 847–861.

Seelig, J. Thermodynamics of lipid-peptide interactions. *Biochim. Biophys. Acta* 1666 (2004): 40–50.

Semrau, S., Idema, T., Schmidt, T., Storm, C. Membrane-mediated interactions measured using membrane domains. *Biophys. J.* 96 (2009): 4906–4915.

Sharp, K. A., Honig, B. Calculating total electrostatic energies with the nonlinear Poisson-Boltzmann equation. *J. Phys. Chem.* 94 (1990): 7684–7692.
Smaby, J. M., Brockman, H. L. Surface dipole moments of lipids at the argon-water interface. Similarities among glycerol-ester-based lipids. *Biophys. J.* 58 (1990): 195–204.
Sriram, I., Singhana, B., Lee, T. R., Schwartz, D. K. Line tension and line activity in mixed monolayers composed of aliphatic and terphenyl-containing surfactants. *Langmuir* 28 (2012): 16294–16299.
Starke-Peterkovic, T., Clarke, R. J. Effect of headgroup on the dipole potential of phospholipid vesicles. *Eur. Biophys. J.* 39 (2009): 103–110.
Stenmark, H., Aasland, R., Toh, B. H., D'Arrigo, A. Endosomal localization of the autoantigen EEA1 is mediated by a zinc-binding FYVE finger. *J. Biol. Chem.* 271 (1996): 24048–24104.
Stock, D., Leslie, A. G., Walker, J. E. Molecular architecture of the rotary motor in ATP synthase. *Science* 286 (1999): 1700–1705.
Tanford, C. *The Hydrophobic Effect: Formation of Micelles and Biological Membranes.* New York: Wiley-Interscience, 1980.
Taylor, D. M. Developments in the theoretical modelling and experimental measurement of the surface potential of condensed monolayers. *Adv. Coll. Int. Sci.* 87 (2000): 183–203.
Teissie, J. Biophysical effects of electric fields on membrane water interfaces: A mini review. *Eur. Biophys. J.* 36 (2007): 967–972.
Tocanne, J. F., Teissié, J. Ionization of phospholipids and phospholipid-supported interfacial lateral diffusion of protons in membrane model systems. *Biochim. Biophys. Acta* 1031 (1990): 111–142.
Tokumasu, F., Jin, A. J., Feigenson, G. W., Dvorak, J. A. Nanoscopic lipid domain dynamics revealed by atomic force microscopy. *Biophys. J.* 84 (2003): 2609–2618.
Toschi, F., Lugli, F., Biscarini, F., Zerbetto, F. Effects of electric field stress on a beta-amyloid peptide. *J. Phys. Chem. B* 113 (2009): 369–376.
Tsong, T. Y. Electrical modulation of membrane proteins: Enforced conformational oscillations and biological energy and signal transductions. *Annu. Rev. Biophys. Biophys. Chem.* 19 (1990): 83–106.
Vega Mercado, F., Maggio, B., Wilke, N. Modulation of the domain topography of biphasic monolayers of stearic acid and dimyristoyl phosphatidylcholine. *Chem. Phys. Lipids* 165 (2012): 232–237.
Vequi-Suplicy, C. C., Riske, K. A., Knorr, R. L., Dimova, R. Vesicles with charged domains. *Biochim. Biophys. Acta* 1798 (2010): 1338–1347.
Villarreal, M. A., Perduca, M., Monaco, H. L., Montich, G. G. Binding and interactions of L-BABP to lipid membranes studied by molecular dynamic simulations. *Biochim. Biophys. Acta* 1778 (2008): 1390–1397.
Voglino, L., McIntosh, T. J., Simon, S. A. Modulation of the binding of signal peptides to lipid bilayers by dipoles near the hydrocarbon-water interface. *Biochemistry* 37 (1998): 12241–12252.
Wang, L. Measurements and implications of the membrane dipole potential. *Annu. Rev. Biochem.* 81 (2012): 615–635.
Warshel, A., Aqvist, J. Electrostatic energy and macromolecular function. *Ann. Rev. Biophys. Biophys. Chem.* 20 (1991): 267–298.
White, S. H., Wimley, W. C. Membrane protein folding and stability: Physical principles. *Ann. Rev. Biophys. Biomol. Struct.* 28 (1999): 319–365.
Wilke, N. Monomolecular films of surfactants with phase-coexistence: Distribution of the phases and their consequences. In *Comprehensive Guide for Nanocoatings Technology*, ed. M. Aliofkhazraei. Hauppauge, NY: Nova Science Publishers, pp. 139–158, 2015.
Zamarreño, F., Herrera, F. E., Córsico, B., Costabel, M. D. Similar structures but different mechanisms: Prediction of FABPs-membrane interaction by electrostatic calculation. *Biochim. Biophys. Acta* 1818 (2012): 1691–1697.
Zhan, H., Lazaridis, T. Influence of the membrane dipole potential on peptide binding to lipid bilayers. *Biophys. Chem.* 161 (2011): 1–7.
Zhao, W., Yang, R. Experimental study on conformational changes of lysozyme in solution induced by pulsed electric field and thermal stresses. *J. Phys. Chem. B* 114 (2010): 503–510.

10 Experimental methods to manipulate cultured cells with electrical and electromagnetic fields

Ze Zhang
Axe Médecine Régénératrice–CHU
Université Laval, Québec, Canada

Shiyun Meng
Chongqing Technology and Business University
Chongqing, China

Mahmoud Rouabhia
Université Laval,
Québec, Canada

Contents

10.1 INTRODUCTION

Electrical (EF) and electromagnetic (EMF) fields are effective tools to manipulate cellular activities for different purposes, including tissue repair and regeneration. Since the early work about bioelectricity and bone growth in the 1960s (Jahn 1968), particularly in the last 20 years, a variety of experimental methods have been developed to modulate the cell activities under the influence of EF or EMF. Based on how a field is established, these methods can be grouped into different categories: electrode-based EF, Helmholtz coil–induced dynamic EMF, static magnetic field (MF), electromagnetically induced EF (transformer-like coupling), capacitively induced EF, conductive substrate–based EF, and other types. This chapter describes these methods, discusses their advantages and disadvantages, and aims to help readers understand the complexity involved in these techniques. The techniques used in classic electrophysiology and electroporation are out of the scope of this chapter.

10.2 ELECTRODE-BASED EF

Because culture medium is ionically conductive, the most convenient way to establish EF in a culture medium is by using electrodes. Electrodes are widely used in research and industry for analysis, synthesis, hydrolysis, plating, and so forth. In these processes, normally there are drastic chemical reactions at the surface of the electrodes and mass transport associated with the movement of charged species in the medium, which are intrinsic to any electrode-based process. However, electrode reactions often produce products that are toxic to cells; mass transport may affect the composition or distribution of the components in the medium. These two phenomena have to be carefully controlled to ensure that the biological observations are really the consequence of EF rather than because of the unwanted processes. As the most widely used method, there are different electrodes and experiment configurations adopted by different research groups, as presented below.

10.2.1 ELECTRODES COUPLED WITH SALT BRIDGE

Since at least the 1970s, salt bridge has become commonly used to avoid contaminations of electrode products in cell culture (Peng and Jaffe 1976). As illustrated in Figure 10.1, two U-shaped glass tubes filled with culture medium agar gel (Steinberg solution or KCl solution may also be used) bridge the culture plate and two beakers, inside of which two electrodes, such as Ag/AgCl electrodes, are inserted and connected to the power source. The beakers contain the same solution as that used to prepare the salt bridges. In such a way, upon the switch-on of the power source, an EF is established through the salt bridges between the two ends of a channel filled with culture medium in a culture plate. Because the electrical circuit in the liquid phase is completed through ionic conduction,

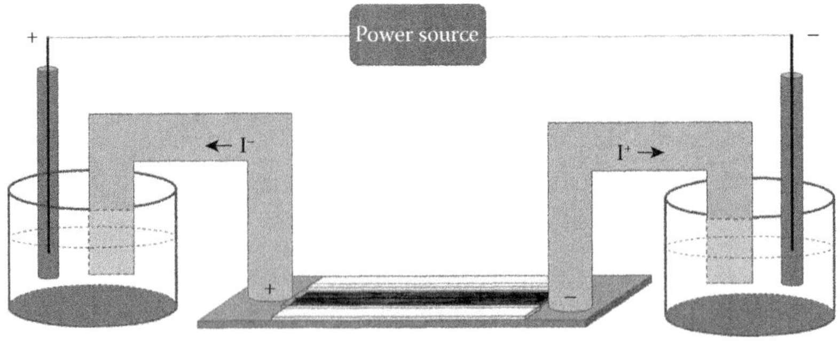

Figure 10.1 Illustration of electrode-based ES through salt bridges, showing cells cultured in a narrow channel to which EF is applied through a pair of Ag/AgCl electrodes and two salt bridges. EF is in culture medium and parallel to the culture plate.

there is an ionic current passing through the culture medium and salt bridges in the form of the movement of cations (I^+) and anions (I^-). Such ionic current, however, was reported to have no detectable effect on cellular behaviors such as neurite orientation. The first proof came from early work by McCaig's group (Hinkle et al. 1981). In that work, the authors designed a perfusion channel allowing culture medium to flow perpendicular to the direction of EF. Such a flow was to neutralize the possible displacement and polarization of nutrients in EF and was found to have no effect on the orientation of the neurites in EF. A large pool of culture medium and a similar composition of the electrolyte in salt bridges may also contribute to keeping the composition of the culture medium in the channel minimally disturbed. Nevertheless, electrophoresis is intrinsic under this configuration and proportional to the strength of the EF, which has to be carefully controlled and taken into account appropriately. The potential contamination to the culture medium by the redox products generated at the electrode surface is effectively blocked by the gel in salt bridges. The salt bridges also significantly increased the resistance of the circuit. The potential gradient to cultured cells should be measured between the ends of the channel.

This setup has been extensively used by McCaig's lab and various groups to study cell electrophysiology (Robinson 1985; McCaig and Zhao 1997). A modified protocol has been published in *Nature Protocols* for the culture of cardiac tissue or three-dimensional culture in the context of tissue engineering (Tandon et al. 2009). A video of the protocol has been published by Zhao's group (Meng et al. 2012).

10.2.2 DIRECT ES WITH ELECTRODE

Probably because of convenience, electrodes are also directly immersed in culture medium to provide electrical stimulation (ES). The setup can be as simple as illustrated in Figure 10.2, where the electrodes are two carbon robs, as reported by Berger et al. (1994). In that work, two parallel carbon robs of 1 cm in diameter were fixed on a specially designed cover of a standard 175 cm^2 culture flask. When the cover was in place, the two carbon electrodes were immersed in culture medium along the long axis of the flask. The authors reported that ES of 15 mA/cm^2 in current density, 0.5 V/cm in voltage gradient, and 0.1–7.0 Hz in frequency preserved the contractile function of

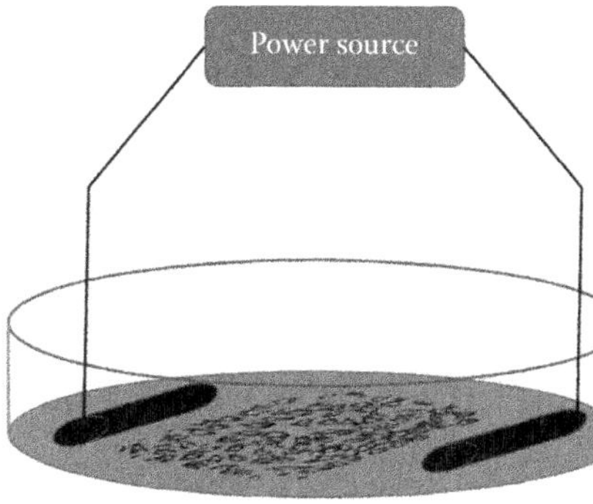

Figure 10.2 Illustration of direct ES through two solid electrodes immersed in culture medium. Cells are plated between the electrodes. EF is in culture medium and parallel to the culture plate.

primary myocytes. The authors selected a carbon electrode of this size because of its low resistivity, stability, and low cost, with respect to a platinum electrode, which is expensive, and a nickel alloy, which is corrosive and generates a large volume of heat. A very similar setup was reported in *Nature Protocols* (Tandon et al. 2009). In fact, a variety of electrodes and electrode configurations have been reported in the literature, ranging from platinum (Schlepüts 2011) to stainless steel, titanium, and titanium nitride (Serena et al. 2009), from two electrodes to three electrodes (i.e., using a conductive substrate as the working electrode, plus counter- and reference electrodes) (Schmidt et al. 1997), and from a single pair of electrodes to integrated electrode arrays (Ahadian et al. 2012; Xiong et al. 2015). Here, we do not include neural prostheses that are highly specialized arrays of microelectrodes for the stimulation and recording of neurons.

While it is easy to fix electrodes in a culture plate, great care must be exercised to avoid harmful redox products and make sure what is observed is indeed the consequence of EF, but the secondary effect of ES. Unlike what is mentioned in previous texts, where salt bridges block redox products from reaching cultured cells, direct ES with electrodes in culture medium represents a risk of generating cytotoxic or potentially stimulatory chemicals. While cytotoxicity can be easily tested, the effect of redox chemicals on cell growth and other activities is unpredictable without knowing the exact nature and quantity of such products. A technique commonly used in neural stimulation called charge-balanced biphasic ES is adopted to reduce the negative effect of the electrode reaction. The principle is to generate two consecutive pulses of equal magnitude in reverse directions (e.g., cathodic followed by anodic). In such a way, after each ES cycle, that is, a pair of cathodic and anodic pulses, there is neither net charge transfer between electrodes nor charge accumulation at the electrodes. While this technique does cancel the net Faradaic current, and hence prevents electrophoresis, irreversible electrode reaction such as water electrolysis can still take place if the voltage is higher than the threshold, that is, about 1.23 V between electrodes. In fact, a pulse as short as 300 ns was reported to generate hydrogen at a platinum electrode (Shimizu et al. 2006). In culture medium, the redox reaction may not be limited to water hydrolysis. A frequent refresh of medium helps minimize the potential change in medium composition. Even so, a control experiment using ES-conditioned culture medium is recommended (Shi et al. 2008). Obviously, direct ES with electrodes is not suggested to provide constant direct current (DC) ES.

To reduce redox contamination, Krauthamer et al. (1991) designed a flow chamber to test the effect of ES on neuroblastoma cells. In that work, two parallel silver wires of 0.5 mm in diameter and 2.5 cm in length were immersed in culture medium separated by a 1 cm space. A constant flow of culture medium at 0.67 ml/min was performed in a 35 mm petri dish, which was expected to eliminate redox contamination. However, in most reported direct ES experiments, no flow chamber had been used. Sato et al. (2008) carried out ES to Müller cells with two needle-type Ag/AgCl electrodes of 0.2 mm in diameter and 2 cm in length inserted in culture medium above the cells in a 35 mm petri dish. A rectangular biphasic train of pulses (1 ms pulse width at 20 Hz and current intensity up to 10 mA) was used. What should be noticed is that the ES was only performed for 30 min. Apparently, a short stimulation time reduces the impact of electrode reactions.

10.3 STIMULATION THROUGH INDUCTIVE EMF

This is a noninvasive approach with respect to the electrode-based ES mentioned previously. Because of its noninvasive nature, EMF is a technique that has been used in the clinic, in orthopedics in particular, to improve wound healing (see Chapter 19 for more information). A basic setup is illustrated in Figure 10.3. Two identical coils of metal wire (Helmholtz coil) with radius *R* are placed in parallel with a separation of *R*. When electrical current passes through the coils, a nearly spatially uniform MF is produced inside and between the coils where the culture plate is placed. With the change of current, a temporally changing EMF is generated accordingly parallel to the coil axis and perpendicular to the culture plate. Different from EF, a geomagnetic field up to about 70 μT is found everywhere if not shielded, and laboratory instruments also generate EMF. These background signals should be considered for both the samples under test and controls. Another important aspect to be considered is the EF that always coexists with a changing MF. A time-varying MF induces electrical current in culture medium in planes perpendicular to the direction of MF, of which the current density can be calculated according to Faraday's law (Tenforde 1996). In the data reported in literature, such induced electrical current and EF strength are relatively weak compared with the EF magnitude reported in other ES methods. For example, an EMF of 2 mT and 75 Hz was reported to induce an EF of 5 mV (Fassina et al. 2006). Heat generation is another factor that one should take into consideration.

Early studies about the effect of EMF on biological systems mostly focused on its safety issues in the context of the increasing presence of electronic and electrical equipment or facilities in close proximity to the human body (Stevens 1996), and the wide availability of medical magnetic resonance imaging (MRI) technology, such as the effect on embryos (Zusman et al. 1990) and lymphocytes (Antonopoulos et al. 1995). Seemingly, Liboff et al. (1984) reported the upregulation of DNA synthesis in fibroblasts treated with an EMF of the same magnitude as the geomagnetic field from 15 Hz to 4 kHz, suggesting possible mutagenesis.

Using EMF to accelerate wound healing probably started from bone repair. In fact, the EBI Bone Healing System from Electro-Biology Inc. was approved by the Food and Drug Administration (FDA) in 1979. EBI's stimulator provides noninvasive pulsed EMF (PEMF) to help the healing of bone nonunion and failed spinal fusion. Study at cellular and molecular

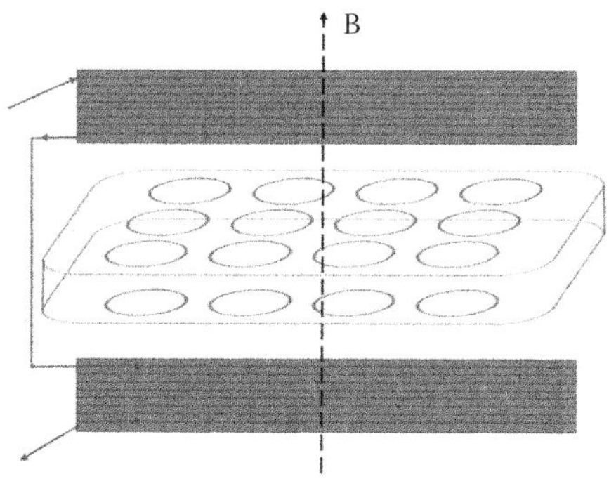

Figure 10.3 Illustration of a 12-well culture plate placed between Helmholtz coils. The induced MF (B) is between the two coils and perpendicular to the culture plate.

levels in the context of tissue repair likely started in the 1980s and early 1990s (Ross 1990). Over the years, the same technique (Helmholtz coil) is still used in studies with little modification (McFarlane et al. 2000; Bai et al. 2013; Falone et al. 2016).

10.4 STIMULATION THROUGH STATIC MF

Static MF generated by magnets was also used to study cell growth (McDonald 1993). However, static MF had been rarely reported in literature until recently, when laboratories equipped with a diamagnetic levitation facility joined the research. In diamagnetic levitation, a large gradient high magnetic field (LGHMF) is used to create a weightless environment to simulate space flight. The field strength in LGHMF can be as high as several tesla, in comparison with the millitesla range normally used in the Helmholtz coil experiment mentioned above. In such a study, cells may experience strong static MF and different gravities (micro, normal, or hypergravity) simultaneously (Di et al. 2012). Unfortunately, such a facility is not readily available.

10.5 ELECTROMAGNETICALLY INDUCED EF

This is a unique design originally reported by Hess et al. (2012a, 2012b) and his colleagues (Thrivikraman et al. 2015), with the objective of establishing a uniform EF in culture medium without any electrode (noninvasive) and, at the same time, without the interference of MF (thus different from EMF stimulation). The design is based on the principle of an electrical transformer, and so being called "transformer-like coupling" by the inventors. The first coil of the transformer is connected with a function generator, which produces a changing MF through the magnetic core. This changing MF induces an electrical current in the second "coil" formed by a loop of culture medium, as illustrated in Figure 10.4. In such a way, an EF is created inside the culture medium through

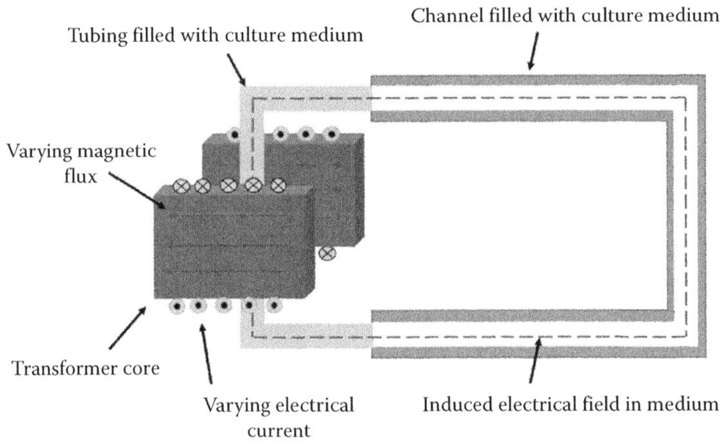

Figure 10.4 Transformer-like coupling, illustrating the primary coil that carries varying electrical current, the transformer core (complete circular core not illustrated), and the varying magnetic flux induced by the electrical current, and the secondary coil formed by a tubing filled with culture medium and the channel-like culture chamber that is also filled with culture medium to complete the loop of the secondary coil. The electrical voltage in culture medium induced by magnetic flux changes according to the electrical current in the primary coil.

electromagnetic induction with the first coil. The form and strength of EF in the culture medium is therefore determined by the EF in the first coil. In Hess's initial work (2012a), a rectangular pulsed ES of 3.6 V/m was used to stimulate human mesenchymal stem cells (MSCs) and was found to up-regulate the expression of osteogenic markers.

The unique feature of this design is the elimination of electrodes, and hence the elimination of any electrode reactions that may complicate the interpretation of experiment data. Since the cell culture portion of the conducting loop is outside the magnetic core, the effect of the MF is negligible. On the other hand, this design is unable to provide a constant DC ES because electromagnetic induction requires a time-varying MF.

10.6 CAPACITIVELY INDUCED ES

This can also be considered a noninvasive ES because the capacitive plates do not directly contact the culture medium. As illustrated in Figure 10.5, this setup is structured like a plate capacitor. The dielectrics between the two metal plates are culture medium plus two pieces of glass or other insulator to which the plates are attached with the help of adhesive. Upon applying an electrical potential between the two plates, an EF is established between the plates, causing polarization of the molecules in the medium, including water and proteins, and those on cell membranes (Finkelstein et al. 2007). An instant displacement current occurs as a result. Compared with the electrode-based ES, the capacitively induced ES neither has any electrode reaction nor causes any continuous ionic flow in culture medium. Because the EF is perpendicular to the culture plate, any potential redistribution of charged particles because of electrophoresis would be the same to all cells. To overcome the low permittivity of glass or other materials used to isolate metal plates from contacting culture medium, a high-voltage power source is required. The glass side of the glass–metal plate on top of the culture dish should be in contact with culture medium to avoid any air. The permittivity of air is more than 10 times lower than that of water. The strength of EF in culture medium has to be calculated by taking into consideration the electrodes, glass cover slip, and conductive glue.

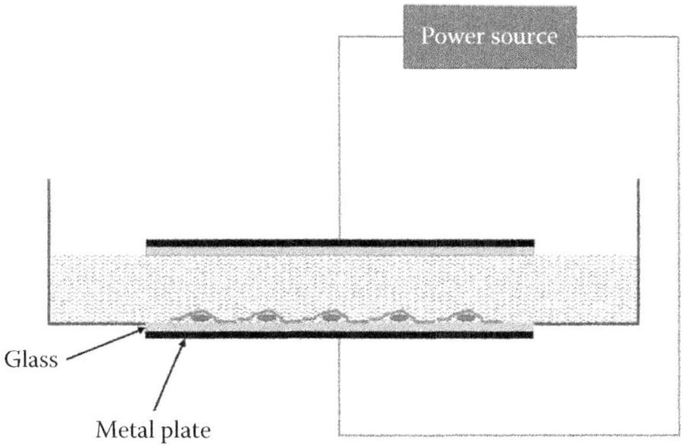

Figure 10.5 ES through capacitively coupled EF. Two metal plates are placed in parallel and in close proximity to culture medium. The metal plates are isolated from the medium by dielectric materials such as glass slides. EF between the two plates induces polarization of the molecules in between.

Korenstein et al. (1984) designed probably the first capacitively coupled ES device using two circular copper electrodes of 54 mm in diameter. Bone cells isolated from rat embryo calvaria were plated in a Petri dish and stimulated with a series of rectangular pulses (25 μs at 3 Hz). They used potential gradients from 10 to 54 V/cm and reported displacement currents of 0.7 to 4.0 A. The ES was found to change the intracellular level of cyclic AMP and enhance DNA synthesis. One may notice the much higher voltage and current reported in this work in comparison with other methods described in this chapter. However, this voltage must not be interpreted as the field strength in culture medium, because of the presence of other dielectrics between the plates. The displacement current is only an instant phenomenon. This technique has been used extensively by Brighton's group since 1984, if not earlier, on bone cells (Brighton et al. 1988a, 1992), chondrocytes (Brighton et al. 1988b, 1989), and tissues of bone (Brighton et al. 1985) and cartilage (Brighton et al. 1984). Mechanistically, they used this technique to study the signaling transduction in bone cells (Brighton et al. 2001), gene and protein expression in cartilage explant (Brighton et al. 2008) and selected gene expression in stimulated bone cells (Clark et al. 2014).

10.7 CONDUCTIVE SUBSTRATE–BASED ES

This method was first reported by Shi et al. (2004) in the context of tissue engineering. The idea is to culture cells on a conductive substrate or scaffold, and then pass electrical current through this substrate. While it is very much ignored in most textbooks, surface charges do exist on a current-carrying circuit (Jackson 1996), and the EF generated by such a static surface charge can be visualized as well (Jacobs et al. 2010). In such a way, the cells in contact with the substrate are exposed to the EF. As illustrated in Figure 10.6, the substrate is integrated as a conductor in a closed circuit. Upon being connected to a power source, current passes through the substrate rather than through the liquid phase. It is important to point out that in the absence of any electrode reaction, there is no ionic flow in the medium even though the medium is conductive.

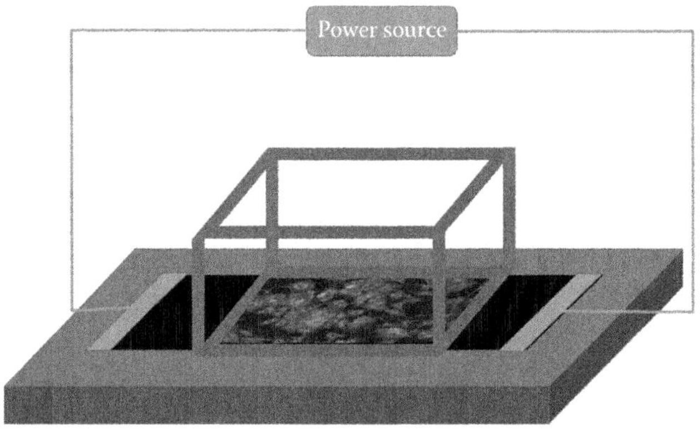

Figure 10.6 ES through conductive substrate. The black conductive substrate is sandwiched between a rectangular plastic culture well and an insulating plate in such a way that there is no leakage of culture medium from the well. The ends of the conductive substrate outside the culture well are connected to a power source. Cells grown on substrate are exposed to the potential gradient along the surface of the substrate.

The materials used to construct the substrate are conductive polymers. Different from metals, conductive polymers are semiconductors, by which one can establish the required potential gradient to stimulate cells (e.g., 10–200 mV/mm) at a very low current density (e.g., microamperes). A high current density is not desirable because of the Joule effect. In this ES model, it is the potential gradient rather than current density that likely plays the principal role (Shi et al. 2004). Another advantage of using conductive polymers is that they can be comprised of other types of polymers to fabricate conductive scaffolds of various surface morphologies, geometries, and mechanical properties. This provides the possibility of stimulating cells or tissues in a geometrically well-defined complex space or scaffold. The drawback of using conductive polymers is their instability in electrical conductivity in an aqueous environment. However, this can be dealt with by selecting appropriate ES protocols, such as short ES time and more stable conductive polymers.

Our group has developed three types of conductive substrates based on polypyrrole (PPy) and poly(3,4-ethylenedioxythiophene) (PEDOT). The first is a composite made of 5% PPy particles and 95% poly-L-lactide (PLLA) (Shi et al. 2004). Because PPy is not degradable, a low content of it is preferred if one wants to use this composite as a temporary implant. When percolation is formed among the PPy particles, this formulation provides a reasonable electrical conductivity (ca. 10^3 ohms/square) while remaining largely bioabsorbable. Importantly, this material is stable enough to stand continuous DC ES for several hours. The conductivity of conductive polymers gradually deteriorates in an aqueous environment, a process mainly because of dedoping that is accelerated by electrical current (Shi et al. 2004). This composite is also relatively rigid and brittle. Obviously, a less rigid, even elastic, composite can be made by mixing PPy particles with other types of polymers. The PPy-coated polyethylene terephthalate (PET) fabric has a mechanical property similar to that of the original fabric. The thin PPy coating, however, requires a pulsed ES because a continuous DC ES quickly reduces its conductivity. A short period of idle time between each pulse allows the thin PPy coating to be quickly redoped by the small anions in culture medium to regain its conductivity (Wang et al. 2013). The third type of substrate we used is a conductive web made of nonwoven PLLA coated with PEDOT (Niu et al. 2015). Compared with the PPy-coated PET fabric, this conductive web is biodegradable and electrically more stable as well. The coating of fabric with conductive polymers appears easy and straightforward. Nevertheless, attention should be paid to avoid an unnecessarily thick coating. A thick coating not only changes the mechanical property of the original fabric but also reduces its porosity and risks delamination.

Through the setup showed in Figure 10.6, we were able to provide both constant and pulsed DC ES to human skin fibroblasts (Shi et al. 2004, 2008a, 2008b; Meng et al. 2008; Wang et al. 2015, 2016; Park et al. 2015; Rouabhia et al. 2013, 2016), osteoblasts (Meng et al. 2011, 2013), and neuron-like cells (Zhang et al. 2007) and nerve tissue (Du et al. 2016). This setup has also been used by Lee et al. (2009) to stimulate PC12 cells cultured on the PPy-coated electrospun poly(lactic-*co*-glycolic acid) (PLGA) meshes, by Nguyen et al. (2014) to stimulate dorsal root ganglia and PC12 cells on polypyrrole-block-polycaprolactone (PPy–PCL), and by Hardy et al. (2015) to stimulate MSCs cultured on a PPy-PSS-coated electrospun PCL mat. An interesting work reported by Hsiao et al. (2013) showed that cardiomyocytes cultured on a conductive electrospun mesh made of PLGA and polyaniline emeraldine base (PANI-EB) were able to synchronize their beating with external ES signals.

10.8 OTHER TYPES OF ES

10.8.1 STIMULATION WITH LIGHT

As a form of EM radiation, light has also been studied to stimulate mammalian cells. One approach was to use photoactive nanoparticles (NPs), as reported by Pappas et al. (2007). In that work, photoactive NPs made of HgTe nanocolloid were stimulated with illumination at 550 nm, which generated a current spike in the magnitude of several microamperes because of the NP to O_2 electron transfer. Such photoelectrical activity was found to activate the electrophysiological activity of NG108 cells (a neuroblastoma glioma cell line). In a different work, a near-infrared laser radiation (NIR) applied to human neural stem cells (hNSCs) cultured on the reduced graphene oxide mesh was found to increase neuronal differentiation (Akhavan et al. 2015). After excluding the contribution of temperature, the authors suggested that the NIR excited the low-energy photoelectron in graphene, which in turn stimulated hNSCs. Light can also be converted into electrical potential to provide ES, as reported by Hsiao et al. (2016). In their work, photovoltaic materials were constructed into power sources that can generate an open-circuit voltage from 0.11 to 0.49 V upon the absorption of NIR. They demonstrated that the generated power was enough to stimulate PC12 cell differentiation with longer neurites (Hsiao et al. 2016). NIR is known by its capability of tissue penetration that is about 4 mm. The authors therefore suggested the potential of using such a device as a wireless implantable power source to provide ES. Notably, this is different from the infrared neural stimulation (INS) that also uses infrared radiation. INS is based on the heat transient converted from infrared to water molecules (Roe et al. 2015).

10.8.2 HYBRID STIMULATION

In addition to EF, cells *in vivo* are also exposed to other conditions that are not accommodated by a conventional monolayer cell culture. These conditions include fluid shear stress to endothelial cells, and mechanical stress to bone and muscle cells. Hybrid stimulation combines ES with other types of designs, such as bioreactors and mechanical stimulation. A work reported by Maidhof et al. (2012) integrated ES with a perfusion bioreactor, and found that under the synergy of ES and perfusion, rat cardiac cells showed increased contraction, DNA content, cell distribution through scaffold, and better tissue morphology. Liu et al. (2013) cultured osteoblasts on a nanostructured PPy membrane and applied both ES and mechanical force to the cells. They reported better cell proliferation, alkaline phosphatase activity, and collagen type I expression upon the dual stimulation of electricity and mechanical force.

10.9 CONCLUDING REMARKS

The variety of methods in electrical and electromagnetic stimulations provide flexibility in setting up experiments based on researchers' expertise and resources. However, this flexibility comes up with a price tag, that is, the complexity when it comes to comparing the data in the literature. While it appears relatively easy to appreciate the difficulty in correlating the biological effects of EF with those of EMF, because their mechanisms of molecular response are not necessarily the same, it may not be much easier if one wants to compare the cellular behaviors in EF of the same strength but generated by different methods.

EF is a vector, having both magnitude and direction. The membrane proteins sensible to EF have both mass and orientation (that may change temporally), a nature similar to that of a vector. EFs parallel (e.g., salt bridge electrodes) or perpendicular (e.g., capacitively coupled EF) to a cell monolayer therefore may not have the same impact on membrane proteins even though they are the same magnitude. The polarization of cell membrane, ECM, and soluble molecules is also sensitive to the magnitude, direction, and frequency of EF. All these tell us that a same set of ES parameters may generate different biological consequences if they are used in different experimental setups. Researchers must be aware of the characteristics and the advantages and disadvantages of the methodologies reported in the literature. Only based on such knowledge may one decide the experimental platform most appropriate for him or her.

REFERENCES

Ahadian, S., Ramón-Azcón, J., Ostrovidov, S., et al. Interdigitated array of Pt electrodes for electrical stimulation and engineering of aligned muscle tissue. *Lab Chip* 12 (2012): 3491–3503.

Akhavan, O., Ghaderi, E., Shirazian, S. A. Near infrared laser stimulation of human neural stem cells into neurons on graphene nanomesh semiconductors. *Colloids Surf B Biointerfaces* 126 (2015): 313–321.

Antonopoulos, A., Yang, B., Stamm, A., Heller, W. D., Obe, G. Cytological effects of 50 Hz electromagnetic fields on human lymphocytes *in vitro*. *Mutat Res* 346 (1995): 151–157.

Bai, W. F., Xu, W. C., Feng, Y., et al. Fifty-hertz electromagnetic fields facilitate the induction of rat bone mesenchymal stromal cells to differentiate into functional neurons. *Cytotherapy* 15 (2013): 961–970.

Berger, H. J., Prasad, S. K., Davidoff, A. J., et al. Continual electric field stimulation preserves contractile function of adult ventricular myocytes in primary culture. *Am J Physiol* 266 (1994): H341–H349.

Brighton, C. T., Jensen, L., Pollack, S. R., Tolin, B. S., Clark, C. C. Proliferative and synthetic response of bovine growth plate chondrocytes to various capacitively coupled electrical fields. *J Orthop Res* 7 (1989): 759–765.

Brighton, C. T., McCluskey, W. P. Response of cultured bone cells to a capacitively coupled electric field: Inhibition of cAMP response to parathyroid hormone. *J Orthop Res* 6 (1988a): 567–571.

Brighton, C. T., Nichols, C. E., 3rd, Arangio, G. A. Amelioration of oxygen-induced osteoporosis in the *in vitro* fetal rat tibia with a capacitively coupled electrical field. *J Orthop Res* 3 (1985): 311–320.

Brighton, C. T., Okereke, E., Pollack, S. R., Clark, C. C. In vitro bone-cell response to a capacitively coupled electrical field. The role of field strength, pulse pattern, and duty cycle. *Clin Orthop Relat Res* 285 (1992): 255–262.

Brighton, C. T., Townsend, P. F. Increased cAMP production after short-term capacitively coupled stimulation in bovine growth plate chondrocytes. *J Orthop Res* 6 (1988b): 552–558.

Brighton, C. T., Unger, A. S., Stanbough, J. L. *In vitro* growth of bovine articular cartilage chondrocytes in various capacitively coupled electrical fields. *J Orthop Res* 2 (1984): 15–22.

Brighton, C. T., Wang, W., Clark, C. C. The effect of electrical fields on gene and protein expression in human osteoarthritic cartilage explants. *J Bone Joint Surg Am* 90 (2008): 833–848.

Brighton, C. T., Wang, W., Seldes, R., Zhang, G., Pollack, S. R. Signal transduction in electrically stimulated bone cells. *J Bone Joint Surg Am* 83A (2001): 1514–1523.

Clark, C. C., Wang, W., Brighton, C. T. Up-regulation of expression of selected genes in human bone cells with specific capacitively coupled electric fields. *J Orthop Res* 32 (2014): 894–903.

Di, S., Tian, Z., Qian, A., et al. Large gradient high magnetic field affects FLG29.1 cells differentiation to form osteoclast-like cells. *Int J Radiat Biol* 88 (2012): 806–813.

Du, Z., Bondarenko, O., Wang, D., Rouabhia, M., Zhang, Z. Ex vivo assay of electrical stimulation to rat sciatic nerves: Cell behaviors and growth factor expression. *J Cell Physiol* 231 (2016): 1301–1312.

Falone, S., Marchesi, N., Osera, C., et al. Pulsed electromagnetic field (PEMF) prevents pro-oxidant effects of H2O2 in SK-N-BE(2) human neuroblastomacells. *Int J Radiat Biol* 4 (2016): 1–6.

Fassina, L., Visai, L., Benazzo, F., et al. Effects of electromagnetic stimulation on calcified matrix production by SAOS-2 cells over a polyurethane porous scaffold. *Tissue Eng* 12 (2006): 1985–1999.

Finkelstein, E. I., Chao, P. H., Hung, C. T., Bulinski, J. C. Electric field-induced polarization of charged cell surface proteins does not determine the direction of galvanotaxis. *Cell Motil Cytoskeleton* 64 (2007): 833–846.

Hardy, J. G., Villancio-Wolter, M. K., Sukhavasi, R. C., et al. Electrical stimulation of human mesenchymal stem cells on conductive nanofibers enhances their differentiation toward osteogenic outcomes. *Macromol Rapid Commun* 36 (2015): 1884–1890.

Hess, R., Jaeschke, A., Neubert, H., et al. Synergistic effect of defined artificial extracellular matrices and pulsed electric fields on osteogenic differentiation of human MSCs. *Biomaterials* 33 (2012b): 8975–8985.

Hess, R., Neubert, H., Seifert, A., Bierbaum, S., Hart, D. A., Scharnweber, D. A novel approach for *in vitro* studies applying electrical fields to cell cultures by transformer-like coupling. *Cell Biochem Biophys* 64 (2012a): 223–232.

Hinkle, L., McCaig, C. D., Robinson, K. R. The direction of growth of differentiating neurones and myoblasts from frog embryos in an applied electric field. *J Physiol* 314 (1981): 121–135.

Hsiao, C. W., Bai, M. Y., Chang, Y., et al. Electrical coupling of isolated cardiomyocyte clusters grown on aligned conductive nanofibrous meshes for their synchronized beating. *Biomaterials* 34 (2013): 1063–1072.

Hsiao, Y. S., Liao, Y. H., Chen, H. L., Chen, P., Chen, F. C. Organic photovoltaics and bioelectrodes providing electrical stimulation for PC12 cell differentiation and neurite outgrowth. *ACS Appl Mater Interfaces* 8 (2016): 9275–9284.

Jackson, D. J. Surface charges on circuit wires and resistors play three roles. *Am J Phys* 64 (1996): 855–870.

Jacobs, R., Salazar, A., Nassar, A. New experimental method of visualizing the electric field due to surface charges on circuit elements. *Am J Phys* 78 (2010): 1432–1433.

Jahn, T. L. A possible mechanism for the effect of electrical potentials on apatite formation in bone. *Clin Orthop Relat Res* 56 (1968): 261–273.

Korenstein, R., Somjen, D., Fischler, H., Binderman, I. Capacitative pulsed electric stimulation of bone cells. Induction of cyclic-AMP changes and DNA synthesis. *Biochim Biophys Acta* 803 (1984): 302–307.

Krauthamer, V., Bekken, M., Horowitz, J. L. Morphological and electrophysiological changes produced by electrical stimulation in cultured neuroblastoma cells. *Bioelectromagnetics* 12 (1991): 299–314.

Lee, J. Y., Bashur, C. A., Goldstein, A. S., Schmidt, C. E. Polypyrrole-coated electrospun PLGA nanofibers for neural tissue applications. *Biomaterials* 30 (2009): 4325–4335.

Liboff, A. R., Williams, T., Jr., Strong, D. M., Wistar, R., Jr. Time-varying magnetic fields: Effect on DNA synthesis. *Science* 223 (1984): 818–820.

Liu, L., Li, P., Zhou, G., et al. Increased proliferation and differentiation of pre-osteoblasts MC3T3-E1 cells on nanostructured polypyrrole membrane under combined electrical and mechanical stimulation. *J Biomed Nanotechnol* 9 (2013): 1532–1539.

Maidhof, R., Tandon, N., Lee, E. J., et al. Biomimetic perfusion and electrical stimulation applied in concert improved the assembly of engineered cardiac tissue. *J Tissue Eng Regen Med* 6 (2012): e12–e23.

McCaig, C. D., Zhao, M. Physiological electrical fields modify cell behaviour. *Bioessays* 19 (1997): 819–826.

McDonald, F. Effect of static magnetic fields on osteoblasts and fibroblasts *in vitro*. *Bioelectromagnetics* 14 (1993): 187–196.

McFarlane, E. H., Dawe, G. S., Marks, M., Campbell, I. C. Changes in neurite outgrowth but not in cell division induced by low EMF exposure: Influence of field strength and culture conditions. *Bioelectrochemistry* 52 (2000): 23–28.

Meng, S., Rouabhia, M., Shi, G., Zhang, Z. Heparin dopant increases the electrical stability, cell adhesion, and growth of conducting polypyrrole/poly(L,L-lactide) composites. *J Biomed Mater Res* 87A (2008): 332–344.

Meng, S., Rouabhia, M., Zhang, Z. Electrical stimulation modulates osteoblast proliferation and bone protein production through heparin-bioactivated conductive scaffolds. *Bioelectromagnetics* 34 (2013): 189–199.

Meng, S., Zhang, S., Rouabhia, M. Accelerated osteoblast mineralization on a conductive substrate by multiple electrical stimulation. *J Bone Miner Metab* 29 (2011): 535–544.

Meng, X., Li, W., Young, F., et al. Electric field-controlled directed migration of neural progenitor cells in 2D and 3D environments. *J Vis Exp* 60 (2012): e3453.

Nguyen, H. T., Sapp, S., Wei, C., et al. Electric field stimulation through a biodegradable polypyrrole-co-polycaprolactone substrate enhances neural cell growth. *J Biomed Mater Res A* 102 (2014): 2554–2564.

Niu, X., Rouabhia, M., Chiffot, N., King, M. W., Zhang, Z. An electrically conductive 3D scaffold based on a nonwoven web of poly(L-lactic acid) and conductive poly(3,4-ethylenedioxythiophene). *J Biomed Mater Res A* 103 (2015): 2635–2644.

Pappas, T. C., Wickramanyake, W. M., Jan, E., Motamedi, M., Brodwick, M., Kotov, N. A. Nanoscale engineering of a cellular interface with semiconductor nanoparticle films for photoelectric stimulation of neurons. *Nano Lett* 7 (2007): 513–519.

Park, H. J., Rouabhia, M., Lavertu, D., Zhang, Z. Electrical stimulation modulates the expression of multiple wound healing genes in primary human dermal fibroblasts. *Tissue Eng* 21 (2015): 1982–1990.

Peng, H. B., Jaffe, L. F. Polarization of fucoid eggs by steady electrical fields. *Dev Biol* 53 (1976): 277–284.

Robinson, K. R. The responses of cells to electrical fields: A review. *J Cell Biol* 101 (1985): 2023–2027.

Roe, A. W., Chernov, M. M., Friedman, R. M., Chen, G. *In vivo* mapping of cortical columnar networks in the monkey with focal electrical and optical stimulation. *Front Neuroanat* 9 (2015): 135.

Rouabhia, M., Park, H. J., Meng, S., Derbali, H., Zhang, Z. Electrical stimulation promotes wound healing by enhancing dermal fibroblast cell growth, migration, growth factor secretion, and alpha smooth muscle actin expression. *PLoS One* 8 (2013): e71660.

Rouabhia, M., Park, H. J., Zhang, Z. Electrically activated primary human fibroblasts improve *in vitro* and *in vivo* skin regeneration. *J Cell Physiol* 231 (2016): 1814–1821.

Ross, S. M. Combined DC and ELF magnetic fields can alter cell proliferation. *Bioelectromagnetics* 11 (1990): 27–36.

Sato, T., Fujikado, T., Lee, T. S., Tano, Y. Direct effect of electrical stimulation on induction of brain-derived neurotrophic factor from cultured retinal Muller cells. *Invest Ophthalmol Vis Sci* 49 (2008): 4641–4646.

Schlepütz, M., Uhlig, S., Martin, C. Electric field stimulation of precision-cut lung slices. *J Appl Physiol* 110 (2011): 545–554.

Schmidt, C. E., Shastri, V. R., Vacanti, J. P., Langer, R. Stimulation of neurite outgrowth using an electrically conducting polymer. *Proc Natl Acad Sci U S A* 94 (1997): 8948–8953.

Serena, E., Figallo, E., Tandon, N., et al. Electrical stimulation of human embryonic stem cells: Cardiac differentiation and the generation of reactive oxygen species. *Exp Cell Res* 315 (2009): 3611–3619.

Shi, G., Rouabhia, M., Meng, S., Zhang, Z. Electrical stimulation enhances viability of human cutaneous fibroblasts on conductive biodegradable substrates. *J Biomed Mater Res A* 84 (2008a): 1026–1037.

Shi, G., Zhang, Z., Rouabhia, M. The regulation of cell functions electrically using biodegradable polypyrrole-polylactide conductors. *Biomaterials* 29 (2008b): 3792–3798.

Shi, G., Rouabhia, M., Wang, Z., Dao, L. H., Zhang, Z. A novel electrically conductive and biodegradable composite made of polypyrrole nanoparticles and polylactide. *Biomaterials* 25 (2004): 2477–2488.

Shimizu, N., Hotta, S., Sekiya, T., Oda, S. A novel method of hydrogen generation by water electrolysis using an ultra-short-pulse power supply. *J Appl Electrochem* 36 (2006): 419–423.

Stevens, R. G. Epidemiological studies of electromagnetic fields and health. In *Handbook of Biological Effects of Electromagnetic Fields*, ed. C. Polk and E. Postow. 2nd ed. Boca Raton, FL: CRC Press, 1996, 275–294.

Tandon, N., Cannizzaro, C., Chao, P. H. et al. Electrical stimulation systems for cardiac tissue engineering. *Nat Protoc* 4 (2009): 155–173.

Tenforde, T. S. Interaction of ELF magnetic fields with living systems. In *Handbook of Biological Effects of Electromagnetic Fields*, eds. C. Polk and E. Postow. 2nd ed. Boca Raton, FL: CRC Press, 1996, 185–230.

Thrivikraman, G., Lee, P. S., Hess, R., Haenchen, V., Basu, B., Scharnweber, D. Interplay of substrate conductivity, cellular microenvironment, and pulsatile electrical stimulation toward osteogenesis of human mesenchymal stem cells *in vitro*. *ACS Appl Mater Interfaces* 7 (2015): 23015–23028.

Wang, Y., Rouabhia, M., Lavertu, D., Zhang, Z. Pulsed electrical stimulation modulates fibroblasts' behavior through Smad signaling pathway. *J Tissue Eng Regen Med* 2015. DOI: 10.1002/term.

Wang, Y., Rouabhia, M., Zhang, Z. PPy-coated PET fabrics and electric pulse-stimulated fibroblasts. *J Mater Chem B* 1 (2013): 3789–3796.

Wang, Y., Rouabhia, M., Zhang, Z. Pulsed electrical stimulation benefits wound healing by activating skin fibroblasts through the TGFβ1/ERK/NFκ axis. *BBA Gen Subj* 1860 (2016): 1551–1559.

Xiong, G. M., Do, A. T., Wang, J. K., Yeoh, C. L., Yeo, K. S., Choong, C. Development of a miniaturized stimulation device for electrical stimulation of cells. *J Biol Eng* 9 (2015): 14.

Zhang, Z., Rouabhia, M., Wang, Z., et al. Electrically conductive biodegradable polymer composite for nerve regeneration: Direct current stimulated neurite outgrowth and axon regeneration. *Artif Organs* 31 (2007): 13–22.

Zusman, I., Yaffe, P., Pinus, H., Ornoy, A. Effects of pulsing electromagnetic fields on the prenatal and postnatal development in mice and rats: *In vivo* and *in vitro* studies. *Teratology* 42 (1990): 157–170.

11 The neurotrophic factor rationale for using brief electrical stimulation to promote peripheral nerve regeneration in animal models and human patients

Tessa Gordon
Peter Gilgan Centre for Research and Learning
Toronto, Ontario, Canada

Contents

11.1 INTRODUCTION

The glial Schwann cells of the peripheral nervous system provide support for the regeneration of lost axons (Fenrich and Gordon 2004). Peripheral nerve injuries that disrupt the continuity of the axons result in the degeneration of the axons distal to the injury and the activation of regenerative programs in the injured neurons (Gordon and Sulaiman 2013; Gordon, 2015). Nerve injuries are sustained in association with trauma, including brachial plexus injuries in babies, and gunshot and knife injuries in children and adults (Midha 2011; Sulaiman et al. 2011; Sulaiman and Gordon 2013).

The dire consequences of delay in surgical repair of transected nerves have directed surgeons to repair the injuries as soon as possible. Where the transection injuries leave nerve stumps on either side of the injury that can be opposed for surgical apposition, the transected stumps may be sutured together or glued together with Tisseel glue (direct repair and neurorrhaphy). Nerve grafts may be required to oppose the transected stumps and, in cases where one or the other of the stumps is not available, nerve transfers and end-to-side repairs may be carried out that oppose the denervated distal stump to a donor intact nerve in order to encourage growth of some axons to the denervated targets (Midha 2011).

Other surgical approaches include transfers of tendons in order to restore required movements (Midha 2011). For all surgical repairs that are carried out whether or not the surgery is performed immediately after the injury, the staggering of axons regenerating across surgical sites and the slow rate of regeneration of 1 mm/day in humans require lengthy periods of time for regenerating nerves to reach and reinnervate denervated targets (Gordon et al. 2003). Because the capacity of the injured neurons to regenerate their axons declines with time when the axons have not yet made functional connections (chronic axotomy), and the Schwann cells in the distal nerve stumps have a relatively short duration of growth support for the regenerating axons, functional recovery after nerve injuries is frequently poor (Fu and Gordon 1997; Gordon and Sulaiman 2013).

This review provides a background of information of the biological events of nerve injury and axon regeneration prior to presentation of the evidence of the key role of neurotrophic factors in nerve regeneration and the efficacy of brief electrical stimulation in accelerating nerve regeneration in animal models and human patients. A brief consideration is also made for using polymer-based materials in nerve grafts.

11.2 PERIPHERAL NERVE INJURY

Following peripheral nerve injury, the axons proximal to the injury remain in continuity with the neuronal cell bodies, while the axons that are disconnected undergo Wallerian degeneration (Fu and Gordon 1997). In the latter process, the distal axons lose their myelin sheaths as the Schwann cells revert from a myelinating phenotype to a growth permissive state, and the axons and myelin sheath progressively disintegrate. The Schwann cells initially ingest or phagocytose the myelin and axon debris prior to entry of macrophages into the nerve as the nerve–blood barrier becomes permeable along the length of the distal axons (Avellino et al. 1995; Beuche and Friede 1984; Gaudet et al. 2011; Hirata and Kawabuchi 2002). The latter cells become the prime cells responsible for the bulk of the removal of the debris, with the process taking at least 3 weeks to complete before the macrophages leave the distal nerve stump and the Schwann cells line the empty endoneurial tubes as the bands of Bungner (You et al. 1997). It is these bands that guide the regenerating axons, with the Schwann cells progressively reverting to the myelinating phenotype on contact with the axons (Arthur-Farraj et al. 2012; Lieberman 1971).

11.2.1 CELL BODY REACTION

The injured neuron undergoes characteristic morphological changes known as chromatolysis (Lieberman 1971). These changes include the dissolution of the Nissl bodies and movement of the nucleus from the center of the neurons to an eccentric position. The nucleolus becomes prominent, with the morphological changes reflecting the changes in gene expression from the normal transmitting phenotype to the growth mode of the regenerating neuron (Gordon 1983). The altered gene response is commonly referred to as the neuronal (or cell body) response with expression of regeneration-associated genes and downregulation of the genes that transcribe proteins associated with chemical transmission (Bisby and Tetzlaff 1992; Fu and Gordon 1997; Gordon 1983; Gordon et al. 2009). The regeneration-associated genes transcribe, among others, the cytoskeletal proteins actin and tubulin (Gordon and Tetzlaff 2015; Gordon et al. 2015). These are normally transported down the axons by the slow component of axonal transport for the normal turnover of the proteins. In regenerating axons, more cytoskeletal proteins are transported for extension of the axons during axon elongation toward denervated targets (Lasek and Hoffman 1976). The rate of regeneration corresponds with the rate of slow transport, consistent with the central role of the cytoskeletal proteins in the axon elongation (McQuarrie and Lasek 1989).

11.2.2 AXON OUTGROWTH FROM INJURED AXONS

At the injury site, the proximal nerve dies back to the first node of Ranvier in response to calcium ions entering through calcium channels, followed by rapid sealing of the disrupted membranes (Gaudet et al. 2011). The ongoing anterograde transport of the cytoskeletal proteins supports the outgrowth of the sealed nerve, but this initial outgrowth is frequently aborted until such time as more proteins arrive at the growth cone. In the same manner that axons send out multiple growth cones *in vitro*, axons *in vivo* may elaborate several growth cones that cross the injury site to enter into the denervated distal nerve stumps (Frizell 1982; McQuarrie and Lasek 1989). The extending axons are guided between the Schwann cells of the bands of Bungner and the basal laminae of the empty endoneurial sheaths (Bunge et al. 1989; Chernousov and Carey 2000; Ide et al. 1983).

The nature of the injury site differs according to the injury type, the integrity of the endoneurial tubes remaining intact after crush injuries whereas the integrity is disrupted by an injury that transects the continuity of the axons (Sulaiman et al. 2011). In both cases, the extracellular matrix is initially disrupted with components such as laminin having to progressively align themselves in parallel arrays (Witzel et al. 2005). Once aligned, the glycoproteins aid in guiding the migration of Schwann cells from the distal nerve stumps and, in turn, allow outgrowing axons to grow across the injury site and into the distal nerve stumps. The processes of the alignment of the extracellular matrix components and the migration of Schwann cells are lengthy, taking more than a week in rats after transection injuries (Witzel et al. 2005). Consequent to this delay, the outgrowth of axons is initially highly disorganized with axons wondering within the injury site prior to their guidance out of the site and into the distal nerve stump. As a result, the entry of regenerating axons into the distal nerve stump is asynchronous with the regenerating axons staggering across the regeneration site; all the neurons regenerate their axons across a transection and surgical repair site over a period of ~28 days in the rat (Brushart et al. 2002). Once the axons enter into the distal nerve stumps, the axons regenerate at the rate of 1–3 mm/day, with the slower rate being characteristic of human nerves and the fast rate more characteristic of rat nerves (Gordon 2009; Gordon et al. 2008; Pfister et al. 2011).

11.2.3 THE GROWTH SUPPORTIVE STATE OF NEURONS AND SCHWANN CELLS IS TRANSIENT

11.2.3.1 Axotomy

Axotomized neurons, namely, those cells whose axons are isolated from and remain without target contact, upregulate regeneration-associated genes, including those that transcribe cytoskeletal proteins that are transported down the axons, and neurotrophic factors and their receptors (Boyd and Gordon 2003b). The changes in gene expression in the injured neurons are a component of the "cell body response," of which the trigger has been the subject of considerable debate and research (Abe and Cavalli 2008; Rishal and Fainzilber 2010). In experiments in which rat sciatic or facial nerves were transected and the proximal and distal nerve stumps were ligated to prevent regeneration, the upregulation of regeneration-associated genes is transient, with the high levels of mRNA for tubulin and actin declining to preoperative levels within months (Gordon and Tetzlaff 2015; Gordon et al. 2015). The finding that a refreshment injury of the stump was sufficient to upregulate the genes again provides strong evidence that the signal is a positive one—the positive signal is related directly to the injury of the neurons themselves—rather than a negative signal resulting from isolation of the injured neurons from factors such as neurotrophic factors from their target contacts (Gordon and Tetzlaff 2015; Gordon et al. 2015). The positive signals may include axonal transcription factors such as (STAT)-3 that is involved in the response of sensory neurons to axotomy (Ben-Yaakov et al. 2012; Rishal and Fainzilber 2014).

11.2.3.2 Schwann cell denervation

Schwann cells undergo a phenotypic change from their myelinating phenotype to a growth-permissive form in response to their loss of axonal contact (Jessen and Mirsky 2008). This represents the conversion to an earlier developmental state when the Schwann cells migrate prior to their contact with outgrowing axons. The cells downregulate their expression of myelin-associated proteins such as Po and peripheral myelin protein 22 (PMP22) concurrent with the upregulation of regeneration-associated genes (Fu and Gordon 1997; Jessen and Mirsky 2005). The latter transcribe, among other proteins, the neurotrophins, nerve growth factor (NGF), and brain-derived neurotrophic factor (BDNF), and their p75 receptor, as well as several other neurotrophic factors and their receptors, including glial-derived neurotrophic factor (GDNF) and pleiotrophin (Hoke et al. 2006). These proteins are elevated within weeks, but their expression, as in the case of the axotomized neurons, is not sustained, declining exponentially with time to reach baseline levels within 1–3 months (Gordon 2014; Hoke et al. 2006).

11.2.4 WINDOW OF OPPORTUNITY FOR NERVE REGENERATION

In lieu of the transient expression of regeneration-associated genes in both injured neurons and denervated Schwann cells, it is perhaps not surprising that the regenerative capacity of axotomized neurons declines exponentially within the same time frame of 4–6 months as does the regenerative support of the Schwann cells in the denervated distal nerve stump (Fu and Gordon 1995a, 1995b; Gordon et al. 2011). The reduced capacity for motor nerve regeneration of 66% due to chronic axotomy alone and of 90% due to the chronic denervation of Schwann cells in the distal nerve stumps provides the basis for the very poor functional outcome of surgical repair of transected nerves or of crushed nerves when the injuries are sustained far from their denervated targets. As a result of the ~6-month period of exponential decline in the regenerative capacities of the injured neurons regenerating their axons within a withering growth environment of the distal nerve stumps, the window of opportunity during which the regenerative capacity is optimal and functional return is likely to be relatively short (Fu and Gordon 1997). Although the denervated muscles undergo rapid atrophy, they are reinnervated by those axons capable of regenerating even after long delays in nerve repair (Fu and Gordon 1995a, 1995b; Gordon et al. 2011). The reinnervated muscles do not, however, fully recover their normal dimensions, likely due to a limited supply of satellite cells in the muscles that fuse with the reinnervated muscle fibers to contribute their nuclei. The number of nuclei in the multinucleated skeletal muscle cells normally increases as a function of the size of the muscle fibers, but as a result of a possible depletion of satellite cells after chronic denervation, the reinnervated muscle fibers do not recover their former size (Fu and Gordon 1995a, 1995b; Gordon et al. 2011).

The rate of regeneration being slow once regenerating axons "stagger" across the injury site and into the denervated distal nerve stumps, it is perhaps not surprising that functional recovery is frequently poor, especially for proximal nerve injuries such as brachial and lumbar plexus injuries, where axons must regenerate over long distances to reinnervate denervated targets. This poor recovery is exacerbated by the delays that are frequently incurred before nerve repair is executed (Sulaiman et al. 2011; Sulaiman and Gordon 2013).

11.3 STRATEGIES TO PROMOTE AXON REGENERATION

11.3.1 CHEMICAL AGENTS INCLUDING *N*-ACETYL-CYSTEINE AND FK506

Many attempts have been made to promote axonal regeneration. Several chemical agents have been explored, many of them having documented effects in animal studies and some being used clinically, including *N*-acetyl-cysteine and FK506 (Chan et al. 2014). However, most of these have a relatively small effect, and the documentation of their efficacy has too frequently been the examination and counting of regenerated axons (Chan et al. 2014).

11.3.2 EXOGENOUS NEUROTROPHIC FACTORS

The discovery of a peptide with the dramatic effect of eliciting neurite outgrowth from sympathetic and dorsal root ganglion neurons *in vitro* was the basis for naming the peptide, NGF. This discovery elicited many studies of whether this and other neurotrophic factors, including BDNF and GDNF, would promote axon regeneration, as the naming of NGF undeniably indicated (Gordon 2009). The *in vitro* studies were generally positive, with the growth factors promoting neurite outgrowth, and the many *in vivo* studies that evaluated the numbers of axons that grow through artificial as well a nerve conduits concluded that many of the factors, including the neurotrophins, NGF, and BDNF, and the trophic factors, GDNF, insulin growth factors 1 and 2, and pleiotrophin, promote axon regeneration (Pfister et al. 2011). However, studies that counted neurons whose regenerated axons were backlabeled with fluorescent dyes noted that these growth factors promoted the outgrowth of several axons from single-parent axons, namely, the sprouting of axons; this response resembled the branching of neurites that grow out from cultured neurons (Boyd and Gordon 2003a).

When neurons that regenerated axons after immediate nerve repair *in vivo* were enumerated, it became apparent that the administration of exogenous growth factors, including BDNF and GDNF, are ineffective in promoting axon regeneration after immediate nerve repair in rats (Boyd and Gordon 2001, 2002, 2003a; Gordon et al. 2003, 2005). In contrast, numbers of both motor and sensory neurons that regenerated their axons after delayed nerve repair were significantly increased when these exogenous factors were administered over a period of a month either via an implanted mini-osmotic pump (Boyd and Gordon 2001, 2002, 2003a; Gordon et al. 2003, 2005, Gordon and Borschel, 2017) or via microspheres placed around the suture site (Wood et al. 2012b, 2013a, 2013b). Moreover, significantly more neurons regenerated their axons through an acellular nerve autograft after immediate nerve repair when GDNF was enclosed within microspheres that were placed at the two suture sites (Tajdaran et al. 2016a).

These studies indicated that the endogenous sources of the growth factors are sufficient to support the regeneration of axons after immediate nerve repair, but their decline during a delay prior to surgery require additional sources of the factors (Furey et al. 2007), as was the case when exogenous sources were provided. Moreover, the appropriate concentrations of the exogenous factors are important because, in the case of the neurotrophic factor BDNF, low-dose infusion is effective in promoting axon regeneration after delayed surgery, while high doses actually reduce regenerative capacity (Boyd and Gordon 2002). These effects of BDNF are mediated by trkB receptors and p75 receptors, respectively. In contrast, GDNF that also binds to two different receptors is effective in promoting regeneration after delayed nerve repair, with the effect being dose dependent. Results such as these led investigators to use retroviral vectors to transfect Schwann cells with neurotrophic factors. While these experiments did demonstrate effective use of exogenous growth factors, including the neurotropic effect of the factors, namely, the growth of axons toward the source of the factors, the axons appeared to be unable to grow past the source of the factors (Mason et al. 2011). This was coined the "candy cane" effect (Eggers et al. 2013; Tannemaat et al. 2009). Interestingly, though, the controlled delivery of exogenous growth factors from microspheres that contained the factors and were placed at the suture site of a transected nerve did not exhibit this "damming" effect (Wood et al. 2013b). The axons grew across the site of delivery, with the neurotrophic factors accelerating the regeneration (Wood et al. 2013b). Similarly, no damming was evident when GDNF was administered via microspheres placed at either end of a 10 mm acellular nerve allograft in rats, with the GDNF promoting excellent regeneration that was equal to the regeneration of nerves through an autograft of the same length (Tajdaran et al. 2016a).

The lack of effect of exogenous growth factors on the regeneration of axons after immediate nerve repair illustrated the fine control exerted by the endogenous concentrations of the neurotrophic factors. However, the problem of the transient expression of the growth factors remains, indicating that methods that are still being developed to sustain the expression of these factors may promote regeneration of axons. These include the use of microspheres in which the neurotrophic factors are contained within the polymers that constitute the microspheres (Tajdaran et al. 2016b; Wood et al. 2013a, 2013b).

11.3.3 ENDOGENOUS NEUROTROPHIC FACTORS UPREGULATED BY ELECTRICAL STIMULATION IN CONCERT WITH ACCELERATED REGENERATION

11.3.3.1 Electrical stimulation in animals *in vivo*

In experiments in which we stimulated the proximal stump of a transected nerve after surgical repair with alternating, charge-balanced, electrical pulses at a frequency of 20 Hz, there was a striking and accelerated upregulation of endogenous BDNF and its trkB receptor (Al-Majed et al. 2000a). This neurotrophic factor and its receptor are normally upregulated in motoneurons after injury, but to a lesser extent and more slowly (Al-Majed et al. 2000a). The stimulation-induced upregulation is followed shortly by an accelerated and a more pronounced upregulation of the cytoskeletal proteins actin and tubulin, as well as the growth-associated gene for GAP-43 with concomitant downregulation of neurofilament protein (Al-Majed et al. 2004). These changes in gene expression accompanied a pronounced acceleration in nerve regeneration (Al-Majed et al. 2000b). We had initially stimulated the proximal femoral nerve stump continuously for 2 weeks after cutting and repairing the nerve. In these experiments in which the femoral nerve was cut and repaired, we were surprised to discover that the motoneurons that regenerated their axons and were backlabeled with retrograde fluorescent dyes 25 mm distal to the surgical site increased progressively over an 8- to 10-week period (Figure 11.1a and b) (Al-Majed et al. 2000b). This lengthy period suggested that regenerating axons might be delayed across the suture site; that was indeed the case, as demonstrated by backlabeling those neurons that had regenerated axons just across the suture site: motoneurons regenerated their axons progressively over a period of ~28 days (Brushart et al. 2002). The electrical stimulation for 2 weeks accelerated this regeneration: *all* the stimulated motoneurons regenerated their axons 25 mm from the surgical site within 3 weeks and across the suture site within a week (Al-Majed et al. 2000b; Brushart et al. 2002; Gordon 2014).

We had adopted a continuous pattern of electrical stimulation in these experiments on the basis of the promising findings that a 2-week period of low-frequency electrical stimulation of the proximal stump of the crushed soleus nerve accelerated the restoration of contractile force in the denervated soleus muscle (Nix and Hopf 1983). The recovery of the reflex foot withdrawal response to tactile stimulation of the foot was also accelerated by electrical stimulation after sciatic nerve crush (Pockett and Gavin 1985). These findings were interpreted as the stimulation accelerating axon regeneration and, in turn, the reinnervation of the denervated muscles. The experiments in which the reflex withdrawal was the outcome measure demonstrated that immediate nerve stimulation appeared to be the most effective, with the effectiveness declining as the period of time

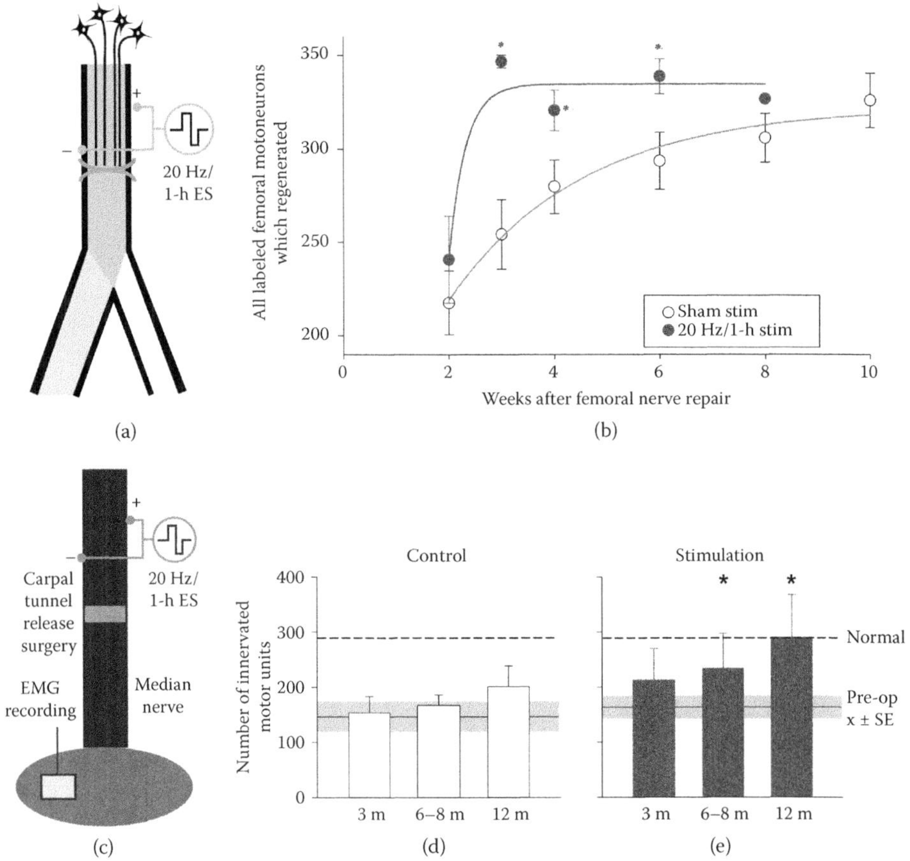

Figure 11.1 Brief (1 h) electrical stimulation immediately after surgical repair of an injured peripheral nerve accelerates nerve regeneration and target reinnervation. (a) Either the rat femoral nerve was subjected to biphasic electrical suprathreshold pulses at 20 Hz for 1 h immediately after cutting and surgically repairing the nerve, or the electrodes were placed but the stimulator was not turned on (sham electrical stimulation). (b) The number of femoral motoneurons that regenerated their axons as a function of the time (weeks) after surgical coaptation of the transected femoral nerve. (c) In patients suffering severe carpal tunnel syndrome, the damaging constriction of the median nerve by the overlying ligament was released by cutting the ligament and stimulating bipolar stainless steel electrodes placed proximal to the site of the release surgery. The median nerve was subjected to the same 20 Hz 1-h electrical stimulation protocol within 15 min of the surgery. (d and e) By determining the number of innervated motor units, that is, the number of median nerves with intact nerve–muscle contacts in the thenar eminence musculature, with a motor unit estimation (MUNE) technique before and 3–12 months after the surgery, it was found that the number did not increase significantly in the patients whose nerves were not electrically stimulated but *did* increase to levels not significantly different from the normal numbers when the median nerve was subjected to 20 Hz stimulation for 1 h.

between the crush injury and the electrical stimulation was prolonged (Pockett and Gavin 1985). Experiments that counted those neurons that regenerated their axons as a function of time after nerve transection and surgical repair established that, indeed, electrical stimulation promotes the regeneration of nerves (Al-Majed et al. 2000b). We determined the optimum period of electrical stimulation by progressively shortening the stimulation time, with a 1-h period of electrical stimulation being as effective as 2 weeks of 20 Hz continuous electrical stimulation (Al-Majed et al. 2000b). Indeed, this period of time was ideal because periods of electrical stimulation that are longer than 1-h downregulate trk receptors on sensory neurons, and the longer periods of electrical stimulation are in turn, ineffective in accelerating sensory nerve regeneration (Geremia et al. 2007).

These findings established that brief electrical stimulation of transected nerves accelerates both motor and sensory nerve regeneration. That the electrical stimulation did not change the rate of slow transport but accelerated axon outgrowth across the suture site and into the distal nerve stump, verified that the stimulation increased the outgrowth of axons across the suture site but did not alter their rate of regeneration once the axons entered into the distal nerve stump (Brushart et al. 2002). These studies elicited many others that verified the effect of the continuous 20 Hz stimulation regime in promoting the regeneration of nerves, including the sciatic nerves (Elzinga et al. 2015; Gordon and English 2015). The manner of electrical stimulation for 1 h remained similar to that of the first report, but experiments of Franz et al. (2008) demonstrated the efficacy of stimulation of the proximal femoral nerve stump via a wick electrode, while others found that, even when the intact nerve was electrically stimulated for an hour at 20 Hz, the stimulation was effective in accelerating axon growth following the subsequent nerve transection and repair (Thompson et al. 2014).

Other stimulation regimes emerged, including daily continuous stimulation of the crushed facial nerve at 20 Hz for 35 min per day (Foecking et al. 2012; Hetzler et al. 2008; Lal et al. 2008; Sharma et al. 2009, 2010b). These experiments noted the same transient upregulation of BDNF as found for the stimulated femoral nerve after transection and repair (Sharma et al. 2010a). In later experiments using transgenic mice that expressed fluorescent neurofilament protein in ~30% of their lumbosacral neurons, English and colleagues used knockout techniques to demonstrate the critical upregulation of neurotrophin-4/5 for promoting outgrowth of axons after a 1-h period of 20 Hz electrical stimulation of the common peroneal nerve (English et al. 2005, 2007). Hence, electrical stimulation is effective in accelerating axon outgrowth, and this effect is directly linked to the expression of neurotrophic factors by the stimulated neurons. The link between the response of the axotomized neurons to the electrical stimulation and the accelerated axon outgrowth was established in 2000, when the effect was eliminated by blocking the conduction of action potentials to the cell bodies using local application of tetrodotoxin to block sodium channels (Al-Majed et al. 2000b).

The use of hetero- and homozygous transgenic mice in which the neurotrophic factors BDNF and NT-4/5 were selectively eliminated in either or both the neurons and the Schwann cells in the denervated distal nerve stumps provided direct evidence for the key involvement of these neurotrophic factors in the enhancement of axon regeneration by electrical stimulation (English et al. 2005, 2007, 2011a). Indeed, the expression of neurotrophic factors by the stimulated neurons is essential for the effect of the electrical stimulation in enhancing nerve regeneration.

11.3.3.2 Electrical stimulation and accelerated nerve regeneration and target reinnervation in animals and humans

Accelerated nerve regeneration following a 1-h period of low-frequency electrical stimulation accelerates functional recovery in both animals and humans. In mice, the stimulation of a transected and surgically repaired femoral nerve accelerated knee extension by the reinnervated quadriceps muscle (Ahlborn et al. 2007). In human patients, the same stimulation paradigm applied to the injured median nerve after carpal tunnel release at the wrist, accelerated both median nerve regeneration and the reinnervation of the muscles of the thenar eminence (Figure 11.1c–e) (Gordon et al. 2010). Sensory nerve regeneration and recovery of sensory function were also accelerated by the same regime of a 1-h period of 20 Hz electrical stimulation after surgical repair of the transected digital nerve (Wong et al. 2015), demonstrating the efficacy of electrical stimulation after nerve injuries in humans, as well as animals. Similar acceleration has also been demonstrated after surgical release of the ulnar nerve at the elbow and after surgical repair of a transected digital nerve (Chan et al. 2016). Ongoing studies in Edmonton and Toronto are currently being extended to more nerve centers internationally.

11.3.3.3 Electrical stimulation and neurite outgrowth *in vitro*

The effectiveness of direct current in promoting bone growth was used as the rationale to determine if the same currents might also promote axon regeneration (Borgens 1982). Isolated dorsal root ganglia or neurons demonstrated neurite outgrowth *in vitro* in response to an electric field created between an applied cathode and anode: neurite outgrowth is enhanced in a current-dependent manner (Zhang et al. 2007). *In vivo*, the corresponding axon outgrowth of axons across a site of a nerve crush or the surgical repair of a transected nerve was more difficult to elicit, possibly due to the difficulties in applying large enough electrical fields (Neumann and Gordon, unpublished observations). This outgrowth required the influx of calcium into the neurons, which in turn induced neurotrophic expression in the neurons (Huang et al. 2010; Wenjin et al. 2011). Hence, neurotrophic factor expression is also a key factor in axon outgrowth both *in vivo* and *in vitro*.

Conducting polymers are currently being developed as tissue scaffolds for nerve regeneration; the repair of several tissues, including muscle and skin; bone fusion; and spinal cord stimulation for pain relief, among many other applications (Hardy et al. 2013). Synthetic polymers such as polypyrrole are conductive for neurite growth, especially when nanoscale fibers are deposited in an orientated manner by electrospinning (Weng et al. 2010; Liu et al. 2010); they support neurite outgrowth from several different neural explants in response to electrical stimulation. The explants include cochlear neural explants (Thompson et al. 2010) and PC 12 cells (Durgam et al. 2010; Gomez and Schmidt 2007; Lee et al. 2009). The effect of the conducting polymer on neurite outgrowth from cochlear neural explants was enhanced by incorporation of neurotrophin-3 and BDNF into the polypyrrole during the electrosynthesis, and the effect was increased significantly more so when the polymer containing the neurotrophins was electrically stimulated. Electrospun fibrous scaffolds also provide contact guidance to Schwann cells, with evidence of electrospun poly(ε-caprolactone) fibers promoting the conversion of growth-supportive Schwann cells to their more mature myelinating phenotype *in vitro* (Chew et al. 2008).

The conducting polymers are currently being explored as a coating on the inner walls of nerve guidance channels, with some evidence of their capacity to support nerve growth through the channels. It is important that the construction of the channels include the incorporation of extracellular matrix glycoproteins such as fibronectin that impart a hydrophilic surface to the biodegradable polymers to facilitate axon growth with eventual replacement with natural functional tissue (Hardy et al. 2013). There is, however, relatively little evidence of improved axon regeneration by *in vivo* application of the polymers in light of problems that include fatigue of the polymers with repeated cycles of electrical stimulation (Guiseppi-Elie 2010), and that many of the conductive polymers, such as polypyrrole, are not inherently biodegradable (Balint et al. 2014). A rat model of tibial nerve transection and surgical repair via a 17 mm gap filled with an aligned polymer fiber–based construct provided immunohistochemical evidence of enhanced longitudinal axon growth and reinnervation of target gastrocnemius muscle (Kim et al. 2008). The study of Huang et al. (2012), who inserted a polypyrrole–chitosan-based 15 mm long conductive conduit between transected stumps of sciatic nerve in rats, used several parameters of enhanced nerve regeneration and target reinnervation to demonstrate enhanced longitudinal axon growth and reinnervation of target gastrocnemius muscle when the conduit was electrically stimulated at 20 Hz for 1 h daily for 2 weeks. The small but significant effect was demonstrated by increased (1) numbers of motor and sensory neurons regenerating their axons through the conduit and 2 mm distal to it; (2) number and size of regenerated axons; (3) amplitude of the gastrocnemius muscle compound action potential; and (4) conduction velocity of the nerves, as well as functional parameters that included reduced time to attend to a "sticky tape on the skin" and an improved sciatic functional index. These effects were demonstrated as early as 4 weeks after the insertion of the conduit and followed over a 12-week period of nerve regeneration and target reinnervation. It is important to note that none of these parameters of regenerative success reached statistical significance unless the conduit was subjected to electrical stimulation. Moreover, the electrical stimulation was not effective if the polypyrrole was not included in the chitosan conduit. The distinct possibility that some of the effect was mediated via the stimulation of Schwann cell differentiation or motility was raised by the authors (Huang et al. 2012). An increased capacity of embryonic Schwann cells to migrate in an electrical field had been demonstrated (McKasson et al. 2008), and was confirmed by findings of increased protein levels of S-100 and BDNF at the site of the conductive scaffold 2 weeks after surgery in the study of Huang et al. (2012). Elevated Par-3 and P0 protein levels that were also reported provided further indications of earlier contact of regenerating axons with Schwann cells and the onset of myelination (Huang et al. 2012).

A study demonstrated improved survival of neurons *in vivo* in response to electrical stimulation of the cochlear in chemically deafened guinea pig ears (Richardson et al. 2009). Neurotrophin-3 was delivered to the cochlear and the cochlear stimulated electrically for 8 h per day for 2 weeks via polypyrrole–*para*-toluene sulfonate applied to cochlear implant electrodes (Richardson et al. 2009). This treatment improved the survival of spiral ganglion neurons whose gradual degeneration compromises hearing outcomes with cochlear implant use, with the polymer coatings of electrodes providing a means of improving the neural tissue–electrode interface to increase the effective lifetime of the implants (Green et al. 2008).

11.3.4 TREADMILL RUNNING AND ENHANCED NERVE REGENERATION

The link between the expression of neurotrophic factors and enhanced nerve regeneration taken together with the evidence of the increased expression of neurotrophic factors by daily exercise was the impetus for examining whether (1) daily exercise implemented within days of a nerve injury and surgical repair may promote axon regeneration (English et al. 2011b; Sabatier et al. 2008) and (2) the combination of brief electrical stimulation with daily exercise might be even more effective than electrical stimulation alone (Asensio-Pinilla et al. 2009; Udina et al. 2011). While the data did not indicate that the combination was necessarily more effective, the exercise was effective in promoting axon regeneration. Of particular interest was that the link between androgens and the daily exercise in promoting this regeneration, especially as electrical stimulation of the facial nerve proximal to a nerve crush was more effective when combined with testosterone in castrated male rats (Foecking et al. 2012; Monaco et al. 2015; Sharma et al. 2010b). Jones (1988) had recognized an important role of androgens in nerve regeneration, after which she and her colleagues demonstrated their importance with evidence of exogenous testosterone-enhancing regeneration of crushed peripheral nerves (Brown et al. 1999; Fargo et al. 2008a, 2008b; Jones et al. 1997, 2000; Monaco et al. 2015; Storer et al. 2002; Tanzer and Jones 1997). The combined treatment of exogenous testosterone and electrical stimulation enhanced the outgrowth of regenerating axons, as well as the rate of regeneration, with the combination promoting a sustained expression of neurotrophic factors (Foecking et al. 2012; Sharma et al. 2010a, 2010b). Endogenous androgens are responsible for the enhanced axonal regeneration observed after daily exercise in male rats and female rats, with the androgens becoming available after continuous exercise in male rats and intermittent exercise in female rats (Thompson et al. 2014; Wood et al. 2012a, Gordon and English, 2016). As found by Jones and colleagues, androgens also enhanced nerve regeneration in combination with electrical stimulation (Thompson et al. 2014).

11.3.5 CONDITIONING LESION AND ACCELERATED NERVE REGENERATION

An important aspect of the combined treatment of electrical stimulation and androgens was that the androgens enhanced both axon outgrowth and the rate of axon regeneration (Sharma et al. 2009). Prior to this demonstration, only a conditioning lesion had been shown to be effective in accelerating the regeneration rate; indeed, the discovery of the efficacy of the conditioning lesion in promoting axon regeneration was a major advance in the 1970s (Bisby and Keen 1984; Carlsen 1983; McQuarrie 1978, 1981; McQuarrie et al. 1977; Oudega et al. 1994). By definition, the conditioning lesion is a crush injury induced prior to or at the same time as the "test" injury, with an injury 3 or more days prior to the test injury being the most effective in accelerating axon regeneration after the test injury itself. This efficacy in promoting regeneration of axons within the peripheral nervous system was also demonstrated within the central nervous system, with a crush of a peripheral nerve promoting regeneration of dorsal root ganglion sensory axons that were transected within the central nervous system (Neumann and Woolf 1999). In contrast, electrical stimulation of the peripheral nerve accelerated axon outgrowth but not the rate of regeneration of the central axons (Gordon et al. 2009). Cyclic AMP (cAMP), a key intermediary in the accelerated rate of regeneration by a conditioning lesion, may also be involved in the mechanism by which electrical stimulation accelerates axon outgrowth. Administration of rolipram that elevates cAMP by inhibiting phosphodiesterase, the enzyme that degrades cAMP, also promoted axon outgrowth across a suture line (Udina et al. 2010). It remains to be seen whether the efficacy of rolipram in promoting axon regeneration involves either or both accelerated axon outgrowth and an increased rate of regeneration.

11.4 CONCLUSIONS AND SIGNIFICANCE

It is becoming clear that endogenous growth factors are involved in the efficacy of both brief low-frequency electrical stimulation and daily exercise in promoting axon regeneration after nerve crush and transection injuries. While other methods are being developed to deliver neurotrophic factors during axon regeneration, the upregulated expression of endogenous neurotrophic factors in response to electrical stimulation or daily exercise indicates a clear and effective means to promote axon regeneration. Clinical evidence for the efficacy of electrical stimulation provides a basis for further exploration and future inclusion in the clinical management of peripheral nerve injuries in human patients.

ACKNOWLEDGMENTS

My thanks are extended to all my many collaborators and colleagues who have contributed their experimental skills to the examination of the efficacy of electrical stimulation in promoting peripheral nerve regeneration. Particular thanks are extended to Dr. Thomas Brushart and my graduate student Dr. Abdul Al-Majed, who initiated the studies on electrical stimulation of transected and surgically repaired rat femoral nerves with me in the 1990s, which in turn has launched many studies in animals and human patients. Our hope is that the extension of this treatment to patients, initiated with Drs. Ming Chan and Jaren Olson and being currently explored actively in children who have suffered nerve injuries by Dr. Gregory Borschel, improves the functional outcomes for many patients who suffer peripheral nerve injuries.

REFERENCES

Abe, N., and Cavalli, V. (2008) Nerve injury signaling. *Curr. Opin. Neurobiol.*, 18, 276–283.

Ahlborn, P., Schachner, M., and Irintchev, A. (2007) One hour electrical stimulation accelerates functional recovery after femoral nerve repair. *Exp. Neurol.*, 208, 137–144.

Al-Majed, A. A., Brushart, T. M., and Gordon, T. (2000a) Electrical stimulation accelerates and increases expression of BDNF and trkB mRNA in regenerating rat femoral motoneurons. *Eur. J. Neurosci.*, 12, 4381–4390.

Al-Majed, A. A., Neumann, C. M., Brushart, T. M., and Gordon, T. (2000b) Brief electrical stimulation promotes the speed and accuracy of motor axonal regeneration. *J. Neurosci.*, 20, 2602–2608.

Al-Majed, A. A., Tam, S. L., and Gordon, T. (2004) Electrical stimulation accelerates and enhances expression of regeneration-associated genes in regenerating rat femoral motoneurons. *Cell. Mol. Neurobiol.*, 24, 379–402.

Arthur-Farraj, P. J., Latouche, M., Wilton, D. K., et al. (2012) c-Jun reprograms Schwann cells of injured nerves to generate a repair cell essential for regeneration. *Neuron*, 75, 633–647.

Asensio-Pinilla, E., Udina, E., Jaramillo, J., and Navarro, X. (2009) Electrical stimulation combined with exercise increase axonal regeneration after peripheral nerve injury. *Exp. Neurol.*, 219, 258–265.

Avellino, A. M., Hart, D., Dailey, A. T., MacKinnon, M., Ellegala, D., and Kliot, M. (1995) Differential macrophage responses in the peripheral and central nervous system during Wallerian degeneration of axons. *Exp. Neurol.*, 136, 183–198.

Balint, R., Cassidy, N. J., and Cartmell, S. H. (2014) Conductive polymers: Towards a smart biomaterial for tissue engineering. *Acta Biomater.*, 10, 2341–2353.

Ben-Yaakov, K., Dagan, S. Y., Segal-Ruder, Y., et al. (2012) Axonal transcription factors signal retrogradely in lesioned peripheral nerve. *EMBO J.*, 31, 1350–1363.

Beuche, W., and Friede, R. L. (1984) The role of non-resident cells in Wallerian degeneration. *J. Neurocytol.*, 13, 767–796.

Bisby, M. A., and Keen, P. (1984) A conditioning lesion increases regeneration rate of rat sciatic-nerve axons containing substance-P. *J. Physiol.*, 354, 56.

Bisby, M. A., and Tetzlaff, W. (1992) Changes in cytoskeletal protein synthesis following axon injury and during axon regeneration. *Mol. Neurobiol.*, 6, 107–123.

Borgens, R. B. (1982) What is the role of naturally produced electric current in vertebrate regeneration and healing. *Int. Rev. Cytol.*, 76, 245–298.

Boyd, J. G., and Gordon, T. (2001) The neurotrophin receptors, trkB and p75, differentially regulate motor axonal regeneration. *J. Neurobiol.*, 49, 314–325.

Boyd, J. G., and Gordon, T. (2002) A dose-dependent facilitation and inhibition of peripheral nerve regeneration by brain-derived neurotrophic factor. *Eur. J. Neurosci.*, 15, 613–626.

Boyd, J. G., and Gordon, T. (2003a) Glial cell line-derived neurotrophic factor and brain-derived neurotrophic factor sustain the axonal regeneration of chronically axotomized motoneurons *in vivo*. *Exp. Neurol.*, 183, 610–619.

Boyd, J. G., and Gordon, T. (2003b) Neurotrophic factors and their receptors in axonal regeneration and functional recovery after peripheral nerve injury. *Mol. Neurobiol.*, 27, 277–324.

Brown, T. J., Khan, T., and Jones, K. J. (1999) Androgen induced acceleration of functional recovery after rat sciatic nerve injury. *Restor. Neurol. Neurosci.*, 15, 289–295.

Brushart, T. M., Hoffman, P. N., Royall, R. M., Murinson, B. B., Witzel, C., and Gordon, T. (2002) Electrical stimulation promotes motoneuron regeneration without increasing its speed or conditioning the neuron. *J. Neurosci.*, 22, 6631–6638.

Bunge, M. B., Bunge, R. P., Kleitman, N., and Dean, A. C. (1989) Role of peripheral nerve extracellular matrix in Schwann cell function and in neurite regeneration. *Dev. Neurosci.*, 11, 348–360.

Carlsen, R. C. (1983) Delayed induction of the cell body response and enhancement of regeneration following a condition/test lesion of frog peripheral nerve at 15 degrees C. *Brain Res.*, 279, 9–18.

Chan, K. M., Gordon, T., Zochodne, D. W., and Power, H. A. (2014) Improving peripheral nerve regeneration: From molecular mechanisms to potential therapeutic targets. *Exp. Neurol.*, 261, 826–835.

Chan, K.M., Power, H., Morhart, M. and Olson, J. (2016) A randomized controlled trial on electrical stimulation to accelerate axon regeneration and functional recovery following cubital tunnel surgery. *Soc. Neurosci.*, 675, 08.

Chernousov, M. A., and Carey, D. J. (2000) Schwann cell extracellular matrix molecules and their receptors. *Histol. Histopathol.*, 15, 593–601.

Chew, S. Y., Mi, R., Hoke, A., and Leong, K. W. (2008) The effect of the alignment of electrospun fibrous scaffolds on Schwann cell maturation. *Biomaterials*, 29, 653–661.

Durgam, H., Sapp, S., Deister, C., et al. (2010) Novel degradable co-polymers of polypyrrole support cell proliferation and enhance neurite out-growth with electrical stimulation. *J. Biomater. Sci. Polym. Ed.*, 21, 1265–1282.

Eggers, R., De Winter, F., Hoyng, S. A., et al. (2013) Lentiviral vector-mediated gradients of GDNF in the injured peripheral nerve: Effects on nerve coil formation, Schwann cell maturation and myelination. *PLoS One*, 8, e71076.

Elzinga, K., Tyreman, N., Ladak, A., Savaryn, B., Olson, J., and Gordon, T. (2015) Brief electrical stimulation improves nerve regeneration after delayed repair in Sprague Dawley rats. *Exp. Neurol.*, 269, 142–153.

English, A. W., Cucoranu, D., Mulligan, A., Rodriguez, J. A., and Sabatier, M. J. (2011a) Neurotrophin-4/5 is implicated in the enhancement of axon regeneration produced by treadmill training following peripheral nerve injury. *Eur. J. Neurosci.*, 33, 2265–2271.

English, A. W., Meador, W., and Carrasco, D. I. (2005) Neurotrophin-4/5 is required for the early growth of regenerating axons in peripheral nerves. *Eur. J. Neurosci.*, 21, 2624–2634.

English, A. W., Schwartz, G., Meador, W., Sabatier, M. J., and Mulligan, A. (2007) Electrical stimulation promotes peripheral axon regeneration by enhanced neuronal neurotrophin signaling. *Dev. Neurobiol.*, 67, 158–172.

English, A. W., Wilhelm, J. C., and Sabatier, M. J. (2011b) Enhancing recovery from peripheral nerve injury using treadmill training. *Ann. Anat.*, 193, 354–361.

Fargo, K. N., Alexander, T. D., Tanzer, L., Poletti, A., and Jones, K. J. (2008a) Androgen regulates neuritin mRNA levels in an *in vivo* model of steroid-enhanced peripheral nerve regeneration. *J. Neurotrauma*, 25, 561–566.

Fargo, K. N., Galbiati, M., Foecking, E. M., Poletti, A., and Jones, K. J. (2008b) Androgen regulation of axon growth and neurite extension in motoneurons. *Horm. Behav.*, 53, 716–728.

Fenrich, K., and Gordon, T. (2004) Canadian Association of Neuroscience review: Axonal regeneration in the peripheral and central nervous systems—Current issues and advances. *Can. J. Neurol. Sci.*, 31, 142–156.

Foecking, E. M., Fargo, K. N., Coughlin, L. M., Kim, J. T., Marzo, S. J., and Jones, K. J. (2012) Single session of brief electrical stimulation immediately following crush injury enhances functional recovery of rat facial nerve. *J. Rehabil. Res. Dev.*, 49, 451–458.

Franz, C. K., Rutishauser, U., and Rafuse, V. F. (2008) Intrinsic neuronal properties control selective targeting of regenerating motoneurons. *Brain*, 131, 1492–1505.

Frizell, M. (1982) The effect of ligation combined with section on anterograde axonal transport in rabbit hypoglossal nerve. *Brain Res.*, 250, 65–69.

Fu, S. Y., and Gordon, T. (1995a) Contributing factors to poor functional recovery after delayed nerve repair: Prolonged axotomy. *J. Neurosci.*, 15, 3876–3885.

Fu, S. Y., and Gordon, T. (1995b) Contributing factors to poor functional recovery after delayed nerve repair: Prolonged denervation. *J. Neurosci.*, 15, 3886–3895.

Fu, S. Y., and Gordon, T. (1997) The cellular and molecular basis of peripheral nerve regeneration. *Mol. Neurobiol.*, 14, 67–116.

Furey, M. J., Midha, R., Xu, Q. G., Belkas, J., and Gordon, T. (2007) Prolonged target deprivation reduces the capacity of injured motoneurons to regenerate. *Neurosurgery*, 60, 723–732.

Gaudet, A. D., Popovich, P. G., and Ramer, M. S. (2011) Wallerian degeneration: Gaining perspective on inflammatory events after peripheral nerve injury. *J. Neuroinflammation*, 8, 110.

Geremia, N. M., Gordon, T., Brushart, T. M., Al-Majed, A. A., and Verge, V. M. (2007) Electrical stimulation promotes sensory neuron regeneration and growth-associated gene expression. *Exp. Neurol.*, 205, 347–359.

Gomez, N., and Schmidt, C. E. (2007) Nerve growth factor-immobilized polypyrrole: Bioactive electrically conducting polymer for enhanced neurite extension. *J. Biomed. Mater. Res. A*, 81, 135–149.

Gordon, T. (1983) Dependence of peripheral nerves on their target organs. In Burnstock, G., O'Brien, R., and Vrbova, G. (eds.), *Somatic and Autonomic Nerve-Muscle Interactions*. Elsevier, Amsterdam, pp. 289–325.

Gordon, T. (2009) The role of neurotrophic factors in nerve regeneration. *Neurosurg. Focus*, 26 (E3), 1–10.

Gordon, T. (2014) Neurotrophic factor expression in denervated motor and sensory Schwann cells: Relevance to specificity of peripheral nerve regeneration. *Exp. Neurol.*, 254, 99–108.

Gordon, T. (2015) The biology, limits, and promotion of peripheral nerve regeneration in rats and humans. In Tubbs, R. S., Rizk, E., Shoja, M., Loukas, M., and Spinner, R. J. (eds.), *Sunderland's Nerves and Nerve Injuries*. Elsevier, Amsterdam, 2, 993–1019.

Gordon, T., Amirjani, N., Edwards, D. C., and Chan, K. M. (2010) Brief post-surgical electrical stimulation accelerates axon regeneration and muscle reinnervation without affecting the functional measures in carpal tunnel syndrome patients. *Exp. Neurol.*, 223, 192–202.

Gordon, T. and Borschel, G. H. (2016) The use of the rat as a model for studying peripheral nerve regeneration and sprouting after complete and partial nerve injuries. Invited review for a special issue on *in vivo* models of neural regeneration. *Exp. Neurol.*, 287, 331–347.

Gordon, T., Boyd, J. G., and Sulaiman, O. A. R. (2006) Experimental approaches to promote functional recovery after severe peripheral nerve injuries. *Eur. Surg.*, 37, 193–203.
Gordon, T., Brushart, T. M., and Chan, K. M. (2008) Augmenting nerve regeneration with electrical stimulation. *Neurol. Res.*, 30, 1012–1022.
Gordon, T., and English, A. W. (2016) Strategies to promote peripheral nerve regeneration: Electrical stimulation and/or exercise. *Eur. J. Neurosci.*, 43, 336–350.
Gordon, T., and Sulaiman, O. A. R. (2013) Nerve regeneration in the peripheral nervous system. In Kettenmann, H., and Ransom, B. R. (eds.), *Neuroglia*. Oxford University Press, Oxford, pp. 701–714.
Gordon, T., Sulaiman, O. A. R., and Boyd, J. G. (2003) Experimental strategies to promote functional recovery after peripheral nerve injuries. *J. Peripher. Nerv. Syst.*, 8, 236–250.
Gordon, T., and Tetzlaff, W. (2015) Regeneration-associated genes decline in chronically injured rat sciatic motoneurons. *Eur. J. Neurosci.*, 42, 2783–2791.
Gordon, T., Tyreman, N., and Raji, M. A. (2011) The basis for diminished functional recovery after delayed peripheral nerve repair. *J. Neurosci.*, 31, 5325–5334.
Gordon, T., Udina, E., Verge, V. M., and de Chaves, E. I. (2009) Brief electrical stimulation accelerates axon regeneration in the peripheral nervous system and promotes sensory axon regeneration in the central nervous system. *Motor Control*, 13, 412–441.
Gordon, T., You, S., Cassar, S. L., and Tetzlaff, W. (2015) Reduced expression of regeneration associated genes in chronically axotomized facial motoneurons. *Exp. Neurol.*, 264, 26–32.
Green, R. A., Lovell, N. H., Wallace, G. G., and Poole-Warren, L. A. (2008) Conducting polymers for neural interfaces: Challenges in developing an effective long-term implant. *Biomaterials*, 29, 3393–3399.
Guiseppi-Elie, A. (2010) Electroconductive hydrogels: Synthesis, characterization and biomedical applications. *Biomaterials*, 31, 2701–2710.
Hardy, J. G., Lee, J. Y., and Schmidt, C. E. (2013) Biomimetic conducting polymer-based tissue scaffolds. *Curr. Opin. Biotechnol.*, 24, 847–854.
Hetzler, L. E., Sharma, N., Tanzer, L., et al. (2008) Accelerating functional recovery after rat facial nerve injury: Effects of gonadal steroids and electrical stimulation. *Otolaryngol. Head Neck Surg.*, 139, 62–67.
Hirata, K., and Kawabuchi, M. (2002) Myelin phagocytosis by macrophages and nonmacrophages during Wallerian degeneration. *Microsc. Res. Tech.*, 57, 541–547.
Hoke, A., Redett, R., Hameed, H., et al. (2006) Schwann cells express motor and sensory phenotypes that regulate axon regeneration. *J. Neurosci.*, 26, 9646–9655.
Huang, J., Lu, L., Zhang, J., et al. (2012) Electrical stimulation to conductive scaffold promotes axonal regeneration and remyelination in a rat model of large nerve defect. *PLoS One*, 7, e39526.
Huang, J., Ye, Z., Hu, X., Lu, L., and Luo, Z. (2010) Electrical stimulation induces calcium-dependent release of NGF from cultured Schwann cells. *Glia*, 58, 622–631.
Ide, C., Tohyama, K., Yokota, R., Nitatori, T., and Onodera, S. (1983) Schwann cell basal lamina and nerve regeneration. *Brain Res.*, 288, 61–75.
Jessen, K. R., and Mirsky, R. (2005) The origin and development of glial cells in peripheral nerves. *Nat. Rev. Neurosci.*, 6, 671–682.
Jessen, K. R., and Mirsky, R. (2008) Negative regulation of myelination: Relevance for development, injury, and demyelinating disease. *Glia*, 56, 1552–1565.
Jones, K. J. (1988) Steroid hormones and neurotrophism: Relationship to nerve injury. *Metab. Brain Dis.*, 3, 1–18.
Jones, K. J., Alexander, T. D., Brown, T. J., and Tanzer, L. (2000) Gonadal steroid enhancement of facial nerve regeneration: Role of heat shock protein 70. *J. Neurocytol.*, 29, 341–349.
Jones, K. J., Durica, T. E., and Jacob, S. K. (1997) Gonadal steroid preservation of central synaptic input to hamster facial motoneurons following peripheral axotomy. *J. Neurocytol.*, 26, 257–266.

Kim, Y. T., Haftel, V. K., Kumar, S., and Bellamkonda, R. V. (2008) The role of aligned polymer fiber-based constructs in the bridging of long peripheral nerve gaps. *Biomaterials*, 29, 3117–3127.

Lal, D., Hetzler, L. T., Sharma, N., et al. (2008) Electrical stimulation facilitates rat facial nerve recovery from a crush injury. *Otolaryngol. Head Neck Surg.*, 139, 68–73.

Lasek, R. J., and Hoffman, P. N. (1976) The neuronal cytoskeleton, axonal transport and axonal growth. In Goldman, R., Pollard, T., and Rosenbaum, J. (eds.), *Cell Motility Book C: Microtubules and Related Proteins*. Cold Spring Harbor Laboratory, Cold Spring Harbor, NY, pp. 1021–1049.

Lee, J. Y., Bashur, C. A., Goldstein, A. S., and Schmidt, C. E. (2009) Polypyrrole-coated electrospun PLGA nanofibers for neural tissue applications. *Biomaterials*, 30, 4325–4335.

Lieberman, A. R. (1971) The axon reaction: A review of the principal features of perikaryal responses to axon injury. *Int. Rev. Neurobiol.*, 14, 49–124.

Liu, X., Chen, J., Gilmore, K. J., Higgins, M. J., Liu, Y., and Wallace, G. G. (2010) Guidance of neurite outgrowth on aligned electrospun polypyrrole/poly(styrene-beta-isobutylene-beta-styrene) fiber platforms. *J. Biomed. Mater. Res. A*, 94, 1004–1011.

Mason, M. R., Tannemaat, M. R., Malessy, M. J., and Verhaagen, J. (2011) Gene therapy for the peripheral nervous system: A strategy to repair the injured nerve? *Curr. Gene Ther.*, 11, 75–89.

McKasson, M. J., Huang, L., and Robinson, K. R. (2008) Chick embryonic Schwann cells migrate anodally in small electrical fields. *Exp. Neurol.*, 211, 585–587.

McQuarrie, I. G. (1978) The effect of a conditioning lesion on the regeneration of motor axons. *Brain Res.*, 152, 597–602.

McQuarrie, I. G. (1981) Acceleration of axonal regeneration in rat somatic motoneurons by using a conditioning lesion. In Gorio, A., Millesi, H., and Mingrino, S. (eds.), *Posttraumatic Peripheral Nerve Regeneration: Experimental Basis and Clinical Implications*. Raven Press, New York, pp. 49–58.

McQuarrie, I. G., Grafstein, B., and Gershon, M. D. (1977) Axonal regeneration in the rat sciatic nerve: Effect of a conditioning lesion and of dbcAMP. *Brain Res.*, 132, 443–453.

McQuarrie, I. G., and Lasek, R. J. (1989) Transport of cytoskeletal elements from parent axons into regenerating daughter axons. *J. Neurosci.*, 9, 436–446.

Midha, R. (2011) Management and repair of peripheral nerve injuries. In Winn, H. R. (ed.), *Youman's Neurological Surgery*. Saunders, Philadelphia, PA, pp. 436–446.

Monaco, G. N., Brown, T. J., Burgette, R. C., et al. (2015) Electrical stimulation and testosterone enhance recovery from recurrent laryngeal nerve crush. *Restor. Neurol. Neurosci.*, 33, 571–578.

Neumann, S., and Woolf, C. J. (1999) Regeneration of dorsal column fibers into and beyond the lesion site following adult spinal cord injury. *Neuron*, 23, 83–91.

Nix, W. A., and Hopf, H. C. (1983) Electrical stimulation of regenerating nerve and its effect on motor recovery. *Brain Res.*, 272, 21–25.

Oudega, M., Varon, S., and Hagg, T. (1994) Regeneration of adult rat sensory axons into intraspinal nerve grafts: Promoting effects of conditioning lesion and graft predegeneration. *Exp. Neurol.*, 129, 194–206.

Pfister, B. J., Gordon, T., Loverde, J. R., Kochar, A. S., Mackinnon, S. E., and Cullen, D. K. (2011) Biomedical engineering strategies for peripheral nerve repair: Surgical applications, state of the art, and future challenges. *Crit Rev. Biomed. Eng.*, 39, 81–124.

Pockett, S., and Gavin, R. M. (1985) Acceleration of peripheral nerve regeneration after crush injury in rat. *Neurosci Lett.*, 59, 221–224.

Richardson, R. T., Wise, A. K., Thompson, B. C., et al. (2009) Polypyrrole-coated electrodes for the delivery of charge and neurotrophins to cochlear neurons. *Biomaterials*, 30, 2614–2624.

Rishal, I., and Fainzilber, M. (2010) Retrograde signaling in axonal regeneration. *Exp. Neurol.*, 223, 5–10.

Rishal, I., and Fainzilber, M. (2014) Axon-soma communication in neuronal injury. *Nat. Rev. Neurosci.*, 15, 32–42.

Sabatier, M. J., Redmon, N., Schwartz, G., and English, A. W. (2008) Treadmill training promotes axon regeneration in injured peripheral nerves. *Exp. Neurol.*, 211, 489–493.

Sharma, N., Coughlin, L., Porter, R. G., et al. (2009) Effects of electrical stimulation and gonadal steroids on rat facial nerve regenerative properties. *Restor. Neurol. Neurosci.*, 27, 633–644.

Sharma, N., Marzo, S. J., Jones, K. J., and Foecking, E. M. (2010a) Electrical stimulation and testosterone differentially enhance expression of regeneration-associated genes. *Exp. Neurol.*, 223, 183–191.

Sharma, N., Moeller, C. W., Marzo, S. J., Jones, K. J., and Foecking, E. M. (2010b) Combinatorial treatments enhance recovery following facial nerve crush. *Laryngoscope,* 120, 1523–1530.

Storer, P. D., Houle, J. D., Oblinger, M., and Jones, K. J. (2002) Combination of gonadal steroid treatment and peripheral nerve grafting results in a peripheral motoneuron-like pattern of beta II-tubulin mRNA expression in axotomized hamster rubrospinal motoneurons. *J. Comp Neurol.*, 449, 364–373.

Sulaiman, O. A. R., Midha, R., and Gordon, T. (2011) Pathophysiology of surgical nerve disorders. In Winn, H. R. (ed.), *Yeomans Neurological Surgery*. Saunders, Philadelphia, PA, pp. 1–12.

Sulaiman, W., and Gordon, T. (2013) Neurobiology of peripheral nerve injury, regeneration, and functional recovery: From bench top research to bedside application. *Ochsner. J.*, 13, 100–108.

Tajdaran, K., Gordon, T., Wood, M. D., Shoichet, M. S., and Borschel, G. H. (2016a) A glial cell line-derived neurotrophic factor delivery system enhances nerve regeneration across acellular nerve allografts. *Acta Biomater.*, 29, 62–70.

Tajdaran, K., Gordon, T., Wood, M. D., Shoichet, M. S., and Borschel, G. H. (2016b) An engineered biocompatible drug delivery system enhances nerve regeneration after delayed repair. *J. Biomed. Mater. Res. A*, 104A, 367–376.

Tannemaat, M. R., Boer, G. J., Eggers, R., Malessy, M. J., and Verhaagen, J. (2009) From microsurgery to nanosurgery: How viral vectors may help repair the peripheral nerve. *Prog. Brain Res.*, 175, 173–186.

Tanzer, L., and Jones, K. J. (1997) Gonadal steroid regulation of hamster facial nerve regeneration: Effects of dihydrotestosterone and estradiol. *Exp. Neurol.*, 146, 258–264.

Thompson, B. C., Richardson, R. T., Moulton, S. E., et al. (2010) Conducting polymers, dual neurotrophins and pulsed electrical stimulation-dramatic effects on neurite outgrowth. *J. Control Release*, 141, 161–167.

Thompson, N. J., Sengelaub, D. R., and English, A. W. (2014) Enhancement of peripheral nerve regeneration due to treadmill training and electrical stimulation is dependent on androgen receptor signaling. *Dev. Neurobiol.*, 74, 531–540.

Udina, E., Cobianchi, S., Allodi, I., and Navarro, X. (2011) Effects of activity-dependent strategies on regeneration and plasticity after peripheral nerve injuries. *Ann. Anat.*, 193, 347–353.

Udina, E., Ladak, A., Furey, M., Brushart, T., Tyreman, N., and Gordon, T. (2010) Rolipram-induced elevation of cAMP or chondroitinase ABC breakdown of inhibitory proteoglycans in the extracellular matrix promotes peripheral nerve regeneration. *Exp. Neurol.*, 223, 143–152.

Weng, B., Shepherd, R. L., Crowley, K., Killard, A. J., and Wallace, G. G. (2010) Printing conducting polymers. *Analyst*, 135, 2779–2789.

Wenjin, W., Wenchao, L., Hao, Z., et al. (2011) Electrical stimulation promotes BDNF expression in spinal cord neurons through Ca(2+)- and Erk-dependent signaling pathways. *Cell. Mol. Neurobiol.*, 31, 459–467.

Witzel, C., Rohde, C., and Brushart, T. M. (2005) Pathway sampling by regenerating peripheral axons. *J. Comp Neurol.*, 485, 183–190.

Wong, J. N., Olson, J. L., Morhart, M. J., and Chan, K. M. (2015) Electrical stimulation enhances sensory recovery: A randomized control trial. *Ann Neurol.*, 77, 996–1006.

Wood, K., Wilhelm, J. C., Sabatier, M. J., Liu, K., Gu, J., and English, A. W. (2012a) Sex differences in the effectiveness of treadmill training in enhancing axon regeneration in injured peripheral nerves. *Dev. Neurobiol.*, 72, 688–698.

Wood, M. D., Gordon, T., Kemp, S. W., et al. (2013a) Functional motor recovery is improved due to local placement of GDNF microspheres after delayed nerve repair. *Biotechnol. Bioeng.*, 110, 1272–1281.

Wood, M. D., Gordon, T., Kim, H., et al. (2013b) Fibrin gels containing GDNF microspheres increase axonal regeneration after delayed peripheral nerve repair. *Regen. Med.*, 8, 27–37.

Wood, M. D., Kim, H., Bilbily, A., et al. (2012b) GDNF released from microspheres enhances nerve regeneration after delayed repair. *Muscle Nerve*, 46, 122–124.

You, S., Petrov, T., Chung, P. H., and Gordon, T. (1997) The expression of the low affinity nerve growth factor receptor in long-term denervated Schwann cells. *Glia*, 20, 87–100.

Zhang, Z., Rouabhia, M., Wang, Z., et al. (2007) Electrically conductive biodegradable polymer composite for nerve regeneration: Electricity-stimulated neurite outgrowth and axon regeneration. *Artif. Organs*, 31, 13–22.

12 *In vitro* modulatory effects of electrical field on fibroblasts

Mahmoud Rouabhia
Université Laval, Québec, Canada

Ze Zhang
Axe Médecine Régénératrice–CHU
Université Laval, Québec, Canada

Contents

12.1 INTRODUCTION

Fibroblasts are primary cells in most connective tissues. They display characteristic phenotypes as to the anatomical region they belong to. Indeed, Castor et al. (1962) demonstrated metabolic differences between mesothelial fibroblasts, fibroblasts of the skin, articular tissues, and periosteum. These were supported by several studies showing site-specific gene expression patterns with a striking division between anterior–posterior, proximal–distal, and dermal–nondermal (Rinn et al. 2008). Furthermore, there is a topographic expression diversity of genes involved in extracellular matrix synthesis, cell growth and differentiation, and cell migration, as well as genes involved in genetic syndromes (Chang et al. 2002; Rinn et al. 2006, 2008). Fibroblasts are also size specific, with palmar fibroblasts being smaller than non-glabrous-derived fibroblasts (Chipev and Simon 2002). In addition to the differences in fibroblasts from different anatomical sites, fibroblasts extracted from a single tissue do not show a homogenous population. As an example, papillary fibroblasts and reticular fibroblasts have a significant difference in extracellular matrix production (Sorrell and Caplan 2004). Indeed, versican is produced at low levels by the papillary fibroblasts compared with reticular fibroblasts. In contrast, a high level of decorin is produced by papillary fibroblast compared with reticular fibroblasts. Interestingly, both the papillary and reticular fibroblast populations produce a comparable amount of collagen type I and collagen type III (Sorrell and Caplan 2004). Thus, fibroblasts are key cells producing and organizing the extracellular matrix. They are also continuously communicating with each other and with different other cell types, such as keratinocytes, regulating the tissue physiology (Ansel et al. 1996; Werner and Smola 2001).

12.2 FIBROBLAST–KERATINOCYTE INTERACTION

The cross talk between fibroblasts and keratinocytes is a vital process for tissue (skin, oral mucosa, etc.), integrity, and physiological functions. Through this cross talk, dermal fibroblasts orchestrate several epidermal biological pathways, including keratinocyte proliferation, differentiation, and stratification, forming a well-structured functional epidermis (Menon et al. 2012). One of the key structures involved in the keratinocyte–fibroblast intercommunication is the basement membrane (BM), providing structural adhesion by attaching the epidermis to the dermis, lending the skin resistance against shearing forces, maintaining a tissue architecture during remodeling and repair, and regulating epithelial–mesenchymal interactions (Figure 12.1). The BM structure includes hemidesmosomes in basal keratinocytes, the anchoring filaments emanating from the hemidesmosomes through the lamina lucida, and the anchoring fibrils extending from

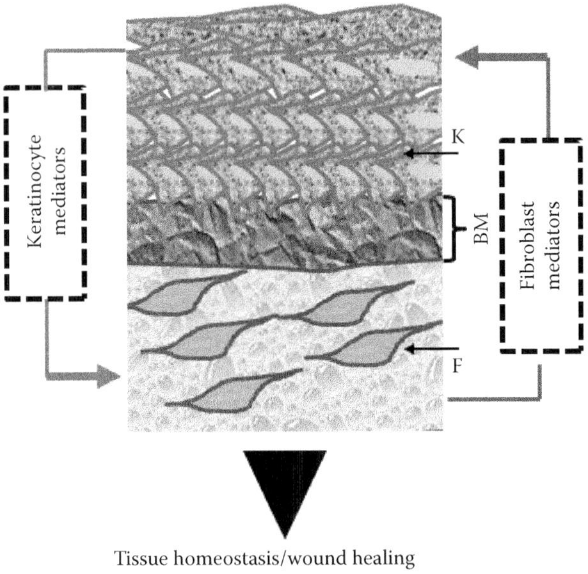

Figure 12.1 Interaction between keratinocytes and fibroblasts through basement membrane. F, fibroblast; K, keratinocyte; BM, basement membrane.

the lamina densa into the dermis (Bruckner-Tuderman and Has 2014). Furthermore, a focal adhesion complex that binds the actin cytoskeleton to the BM was demonstrated (Tsuruta et al. 2011). *In vitro*, studies demonstrated that keratinocyte and fibroblast interact, leading to significant extracellular matrix contraction (Souren et al. 1989). Keratinocytes can regulate fibroblast phenotype contributing to tissue remodeling in addition to extracellular matrix production and deposition, promoting wound healing (Souren et al. 1989; Harrison et al. 2006). Interaction between keratinocytes and fibroblasts is also established through different mediators (fibroblast- and keratinocyte-secreted factors), maintaining epidermal homeostasis and promoting regeneration during wound healing (Cabrijan and Lipozencic 2011; Wojtowicz et al. 2014) (Figure 12.1).

12.3 ROLE OF GROWTH FACTORS MEDIATING FIBROBLAST–KERATINOCYTE INTERACTIONS

It is well established that keratinocytes and fibroblasts communicate with each other via double-paracrine signaling loops, known as cross talk or dynamic reciprocity, which coordinate their actions to maintain and, if needed, restore normal tissue homeostasis after wounding (Tomic-Canic 2005; Schultz et al. 2011). Fibroblasts communicate with keratinocytes through secreted paracrine factors, such as basic fibroblast growth factor (bFGF/FGF-2) (Barrientos et al. 2008; Tomic-Canic 2005), keratinocyte growth factor (KGF-1/FGF-7) (Lee et al. 2014; Werner et al. 2007), vascular endothelial growth factor A (VEGF-A) (Barrientos et al. 2008), and insulin-like growth factor-1 (IGF-1) (Kilani et al. 2007; Talebpour Amiri et al. 2014). Following interaction with keratinocytes, fibroblasts synthesize collagen and promote cross-linking to form an extracellular matrix and differentiate into a myofibroblastic phenotype to facilitate wound closure (Glim et al. 2013; Pastar et al. 2014). FGF-2, or basic FGF, contributes to granulation tissue formation, reepithelialization, and tissue remodeling (Abe et al. 2012). On the other hand, FGF-7 stimulates keratinocyte proliferation and migration, contributing to tissue reepithelialization (Peng et al. 2011). Angiogenesis is very important to wound healing. VEGF-A plays an important role in promoting tissue vascularization, particularly through endothelial cell migration (Senger et al. 1996). Keratinocytes also contribute to tissue vascularization by secreting different key factors such as VEGF (Barrientos et al. 2008) and platelet-derived growth factor (PDGF) (Werner et al. 2007). These factors contribute to endothelial cell migration and wound bed vascularization (Bao et al. 2009). Furthermore, interleukin-1 (IL-1) secreted by keratinocytes upregulates KGF production by fibroblasts, which in turn stimulates keratinocyte proliferation and migration (Maas-Szabowski et al. 2001). These growth factors and keratinocyte–fibroblast interaction during the wound healing process are major players in maintaining tissue homeostasis.

12.4 ROLE OF FIBROBLASTS IN WOUND HEALING

Fibroblasts are a key player in epithelium restoration after injury. In a normal wound healing process, fibroblasts directly or indirectly, through mediators, play key roles in inflammation, the formation of granulation tissue, tissue reepithelialization, and remodeling. Involved in the wound healing process, fibroblasts change phenotypes to myofibroblasts expressing alpha-smooth muscle actin (α-SMA), an early differentiation marker of smooth muscle cells (Hinz and Gabbiani 2003). Myofibroblasts are found in the early phase of granulation tissue formation (Marangoni et al. 2015), become most abundant during the proliferation phase of wound healing, and gradually disappear during the remodeling phase, presumably through an apoptotic process (Ross et al. 1970). However, scars can occur during wound healing. In nonhealing wounds, fibroblasts become dysfunctional, producing less much-needed growth factors, including bFGF (Akita et al. 2013), PDGF (Demaria et al. 2014), and VEGF (Bao et al. 2009). This suggests possible clinical strategies that include modulating one or multiple mediators to restore the wound healing process. However, this may not be possible without restoring the normal function of fibroblasts and keratinocytes. Thus, there still a critical need for new strategies to manage nonhealing wounds, such as diabetic foot ulcers, pressure ulcers, and chronic venous ulcers, which represent a major healthcare burden all over the world. Stimulating cells, including fibroblasts, during the wound healing process could be a valuable way to restore cell functions and contribute to the clinical management of nonhealing wounds.

12.5 PROMOTING WOUND HEALING PROCESS

Multiple strategies have been used to promote wound healing, including vacuum-assisted wound closure (Sinha et al. 2013). Through this process, the authors showed that exposure to vacuum significantly reduced soft tissue defects compared with standard wound therapy (Sinha et al. 2013). This suggests that vacuum-assisted wound therapy was able to facilitate the rapid formation of healthy granulation tissue on open wounds, thus shortening the healing period and contributing to the patient well-being. Extracorporeal shock wave therapy (ESWT) (Moretti et al. 2009) is physiotherapy to promote wound healing. Following several weeks of treatment with standard care or ESWT, more than 50% of the ESWT-treated patients showed complete wound closure, compared with about 33% of the control patients. This was supported by a high (2.97 mm^2/die) index of the reepithelization in the ESWT group compared with the control group (1.30 mm^2/die) (Moretti et al. 2009). Wound healing may also be promoted by local warming as reported by Alvarez et al. (2003), showing that ulcers treated with noncontact normothermic wound therapy had greater healing during the first 4 months. Later on, 70% of the wounds treated with noncontact normothermic wound therapy were healed, compared with 40% in the control group. Electrical stimulation (ES) was also suggested as a wound healing–promoting agent. The subsequent sections of this review focus on ES promoting wound healing through fibroblast modulations.

12.6 EFFECT OF ELECTRICAL FIELD ON FIBROBLAST ADHESION

Cell adhesion is essential in all aspects of development. As a key primary process, cell adhesion is tightly linked to different signaling pathways that are also involved in cell division, cell migration, cell differentiation, and apoptosis. Adhesive properties are also important in cell–cell or cell–substrate binding and interaction (Ramjaun and Hodivala-Dilke 2009). The cell adhesive function is attributed to the membrane molecules called cell adhesion molecules (CAMs) (Hatzfeld 2005). Among these molecules, E-cadherin and β1-integrin have critical roles in early development (Shirazi et al. 2015; Toh et al. 2015). Cell adhesion is linked to a high density of focal contacts promoting the rearrangement of the cytoskeletal structures (Block et al. 2008). Cell orientation determines the alignment of collagenous matrix in healing tissue after injury (Smitha and Donoghue 2011). Cell adhesion can be modulated by different exogenous stimuli (Rico et al. 2010). Indeed, a mechanical process such as stretching was found to modulate cell adhesion and orientation (Balcioglu et al. 2015; Yamada and Ando 2007). Physical stimuli such as the application of electrical stimulus were also found to modulate different cell properties, including adhesion and orientation (Sun et al. 2006). Exposure of human dermal fibroblasts to 100 mV/mm for 60 min led to an increased expression of adhesion genes, such as myosin regulatory light-chain interacting protein, protocadherin-9, and CD44 (Sun et al. 2006). Also, when electrical current ranging from 0 to 200 mA was applied with a direct current (DC) power supply, fibroblast adhesion and growth were upregulated (Jeong et al. 2008). ES (100 mV/mm) through conductive membrane was able to promote cell adhesion and growth (Shi et al. 2008a). DC at a constant electrical field (EF) strength (0–6 V/cm for sparse cells, 0–2 V/cm for monolayers, 37°C) was reported to mediate fibroblast attachment to substrata and fibroblast migration via the participation of microtubules and adhesive components (Finkelstein et al. 2004). Overall, available data demonstrated that ES can improve fibroblast adhesion through the activation of several adhesion genes and proteins. These are the keys for subsequent cell evolution, such as cell proliferation and tissue formation.

12.7 EFFECT OF ELECTRICAL FIELD ON FIBROBLAST PROLIFERATION

Given the contribution of fibroblasts to tissue restoration and functionality, it is important that these cells can proliferate to increase their number, and thus provide adequate protein synthesis and interactions with other cell types, including keratinocytes. Fibroblast proliferation can be promoted by different endogenous and exogenous mediators (Kendall and Feghali-Bostwick 2014; Majtan 2014). Among the exogenous mediators, ES was reported to upregulate fibroblast proliferation (Dubey et al. 2013; Rouabhia et al. 2013). We have previously reported that ES delivered through conductive polypyrrole–polylactide (PPy/PLA) substrate was able to upregulate primary human fibroblast viability (Shi et al. 2008a) and proliferation (Shi et al. 2008b). ES was suggested to promote cell passage by entering from the G0 to G1 phase earlier, compared with the nonstimulated fibroblasts (Cheng and Goldman 1998), confirming the effect of ES on cell proliferation. This ES-mediated cell growth can be through the activation of multiple growth and proliferation genes (Jennings et al. 2008). Furthermore, ES-activated signaling goes through EF receptors on the cell membrane, leading to the transduction of the signal determining cell fate and response. The family of EF-sensitive receptors includes epidermal growth factor receptor (EGFR) and acetylcholine receptor (AchR) (H. F. Tsai et al. 2013). By activating these receptors, different intracellular pathways, such as phosphatidylinositol-3-OH kinase (PI3K)/Akt may be involved in modulating cell growth (Wang et al. 2013). Recently, we demonstrated that ES stimulated fibroblasts through the Smad signaling pathway, as we observed a higher level of phosphorylated Smad2 and Smad3 in the stimulated than in the non-ES-stimulated fibroblasts (Wang et al. 2015). These data suggest that ES modulates fibroblast activities through the Smad signaling pathway, thus providing new mechanistic insights related to the use of ES to promote wound healing in humans.

12.8 EFFECT OF ELECTRICAL FIELD ON FIBROBLAST MOBILITY AND MIGRATION

To play an active role in wound healing, resting fibrocytes or fibroblasts are recruited and activated to migrate from one site to another, with upregulated extracellular matrix secretion and deposition promoting the tissue homeostasis. Fibroblast migration can be modulated by not only chemotaxis growth factors, but also an endogenous electric signal following the wound (Martins et al. 2012). This wound EF was reported to promote wound healing by increasing fibroblast migration (Guo et al. 2010). Lateral EFs of approximately 100 mV/mm have been measured in mammalian skin wounds (Illingworth and Barker 1980) and are thought to be an important mechanism in guiding cell migration toward the wound. *In vitro* experiments showed that EF guides the migration of different cell types, including epithelial cells, embryonic neurons, and fibroblasts (Zhao et al. 1999). As a key process in the wound healing, the migration of fibroblasts stimulated by EF has attracted different groups. Previous work has shown that murine fibroblasts (Yang et al. 1984), embryonic chick heart fibroblasts (Harris et al. 1990), human gingival fibroblasts (Ross et al. 1989), ligament fibroblasts (Chao et al. 2007), and dermal fibroblasts (Guo et al. 2010) migrate toward the cathode, in response to an EF established between two electrodes. In a different setting not involving electrodes, our recent work demonstrated that ES promoted dermal fibroblast migration following wound creation in a confluent monolayer culture. Different from the directional migration toward the cathode, the fibroblasts exposed to EF on the conductive polymer surface and then replated recorded higher migration activity without a preferred direction. Apparently, different mechanisms must be involved in the two migration behaviors. Indeed, wounds were created in skin fibroblast monolayers, and cell migration from the edge of the scratch toward the center, namely, wound closure. This process was recorded at 6, 12, and 24 h postwounding. The ES-exposed fibroblasts actively migrated from both edges and had closed the entire wound at 24 h. Further analysis showed that compared with the nonexposed controls, a significant ($p < 0.01$) reduction of wound distance was observed at 12 and 6 h for the cells treated with 50 and 200 mV/mm, respectively, with a higher migration rate observed for the cells treated at 200 mV/mm (Figure 12.2). We also noticed that the migration rate was linear. Cell migration was comparable between those different stimulation periods (2, 4, or 6 h) (Rouabhia et al. 2013).

Cell migration involves different specific molecules, such as integrins. These are transmembrane receptors contributing to the interaction of the extracellular matrix with the actin cytoskeleton. Integrins are considered mechanoreceptors that, upon activation by binding to extracellular matrix ligands, trigger different signals that mediate cell growth, survival, and migration (Liu et al. 2014). Through the cytoplasmic domain, integrins interact with key cytoplasmic signaling proteins, such as focal adhesion kinase (FAK), to regulate cell adhesion and migration (Wolfenson et al. 2009). The ES-promoted fibroblast migration was therefore thought to be related to integrin modulations. Following exposure to EF, porcine ligament fibroblasts showed integrin polarization in the direction of cell migration (C. H. Tsai et al., 2013). Such polarization,

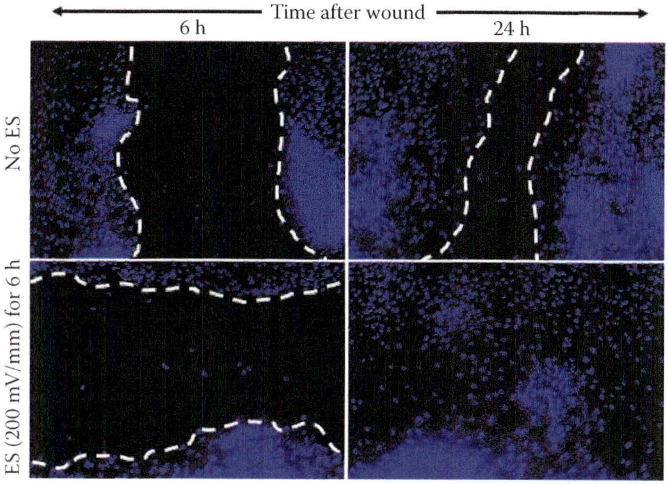

Figure 12.2 ES promotes the migration of primary human dermal fibroblasts after scratch. Cells were stained with Hoechst (blue staining), photographed, and then presented. (Reproduced from Rouabhia, M., et al., *PLoS One*, 19, e71660, 2013. With permission.)

however, was not significant in the groups of high migration speed but little orientation. The authors hypothesized that EF-induced integrin aggregation contributed to the directionality of cell migration. Nevertheless, the integrin polarization does not explain the high migration speed. This increased cell migration may have involved growth factors such as FGFs.

12.9 EFFECT OF ELECTRICAL FIELD ON GROWTH FACTOR RELEASE

Fibroblast is a source of different cytokines and chemokines, such as IL-1β, IL-6, IL-8, IL-33, transforming growth factor β1 (TGFβ1), and CXC and CC chemokines (Feghali and Wright 1997; Gharaee-Kermani et al. 2012). These mediators have different roles. As an example, TGFβ promotes cell proliferation, migration, extracellular matrix production, and fibroblast differentiation to myofibroblasts. Inhibition of TGFβ secretion prevents fibrosis (Bonniaud et al. 2005). As another example, fibroblasts produce IL-1β, a potent pro-inflammatory cytokine (Feghali and Wright 1997). IL-1β is involved in the production of profibrotic cytokines such as PDGF and TGFβ. Fibroblasts play an important role in angiogenesis by producing VEGF, which acts on VEGF receptors expressed on endothelial cells to promote angiogenesis by stimulating mitogenesis and migration, for example (Kajihara et al. 2013; Newman et al. 2011). Fibroblasts regulate wound healing through the secretion of matrix metalloproteinases (MMPs), including MMP-1 and MMP-3 (Zhang et al. 2014). Growth factors such as FGF1 and FGF2 are required for fibroblast growth and interaction with neighbor cells. Growth factors and cytokines are therefore critical players in tissue repair (Powers et al. 2000; Komi-Kuramochi et al. 2005).

The production of these different mediators (FGF, VEGF, MMPs, etc.) can be modulated by exogenous stimuli such as ES (Rouabhia et al. 2013; Asadi et al. 2013). Exposure of primary human fibroblasts to ES of 50 or 200 mV/mm upregulated FGF-1 and FGF-2 secretion by the skin fibroblasts (Rouabhia et al. 2013). In a full-thickness wound, high levels of growth factors, including FGF-2, are readily found in wound fluid (Fei et al. 2013). A recent work (Asadi et al. 2013) studied the secretion of FGF-2 in a full-thickness wound following ES using a cathode directly over the wound. It demonstrated that FGF-2 levels in the sensory ES group (600 μA DC, 1 h/day, every other day for 3 or 7 days) were significantly greater than those in the motor ES (2.5 mA monophasic pulse of 300 μs, 100 Hz, 1 h/day for 3 or 7 days) and control groups. This suggests that application of the sensory ES to incisional wounds induces the expression of FGF-2 in the early phase of wound healing. Human cutaneous fibroblasts exposed to an EF of 50 mV/mm showed higher gene and protein expressions of IL-6 and IL-8 (Shi et al. 2008b). Recently, we demonstrated that exposure of primary dermal fibroblast to ES modulated gene and protein expression of different MMPs and their inhibitors, including MMP-2 and TIMP-2 (Park et al. 2015). Altogether, these studies suggest the implication of ES in fibroblast physiology through fibroblast-secreted mediators. Further studies are needed to broaden our understanding of the roles of ES in fibroblast-produced growth factors and their impact on tissue homeostasis and wound healing.

12.10 FIBROBLAST DIFFERENTIATION TO MYOFIBROBLASTS UNDER ELECTRICAL STIMULATION

During the wound healing process, high numbers of fibroblasts are activated, adopting a myofibroblast phenotype and playing an active role in tissue repair (Darby et al. 2014). Myofibroblast is a cell type phenotypically between fibroblast and smooth muscle cell because it expresses α-SMA, an early differentiation marker of smooth muscle cells (Hinz et al. 2007). Myofibroblasts are found in the early phase of granulation tissue formation, become most abundant during the proliferation phase of wound healing, and gradually disappear during the final stages of healing, presumably through an apoptotic process (Kis et al. 2011). These cells exert a high traction force promoting wound contraction and closure (Tomasek et al. 2012). Promoting wound closure will help prevent infection, chronic inflammation, and scars. Because ES has been shown to promote fibroblast adhesion, migration, and growth (Kloth and Feedar 1991; Thakral et al. 2015), this suggests that ES may also promote wound healing by promoting fibroblast-to-myofibroblast differentiation. In a recent study, we demonstrated that fibroblasts exposed to ES were able to contract a 3D collagen matrix (Wang et al. 2015), supporting the animal studies showing that EF increased collagen synthesis and wound closure and contraction (Cinar et al. 2009). The ES-enhanced fibroblast contraction to collagen matrix we demonstrated could be due to a high level of expression of α-SMA protein by the fibroblasts following ES exposure (Rouabhia et al. 2013) (Figure 12.3). Expression of α-SMA mediated by ES was previously reported in vascular smooth muscle cells (VSMCs) (Rowlands and Cooper-White 2008). These data suggest that one of a number of ways that ES contributes to wound healing can be by favoring fibroblast differentiation to myofibroblasts. However, detailed study about the mechanisms leading to myofibroblast activation has yet to be investigated.

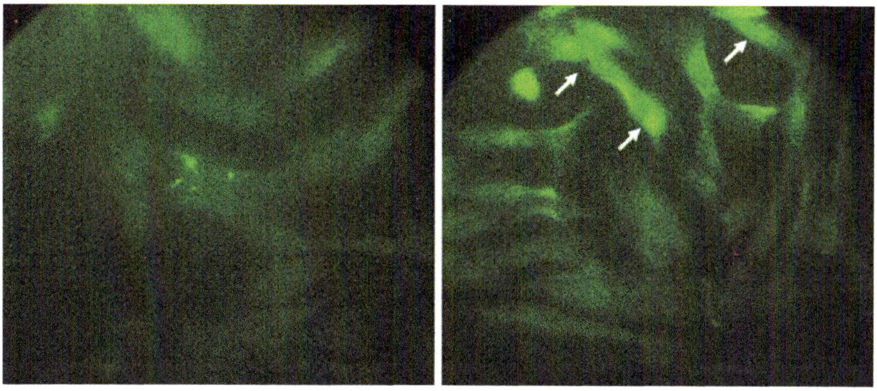

Figure 12.3 ES promotes the expression of α-SMA by primary human dermal fibroblasts. Arrow, α-SMA-positive cells. (Reproduced from Rouabhia, M., et al., *PLoS One*, 19, e71660, 2013. With permission.)

12.11 CONCLUSION

Available studies reported some benefits of ES on bone regeneration, cardiac activities, nerve regeneration in sensory and motor neurons after injury, and tissue wound healing. However, basic and clinical research is still needed to shed light on the mechanisms of how ES affects cell responses. As a potential wound healing modulator, ES may affect different cell types, including fibroblasts. Through this chapter, we demonstrated the role of ES on fibroblast, but more investigations are needed to understand how ES modulates fibroblast behaviors and functions during wound healing. ES may also modulate the specific interactions of fibroblasts with other cell types, such as keratinocytes, endothelial cells, and inflammatory cells, during wound healing. Thus, future studies should include cell signaling mechanisms involved in the cell response to ES. We should also investigate the effect of ES on fibroblast interactions with neighbor cells.

REFERENCES

Abe, M., Yokoyama, Y., and Ishikawa, O. A possible mechanism of basic fibroblast growth factor-promoted scarless wound healing: the induction of myofibroblast apoptosis. *Eur J Dermatol.* 22 (2012): 46–53.

Akita, S., Akino, K., and Hirano, A. Basic fibroblast growth factor in scarless wound healing. *Adv Wound Care (New Rochelle)* 2 (2013): 44–49.

Alvarez, O. M., Rogers, R. S., Booker, J. G., and Patel, M. Effect of noncontact normothermic wound therapy on the healing of neuropathic (diabetic) foot ulcers: An interim analysis of 20 patients. *J Foot Ankle Surg* 42 (2003): 30–35.

Ansel, J. C., Kaynard, A. H., Armstrong, C. A., Olerud, J., Bunnett, N., and Payan, D. Skin-nervous system interactions. *J Invest Dermatol* 106 (1996): 198–204.

Asadi, M. R., Torkaman, G., Hedayati, M., and Mofid, M. Role of sensory and motor intensity of electrical stimulation on fibroblastic growth factor-2 expression, inflammation, vascularization, and mechanical strength of full-thickness wounds. *J Rehabil Res Dev* 50 (2013): 489–498.

Balcioglu, H. E., van Hoorn, H. V., Donato, D. M., Schmidt, T., and Danen, E. H. Integrin expression profile modulates orientation and dynamics of force transmission at cell matrix adhesions. *J Cell Sci* 128 (2015): 1316–1326.

Bao, P., Kodra, A., Tomic-Canic, M., Golinko, M. S., Ehrlich, H. P., and Brem, H. The role of vascular endothelial growth factor in wound healing. *J Surg Res* 153 (2009): 347–358.

Barrientos, S., Stojadinovic, O., Golinko, M. S., Brem, H., and Tomic-Canic, M. Growth factors and cytokines in wound healing. *Wound Repair Regen* 16 (2008): 585–601.

Block, M. R., Badowski, C., Millon-Fremillon, A., et al. Podosome-type adhesions and focal adhesions, so alike yet so different. *Eur J Cell Biol* 87 (2008): 491–506.

Bonniaud, P., Margetts, P. J., Kolb, M., Schroeder, J. A., Kapoun, A. M., and Damm, D. Progressive transforming growth factor beta1-induced lung fibrosis is blocked by an orally active ALK5 kinase inhibitor. *Am J Respir Crit Care Med* 171 (2005): 889–898.

Bruckner-Tuderman, L., and Has, C. Disorders of the cutaneous basement membrane zone—The paradigm of epidermolysis bullosa. *Matrix Biol* 33 (2014): 29–34.

Cabrijan, L., and Lipozencic, J. Adhesion molecules in keratinocytes. *Clin Dermatol* 29 (2011): 427–431.

Castor, C. W., Prince, R. K., and Dorstewitz, E. L. Characteristics of human fibroblasts cultivated *in vitro* from different anatomical sites. *Lab Invest* 11 (1962): 703–713.

Chang, H. Y., Chi, J. T., Dudoit, S., Bondre, C., Van de Rijn, M., and Botstein, D. Diversity, topographic differentiation, and positional memory in human fibroblasts. *Proc Natl Acad Sci* U S A 99 (2002): 12877–12882.

Chao, P. H., Lu, H. H., Hung, C. T., Nicoll, S. B., and Bulinski, J. C. Effects of applied DC electric field on ligament fibroblast migration and wound healing. *Connect Tissue Res* 48 (2007): 188–197.

Cheng, K., and Goldman, R. J. Electric fields and proliferation in a dermal wound model: Cell cycle kinetics. *Bioelectromagnetics* 19 (1998): 68–74.

Chipev, C. C., and Simon, M. Phenotypic differences between dermal fibroblasts from different body sites determine their responses to tension and TGFbeta1. *BMC Dermatol* 21 (2002): 13.

Cinar, K., Comlekci, S., and Senol, N. Effects of a specially pulsed electric field on an animal model of wound healing. *Lasers Med Sci* 24 (2009): 735–740.

Darby, I. A., Laverdet, B., Bonté, F., and Desmoulière, A. Fibroblasts and myofibroblasts in wound healing. *Clin Cosmet Investig Dermatol* 7 (2014): 301–311.

Demaria, M., Ohtani, N., Youssef, S. A., et al. An essential role for senescent cells in optimal wound healing through secretion of PDGF-AA. *Dev Cell* 31 (2014): 722–733.

Dubey, A. K., Agrawal, P., Misra, R. D., and Basu, B. Pulsed electric field mediated *in vitro* cellular response of fibroblast and osteoblast-like cells on conducting austenitic stainless steel substrate. *J Mater Sci Mater Med* 24 (2013): 1789–1798.

Feghali, C. A., and Wright, T. M. Cytokines in acute and chronic inflammation. *Front Biosci* 2 (1997): 12–26.

Fei, Y., Gronowicz, G., and Hurley, M. M. Fibroblast growth factor-2, bone homeostasis and fracture repair. *Curr Pharm Des* 19 (2013): 3354–3363.

Finkelstein, E., Chang, W., Chao, P. H., Gruber, D., Minden, A., Hung, C. T., and Bulinski, J. C. Roles of microtubules, cell polarity and adhesion in electric-field-mediated motility of 3T3 fibroblasts. *J Cell Sci* 117 (2004): 1533–1545.

Gharaee-Kermani, M., Kasina, S., Moore, B. B., Thomas, D., Mehra, R., and Macoska, J. A. CXC-type chemokines promote myofibroblast phenoconversion and prostatic fibrosis. *PLoS One* 7 (2012): e49278.

Glim, J. E., van Egmond, M., Niessen, F. B., Everts, V., and Beelen, R. H. Detrimental dermal wound healing: What can we learn from the oral mucosa? *Wound Repair Regen* 21 (2013): 648–660.

Guo, A., Song, B., Reid, B., Gu, Y., Forrester, J. V., Jahoda, C. A., and Zhao, M. Effects of physiological electric fields on migration of human dermal fibroblasts. *J Invest Dermatol* 130 (2010): 2320–2327.

Harris, A. K., Pryer, N. K., and Paydarfar, D. Effects of electric fields on fibroblast contractility and cytoskeleton. *J Exp Zool* 253 (1990): 163–176.

Harrison, C., Gossiel, F., Bullock, A., Sun, T., Blumsohn, A., and Mac Neil, S. Investigation of keratinocyte regulation of collagen I synthesis by dermal fibroblasts in a simple *in vitro* model. *Br J Dermatol* 154 (2006): 401–410.

Hatzfeld, M. The p120 family of cell adhesion molecules. *Eur J Cell Biol* 84 (2005): 205–214.

Hinz, B., and Gabbiani, G. Cell-matrix and cell-cell contacts of myofibroblasts: Role in connective tissue remodeling. *Thromb Haemost* 90 (2003): 993–1002.

Hinz, B., Phan, S. H., Thannickal, V. J., Galli, A., Bochaton-Piallat, M. L., and Gabbiani, G. The myofibroblast: One function, multiple origins. *Am J Pathol* 170 (2007): 1807–1816.

Illingworth, C. M., and Barker, A. T. Measurement of electrical currents emerging during the regeneration of amputated fingertip in children. *Clin Phys Physiol Meas* 1 (1980): 87–89.

Jennings, J., Chen, D., and Feldman, D. Transcriptional response of dermal fibroblasts in direct current electric fields. *Bioelectromagnetics* 29 (2008): 394–405.

Jeong, S. I., Jun, I. D., Choi, M. J., Nho, Y. C., Lee, Y. M., and Shin H. Development of electroactive and elastic nanofibers that contain polyaniline and poly(L-lactide-co-epsilon-caprolactone) for the control of cell adhesion. *Macromol Biosci* 8 (2008): 627–637.

Kajihara I., Jinnin, M., Honda, N., Makino, K., Makino, T., and Masuguchi, S. Scleroderma dermal fibroblasts overexpress vascular endothelial growth factor due to autocrine transforming growth factor beta signalling. *Mod Rheumatol* 23 (2013): 516–524.

Kendall, R. T., and Feghali-Bostwick, C. A. Fibroblasts in fibrosis: Novel roles and mediators. *Front Pharmacol* 27 (2014): 123.

Kilani, R. T., Guilbert, L., Lin, X., and Ghahary, A. Keratinocyte conditioned medium abrogates the modulatory effects of IGF-1 and TGF-beta1 on collagenase expression in dermal fibroblasts. *Wound Repair Regen* 15 (2007): 236–244.

Kis, K., Liu, X., and Hagood, J. S. Myofibroblast differentiation and survival in fibrotic disease. *Expert Rev Mol Med* 13 (2011): e27.

Kloth, L. C., and Feedar, J. A. Acceleration of wound healing with high voltage, monophasic, pulsed current. *Phys Ther* 71 (1991): 433–442.

Komi-Kuramochi, A., Kawano, M., Oda, Y., Asada, M., Suzuki, M., Oki, J., and Imamura, T. Expression of fibroblast growth factors and their receptors during full-thickness skin wound healing in young and aged mice. *J Endocrinol* 186 (2005): 273–289.

Lee, D. H., Choi, K. H., Cho, J. W., et al. Recombinant growth factor mixtures induce cell cycle progression and the upregulation of type I collagen in human skin fibroblasts, resulting in the acceleration of wound healing processes. *Int J Mol Med* 33 (2014): 1147–1152.

Liu, L., Zong, C., Li, B., et al. The interaction between β1 integrins and ERK1/2 in osteogenic differentiation of human mesenchymal stem cells under fluid shear stress modelled by a perfusion system. *J Tissue Eng Regen Med* 8 (2014): 85–96.

Maas-Szabowski, N., Szabowski, A., Stark, H. J., Andrecht, S., Kolbus, A., and Schorpp-Kistner, M. Organotypic cocultures with genetically modified mouse fibroblasts as a tool to dissect molecular mechanisms regulating keratinocyte growth and differentiation. *J Invest Dermatol* 116 (2001): 816–820.

Majtan, J. Honey: An immunomodulator in wound healing. *Wound Repair Regen* 2 (2014): 187–192.

Marangoni, R. G., Korman, B., Wei, J., et al. Myofibroblasts in cutaneous fibrosis originate from adiponectin-positive intradermal progenitors. *Arthritis Rheumatol* 67 (2015): 1062–1073.

Martins, M., Warren, S., Kimberley, C., et al. Activity of PLCε contributes to chemotaxis of fibroblasts towards PDGF. *J Cell Sci* 125 (2012): 5758–5769.

Menon, S. N., Flegg, J. A., McCue, S. W., Schugart, R. C., Dawson, R. A., and McElwain, D. L. Modelling the interaction of keratinocytes and fibroblasts during normal and abnormal wound healing processes. *Proc Biol Sci* 279 (2012): 3329–3338.

Moretti, B., Notarnicola, A., Maggio, G., Moretti, L., and Pascone, M. The management of neuropathic ulcers of the foot in diabetes by shock wave therapy. *BMC Musculoskelet Disord* 10 (2009): 54.

Newman, A. C., Nakatsu, M. N., Chou, W., Gershon, P. D., and Hughe, C. C. The requirement for fibroblasts in angiogenesis: Fibroblast-derived matrix proteins are essential for endothelial cell lumen formation. *Mol Biol Cell* 22 (2011): 3791–3800.

Park, H. J., Rouabhia, M., Lavertu, D., and Zhang, Z. Electrical stimulation modulates the expression of multiple wound healing genes in primary human dermal fibroblasts. *Tissue Eng Part A* 21 (2015): 1982–1990.

Pastar, I., Stojadinovic, O., Yin, N. C., et al. Epithelialization in wound healing: A comprehensive review. *Adv Wound Care (New Rochelle)* 3 (2014): 445–464.

Peng, C., Chen, B., Kao, H. K., Murphy, G., Orgill, D. P., and Guo, L. Lack of FGF-7 further delays cutaneous wound healing in diabetic mice. *Plast Reconstr Surg* 128 (2011): 673e–684e.

Powers, C. J., McLeske, S. W., and Wellstein, A. Fibroblast growth factors, their receptors and signaling. *Endocr Relat Cancer* 7 (2000): 165–197.

Ramjaun, A. R., and Hodivala-Dilke, K. The role of cell adhesion pathways in angiogenesis. *Int J Biochem Cell Biol* 41 (2009): 521–530.

Rico, F., Chu, C., Abdulreda, M. H., Qin, Y., and Moy, V. T. Temperature modulation of integrin-mediated cell adhesion. *Biophys J* 99 (2010): 1387–1396.

Rinn, J. L., Bondre, C., Gladstone, H. B., Brown, P. O., and Chang, H. Y. Anatomic demarcation by positional variation in fibroblast gene expression programs. *PLoS Genet* 2 (2006): 119.

Rinn, J. L., Wang, J. K., Liu, H., Montgomery, K., Van de Rijn, M., and Chang, H. Y. A systems biology approach to anatomic diversity of skin. *J Invest Dermatol* 128 (2008): 776–782.

Ross, R., Everett, N. B., and Tyler, R. Wound healing and collagen formation. VI. The origin of the wound fibroblast studied in parabiosis. *J Cell Biol* 44 (1970): 645–654.

Ross, S. M., Ferrier, J. M., and Jaubin, J. E. Studies on the alignment of fibroblasts in uniform applied electrical fields. *Bioelectromagnetics* 10 (1989): 371–384.

Rouabhia, M., Park, H. J., Meng, S., Derbali, H., and Zhang, Z. Electrical stimulation promotes wound healing by enhancing dermal fibroblast activity and promoting myofibroblast transdifferentiation. *PLoS One* 19 (2013): e71660.

Rowlands, A. S., and Cooper-White, J. J. Directing phenotype of vascular smooth muscle cells using electrically stimulated conducting polymer. *Biomaterials* 29 (2008): 4510–4520.

Schultz, G. S., Davidson, J. M., Kirsner, R. S., Bornstein, P., and Herman, I. M. Dynamic reciprocity in the wound microenvironment. *Wound Repair Regen* 19 (2011): 134–148.

Senger, D. R., Ledbetter, S. R., Claffey, K. P., Papadopoulos-Sergiou, A., Peruzzi, C. A., and Detmar, M. Stimulation of endothelial cell migration by vascular permeability factor/vascular endothelial growth factor through cooperative mechanisms involving the alphavbeta3 integrin, osteopontin, and thrombin. *Am J Pathol* 149 (1996): 293–305.

Shi, G., Rouabhia, M., Meng, S., and Zhang, Z. Electrical stimulation enhances viability of human cutaneous fibroblasts on conductive biodegradable substrates. *J Biomed Mater Res A* 84 (2008a): 1026–1037.

Shi, G., Zhang, Z., and Rouabhia, M. The regulation of cell functions electrically using biodegradable polypyrrole-polylactide conductors. *Biomaterials* 9 (2008b): 3792–3798.

Shirazi, A., Heidari, M., Shams-Esfandabadi, N., Momeni, A., and Derafshian, Z. Overexpression of signal transducers and activators of transcription in embryos derived from vitrified oocytes negatively affect E-cadherin expression and embryo development. *Cryobiology* 70 (2015): 239–245.

Sinha, K., Chauhan, V. D., Maheshwari, R., Chauhan, N., Rajan, M., and Agrawal, A. Vacuum assisted closure therapy versus standard wound therapy for open musculoskeletal injuries. *Adv Orthop* 2013 (2013): 245940.

Smitha, B., and Donoghue, M. Clinical and histopathological evaluation of collagen fiber orientation in patients with oral submucous fibrosis. *J Oral Maxillofac Pathol* 15 (2011): 154–60.

Sorrell, J. M., and Caplan, A. I. Fibroblast heterogeneity: More than skin deep. *J Cell Sci* 117 (2004): 667–675.

Souren, J. E., Ponec M. M., and Van Wijk, R. Contraction of collagen by human fibroblasts and keratinocytes. *In Vitro Cell Dev B* 25 (1989): 1039–1045.

Sun, S., Titushkin, I., and Cho, M. Regulation of mesenchymal stem cell adhesion and orientation in 3D collagen scaffold by electrical stimulus. *Bioelectrochemistry* 69 (2006): 133–141.

Talebpour Amiri, F., Fadaei Fathabadi, F., Mahmoudi Rad, M., et al. The effects of insulin-like growth factor-1 gene therapy and cell transplantation on rat acute wound model. *Iran Red Crescent Med J* 16 (2014): e16323.

Thakral, G., La Fontaine, J., Kim, P., Najafi, B., Nichols, A., and Lavery, L. A. Treatment options for venous leg ulcers: Effectiveness of vascular surgery, bioengineered tissue, and electrical stimulation. *Adv Skin Wound Care* 28 (2015): 164–172.

Toh, Y. C., Xing, J., and Yu, H. Modulation of integrin and E-cadherin-mediated adhesions to spatially control heterogeneity in human pluripotent stem cell differentiation. *Biomaterials* 50 (2015): 87–97.

Tomasek, J. J., Gabbiani, G., Hinz, B., Chaponnier, C., and Brown, A. Myofibroblasts and mechano-regulation of connective tissue remodelling. *Nat Rev Mol Cell Biol* 3 (2012): 349–363.

Tomic-Canic, M. Keratinocyte cross-talks in wounds. *Wounds* 17 (2005): S3–S6.

Tsai, C. H., Lin, B. J., and Chao, P. H. $\alpha 2\beta 1$ integrin and RhoA mediates electric field-induced ligament fibroblast migration directionality. *J Orthop Res* 31 (2013): 322–327.

Tsai, H. F., Huang, C. W., Chang, H. F., Chen, J. J., Lee, C. H., and Cheng, J. Y. Evaluation of EGFR and RTK signaling in the electrotaxis of lung adenocarcinoma cells under direct-current electric field stimulation. *PLoS One* 8 (2013): e73418.

Tsuruta, D., Hashimoto, T., Hamill, K. J., and Jones, J. C. Hemidesmosomes and focal contact proteins: Functions and cross-talk in keratinocytes, bullous diseases and wound healing. *J Dermatol Sci* 1 (2011): 1–7.

Wang, M., Li, P., Liu, M., Song, W., Wu, Q., and Fan, Y. Potential protective effect of biphasic electrical stimulation against growth factor-deprived apoptosis on olfactory bulb neural progenitor cells through the brain-derived neurotrophic factor-phosphatidylinositol 3′-kinase/Akt pathway. *Exp Biol Med (Maywood)* 238 (2013): 951–959.

Wang, Y., Rouabhia, M., Lavertu, D., and Zhang, Z. Pulsed electrical stimulation modulates fibroblasts' behaviour through the Smad signalling pathway. *J Tissue Eng Regen Med* (2015): DOI: 10.1002/term.2014.

Werner, S., Krieg, T., and Smola, H. Keratinocyte-fibroblast interactions in wound healing. *J Invest Dermatol* 127 (2007): 998–1008.

Werner, S., and Smola, H. Paracrine regulation of keratinocyte proliferation and differentiation. *Trends Cell Biol* 11 (2001): 143–146.

Wojtowicz, A. M., Oliveira, S., Carlson, M. W., Zawadzka, A., Rousseau, C. F., and Baksh, D. The importance of both fibroblasts and keratinocytes in a bilayered living cellular construct used in wound healing. *Wound Repair Regen* 22 (2014): 246–255.

Wolfenson, H., Henis, Y., Geiger, B., and Bershadsky, A. D. The heel and toe of the cell's foot: A multifaceted approach for understanding the structure and dynamics of focal adhesions. *Cell Motil Cytoskel* 66 (2009): 1017–1029.

Yamada, H., and Ando, H. Orientation of apical and basal actin stress fibers in isolated and subconfluent endothelial cells as an early response to cyclic stretching. *Mol Cell Biomech* 4 (2007): 1–12.

Yang, W. P., Onuma, E. K., and Hui, S. W. Response of C3H/10T1/2 fibroblasts to an external steady electric field stimulation. Reorientation, shape change, ConA receptor and intramembranous particle distribution and cytoskeleton reorganization. *Exp Cell Res* 155 (1984): 92–104.

Zhang, Y., Lin, Z., Foolen, J., Schoen, I., Santoro, A., Zenobi-Wong, M., and Vogel, V. Disentangling the multifactorial contributions of fibronectin, collagen and cyclic strain on MMP expression and extracellular matrix remodeling by fibroblasts. *Matrix Biol* 40 (2014): 62–72.

Zhao, M., Forrester, J. V., and McCaig, C. D. A small, physiological electric field orients cell division. *Proc Natl Acad Sci USA* 96 (1999): 4942–4946.

13 The role of electrical field on neurons: *In vitro* studies

A. Lee Miller II, Huan Wang, Michael J. Yaszemski, and Lichun Lu
Mayo Clinic College of Medicine
Rochester, Minnesota

Contents

13.1 INTRODUCTION

13.1.1 BIOLOGY AND ORGANIZATION OF NEURONS

Cells that make up the nervous system are divided into two broad groups: neurons, which are the primary signaling cells, and glia, which support neurons. In the late nineteenth century, Spanish neuroanatomist Santiago Ramón y Cajal applied Italian physician Camillo Golgi's black staining technique to nervous tissues obtained from different animals and humans. This application produced detailed sketches of discrete cells of the nervous system. These cells were later dubbed neurons by the German anatomist

Wilhelm Waldeyer (Bentivoglio 1998; Costandi 2006). Together, Cajal and Golgi were awarded the Nobel Prize in Physiology or Medicine in 1906.

There is a great deal of variability of neurons. They vary in function, size (small, medium, large, and giant), and shape (spherical, conical, star shaped, fusiform, polyhedral, and pyramidal). Neurons can be classified by their morphology or function. A broad and basic classification of neurons categorizes them based on their functions into motor, sensory, or interneurons. Motor neurons carry information from the central nervous system to organs, glands, and muscles (hence they are also termed efferent neurons). Sensory neurons send information to the central nervous system from internal organs or from external stimuli (hence they are also termed *afferent neurons*). Interneurons relay signals between motor and sensory neurons. They communicate with each other via chemical and electrical synapses in a process known as synaptic transmission (Shepherd 2000).

The majority of neurons belong to the central nervous system; one example is the motoneurons located in the anterior horn of the spinal cord. Other neurons reside in peripheral ganglia, such as sensory neurons located in the dorsal root ganglia. Some sensory neurons are situated in sensory organs such as the cochlea and retina. A typical neuron has three main structures: the cell body (soma), the axon, and the dendrites (Lodish et al. 2013). The cell body contains the nucleus and is where virtually all neuronal proteins and membranes are synthesized. The anatomy of a neuron can be seen in Figure 13.1. Almost every neuron has a single axon, specialized for conducting action potentials (outward for efferent neurons and inward for afferent neurons) away from the cell body toward the axon terminus. Most neurons have multiple dendrites, which extend outward from the cell body and are specialized to receive chemical signals from the axon terminals of other neurons. Synaptic signals from other neurons are received by the dendrites and soma, whereas signals sent to other neurons are transmitted by the axon. In addition to the typical axodendritic and axosomatic synapses, there are axoaxonic (axon-to-axon) (Ashrafi et al. 2014; Hermel et al. 2006) and dendrodendritic (dendrite-to-dendrite) synapses (Shepherd 2009).

13.1.2 NEURITE DIFFERENTIATION

Mature neurons are highly polarized cells, exhibiting morphological, biochemical, and functional compartmentalization. Axonal and dendritic domains are distinct. The ion channels and receptors in dendrite membranes are different from those in axon membranes. This differentiation sustains the proper flux of information in the nervous system

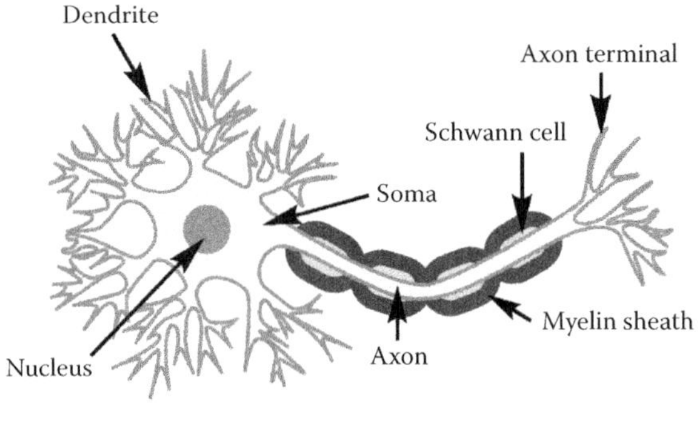

Figure 13.1 Structure of a typical mammalian peripheral motoneuron.

(Villarroel-Campos et al. 2014). The term *neurite* refers to any projection from the cell body of a neuron. This projection can be either a dendrite or an axon. Because it can be difficult to tell axons from dendrites before differentiation is complete, this term is frequently used when speaking of developing or immature neurons, especially of cells in culture.

The axon leaves the neuronal cell body at an enlarged region called the axon hillock (Fuortes et al. 1957). It travels away from the soma for a greater distance before it becomes axon terminals, tiny branches that form synapses with other cells. An action potential propagates along the axon, reaches axon terminals, and induces a localized rise in the level of Ca^{2+}, causing vesicles in the axon terminal to release neurotransmitters into the synaptic cleft. Dendrites receive chemical signals from the axon termini of other neurons. They convert these signals into small electric impulses and transmit them inward toward the cell body.

13.2 *IN VITRO* CELL STUDIES

13.2.1 CELLS

Primary neuronal cultures and neuronal cell lines derived from rodents are widely used to study basic physiological properties of neurons. Neurons from the hippocampus, cerebella, and striatum (Giordano and Costa 2011); dorsal root ganglion (DRG) neurons (Runge et al. 2010; Koppes et al. 2014; Nguyen et al. 2014); neural progenitor cells (Sudwilai et al. 2014); neuroblastoma cells (Neuro2a [N2a]) (Jain et al. 2013; Valero et al. 2010); neural stem cell lines (Bechara et al. 2011); primary cortical neurons (Zhou et al. 2012); olfactory ensheathing cells (Qi et al. 2013); and pheochromocytoma (PC12) cells have been used. PC12 is the most widely used cell line (Bettinger et al. 2009; Runge et al. 2010; Cen et al. 2004; Cho and Borgens 2010; Jin et al. 2012; Kang et al. 2011; Liu et al. 2011; Moroder et al. 2011; Schmidt et al. 1997; Shi et al. 2014; Zeng et al. 2013) in studying conductive polymers.

Culturing these cells must meet certain growth requirements, yet the requirements for each cell or cell line are different. The culture conditions of these cells are summarized in Table 13.1.

13.2.2 CELL VIABILITY

Cell viability is evaluated to assess toxicity and nerve cell affinity toward test substrates. Cell viability can be evaluated by the trypan blue exclusion test (Zhang et al. 2007), a live–dead assay using a commercial kit (Muller et al. 2013). However, MTT assay is the most common way to quantify cell viability and proliferation (Cho and Borgens 2010; Valero et al. 2010; Jain et al. 2013). The tetrazolium dye MTT reacts with dehydrogenase enzymes in the mitochondria of metabolically active cells and forms a purple color formazan crystal. The density of formazan crystals is directly related to the amount of viable cells. The formed formazan crystals are dissolved in dimethyl sulfoxide, and the optical density (OD) is recorded at 540 nm. By plotting the OD value against a standard curve, cell numbers can be calculated.

13.2.3 CELL ATTACHMENT AND PROLIFERATION

One of the key requirements for biomaterials, conductive or not, is biocompatibility. Ideally, the conductive polymer substrate would support cell growth even without the effects of electrical stimulation. Whether a biomaterial contributes to cell growth is usually tested using cell proliferation assay. A cell line close to the biology of the tissue

Table 13.1 **Types of neuronal cells and culture conditions**

CELL TYPE	PRIMARY OR CELL LINE	SOURCE	MORPHOLOGY	CULTURE MEDIUM AND/OR SUBSTRATE	DIFFERENTIATION STATUS
Hippocampus neuron	Primary culture; do not divide in culture; form synapses; electrically active; mortal	Hippocampus region of the brain; rat or mouse fetuses of gestation day 20	Pyramidal cell; triangular-shaped soma; a single axon, a large apical dendrite, and multiple basal dendrites with dendritic spines	Poly-D-lysine-coated culture substrate; neurobasal medium with B27 supplement; serum; pH 7.2–7.5	Already differentiated; maintained for up to 4 weeks (Beaudoin 2012)
Cerebella neuron	Primary culture; do not divide in culture; form synapses; electrically active; mortal	Cerebella; 6- to 8-day-old rat or mouse pups	Granule cell; most populous neuron in the brain; small; most of soma occupied by nucleus	Poly-D-lysine-coated culture substrate; neurobasal medium with B27 supplement; serum; KCl; pH 7.2–7.5	Already differentiated; maintained for up to 2 weeks (Lee 2009)
Striatum neuron	Primary culture; do not divide in culture; form synapses; electrically active; mortal	Subcortical area of brain; rat or mouse fetuses of gestation day 18–19	Medium spiny neuron; medium-sized; complex dendritic arbors; high density of dendritic spines	Poly-D-lysine-coated culture substrate; neurobasal medium with B27 supplement; serum or serum-free; KCl; pH 7.2–7.5	Already differentiated; maintained for up to 8 weeks (Sebben 1990)

(Continued)

Table 13.1 *(Continued)* **Types of neuronal cells and culture conditions**

CELL TYPE	PRIMARY OR CELL LINE	SOURCE	MORPHOLOGY	CULTURE MEDIUM AND/OR SUBSTRATE	DIFFERENTIATION STATUS
DRG neuron	Primary culture; do not divide in culture; used as DRG explants or dissociated cells; die	Dorsal root ganglia; spinal cord of E 15 rats or adult rats	Primary somatic and visceral afferent neurons; medium-sized; bright, round soma with clear nuclei and sometimes nucleoli	DMEM, serum	Requires NGF for survival and neurite extension
Neuroblastoma cells	Cell line; rapid cell division after plating; differentiate after confluency; immortal	Mouse neural crest-derived cell line	Neuronal/ amoeboid-like	DMEM, high glucose, w/L-gluatamine, serum	cAMP (3',5'-cyclic AMP) sodium salt; retinoic acid; low serum medium
PC12 cells	Cell line; rapid cell division after plating; differentiate after confluency; immortal	Rat pheochromocytoma-derived cell line	Small irregularly shaped	DMEM, serum	NGF

structure that the biomaterial is intended to replace or reconstruct is chosen for the assay, for example, osteoblasts for bone tissue engineering. For neural applications, the PC12 cell line is widely used due to its extreme versatility for pharmacological manipulation, ease of culture, the large amount of information on their proliferation and differentiation, and their response to extracellular cues such as electrical stimulation (Westerink and Ewing 2008). Since neuron-like cells are anchoring cells, they attach to the culture surface prior to proliferation and extending neurites. By allowing time for the cells to attach before changing the medium (to get rid of floating cells) and staining the cell nucleus and actin cytoskeleton, anchored cells are well visualized for the assessment of cell attachment. Neuronal cell density and the cell coverage of the substrate surface can both reflect the affinity of biomaterials to their biological counterparts. Therefore, the cell attachment assay is generally used to investigate various surface properties of the materials (Bettinger et al. 2009; Bhang et al. 2012; Runge et al. 2010; Lee et al. 2009; Li et al. 2007; Li et al. 2005; Muller et al. 2013; Stauffer and Cui 2006).

For cell proliferation, the adherent cells can be detached with trypsin, resuspended in medium, sampled, and counted under the microscope using a hemocytometer to calculate the total number of cells that reflect cell proliferation. Cell proliferation can also be quantified using an MTT assay as described in Section 13.2.2.

Another important aspect of cell attachment and the proliferation assay is to test the stability of the conductive polymer upon electrical stimulation. A biocompatible material can potentially become toxic upon electrical stimulation if there is toxic leaching. Cell death due to cytotoxicity of the leaching products will result in cell detachment (Cullen et al. 2008; Durgam et al. 2010).

13.2.4 NEURITE GROWTH, LENGTH, AND ORIENTATION

During nerve regeneration, the axon growth cone continuously moves toward the reinnervation target. Regenerating axons find their way into endoneurium tubes and eventually reconnect with the terminal organ. The ones that are not successful in reestablishing connection with their targets will retract and disappear, known as the pruning process. It is apparent that both neurite outgrowth and directionality are important aspects, and a treatment that promotes both would be ideal.

Neurite growth is an indication of neuronal differentiation that can reflect the quality of the test substrate and the effect of electric stimulation. Neurite growth can be directly determined using phase contrast microscopy and computer-based quantitative image analysis. The cells can also be fixed and stained with actin filaments or with neurofilament to visualize the cell cytoskeleton. The number of cells bearing neurites, the number of neurites per cell, and the neurite length provide quantitative information of cell differentiation (Figure 13.2). The directionality of neuritis has also been investigated upon electrical field application (Koppes et al. 2014; Moroder et al. 2011).

Neurite length is the most widely adopted criterion for quantifying neurite outgrowth. It is defined as the distance from the tip of the neurite to the junction between the cell body and neurite base. Most studies reported that subjecting cells to electrical stimulation on conductive polymer substrates led to longer neurites (Durgam et al. 2010; Ghasemi-Mobarakeh et al. 2009; Gomez and Schmidt 2007; Koppes et al. 2014; Lee et al. 2009; Moroder et al. 2011; Nguyen et al. 2014; Prabhakaran et al. 2011; Schmidt et al. 1997; Shi et al. 2014; Zeng et al. 2013), whereas high field strength (10 V/cm) could also damage neurites (Jain et al. 2013).

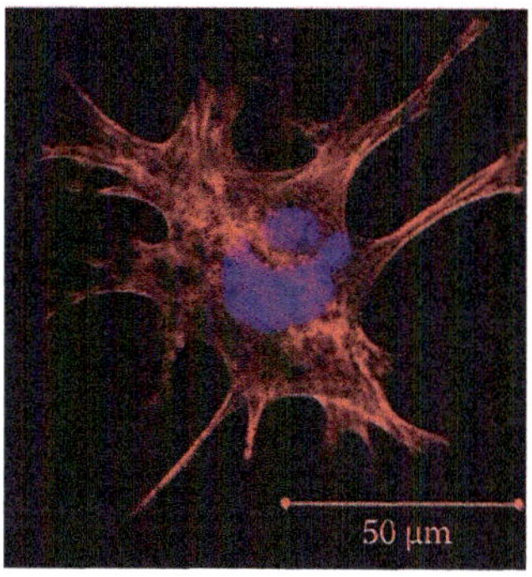

Figure 13.2 Fluorescence microscopy of an exemplary image of a cultured PC12 cell under electrical stimulation bearing multiple neurites. The average number of neurites per cell could be quantified. (Reprinted from Moroder, P., et al., *Acta Biomater.*, 7, 944–953, 2011. With permission.)

Work by Erskine et al. (1995) showed that *Xenopus* neurites extend randomly in the absence of an electrical field. In contrast, the presence of a weak direct current electric field directs the orientation of the extending neurites to the cathode, and the amount of alignment was directly dependent on the applied field strength. Similar observations were reported using N2a cells where there was no directional growth of neurites at 0 V/cm and greater neurite growth parallel to the lateral electric field direction (Jain et al. 2013). Analyzing the orientation of neurite extension of PC12 cells cultured on polycaprolactone fumarate–polypyrrole (PCLF-PPy) substrate also revealed that neurites align with the direction of current (Moroder et al. 2011).

13.3 ELECTRICAL CONDUCTING POLYMERS

13.3.1 ROLE OF DOPANTS ON CONDUCTIVITY

The method for preparing a conductive polymer is typically through the oxidation of its analogous monomer. An oxidizing dopant can change the band structure by p-doping. P-doping is performed by removing electrons from the valence band and transferring them to the conduction band. N-doping can also take place by adding electrons directly to the conduction band. By adding electrons to the lowest unoccupied molecular orbital (LUMO) and partially filling the conduction band, a radical anion called a polaron is formed. The polaron initiates the injection of states from the bottom of the conduction band and the top of the valence band through the band gap, resulting in the polymer being a conductor. For a detailed explanation of how conducting polymers originated and are synthesized, refer to Chapters 1 and 2.

Dopants are characterized based on molecular size and are termed either small dopants (such as Cl) or large dopants (such as polystyrene sulfonate [PSS]). The polymer conductivity and structural properties are affected similarly by using either large or small dopants. However, large dopants have more effects on the material properties of the polymer, such as density. By using a large dopant, the dopant is more integrated into the polymer itself and will not leach out as easily, leading to higher stability (Balint et al. 2014).

13.3.2 CONDUCTIVE POLYMERS FOR NEURONAL GROWTH

Numerous techniques for incorporating electrically conductive materials in order to restore nerve function include the attachment of metal electrodes to proximal and distal nerve stumps, scaffolds coated with nanoparticles, and electrically conductive polymers (Runge et al. 2010; Balint et al. 2014). Electrically conductive polymers used for restoration of nerve function include polyaniline (PANI) (Ghasemi-Mobarakeh et al. 2011; Green et al. 2008; Li et al. 2007; Qazi et al. 2014), PPy (Ghasemi-Mobarakeh et al. 2011; Gomez and Schmidt 2007; Green et al. 2008; Keohan et al. 2007; Stauffer and Cui 2006; Thompson et al. 2011; Wadhwa et al. 2006), polythiophenes (PTs) (Abidian et al. 2010; Green et al. 2008; Jeong et al. 2011; Moral-Vico et al. 2014; Srivastava et al. 2014), electroconductive hydrogels (ECHs) (Guiseppi-Elie 2010), and copolymers (Bechara et al. 2011; Runge et al. 2010; Durgam et al. 2010; Ghasemi-Mobarakeh et al. 2011; Lee et al. 2009; Moroder et al. 2011). A summary of conducting polymers and their dopants, cell types, advantages, and limitations can be found in Table 13.2.

13.3.2.1 Polyaniline

PANI, being the first electrically conductive polymer discovered, has found numerous applications in stimulating the growth of neurons. PANI in its pure form has been shown to have good biocompatibility *in vivo*, as well as having good conductivity and high surface area, and is easy to synthesize (Qazi et al. 2014). For details on PANI synthesis, refer to Chapters 1 and 2.

PANI is an electrically conductive polymer that can exist in three different oxidation states (Qazi et al. 2014). It should be noted that the three oxidation states of PANI are not inherently conductive. In order to add electrical conductivity, the emeraldine salt must be produced (Qazi et al. 2014). The formation of the emeraldine salt is important because the emeraldine base of PANI has been shown to have a low conductivity of approximately 10^{-10} S/cm, whereas the emeraldine salt has been shown to have a conductivity as high as 30 S/cm (Balint et al. 2014). Producing emeraldine salt occurs through a process called doping by using either oxidized (p-doping) or reduced (n-doping) dopants (Green et al. 2008; Li et al. 2007). Typical dopants used for PANI include strong acids, the most common being HCl.

PANI is an attractive conductive polymer due to its low cost, high biocompatibility, high conductivity, and flexibility, and the ability to support cell attachment and proliferation (Ghasemi-Mobarakeh et al. 2011).

13.3.2.2 Polypyrrole

PPy is an electrically conductive polymer that has found numerous applications in nerve regeneration. It has a highly conjugated backbone, which makes it extremely rigid. Its insolubility in many solvents leads to poor processability. PPy itself is not conductive and thus must be doped. The dopant anion in PPy has been shown to play a critical role in determining the chemical and physical properties of the polymer (Thompson et al. 2011). PPy has been shown to possess a conductivity of up to 7.5×10^3 S/cm, which can be controlled by varying dopants, conjugation length, the polaron, and charge transfer to adjacent molecules (Balint et al. 2014). PPy is typically synthesized electrochemically or through oxidative polymerization, followed by doping to make it conductive. Thompson et al. (2011) have shown that smaller dopants, such as *para*-toluene sulfonate (PTS) and dodecylbenzene sulfonate (DBS), are found to produce PPy films that have superior

Table 13.2 **Summary of conducting polymers, dopants, cell types, advantages, and limitations**

POLYMER TYPE	DOPANTS/ ADDITIVES	CELL TYPES	ADVANTAGES	LIMITATIONS	REFERENCES
Polypyrrole (PPy)	NSA, DBSA, DOSS, I, lysine, laminin, chondroitin sulfate	PC12 cells, Rat DRG neurons, neuroblastoma cells, murine glia cells, Schwann cells	High biocompatibility and conductivity, ease of synthesis, high surface area	Poor porosity, poor processability, poor mechanical properties, and biodegradability	(Stauffer and Cui 2006) (Thompson et al. 2011) (Wadhwa et al. 2006) (Keohan et al. 2007) (Gomez and Schmidt 2007) (Green et al. 2008) (Ghasemi-Mobarakeh et al. 2009)
Polyaniline (PANI)	Strong acids	PC12 cells, rat nerve stem cells, chick embryo DRG neurons, neural crest cells	Semiflexible, high conductivity	Poor cell growth and adhesion	(Qazi et al. 2014) (Green et al. 2008) (Ghasemi-Mobarakeh et al. 2009) (Li et al. 2007)
Polythiophenes/ PEDOT	Organic acids	DRG cells, PC12 cells, Cortical neuron cells, SGNs, neuronal progenitor cells	High biocompatibility, conductivity, and optical properties	No covalent bound materials, poor solubility	(Abidian et al. 2010) (Jeong et al. 2011) (Moral-Vico et al. 2014) (Green et al. 2008) (Srivastava et al. 2014)
PPy copolymers with PCLF, PCL, PLGA, and PECA	NSA, DBSA	PC12 cells, DRG explants, Schwann cells, C17.2 murine neural stem cells	Increased cell adhesion, good biodegradability	More complicated synthesis	(Runge et al. 2010) (Moroder et al. 2011) (Lee et al. 2009)(Durgam et al. 2010) (Bechara et al. 2011) (Ghasemi-Mobarakeh et al. 2011)
Electroconductive hydrogels	Same as PPy and PANI	PC12 cells, primary chicken sciatic nerve explants	Facile small molecule transport, high hydration levels, and biocompatibility	Poor mechanical strength	(Guiseppi-Elie 2010)

biocompatibility with the cultured neural tissue. It has also been shown by Stauffer and Cui (2006) that PPy doped with the laminin fragment p20 peptide produced significantly longer neurites than other doped PPy polymer surfaces.

Gomez and Schmidt (2007) successfully immobilized nerve growth factor (NGF) onto PPy and subsequently reported that PC12 cells successfully grew neurites on the polymer–NGF matrix. PPy has also been shown to possess good biocompatibility with rat peripheral nerve tissue, and also is an appropriate substrate for bridging the peripheral nerve gap in rats (Ghasemi-Mobarakeh et al. 2011). PPy has successfully been used to make coatings that electrically control and locally deliver the anti-inflammatory drug dexamethasone (Wadhwa et al. 2006).

13.3.2.3 Polythiophenes

Due to their electronic and ionic conductivity, PT and its derivatives have shown remarkable potential as microelectrodes for transducing biological signals to and from electronic signals. PTs are relatively soft and flexible, have low cytotoxicity, conduct electrons and ions well, and are easily chemically modified. In addition, electrodes coated with PTs show reduced electrical impedance and allow for an appropriate connection between devices and biological tissues (Abidian et al. 2010). PTs can be synthesized either electrochemically or chemically, either through oxidation or by using cross-coupling metal catalysts. Typical dopants used with PTs are organic acids such as trifluoroacetic acid, propionic acid, and trifluoromethanesulfonic acid. One of the most widely used PTs for neuron stimulation is poly(3,4-ethylenedioxythiophene) (PEDOT) (Abidian et al. 2010). PEDOT has the distinct advantage of being optically transparent in its conducting state, as well as being highly stable and having a moderate band gap and low redox potential. However, it is poorly soluble. The solubility can be increased by using PSS (Srivastava et al. 2014). This combination results in a cationic PEDOT polymer and an anionic PSS polymer dopant. PEDOT:PSS has recently been utilized by Srivastava et al. (2014) and has shown that varying morphologies and applied strains can affect the differentiation and neuron cellular distribution. Cui and Martin (2003) have also been able to electrochemically deposit PEDOT:PSS on microelectrodes of neural probes from an aqueous solution, resulting in an impedance modulus decrease of almost two orders of magnitude. Baytron®, previously known as Clevios™, is a commercially available aqueous version of PEDOT:PSS that achieves conductivities of 1–2 S/cm. PT and PEDOT were also deposited onto silica nanoparticles and the cellular uptake monitored via transmission electron microscopy (Jeong et al. 2011). PEDOT–iridium oxide nanocomposites have also been used to show biocompatibility for the growing and differentiation of neurons (Moral-Vico et al. 2014).

13.3.2.4 Electroconductive hydrogels

In order to combine the redox switching and conductive properties of conductive polymers and allow for facile small-molecule transport, Guiseppi-Elie (2010) developed a series of ECHs. ECHs were developed from poly(hydroxyethyl)methacrylate (HEMA)-based hydrogels with either PANI or PPy to facilitate conductivity. Similar dopants that are used for PANI and PPy are typically used when making ECHs. Poly(HEMA) is a hydrophobic acrylate polymer, but it swells in the presence of water due to the hydrophilic end group.

Figure 13.3 (Guiseppi-Elie 2010) shows the typical synthetic routes of ECHs. The hydrogel precursors are mixed with the conducting polymer precursor monomer, and after being cast into thin films, the hydrogel network is photopolymerized, prepared

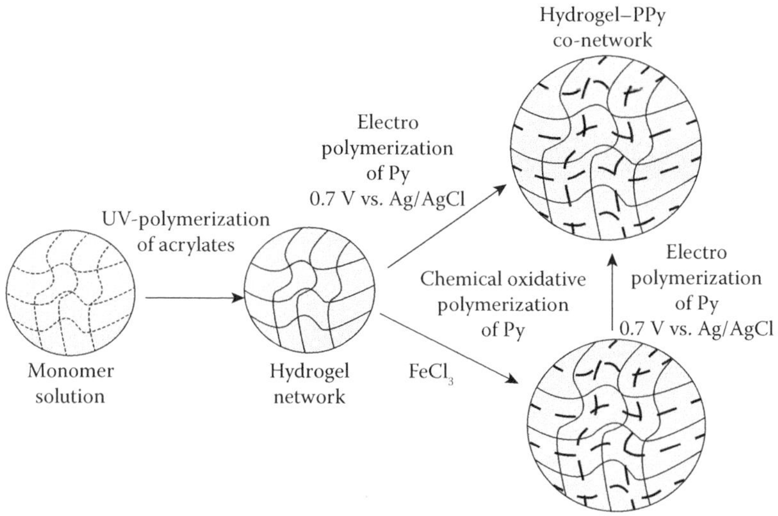

Figure 13.3 Synthetic scheme toward electroconductive hydrogels. (Reprinted from Guiseppi-Elie, A., *Biomaterials*, 31, 2701–2716, 2010. With permission.)

into microspheres, spun as fibers, or applied to electrodes (Guiseppi-Elie 2010). Once the hydrogel network is polymerized, the conducting monomer can be polymerized either electrochemically or oxidatively to yield the ECH network.

ECHs are distinctive platforms that combine the responsive properties of both incorporated materials within an aqueous environment that is suitable for biological molecules such as neurons. ECHs have been utilized to make a new class of devices that have low interfacial impedances. These materials are reportedly suitable for neural prosthetic devices that can be applied to electrodes for deep brain stimulation, electrically stimulated drug release devices, and *in vivo* biocompatibility in implantable biosensors (Guiseppi-Elie 2010). The facile molecule transport, high hydration levels, and increased biocompatibility are promising properties for ECHs. One downside of ECHs is their poor mechanical stability; however, in certain applications, high mechanical strength is not needed.

13.3.2.5 Copolymers of PPy

It is possible to utilize the conductive capacity of PPy while minimizing PPy's disadvantages. This can be accomplished by using a copolymer; numerous copolymers have been developed using various polymers to add processability and better mechanical properties. PCLF has successfully been used (Runge et al. 2010; Moroder et al. 2011) to create electrically conductive polymer nanocomposites of PCLF-PPy capable of supporting both PC12 cells and dorsal root ganglia neurite extension (Runge et al. 2010). Dopants, including naphthalene-2-sulfonic acid (NSA) sodium salt and dodecylbenzenesulfonic acid (DBSA) sodium salt, have been shown to support cell attachment, proliferation, and neurite extension (Runge et al. 2010). Having such promising material properties, several studies involving electrical stimulation were conducted. One notable finding from these studies is the ability to form nerve conduits. PCLF-PPy materials have increased flexibility compared with normal PPy, while still maintaining good mechanical strength (Moroder et al. 2011). Additionally, it was shown that neurites can grow parallel to the applied current, leading to the ability to direct the extension of the neurites (Moroder et al. 2011).

Lee et al. (2009) developed a different copolymer system that consisted of PPy-coated electrospun PLGA nanofibers that were used for neural tissue applications. These materials are biocompatible, and the addition of nanofibers provides a high ratio of surface area to volume, interconnecting pores, and nanofibrous topographies (Lee et al. 2009). Neurite growth of PC12 cells was improved by using electrical stimulation compared with the neurite growth of nonstimulated PC12 cells (Lee et al. 2009). This work can lead to the design of neuronal tissue interfaces that are integrated with electrical and topographical cues for use in nerve tissue scaffolds and interfaces (Lee et al. 2009).

Another electroconductive copolymer system that has been used to facilitate c17.2 neural stem cell line proliferation, adhesion, and differentiation is a mixture of PPy and polycaprolactone (PCL) (Bechara et al. 2011; Durgam et al. 2010; Ghasemi-Mobarakeh et al. 2011). Bechara et al. (2011) developed nanowires made of PCL that were subsequently coated with PPy. The nanowire surfaces were fabricated as shown in Figure 13.4 utilizing template synthesis. Commercially available nanoporous Al_2O_3 membranes with a 20 nm pore size were used. PCL disks were subsequently added to the

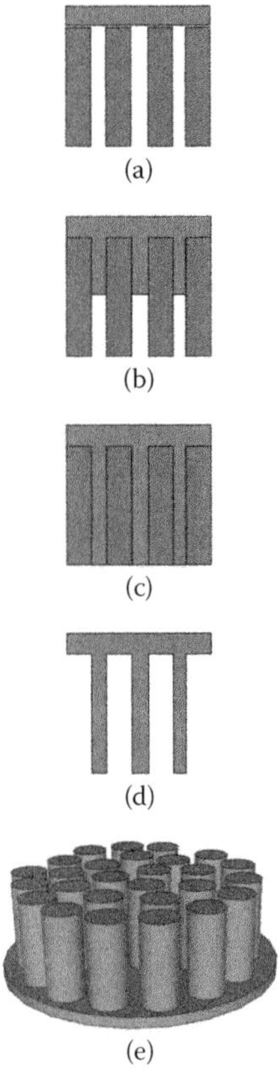

Figure 13.4 Schematic of PCL nanowire fabrication. (a) A PCL pellet is placed on top of the alumina nanoporous membrane. (b and c) The polymer is extruded through the nanoporous membrane in a vacuum oven. (d) The alumina nanoporous membrane is dissolved in NaOH to release the nanowires. (e) PCL nanowire surfaces. (Reprinted from Bechara, S., et al., *Acta Biomater.*, 7, 2892–2901, 2011. With permission.)

surface of the membranes and extruded at 115°C. The nanowires were released from the membranes by dissolving the membranes in 1 M NaOH. The nanowire surfaces were then rinsed with deionized (DI) water and dried. The PPy coating was applied to the nanowire surface by sonication in a 14 mM pyrrole and 14 mM sodium PTS solution. The surface was then incubated for 1 h at 4°C, and the polymerization was performed by addition of 38 mM $FeCl_3$ as oxidant and incubated at 4°C for 24 h. The PPy–nanowire surfaces were finally sonicated in DI water and dried to prepare for use (Bechara et al. 2011). The addition of the PPy reduced the electrical resistivity of the PCL nanowires to a functional level, allowing electrical stimulation to be applied and resulting in the growth of neurons (Bechara et al. 2011). The PPy nanowire surfaces had an average resistivity of 3.68 kΩ/square, compared with 40 MΩ/square for the nanowire surface alone, a 13,000-fold decrease (Bechara et al. 2011).

Durgam et al. (2010) utilized two different copolymer systems of PPy and compared how they supported cell proliferation and enhanced neurite growth using electrical stimulation. They compared PPy–PCL with PPy–poly(ethyl 2-cyanoacrylate) (PECA) and determined that the films made of these biomaterials have high electrical conductivity (20–30 S/cm) and allow for the electrical stimulation of cells. No toxic effects occurred during *in vitro* studies. Neurite-bearing cells and long neurites were observed to occur on the PPy–PCL films compared with the PPy–PECA or unstimulated control (Durgam et al. 2010). By incorporating copolymers with better processability and different mechanical properties, many new materials have been developed to overcome the challenges of using pure PPy systems for neuronal growth.

13.4 ELECTRICAL STIMULATION

For neural stimulation, typical charge-balanced biphasic current pulses have been used (Cogan 2008). Most of the works on field-induced neural stimulation apply a lateral electric field parallel to the substrate surface by the use of electrodes flush with the substrate (Moroder et al. 2011; Schmidt et al. 1997; Zhang et al. 2007). The electric field has also been applied perpendicular to the substrate while growing cells *in vitro* (Jain et al. 2013). The vertically applied electrical field is thought to be less disturbing to cell morphology and cell transport.

Various settings have been used to deliver electrical stimulation to the conductive polymer substrates in a cell culture environment. Borosilicate cover glass chambers consisting of three electrodes, a working electrode (the conductive polymer of interest), a reference electrode (Ag/AgCl), and a counterelectrode (Pt), have been used to culture cells in which they receive electrical stimulation (Cho and Borgens 2010). A similar setting used the conductive polymer of interest as the anode, a gold wire placed at the opposite end as the cathode, and a silver wire as the reference electrode (Gomez and Schmidt 2007; Schmidt et al. 1997). A snug-fitting conductive polymer scaffold into a cell culture well and delivering electrical stimulation via platinum wires in tight contact with the scaffold has been shown to be an effective method. The wires are insulated from the culture medium using grease sealant (Moroder et al. 2011; Zhang et al. 2007). A different setting is to have the electrical stimulation system outside the cell culture dish. Attaching conductive polymer film to the bottom of the cell culture dish and running electrical currents across the film generates a lateral electric field for stimulation of the cells (Durgam et al. 2010).

There is a wide variation of electrical stimulation parameters used throughout the literature, which is summarized in Table 13.3. Either constant currents (Cho and Borgens 2010; Ghasemi-Mobarakeh et al. 2009; Gomez and Schmidt 2007; Koppes et al. 2014; Prabhakaran et al. 2011; Schmidt et al. 1997; Zeng et al. 2013), alternate currents (Thompson et al. 2011; Jain et al. 2013), or both (Moroder et al. 2011) have been used. Simulation intensity and field strength vary from study to study as well, with varying results. An increase in the field strength from 0 to 2.5 V/cm increases neurite length, while field strength greater than 5 V/cm inhibits neurite length (Jain et al. 2013). The following parameters have been reported to be helpful: 100 mV for 2 h (Gomez and Schmidt 2007; Nguyen et al. 2014; Schmidt et al. 1997), 10 μA for 1 h (Moroder et al. 2011), 1.7–8.4 μA/cm (Zhang et al. 2007), 10 mV/cm for 2 h (Lee et al. 2009), 100 μA for 2 h (Durgam et al. 2010), 1.5 V for 1 h (Ghasemi-Mobarakeh et al. 2009), 100 mV/mm for 1 h (Prabhakaran et al. 2011; Qi et al. 2013), 100 mV/cm for 2 h (Nguyen et al. 2014; Zeng et al. 2013), and 100 mV/mm for 4 h (Huang et al. 2010).

Current frequency is a variable parameter that is likely to be one of the most important factors affecting how neurons respond to electrical stimulation. The optimal current frequency for stimulating regeneration is likely to be different between motor and sensory nerves. Pulsed stimulation at a current frequency of 50 Hz had been shown to have the largest effect on dorsal column nuclei (Qin et al. 2009). For motor neurons, 20 Hz frequency might be a better choice since motoneurons fire at this frequency (Gordon et al. 2009).

Pulse duration is another parameter of concern. It is known that response of neurite extension and guidance to electrical stimulation are partly controlled by voltage-gated Ca^{2+} channels. However, overstimulation can cause extensive cell depolarization, resulting in the activation of multiple Ca^{2+} channels and the retraction of extending neurites (Wood and Willits 2009). Since Ca^{2+} channels can be directly activated by electrical stimulation, the amount of time that the electric field is applied may be critical for obtaining and maintaining enhanced neurite outgrowth. Ten minutes of electrical stimulation had been shown to have produced the same result as 100 min of stimulation when cells were cultured on collagen surfaces. On the other hand, increased stimulation time further enhanced neurite length on a laminin substrate (Wood and Willits 2009). Exposure to a constant voltage of 0.1 V for 6 h led to slightly enhanced expansion of PC12 cell neurites in a separate study (Cho and Borgens 2010).

13.5 SUMMARY

The incidence of nerve injuries has increased recently due to accidents, traumatic injuries, systematic diseases, and compression syndromes (Bennet and Kim 2011). Neurological damage to the central or peripheral nervous system occurs frequently as a result of these injuries. More than 250,000 occurrences each year are diagnosed in the United States alone (Runge et al. 2010). The most common injuries are peripheral nerve injuries (PNIs), which normally require therapies to restore the loss of nerve function (Runge et al. 2010). Conductive polymers are a class of materials that are increasingly important in neural tissue engineering applications due to their versatility. Their properties can be modulated by varying the polymer type, molecular weight, and choice of dopant; modifying the surfaces; and forming copolymers or composites. The polymer properties, as well as the methods and regimes of electrical stimulation, will ultimately influence the cell behavior *in vitro*.

Table 13.3 **Electrical stimulation conditions**

STIMULATION PARAMETERS	TYPE	CELL TYPES	VALUE	COMMENTS	REFERENCES
Electric field orientation	Lateral	PC12 cells, Schwann cells	Parallel to substrate	Most commonly used	(Moroder et al. 2011) (Schmidt et al. 1997) (Zhang et al. 2007)
	Vertical	N2a cells	Perpendicular to substrate	Less disturbing to cell morphology and transport	(Jain et al. 2013)
Current frequency	Motor neurons	Schwann cells	20 Hz	Motoneurons fire at this frequency	(Gordon et al. 2009)
	Sensory neurons	Dorsal column nuclei	50 Hz	Largest effect of dorsal column nuclei	(Qin et al. 2009)
Pulse duration	Short	PC12 cells, chick embryo DRG neurons	10–100 min	Same neurite growth	(Gomez and Schmidt 2007) (Wood and Willits 2009)
	Long	Chick embryo DRG neurons, PC12 cells	6 h	Slightly enhanced expansion of PC12 cell neurites	(Wood and Willits 2009) (Cho and Borgens 2010)
Field strength	Low	N2a cells, PC12 cells, Schwann cells, Neural crest cells, chick embryo DRG neurons	0–2.5 V/cm	Increased neurite length	(Jain et al. 2013) (Nguyen et al. 2014) (Schmidt et al. 1997) (Moroder et al. 2011) (Zhang et al. 2007) (Lee et al. 2009) (Ghasemi-Mobarakeh et al. 2009) (Prabhakaran et al. 2011) (Huang et al. 2010)
	High	N2a cells, PC12 cells, Schwann cells, Neural crest cells, chick embryo DRG neurons	> 5 V/cm	Inhibited neurite length	(Gomez and Schmidt 2007) (Jain et al. 2013) (Schmidt et al. 1997) (Moroder et al. 2011) (Zhang et al. 2007) (Lee et al. 2009) (Ghasemi-Mobarakeh et al. 2009) (Prabhakaran et al. 2011) (Huang et al. 2010)

REFERENCES

Abidian, M. R., J. M. Corey, D. R. Kipke, and D. C. Martin. Conducting-polymer nanotubes improve electrical properties, mechanical adhesion, neural attachment, and neurite outgrowth of neural electrodes. *Small* 6 (2010): 421–429.

Ashrafi, S., J. N. Betley, J. D. Comer, et al. Neuronal Ig/Caspr recognition promotes the formation of axoaxonic synapses in mouse spinal cord. *Neuron* 81 (2014): 120–129.

Balint, R., N. J. Cassidy, and S. H. Cartmell. Conductive polymers: Towards a smart biomaterial for tissue engineering. *Acta Biomater.* 10 (2014): 2341–2353.

Bechara, S., L. Wadman, and K. C. Popat. Electroconductive polymeric nanowire templates facilitates in vitro C17.2 neural stem cell line adhesion, proliferation and differentiation. *Acta Biomater.* 7 (2011): 2892–2901.

Bennet, D., and S. Kim. Implantable microdevice for peripheral nerve regeneration: materials and fabrications. *J. Mater. Sci.* 46 (2011): 4723–4740.

Bentivoglio, M. Life and discoveries of Santiago Ramon y Cajal. Nobelprize.org, 1998. http://www.nobelprize.org/nobel_prizes/medicine/laureates/1906/cajal-article.html (accessed on April 20, 1998).

Bettinger, C. J., J. P. Bruggeman, A. Misra, J. T. Borenstein, and R. Langer. Biocompatibility of biodegradable semiconducting melanin films for nerve tissue engineering. *Biomaterials* 30 (2009): 3050–3057.

Bhang, S. H., S. I. Jeong, T.-J. Lee, et al. Electroactive electrospun polyaniline/poly[(L-lactide)-co-(ε-caprolactone)] fibers for control of neural cell function. *Macromol. Biosci.* 12 (2012): 402–411.

Cen, L., K. G. Neoh, Y. Li, and E. T. Kang. Assessment of in vitro bioactivity of hyaluronic acid and sulfated hyaluronic acid functionalized electroactive polymer. *Biomacromolecules* 5 (2004): 2238–2246.

Cho, Y., and R. B. Borgens. The effect of an electrically conductive carbon nanotube/collagen composite on neurite outgrowth of PC12 cells. *J. Biomed. Mater. Res. A* 95A (2010): 510–517.

Cogan, S. F. Neural stimulation and recording electrodes. *Annu. Rev. Biomed. Eng.* 10 (2008): 275–309.

Costandi, M. The discovery of the neuron. *Neurophilosophy* 2006. http://neurophilosophy.wordpress.com/2006/08/29/the-discovery-of-the-neuron (accessed on August 29, 2006).

Cui, X., and D. C. Martin. Electrochemical deposition and characterization of poly(3,4-ethylenedioxythiophene) on neural microelectrode arrays. *Sens. Actuators B* 89 (2003): 92–102.

Cullen, D. K., R. P. Ankur, J. F. Doorish, D. H. Smith, and B. J. Pfister. Developing a tissue-engineered neural-electrical relay using encapsulated neuronal constructs on conducting polymer fibers. *J. Neural Eng.* 5 (2008): 374–384.

Durgam, H., S. Sapp, C. Deister, et al. Novel degradable co-polymers of polypyrrole support cell proliferation and enhance neurite out-growth with electrical stimulation. *J. Biomater. Sci. Polym. Ed.* 21 (2010): 1265–1282.

Erskine, L., R. Stewart, and C. D. McCaig. Electric field-directed growth and branching of cultured frog nerves: Effects of aminoglycosides and polycations. *J. Neurobiol.* 26 (1995): 523–536.

Fuortes, M. G., K. Frank, and M. C. Becker. Steps in the production of motoneuron spikes. *J. Gen. Physiol.* 40 (1957): 735–752.

Ghasemi-Mobarakeh, L., M. P. Prabhakaran, M. Morshed, et al. Application of conductive polymers, scaffolds and electrical stimulation for nerve tissue engineering. *J. Tissue Eng. Regener. Med.* 5 (2011): e17–e35.

Ghasemi-Mobarakeh, L., M. P. Prabhakaran, M. Morshed, M. H. Nasr-Esfahani, and S. Ramakrishna. Electrical stimulation of nerve cells using conductive nanofibrous scaffolds for nerve tissue engineering. *Tissue Eng. Part A* 15 (2009): 3605–3619.

Giordano, G., and L. G. Costa. Primary neurons in culture and neuronal cell lines for in vitro neurotoxicological studies. *Methods Mol. Biol.* 758 (2011): 13–27.

Gomez, N., and C. E. Schmidt. Nerve growth factor-immobilized polypyrrole: Bioactive electrically conducting polymer for enhanced neurite extension. *J. Biomed. Mater. Res. Part A* 81A (2007): 135–149.

Gordon, T., K. M. Chan, O. A. R. Sulaiman, et al. Accelerating axon growth to overcome limitations in functional recovery after peripheral nerve injury. *Neurosurgery* 65 (2009): A132–A144.

Green, R. A., N. H. Lovell, G. G. Wallace, and L. A. Poole-Warren. Conducting polymers for neural interfaces: Challenges in developing an effective long-term implant. *Biomaterials* 29 (2008): 3393–3399.

Guiseppi-Elie, A. Electroconductive hydrogels: Synthesis, characterization and biomedical applications. *Biomaterials* 31 (2010): 2701–2716.

Hermel, E. E. S., M. C. Faccioni-Heuser, S. Marcuzzo, A. A. Rasia-Filho, and M. Achaval. Ultrastructural features of neurons and synaptic contacts in the posterodorsal medial amygdala of adult male rats. *J. Anat.* 208 (2006): 565–575.

Huang, J., X. Hu, L. Lu, et al. Electrical regulation of Schwann cells using conductive polypyrrole/chitosan polymers. *J. Biomed. Mater. Res. Part A* 93A (2010): 164–174.

Jain, S., A. Sharma, and B. Basu. Vertical electric field stimulated neural cell functionality on porous amorphous carbon electrodes. *Biomaterials* 34 (2013): 9252–9263.

Jeong, Y. S., W.-K. Oh, S. Kim, and J. Jang. Cellular uptake, cytotoxicity, and ROS generation with silica/conducting polymer core/shell nanospheres. *Biomaterials* 32 (2011): 7217–7225.

Jin, L., Z.-Q. Feng, M.-L. Zhu, et al. A novel fluffy conductive polypyrrole nano-layer coated PLLA fibrous scaffold for nerve tissue engineering. *J. Biomed. Nanotechnol.* 8 (2012): 779–785.

Kang, G., R. B. Borgens, and Y. Cho. Well-ordered porous conductive polypyrrole as a new platform for neural interfaces. *Langmuir* 27 (2011): 6179–6184.

Keohan, F., X. F. Wei, A. Wongsarnpigoon, et al. Fabrication and evaluation of conductive elastomer electrodes for neural stimulation. *J. Biomater. Sci. Polym. Ed.* 18 (2007): 1057–1073.

Koppes, A. N., N. W. Zaccor, C. J. Rivet, et al. Neurite outgrowth on electrospun PLLA fibers is enhanced by exogenous electrical stimulation. *J. Neural Eng.* 11 (2014): 046002.

Lee, J. Y., C. A. Bashur, A. S. Goldstein, and C. E. Schmidt. Polypyrrole-coated electrospun PLGA nanofibers for neural tissue applications. *Biomaterials* 30 (2009): 4325–4335.

Li, M.-Y., P. Bidez, E. Guterman-Tretter, et al. Electroactive and nanostructured polymers as scaffold materials for neuronal and cardiac tissue engineering. *Chin. J. Polym. Sci.* 25 (2007): 331–339.

Li, Y., K. G. Neoh, L. Cen, and E. T. Kang. Porous and electrically conductive polypyrrole-poly(vinyl alcohol) composite and its applications as a biomaterial. *Langmuir* 21 (2005): 10702–10709.

Liu, X., Z. Yue, M. J. Higgins, and G. G. Wallace. Conducting polymers with immobilized fibrillar collagen for enhanced neural interfacing. *Biomaterials* 32 (2011): 7309–7317.

Lodish, H. J., Berk, A., Kaiser, C. A., Krieger, M., Scott, M. P., Bretscher, A., Ploegh, H., and Matsudaira, P. Molecular Cell Biology. In *Overview of Neuron Structure and Function.* New York: W. H. Freeman and Company, 2013, 1019–1057.

Moral-Vico, J., S. Sanchez-Redondo, M. P. Lichtenstein, C. Sunol, and N. Casan-Pastor. Nanocomposites of iridium oxide and conducting polymers as electroactive phases in biological media. *Acta Biomater.* 10 (2014): 2177–2186.

Moroder, P., M. B. Runge, H. Wang, et al. Material properties and electrical stimulation regimens of polycaprolactone fumarate-polypyrrole scaffolds as potential conductive nerve conduits. *Acta Biomater.* 7 (2011): 944–953.

Muller, D., J. P. Silva, C. R. Rambo, et al. Neuronal cells' behavior on polypyrrole coated bacterial nanocellulose three-dimensional (3D) scaffolds. *J. Biomater. Sci. Polym. Ed.* 24 (2013): 1368–1377.

Nguyen, H. T., S. Sapp, C. Wei, et al. Electric field stimulation through a biodegradable polypyrrole-co-polycaprolactone substrate enhances neural cell growth. *J. Biomed. Mater. Res. A* 102A (2014): 2554–2564.

Prabhakaran, M. P., L. Ghasemi-Mobarakeh, G. Jin, and S. Ramakrishna. Electrospun conducting polymer nanofibers and electrical stimulation of nerve stem cells. *J. Biosci. Bioeng.* 112 (2011): 501–507.

Qazi, T. H., R. Rai, and A. R. Boccaccini. Tissue engineering of electrically responsive tissues using polyaniline based polymers: A review. *Biomaterials* 35 (2014): 9068–9086.

Qi, F., Y. Wang, T. Ma, et al. Electrical regulation of olfactory ensheathing cells using conductive polypyrrole/chitosan polymers. *Biomaterials* 34 (2013): 1799–1809.

Qin, C., X. Yang, M. Wu, et al. Modulation of neuronal activity in dorsal column nuclei by upper cervical spinal cord stimulation in rats. *Neuroscience* 164 (2009): 770–776.

Runge, M. B., M. Dadsetan, J. Baltrusaitis, et al. The development of electrically conductive polycaprolactone fumarate-polypyrrole composite materials for nerve regeneration. *Biomaterials* 31 (2010): 5916–5926.

Schmidt, C. E., V. R. Shastri, J. P. Vacanti, and R. Langer. Stimulation of neurite outgrowth using an electrically conducting polymer. *Proc. Natl. Acad. Sci.* U S A 94 (1997): 8948–8953.

Shepherd, G. M. Principles of Neural Science, Fourth Edition by Eric R. Kandel, James H. Schwartz, and Thomas M. Jessell. *Neuron* 26 (2000): 581–582.

Shepherd, G. M. Symposium overview and historical perspective: Dendrodendritic synapses: Past, present, and future. *Ann. N.Y. Acad. Sci.* 1170 (2009): 215–223.

Shi, Z., H. Gao, J. Feng, et al. In situ synthesis of robust conductive cellulose/polypyrrole composite aerogels and their potential application in nerve regeneration. *Angew. Chem. Int. Ed.Engl.* 53 (2014): 5380–4.

Srivastava, N., K. S. Narayan, and J. James. Morphology and electrostatics play active role in neuronal differentiation processes on flexible conducting substrates. *Organogenesis* 10 (2014): 1–5.

Stauffer, W. R., and X. T. Cui. Polypyrrole doped with 2 peptide sequences from laminin. *Biomaterials* 27 (2006): 2405–2413.

Sudwilai, T., J. J. Ng, C. Boonkrai, et al. Polypyrrole-coated electrospun poly(lactic acid) fibrous scaffold: Effects of coating on electrical conductivity and neural cell growth. *J. Biomater. Sci. Polym. Ed.* 25 (2014): 1240–1252.

Thompson, B. C., S. E. Moulton, R. T. Richardson, and G. G. Wallace. Effect of the dopant anion in polypyrrole on nerve growth and release of a neurotrophic protein. *Biomaterials* 32 (2011): 3822–3831.

Valero, T., G. Moschopoulou, S. Kintzios, et al. Studies on neuronal differentiation and signalling processes with a novel impedimetric biosensor. *Biosens. Bioelectron.* 26 (2010): 1407–1413.

Villarroel-Campos, D., L. Gastaldi, C. Conde, A. Caceres, and C. Gonzalez-Billault. Rab-mediated trafficking role in neurite formation. *J. Neurochem.* 129 (2014): 240–248.

Wadhwa, R., C. F. Lagenaur, and X. T. Cui. Electrochemically controlled release of dexamethasone from conducting polymer polypyrrole coated electrode. *J. Control. Release* 110 (2006): 531–541.

Westerink, R. H. S., and A. G. Ewing. The PC12 cell as model for neurosecretion. *Acta Physiol.* 192 (2008): 273–285.

Wood, M. D., and R. K. Willits. Applied electric field enhances DRG neurite growth: Influence of stimulation media, surface coating and growth supplements. *J. Neural Eng.* 6 (2009): 046003.

Zeng, J., Z. Huang, G. Yin, et al. Fabrication of conductive NGF-conjugated polypyrrole-poly(L-lactic acid) fibers and their effect on neurite outgrowth. *Colloids Surf. B* 110 (2013): 450–457.

Zhang, Z., M. Rouabhia, Z. Wang, et al. Electrically conductive biodegradable polymer composite for nerve regeneration: Electricity-stimulated neurite outgrowth and axon regeneration. *Artif. Organs* 31 (2007): 13–22.

Zhou, K., G. A. Thouas, C. C. Bernard, et al. Method to impart electro- and biofunctionality to neural scaffolds using graphene-polyelectrolyte multilayers. *ACS Appl. Mater. Interfaces* 4 (2012): 4524–4531.

14 Modulation of bone cell activities *in vitro* by electrical and electromagnetic stimulations

Ze Zhang
Axe Médecine Régénératrice–CHU
Université Laval, Québec, Canada

Mahmoud Rouabhia
Université Laval, Québec, Canada

Contents

14.1 INTRODUCTION

The integrity and quality of bone are maintained by the coordinated actions of osteoblasts that lay down new bone and osteoclasts that resorb bone. This dynamic process occurs all the time to adapt the environment change, such as gravity and mechanical force, and also to repair damaged bones. The load-bearing function of human skeleton makes it easy to accept that bone cells, that is, osteocytes, osteoblasts, and osteoclasts, are responsive to mechanical loading, which is true and has been widely recognized (Nelson et al. 2003). What is less recognized is the fact that bone cells, along with neurons, are probably the most intensively studied cell type in electrical (EF) and electromagnetic (EMF) fields. Bassett and Becker in 1962 revealed that bone generates EF under mechanical stress, that is, stress potential, and such potential has polarity depending on compression or stretching. Probably inspired by this relationship between bone tissue and electricity, research interests had been developed in bone cell electrophysiology, and such interests have continued. A Medline search using the combination "(bone OR osteoblast OR osteoclast OR osteocyte) AND (electrical OR electromagnetic)" shows that more than 10 times the articles were published between 2000 and 2009 than between 1960 and 1969 (4071 vs. 347). This chapter focuses on the *in vitro* behaviors of bone cells, mostly osteoblasts, under the influence of EF and EMF. Information about bone healing in conjunction with electro- or electromagnetic therapy can be found in Chapter 19.

14.2 EFFECT ON DIRECTIONAL MIGRATION

The galvanotaxis or electrotaxis of osteogenic cells in EF was studied by Ferrier using time-lapse photography (Ferrier et al. 1986). They first cultured primary osteoclasts and osteoblasts on tissue culture petri dishes to let cells adhere and spread. An EF of 0.1 and 1.0 V/mm was established between two salt bridges positioned at the opposite sides of a narrow and shallow channel formed by two strips of separated glass covered by a coverslip. This setup is classic and used by many researchers (see Chapters 10 and 17). What was found most interesting was that the cells were moving toward opposite directions, with osteoclasts to the positive electrode (anode) and osteoblasts to the negative electrode (cathode). The authors tried to explain this phenomenon using various possibilities, including the depolarization and hyperpolarization of the channels of calcium, potassium, and sodium ions. A much more recent work revisited this phenomenon by looking at osteoblasts from rat calvarial and osteoblast-like SaOS-2 cells (Ozkucur et al. 2009). The authors exposed the cells to similar EFs, that is, 1.0–1.5 V/mm, and found that the calvarial osteoblasts moved toward the cathode, but SaOS-2 cells moved toward the anode. The authors carried out a delicate experiment to investigate the intracellular calcium wave, revealing that intracellular calcium elevation started at the anode-facing side and then propagated to the cathode-facing side for calvarial osteoblasts. For SaOS-2 cells, however, the calcium wave traveled in the opposite direction (Figure 14.1). Blocking calcium channels prevented the directional cell movement, confirming the key role of calcium ions. To study the mechanism, focus has been on the essential steps of cell crawling and the relationship between Ca^{2+} and actin distribution in the cell cortex. Ozkucur et al. (2009) suggested that an elevated Ca^{2+} would activate myosin light-chain kinase and then actin-activated myosin ATPase, resulting in elevated membrane contraction. Membrane contraction is a necessary step in cell crawling leading to cell membrane detachment opposite to the cell migration direction. Another work by Ozkucur et al. (2011) demonstrated that sodium–potassium pump Na^+/K^+ ATPase and sodium–hydrogen exchangers 1 and 3 (NHE-1 and NHE-3) functioned as direction sensors. In this work, Na^+/K^+ ATPase was found elevated and colocalized with the focal adhesion protein vinculin at the leading edge of cell migration regardless of the direction, that is, both cathode- and anode-directed migrations. Since an extra cell membrane is required to form filopodia and lamellipodia in the direction of migration, the distribution and polymerization or depolymerization of actin, the critical component of cytoskeleton, may provide a clue. As an actin-binding protein known to sever actin and also to be Ca^{2+} dependent (Ono 2007), gelsolin was suggested to downregulate actin synthesis following Ca^{2+} influx and consequently favor the detachment of the cell membrane that is depolarized in the first place (Borys 2012). A mathematical model was established to describe the electrotaxis of osteoblasts (Vanegas-Acosta et al. 2012).

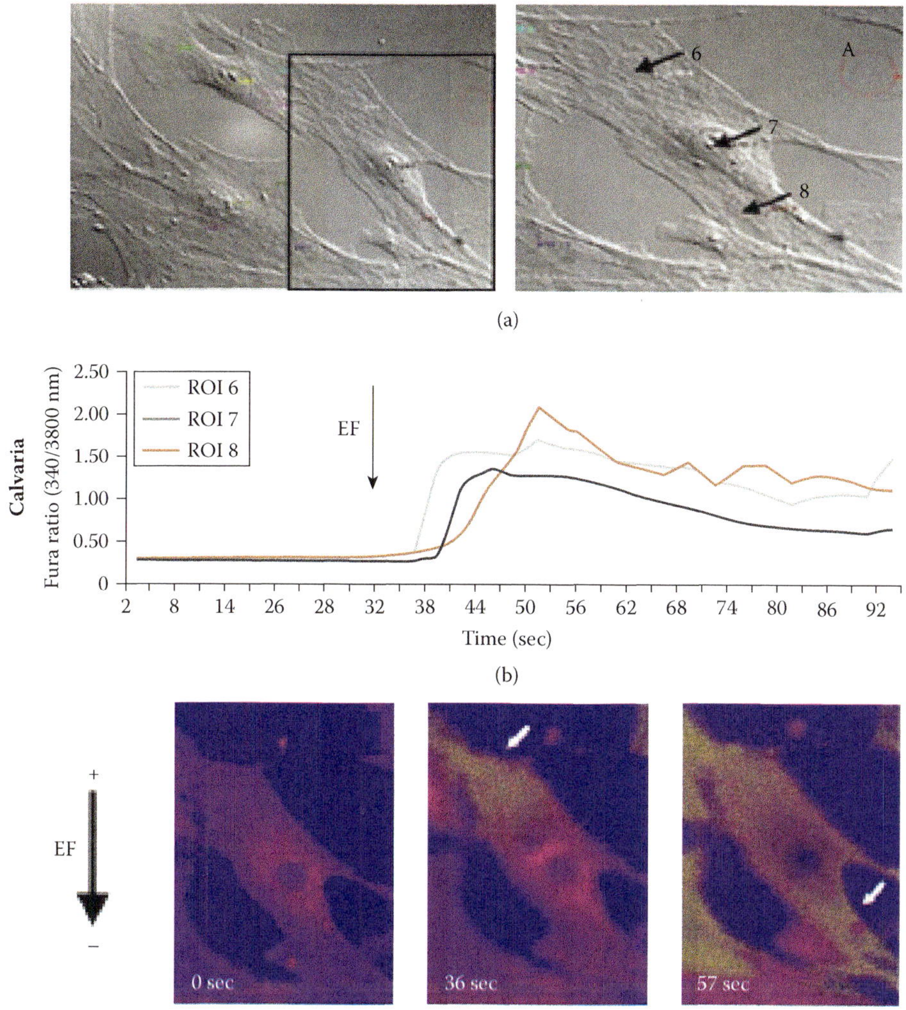

Figure 14.1 Local and opposite Ca^{2+} initials in Fura-loaded calvaria and SaOS-2 cells exposed to strong DC EF (14 V/cm). (a [A]) Differential interference contrast images of calvaria cells with regions of interest (ROI_s) within colored circles. (b) One-cell Fura kinetics representing the sequential elevation of $[Ca^{2+}]$i first at the anode-facing side (ROI 6, grey circle) and its propagation through the cell (ROI 7, black circle) to the cathode-facing side (ROI 8, dark red circle) of a calvaria cell. (c) False color time-lapse frames showing the contrary initiation and propagation (white arrows) of $[Ca^{2+}]$i elevation in calvaria cells. From 6–10 independent experiments, 15–17 cells from a total of 22–23 were scored. (From Ozkucur, N., et al., *PLoS One*, 4, e6131, 2009. With permission.) *(Continued)*

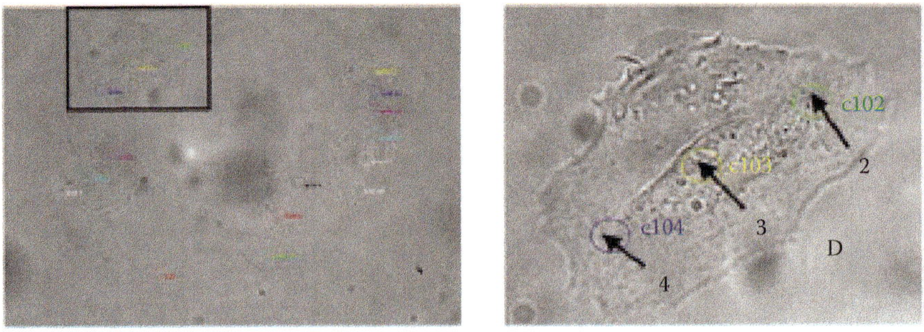

(d)

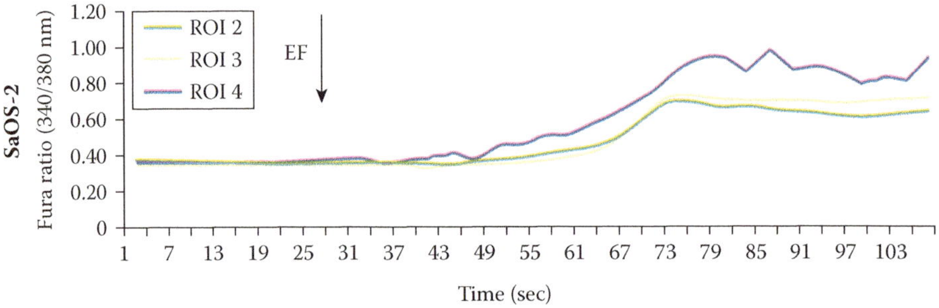

(e)

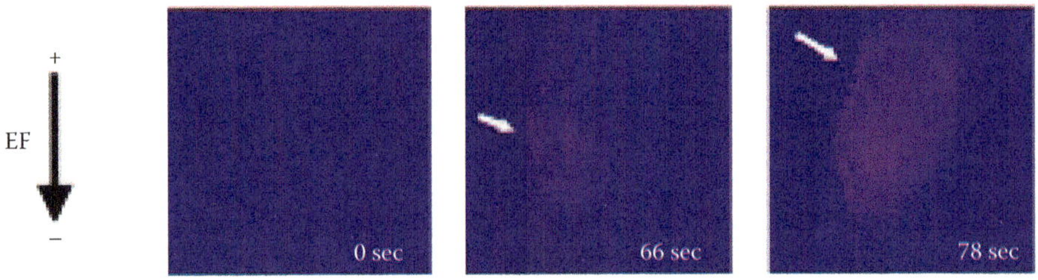

(f)

Figure 14.1 (Continued) Local and opposite Ca^{2+} initials in Fura-loaded calvaria and SaOS-2 cells exposed to strong DC EF (14 V/cm). (d [D]) Differential interference contrast images of SaOS-2 cells with regions of interest (ROI_s) within colored circles. (e) The same pattern but opposite direction in a SaOS-2 cell. (f) False color time-lapse frames showing the contrary initiation and propagation (white arrows) of $[Ca^{2+}]i$ elevation in SaOS-cells. From 6–10 independent experiments, 15–17 cells from a total of 22–23 were scored. (From Ozkucur, N., et al., *PLoS One*, 4, e6131, 2009. With permission.)

This model, however, does not include the biological difference between normal osteoblasts and cancel cells like SaOS-2, or between osteoblasts and osteoclasts, and therefore could not fully explain the differential electrotaxis.

14.3 EFFECT ON GENES AND PROTEINS

A variety of growth factors are involved in bone formation, including insulin-like growth factors (IGFs), transforming growth factors (TGFs), and bone morphogenetic growth factors (BMPs). As the most abundant growth factors in bone tissue, IGFs are potent mitogens that interact with type I and II receptors called IGFR-I and IGFR-II (Giustina et al. 2008). An early work by Fitzsimmons et al. (1995) tested the reaction of human osteosarcoma–derived osteoblast-like cells (TE-85) to a magnetic field (MF) of 40 (alternating current [AC] field) and 20 (direct current [DC] field) μT, and reported the maximum upregulation of IGFR-II at 15.3 Hz. As one can expect, an EF was also induced by the changing MF in the electrically conductive culture medium. This EF was calculated to be 2×10^{-5} V/m, which is extremely weak if one compares it with the biologically effective strength of EF reported in the literature, normally ranging from tens to hundreds of millivolts per millimeter. The authors therefore suggested that a high frequency was necessary to couple the EMF with the "biochemical transductive process(es)" (Fitzsimmons et al. 1995). Notably, the MF in that work was roughly the same magnitude of geomagnetic field, that is, from 25 to 65 μT. The biological effect of an MF strength lower than 1 mT was criticized because this magnitude was not considered physically possible to overcome the ubiquitous noise caused by normal fluctuation of membrane potential and thermal molecular movement. A carefully designed experiment was performed to address this critical argument by canceling ambient MF and looking at the cystolic calcium spike of individual primary bone cells (from rat calvaria) during the stimulation process (real time) (Reinbold and Pollack 1997). Their work confirmed the effect of a weak MF (60 μT to ≤1 mT, 15–100 Hz) and at same time reported the critical role of serum; that is, the effect became significant only in the presence of 2% of serum. The induced EF and current density were between 4.0 μV/m and 3.1 mV/m and 6.9 $\mu A/m^2$ and 5.4 mA/m^2, respectively. Such an EF and current density are much lower than those reported in the literature and so should not be responsible for the observed phenomena. Their work also revealed the heterogeneous response of primary bone cells to MF; that is, only a portion of cells were responsive to MF stimulation. Such a heterogeneous nature of cellular response to EMF could become an important issue if these activated cells would maintain their phenotypical changes while proliferating. Between 2000 and 2003, Lohmann et al. published two works using osteoblast-like MG 63 from human osteosarcoma (Lohmann et al. 2000), rat osteosarcoma cell ROS 17/2.8, and osteocyte MLO-Y4 from transgenic mice (Lohmann et al. 2003). The pulsed EMF they used was the same as that used in the clinic, containing a train of 20 pulses at 15 Hz and a magnitude of 1.6–1.8 mT (16–18 G, 1 G = 10^{-4} T). For the MG 63 cells, they reported suppressed cell proliferation and enhanced differentiation marked by a higher level of alkaline phosphatase (ALP), osteocalcin, collagen, and TGFβ. A much higher frequency was adopted by Fassina et al. (2006) to SaOS-2 cells cultured in a polyurethane scaffold. At 2 mT and 75 Hz, the cells proliferated at a double speed. The transcription of TGFβ was found upregulated. At the protein level, five types of important matrix proteins were found significantly increased in the stimulated group. Among them, the production of collagen types I and III, osteocalcin, and osteopontin, increased more than 10-fold. A setup similar to that used in Fassina's work can be found in Chapter 10, with the MF and EF being perpendicular and parallel to the culture plate, respectively.

A similar setup was used in a comprehensive gene activation screen experiment reported by Sollazzo et al. (2010), in which 19,000 genes were analyzed using a cDNA microarray technology. Using stimulation parameters and a cell line similar to those used by Fassina et al. (2006), that is, 2 mT at 75 Hz and osteosarcoma cell line MG 63, Solazzo et al. (2010) reported 268 upregulated and 277 downregulated genes. These genes are involved in cell differentiation, proliferation, and osteogenesis. In this system, an induced electrical tension of 5 mV was reported. A most recent work focused on 10 important genes and found the upregulation of *Runx2* at the proliferation phase, and *OCN*, *ALP*, *Runx2*, and *COL1* at the differentiation phase (Zhai et al. 2016). At the protein level, *ALP* was downregulated at the proliferation phase, and *ALP*, *OCN*, and *Runx2* were upregulated at the differentiation phase. The authors also reported the activation of the Wnt signaling pathway. In this work, the cell line was MC3T3-E1, and the most effective stimulation occurred at 2 mT and 15.4 Hz. In fact, 2 mT is the highest intensity in their protocol, which tested 0.5, 1, and 2 mT (peak value) in a 2 h daily stimulation (2 days for proliferation, 5–7 days for differentiation). Polyamines and proto-oncogenes were also reportedly affected by a pulsed EMF of 2.3 mT at 75 Hz (De Mattei et al. 2005).

With respect to the dynamic EMF generated with Helmholtz coils at a certain frequency, as discussed above, static MF around a magnet was also studied. In the literature, the strength of the static MF used in biological experiments is often much higher than that of the dynamic MF. In an early work, fibroblasts and osteoblasts were cultured near magnets that produced an average of 0.45 T MF (McDonald 1993). While fibroblasts were found to proliferate faster, accompanied with higher metabolic activity, osteoblasts were merely affected. A more recent work exposed osteoclasts to 5 T MF and found that such exposure reduced cell resorption activity (Sun et al. 2014). Long-time exposure of experimental rats in an MRI machine (1.5 T) was reported to cause osteoporosis (Gungor et al. 2015).

In addition to EMF, EF also modulates bone cell gene activation and protein production. Using capacitively coupled ES, Zhuang et al. (1997) from Brighton's group reported elevated proliferation and a high level of TGFβ1 mRNA of osteoblastic cells (MC3T3-E1) following 30 min to 24 h of exposure to a sine wave signal (20 mV/cm, 300 $\mu A/cm^2$) at 65 kHz. They found that both cell proliferation and TGFβ1 mRNA upregulation were calcium channel dependent, which was stopped by calcium blocking chemicals verapamil and W-7. Based on these data, they suggested that the calcium or calmodulin pathway might have been involved in the capacitive ES-induced osteoblast activation. In a beautifully executed work, using similar field strength (i.e., 20 mV/cm and 300 $\mu A/cm^2$) and ES signal (sine wave at 60 kHz), Wang et al. (2004) studied the dose effect of the capacitive ES on chondrocytes and reported the optimal conditions to specifically upregulate collagen type II and aggrecan, the most important extracellular matrix molecules in cartilage. The individual modulation of these two proteins was achieved through different stimulation times and duty cycles. Using the same approach, Clark et al. (2014) further optimized the stimulation duration, duty cycle, frequency, and amplitude. They reported the upregulation of BMP2 and BMP4, TGFβ1 and TGFβ2, FGF-2, osteocalcin, and ALP. This capacitive ES, explored by Brighton in the 1980s (Brighton et al. 1985, 1989, 1992), is noninvasive and has been used in the clinical setting. Brighton's group has published extensively about how to use ES to help bone growth, particularly in the case of nonunion (Nelson et al. 2003).

Osteocytes, the most abundant cell type in bone, were less studied but still found to be responsive to ES. Lohmann et al. (2003) used pulsed EMF to stimulate osteocyte-like cells isolated from the long bone of transgenic mice (MLO-Y4 cells). The stimulation generated with Helmholtz coils was a train of 4.5 ms bursts consisting of 20 pulses of 16 G maximum MF. The stimulation was 8 h a day for 1, 2, and 4 days. The MLO-Y4 cells showed upregulated ALP, TGFβ1, and prostaglandin, but reduced osteocalcin and connexin 43. Cell numbers were found to be unaffected even though the data showed a significant increase at day 4. Interestingly, the response was different from osteoblasts (ROS 17/2.8 cells).

14.4 EFFECT ON MINERALIZATION

ES and bone mineralization are a complex phenomenon involving both cellular and noncellular processes. Ion enrichment caused by electrophoresis and surface charge are important factors of cathodal apatite formation (Jahn 1968). This noncellular process, however, is not the subject of this chapter. This chapter only deals with the mineralization caused by cellular processes under the influence of well-defined ES, which is mostly *in vitro*.

Using bursts of 20 trains of 5 ms wide pulsed square waves of EMF (PEMF) at 15 Hz and 7 mT (70 mV of induced EF) peak magnetic flux, generated with Helmholtz coils, Diniz et al. (2002) looked at the proliferation (DNA assay), differentiation (ALP assay), and mineralization (von Kossa stain) of MC3TC-E1 cells. While they found that the PEMF accelerated the proliferation and differentiation of osteoblasts at both the proliferative and differentiation phases, the mineralization of the cells at the mineralization phase was reduced. These findings led to the conclusion that the PEMF accelerated bone mineralization through promoting osteoblast differentiation. A similar work was reported in 2004 by Chang et al. with the conclusion that PEMF accelerated bone tissue formation through enhanced cell proliferation rather than differentiation. In this work, the ALP activity was found to be reduced by PEMF. The difference in conclusions of the abovementioned two reports could be related to the difference in cell lines (MC3TC vs. primary osteoblasts) and stimulation strength (7 mT vs. 0.1 mT). The Helmholtz coils used by Diniz et al. and the single solenoid used by Chang et al. may also have contributed to the difference. In Chang's work, the experimental and control groups were cultured in different incubators to avoid the interference of environmental EMF to the control cultures.

An interesting work by Lin and Lin (2011) studied the coculture of osteoblasts and macrophages under a PEMF of 1.5 mT and 75 Hz. This design was to simulate the inflammatory environment upon implantation. The nitric oxide (NO) produced by the coculture was found to be significantly elevated by the PEMF, even though macrophage viability remained unchanged upon PEMF. In comparison, the osteoblasts showed an elevated viability and number of cells, together with an upregulated collagen type I and downregulated osteocalcin and ALP. This work therefore supports Chang's conclusion that PEMF stimulation promotes osteoblast proliferation but differentiation.

One challenging issue in ES and wound healing is the diversity in methodology, making it difficult to compare the data generated from different experiment setups. One may agree that it is the field strength in the culture medium or tissue, rather than the nominal ES parameters, that should be compared. However, the following example shows the complexity. Griffin et al. (2013) compared two models of ES with degenerative waves: SaOS-2 cells cultured on coverslips and stimulated with salt bridge electrodes inserted in culture medium, and the same cultured coverslips stimulated with capacitive plates that were isolated from the culture medium. The key point in this design was the same field strength in culture medium in both configurations: 10 mV/mm and 16 Hz of degenerate electrical wave. The outcomes were different, despite the same field strength, frequency, and waveform. The capacitive ES resulted in higher cytotoxicity (lactate dehydrogenase [LDH] assay), while the electrode ES came up with the upregulation of differentiation (ALP, collagen type I, osteocalcin, etc.) and mineralization (Alizarin Red S and von Kossa stains). There must be other factors that caused the difference. Those factors could be the orientation of the EF (parallel vs. perpendicular to cell monolayer) and the invasive (electrodes) and noninvasive (capacitor) nature of the methods. Therefore, a great caution should be taken when one wants to compare the behaviors of the cells stimulated in different configurations.

14.5 SIGNALING PATHWAYS

Calcium ions and calcium channels have been in the center of the mechanistic studies of electrically stimulated bone cells. This is logical because of the importance of calcium ions in cell physiology and the fact that EF is known to impact the voltage-gated ion channels, membrane potential, and mobility of the charged molecules. An early work by Korenstein et al. (1984) showed elevated cAMP and DNA synthesis in bone cells isolated from rat embryo calvaria 24 h following a capacitive ES. They suggested that a calcium influx might have regulated cAMP, which in turn triggered cell proliferation (DNA synthesis assay). In that work, however, neither calcium ions nor relevant channels were investigated. A few years later, a study using a similar experiment setup and MC3T3-E1 cells (osteoblast-like cells) confirmed the effect of ES on DNA synthesis and the influx of calcium ions. Contrary to the previous work, the cAMP was found to be not affected by ES (Ozawa et al. 1989). A careful study by Lorich et al. (1998) used verapamil to block calcium channels, W-7 to deactivate calmodulin-dependent phosphodiesterase, and indocin to inhibit prostaglandin synthesis. Their data convincingly demonstrated that the capacitively stimulated osteoblast proliferation was initiated by a calcium influx and through the Ca^{2+}/calmodulin pathway. Their work also suggested that the different pathways triggered by EF or mechanical force all involved calmodulin activation. This work was further expanded to include stimulations of inductive coupling and combined EMFs, in comparison with capacitive ES (Brighton et al. 2001). While having confirmed the Ca^{2+} influx through the voltage-gated channels upon capacitive ES, the authors were also able to demonstrate that both the inductive coupling and the combined EMFs increased cytosolic Ca^{2+} by releasing it from intracellular stores rather than from an extracellular source through calcium channels. This points out the possibility of electrically activating intracellular components without disturbing membrane proteins. The pathways are illustrated in Figure 14.2.

Cell membrane receptors may be sensitive to ES. Fitzsimmons et al. (1995) studied the effect of combined MF on the membrane receptors of IGF-I and IGF-II. The MF was reported to be as low as 40 and 20 μT for the AC and DC modes, respectively. The osteosarcoma-derived osteoblast-like cells TE-85 express membrane receptors of both IGF-I and IGF-II. The results showed that the amount of the IGF-II receptors increased

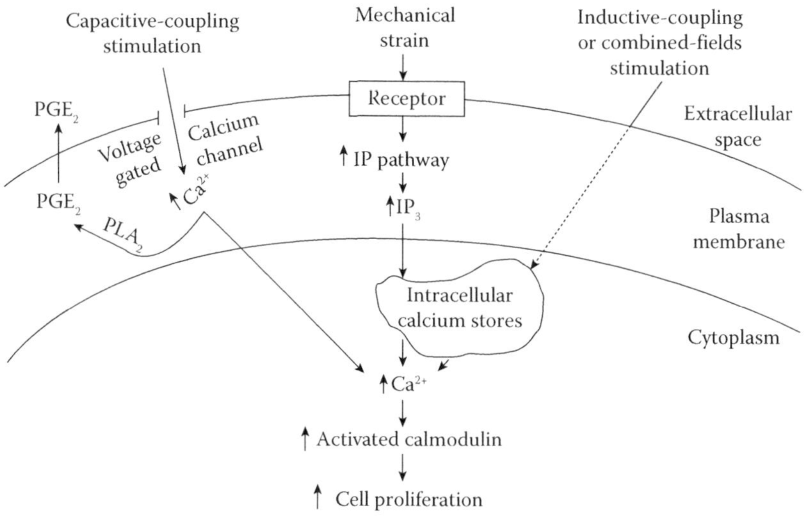

Figure 14.2 Schematic drawing showing the signal transduction pathways, followed by the three forms of electrical stimulation compared with that followed by mechanical strain (cyclic, biaxial, 0.17% strain at 1 H3). PGE2, prostaglandin E2; PLA2, phospholipase A2; IP, inositol phosphate; IP3, inositol triphosphate. (From Brighton, C. T., et al., *J. Bone Joint Surg. Am.*, 83A, 1514–1523, 2001. With permission.)

in a frequency-dependent manner, while the IGF-I receptors were not affected. The authors suggested that the increased receptors were translocated from the intracellular pool to the membrane surface. An interesting work by Schnoke and Midura (2007) compared the effect of pulsed EMF on the phosphorylation of three intracellular signaling molecules in osteoblastic cells (UMR-106) with that of insulin and parathyroid hormone (PTH). Insulin and PTH are two anabolic agents known to cause the phosphorylation of insulin receptor substrate-1 (IRS-1), endothelial nitric oxide synthase (eNOS), and S6 ribosomal subunit (S6). Their data demonstrated an immediate (10–30 min) effect of the EMF on the phosphorylation of these three molecules, as effective as that of insulin or PTH. However, a pulsed EMF stimulation failed to activate Erk 1/2 and CREB pathways as insulin and PTH did, suggesting different downstream pathways. The authors suggested that the field-activated phosphorylation might be initiated by acting on plasma membrane receptors. However, they did not make the effort to test this hypothesis. Clear evidence of plasma membrane receptor activation by EMF was demonstrated by Vincenzi et al. (2013). They used a pulsed EMF of 1.3 ms pulse width at 75 Hz, generating 1.5 mT to treat T/C-28a2 chondrocytes and 2.5 mT to treat hFOB 1.19 osteoblasts. They identified upregulated mRNA transcription and protein expression of A_{2A} and A_3 adenosine receptors (ARs), and verified the findings with their respective antagonists. A_{2A} and A_3 ARs are known to modulate inflammation and bone remodeling. Nevertheless, the mechanism of the activation of A_{2A} and A_3 ARs is still unknown.

The sensitivity of membrane receptors to EF has also been demonstrated in other types of cells. The receptors of low-density lipoprotein in fibroblasts were found to redistribute in an EF of 10.0 V/cm using electrophoresis equipment (Tank et al. 1985).

Acetylcholine receptors (AChRs) of muscle cells formed clusters in EF, which was thought to be through the aggregation of tyrosine kinase in the EF (Peng et al. 1993; Zhang and Peng 2011).

14.6 CONDUCTIVE POLYMER–MEDIATED ES TO BONE CELLS

The idea of using conductive polymers as a substrate to culture cells was initiated by their tunable surface charges that were thought to affect protein absorption and cell adhesion (Wong et al. 1993, 1994), and then because of its good compatibility with neurons and the ability to conduct ES (Schmidt et al. 1997). The polypyrrole (PPy)-coated conductive fabrics were then studied as a new class of biomaterial (Jakubiec et al. 1998). Later, the conductive biodegradable composite was developed as an electrically conductive scaffolding material (Shi et al. 2004). By using the conductive composite as a substrate for cell culture, ES can be readily delivered without using electrodes, Helmholtz coils, or capacitors. Considering the potential advantages of the conductive scaffold in bone regeneration, Meng et al. (2013) used a PPy and polylactide (PLLA) composite to test if such a conductive scaffold was capable of supporting electrically stimulated bone cell growth. The composite was made of 95% (w/w) PLLA and 5% PPy particles doped by chlorine anions and heparin, and had a surface resistivity of 10^2 to 10^3 ohm/square (Meng et al. 2008). Heparin is a cell adhesion molecule and also a polyanion, which helps cell adhesion to the conductive composite and at same time makes the electrical conductivity of the composite more stable in an aqueous environment. Taking advantage of both the pyrrole monomers and heparin being water soluble, heparin can be easily incorporated into PPy during the polymerization of pyrrole, both in the form of a dopant and through physical mixing. Other types of polysaccharides, such as hyaluronic acid (HA) and chondroitin-4-sulfate (CSA), were also doped into PPy through electrochemical synthesis, showing adequate surface property in supporting osteoblast growth and the normal expression of key proteins (Serra Moreno et al. 2009). In fact, the ability to incorporate negatively charged molecules as dopants is a powerful approach to render PPy compatible with cells, because many proteins and polysaccharides are negatively charged. In Meng et al.'s work (2008), the conductive substrates were directly connected as conductors (see Chapter 10) in a DC circuit. When electrical current passed the substrate, cells on the substrate were exposed to the EF on the substrate surface. Under this configuration, the intensity of EF is calculated as the potential gradient along the length of the substrate, that is, the voltage (millivolts) across the substrate membrane divided by its length (millimeters). The osteoblast-like cells (Soas-2) cultured on the conductive substrate were exposed to ES of 100, 200, 300, and 400 mV/mm, to find out that 200 mV/mm was the most effective in promoting cell proliferation. High EF strength (e.g., 400 mV/mm) was found to suppress cell growth. Upregulated ALP and osteocalcin secretion relative to the nonstimulated controls was also identified. After 6 days of culture in mineralization medium in conjunction with a daily ES of 6 h at 200 mV/mm, Soas-2 cells were found to be mineralized more intensively than controls, together with a high expression of the factors implicated in osteoblast differentiation and mineralization, such as ALP, BMP-2, osteocalcin, and Runx2 (Meng et al. 2011). Chitosan, a naturally occurred biodegradable polymer, has also been composed with PPy nanoparticles to form a conductive membrane that was immobilized with BMP-2 by Zhang et al. (2013). They reported elevated metabolic activity, ALP production, and calcium deposit, and increased transcription of important osteogenic genes (*Runx2, OPN, OC, and OSX*).

PPy has good compatibility with osteoblasts (Fahlgren et al. 2015) and can be readily coated to the metal surface of complex geometry through electropolymerization, making PPy a material of choice in the surface modification of orthopedic implants (De Giglio et al. 1999). Water-soluble bioactive molecules such as cell adhesion peptides can be added into electrolyte and deposited to the metal surface during electropolymerization through doping if they are negatively charged, or through physical trapping or entanglement when PPy deposits to the electrode surface. Evidently, one has little control about how these biomolecules interact with PPy, which eventually affects the bioactivity of those molecules. The amount of the biomolecules accessible at the PPy surface is only a small portion of the total biomolecules in PPy. Nevertheless, this relatively simple process does have great potential to render the surface of electrodes bioactive.

Depending on the surface property of a substrate, PPy or any other conductive polymers can also be polymerized to the surface of insulating plastics. One key issue is to resolve how to form a strong adhesion of PPy to the substrate. Knowing oxidized PPy is positively charged, a negatively charged substrate is expected to form a strong ionic bond with PPy, preventing the delamination of the normally brittle and less adhesive PPy coating. A simple coating of PPy on a plastic surface was reported to upregulate the differentiation of bone marrow stromal cells into the osteogenic phenotype, and ES in the form of DC further accelerated calcium deposition to the extracellular matrix (Hu et al. 2014). PPy was also reported to deposit to the biodegradable polymer substrate, together with hydroxyapatite, to form an osteoconductive and electrically conductive surface (Runge et al. 2011), allowing ES to a hydroxyapatite scaffold. The convenience to coat polymer membranes and to compose other elastic polymers with PPy makes it feasible to simultaneously stimulate osteoblasts electrically and mechanically (Liu et al. 2013).

14.7 CONCLUDING REMARKS

Convincing and mounting evidence has demonstrated the feasibility of modulating bone cell physiology with EF and EMF. An exogenous field acts on plasma membrane proteins or intracellular compartments, triggering downstream pathways. A wide range of cellular behaviors can be modified, of which some are osteogenic and beneficial for bone repair. However, major challenges must be resolved before EF and EMF could be accepted as standard techniques in a cell culture laboratory and in clinic management. These challenges include knowing the molecular mechanisms of how EF or EMF interacts with the binding sites of plasma membrane receptors, probably with the help of molecular modeling, such as the protein–ligand docking technique. For example, is the weak EF reported in literature able to directly impact the conformation of the active center of a receptor, or does it work together with soluble ions and even water molecules? Do DC EF and EMF act on cells differently, and if they do, how? Another important challenge is to identify specific combinations of ES parameters leading to foreseeable and reproducible cellular consequence. EF and EMF cannot become a standard technique without accuracy. Clinical research must consider the complexity of tissue dielectric properties that vary among individuals and govern the magnitude and distribution of EF in tissue.

A conductive scaffold may provide a uniform two- or three-dimensional space of EF for the cells cultured on or in it. However, the feasibility and efficacy of *in vivo* use of the conductive scaffold remain to be demonstrated.

REFERENCES

Bassett, C. A., Becker, R. O. Generation of electric potentials by bone in response to mechanical stress. *Science* 137 (1962): 1063–4.

Borys, P. On the biophysics of cathodal galvanotaxis in rat prostate cancer cells: Poisson-Nernst-Planck equation approach. *Eur Biophys J* 41 (2012): 527–34.

Brighton, C. T., Hozack, W. J., Brager, M. D., et al. Fracture healing in the rabbit fibula when subjected to various capacitively coupled electrical fields. *J Orthop Res* 3 (1985): 331–40.

Brighton, C. T., Jensen, L., Pollack, S. R., Tolin, B. S., Clark, C. C. Proliferative and synthetic response of bovine growth plate chondrocytes to various capacitively coupled electrical fields. *J Orthop Res* 7 (1989): 759–65.

Brighton, C. T., Okereke, E., Pollack, S. R., Clark, C. C. In vitro bone-cell response to a capacitively coupled electrical field. The role of field strength, pulse pattern, and duty cycle. *Clin Orthop Relat Res* 285 (1992): 255–62.

Brighton, C. T., Wang, W., Seldes, R., Zhang, G., Pollack, S. R. Signal transduction in electrically stimulated bone cells. *J Bone Joint Surg Am* 83A (2001): 1514–23.

Clark, C. C., Wang, W., Brighton, C. T. Up-regulation of expression of selected genes in human bone cells with specific capacitively coupled electric fields. *J Orthop Res* 32 (2014): 894–903.

Chang, W. H., Chen, L. T., Sun, J. S., Lin, F. H. Effect of pulse-burst electromagnetic field stimulation on osteoblast cell activities. *Bioelectromagnetics.* 25 (2004): 457–65.

De Giglio, E., Sabbatini, L., Zambonin, P. G. Development and analytical characterization of cysteine-grafted polypyrrole films electrosynthesized on Pt- and Ti-substrates as precursors of bioactive interfaces. *J Biomater Sci Polym Ed* 10 (1999): 845–58.

De Mattei, M., Gagliano, N., Moscheni, C., et al. Changes in polyamines, c-myc and c-fos gene expression in osteoblast-like cells exposed to pulsed electromagnetic fields. *Bioelectromagnetics* 26 (2005): 207–14.

Diniz, P., Shomura, K., Soejima, K., Ito, G. Effects of pulsed electromagnetic field (PEMF) stimulation on bone tissue like formation are dependent on the maturation stages of the osteoblasts. *Bioelectromagnetics* 23 (2002): 398–405.

Fahlgren, A., Bratengeier, C., Gelmi, A., et al. Biocompatibility of polypyrrole with human primary osteoblasts and the effect of dopants. *PLoS One* 10 (2015): e0134023.

Fassina, L., Visai, L., Benazzo, F., et al. Effects of electromagnetic stimulation on calcified matrix production by SAOS-2 cells over a polyurethane porous scaffold. *Tissue Eng* 12 (2006): 1985–99.

Ferrier, J., Ross, S. M., Kanehisa, J., Aubin, J. E. Osteoclasts and osteoblasts migrate in opposite directions in response to a constant electrical field. *J Cell Physiol* 129 (1986): 283–8.

Fitzsimmons, R. J., Ryaby, J. T., Magee, F. P., Baylink, D. J. IGF-II receptor number is increased in TE-85 osteosarcoma cells by combined magnetic fields. *J Bone Miner Res* 10 (1995): 812–19.

Giustina, A., Mazziotti, G., Canalis, E. Growth hormone, insulin-like growth factors, and the skeleton. *Endocr Rev* 29 (2008): 535–59.

Griffin, M., Sebastian, A., Colthurst, J., Bayat, A. Enhancement of differentiation and mineralisation of osteoblast-like cells by degenerate electrical waveform in an in vitro electrical stimulation model compared to capacitive coupling. *PLoS One* 8 (2013): e72978.

Gungor, H. R., Akkaya, S., Ok, N., et al. Chronic exposure to static magnetic fields from magnetic resonance imaging devices deserves screening for osteoporosis and vitamin D levels: A rat model. *Int J Environ Res Public Health* 12 (2015): 8919–32.

Hu, W. W., Hsu, Y. T., Cheng, Y. C., et al. Electrical stimulation to promote osteogenesis using conductive polypyrrole films. *Mater Sci Eng C Mater Biol Appl* 37 (2014): 28–36.

Jahn, T. L. A possible mechanism for the effect of electrical potentials on apatite formation in bone. *Clin Orthop Relat Res* 56 (1968): 261–73.

Jakubiec, B., Marois, Y., Zhang, Z., et al. In vitro cellular response to polypyrrole-coated woven polyester fabrics: Potential benefits of electrical conductivity. *J Biomed Mater Res* 41 (1998): 519–26.

Korenstein, R., Somjen, D., Fischler, H., Binderman, I. Capacitative pulsed electric stimulation of bone cells. Induction of cyclic-AMP changes and DNA synthesis. *Biochim Biophys Acta* 803 (1984): 302–7.

Lin, H. Y., Lin, Y. J. In vitro effects of low frequency electromagnetic fields on osteoblast proliferation and maturation in an inflammatory environment. *Bioelectromagnetics* 32 (2011): 552–60.

Liu, L., Li, P., Zhou, G., et al. Increased proliferation and differentiation of pre-osteoblasts MC3T3-E1 cells on nanostructured polypyrrole membrane under combined electrical and mechanical stimulation. *J Biomed Nanotechnol* 9 (2013): 1532–9.

Lohmann, C. H., Schwartz, Z., Liu, Y., et al. Pulsed electromagnetic field stimulation of MG63 osteoblast-like cells affects differentiation and local factor production. *J Orthop Res* 18 (2000): 637–46.

Lohmann, C. H., Schwartz, Z., Liu, Y., et al. Pulsed electromagnetic fields affect phenotype and connexin 43 protein expression in MLO-Y4 osteocyte-like cells and ROS 17/2.8 osteoblast-like cells. *J Orthop Res* 21 (2003): 326–34.

Lorich, D. G., Brighton, C. T., Gupta, R., et al. Biochemical pathway mediating the response of bone cells to capacitive coupling. *Clin Orthop Relat Res* 50 (1998): 246–56.

McDonald, F. Effect of static magnetic fields on osteoblasts and fibroblasts in vitro. *Bioelectromagnetics* 14 (1993): 187–96.

Meng, S., Rouabhia, M., Shi, G., Zhang, Z. Heparin dopant increases the electrical stability, cell adhesion, and growth of conducting polypyrrole/poly(L,L-lactide) composites. *J Biomed Mater Res* 87A (2008): 332–44.

Meng, S., Rouabhia, M., Zhang, Z. Electrical stimulation modulates osteoblast proliferation and bone protein production through heparin-bioactivated conductive scaffolds. *Bioelectromagnetics* 34 (2013): 189–99.

Meng, S., Zhang, Z., Rouabhia, M. Accelerated osteoblast mineralization on a conductive substrate by multiple electrical stimulation. *J Bone Miner Metab* 29 (2011): 535–44.

Nelson, F. R., Brighton, C. T., Ryaby, J., et al. Use of physical forces in bone healing. *J Am Acad Orthop Surg* 11 (2003): 344–54.

Ono, S. Mechanism of depolymerization and severing of actin filaments and its significance in cytoskeletal dynamics. *Int Rev Cytol* 258 (2007): 1–82.

Ozawa, H., Abe, E., Shibasaki, Y., Fukuhara, T., Suda, T. Electric fields stimulate DNA synthesis of mouse osteoblast-like cells (MC3T3-El) by a mechanism involving calcium ions. *J Cell Physiol* 138 (1989): 477–83.

Ozkucur, N., Monsees, T. K., Perike, S., Do, H. Q., Funk, R. H. Local calcium elevation and cell elongation initiate guided motility in electrically stimulated osteoblast-like cells. *PLoS One* 4 (2009): e6131.

Ozkucur, N., Perike, S., Sharma, P., Funk, R. H. Persistent directional cell migration requires ion transport proteins as direction sensors and membrane potential differences in order to maintain directedness. *BMC Cell Biol* 12 (2011): 4.

Peng, H. B., Baker, L. P., Dai, Z. A role of tyrosine phosphorylation in the formation of acetylcholine receptor clusters induced by electric fields in cultured *Xenopus* muscle cells. *J Cell Biol* 120 (1993): 197–204.

Reinbold, K. A., Pollack, S. R. Serum plays a critical role in modulating [Ca^{2+}]c of primary culture bone cells exposed to weak ion-resonance magnetic fields. *Bioelectromagnetics* 18 (1997): 203–14.

Runge, M. B., Dadsetan, M., Baltrusaitis, J., Yaszemski, M. J. Electrically conductive surface modifications of three-dimensional polypropylene fumarate scaffolds. *J Biol Regul Homeost Agents* 25 (2011): S15–23.

Schmidt, C. E., Shastri, V. R., Vacanti, J. P., Langer, R. Stimulation of neurite outgrowth using an electrically conducting polymer. *Proc Natl Acad Sci USA* 94 (1997): 8948–53.

Schnoke, M., Midura, R.J. Pulsed electromagnetic fields rapidly modulate intracellular signaling events in osteoblastic cells: Comparison to parathyroid hormone and insulin. *J Orthop Res* 25 (2007): 933–40.

Serra Moreno, J., Panero, S., Materazzi, S., et al. Polypyrrole-polysaccharide thin films characteristics: Electrosynthesis and biological properties. 88 (2009): 832–40.

Shi, G., Rouabhia, M., Wang, Z., Dao, L. H., Zhang, Z. A novel electrically conductive and biodegradable composite made of polypyrrole nanoparticles and polylactide. *Biomaterials* 25 (2004): 2477–88.

Sollazzo, V., Palmieri, A., Pezzetti, F., Massari, L., Carinci, F. Effects of pulsed electromagnetic fields on human osteoblastlike cells (MG-63): a pilot study. *Clin Orthop Relat Res.* 468 (2010): 2260–77.

Sun, Y. L., Chen, Z. H., Chen, X. H., et al. Diamagnetic levitation promotes osteoclast differentiation from RAW264.7 cells. *IEEE Trans Biomed Eng* 62 (2014): 900–8.

Tank, D. W., Fredericks, W. J., Barak, L. S., Webb, W. W. Electric field-induced redistribution and postfield relaxation of low density lipoprotein receptors on cultured human fibroblasts. *J Cell Biol* 101 (1985): 148–57.

Vanegas-Acosta, J. C., Garzón-Alvarado, D. A., Zwamborn, A. P. Mathematical model of electrotaxis in osteoblastic cells. *Bioelectrochemistry* 88 (2012): 134–43.

Vincenzi, F., Targa, M., Corciulo, C., et al. Pulsed electromagnetic fields increased the anti-inflammatory effect of A_{2A} and A_3 adenosine receptors in human T/C-28a2 chondrocytes and hFOB 1.19 osteoblasts. *PLoS One* 8 (2013): e65561.

Wang, W., Wang, Z., Zhang, G., Clark, C. C., Brighton, C. T. Up-regulation of chondrocyte matrix genes and products by electric fields. *Clin Orthop Relat Res* 427 (2004): S163–73.

Wong, J. Y., Langer, R. S., Ingber, D. E. Cell attachment and protein adsorption to polypyrrole-thin films. *Mater Res Soc Symp Proc* 293 (1993): 179–84.

Wong, J. Y., Langer, R., Ingber, D. E. Electrically conducting polymers can noninvasively control the shape and growth of mammalian cells. *Proc Natl Acad Sci USA* 91 (1994): 3201–4.

Zhai, M., Jing, D., Tong, S., et al. Pulsed electromagnetic fields promote in vitro osteoblastogenesis through a Wnt/β-catenin signaling-associated mechanism. *Bioelectromagnetics* (2016). DOI: 10.1002/bem.21961.

Zhang, H. L., Peng, H. B. Mechanism of acetylcholine receptor cluster formation induced by DC electric field. *PLoS One* 6 (2011): e26805.

Zhang, J., Neoh, K. G., Hu, X., Kang, E. T., Wang, W. Combined effects of direct current stimulation and immobilized BMP-2 for enhancement of osteogenesis. *Biotechnol Bioeng* 110 (2013): 1466–75.

Zhuang, H., Wang, W., Seldes, R. M., Tahernia, A. D., Fan, H., Brighton, C. T. Electrical stimulation induces the level of TGF-beta1 mRNA in osteoblastic cells by a mechanism involving calcium/calmodulin pathway. *Biochem Biophys Res Commun* 237 (1997): 225–9.

15

Electrical stimulation of cells derived from muscle

Anita F. Quigley
University of Wollongong
Wollongong, New South Wales, Australia

Justin L. Bourke
University of Melbourne
Melbourne, Victoria, Australia

Robert M. I. Kapsa
University of Wollongong
Wollongong, New South Wales, Australia

Contents

15.1 INTRODUCTION

Electricity is a fundamental force in nature and is increasingly being recognized as an influencing factor in the regulation of cellular differentiation and activity. This chapter explores the effects of electrical stimulation on muscular tissues at the tissue, cellular, and molecular levels. In addition, this chapter describes the use of conducting polymers for influencing cardiac and skeletal muscle differentiation and regeneration and recent developments in the use of conducting polymers for electroactive tissue scaffolds.

15.2 BIOELECTRICAL STIMULATION: FROM THE EGYPTIANS TO DUCHENNE AND BEYOND

Bioelectricity is the term used to describe the electrical currents and potentials produced by, or occurring within, living organisms. These currents and potentials are created by the flux and active transport of positively and negatively charged ions across biological membranes and through tissues. The first observation of bioelectricity predates any modern experimentation. The Egyptians made reference to electric fish, describing the strange properties of the electric eel and Nile catfish (*Malapterurus electricus*), on Egyptian murals (Howes 1984; Wu 1984). The Greek philosophers were also familiar with the electrical activity of the Mediterranean ray (*Torpedo ocellata*); it was recognized that certain kinds of fish had the ability to generate a force that could "stun." This phenomenon was put to use by Scribonius Largus (court physician to the Roman emperor Claudius, ca. 47 AD), who advocated "piscine electrotherapy" for the relief of pain (Kane and Taub 1975; Kellaway 1946). However, it was in Galvani's age that the knowledge of bioelectricity, as it is known today, was in its very beginnings. The phenomenon of bioelectricity and its effects on the movement of muscle was described by Galvani in *De viribus electricitatis in motu musculari commentaries*, whose experiments explore a direct relationship between electricity and tissue response. Galvani's famous frog leg experiments on "bioelectricity" were performed at the University of Bologna (Italy). In these well-known experiments, he applied electrical impulses to the legs, making them twitch and jump, demonstrating what we now understand to be the electrical basis of nerve impulses and muscle contraction. In brief, the findings of Galvani were summarized as the discovery of an "animal electricity" that is produced within the body. As a result of these early experiments, Galvani was posthumously given the unofficial title of "father of bioelectricity."

There were a number of pre-Galvanian "electrophysiologists" that provided a wealth of background to inspire Galvani, including Croone in the seventeenth century and Langrish in the eighteenth (Walker 1937); these early researchers are relatively unrecognized for their achievements. These early electrophysiologists no doubt provided some inspiration for Galvani to conduct his famous experiments that would serve as the foundation for modern electrophysiology. An excellent summary of Galvani's work and the work of pre-Galvanian electrophysiologists was made by Holt in 1936.

The concept and study of bioelectricity was carried on by Galvani's nephew, Giovanni Aldini (born in Bologna, Italy, in 1762), who also carried out scientific work associated with Galvanism. Aldini took bioelectricity to the masses and traveled Europe and the United Kingdom, putting on theatrical public displays of electrifying (or reanimating) animal and human bodies (Figure 15.1). Many of these experiments were performed in front of surgeons and other notables, including the Prince of Wales (Sleigh 1998). Aldini's most famous public experiments were carried out at the Royal College of Surgeons in London in 1803, where he conducted experiments on a hanged man named George Forster. Aldini's experiments gained a large amount of public interest and were reported in the *Times* (London newspaper) (Parent 2004). It is widely believed that the

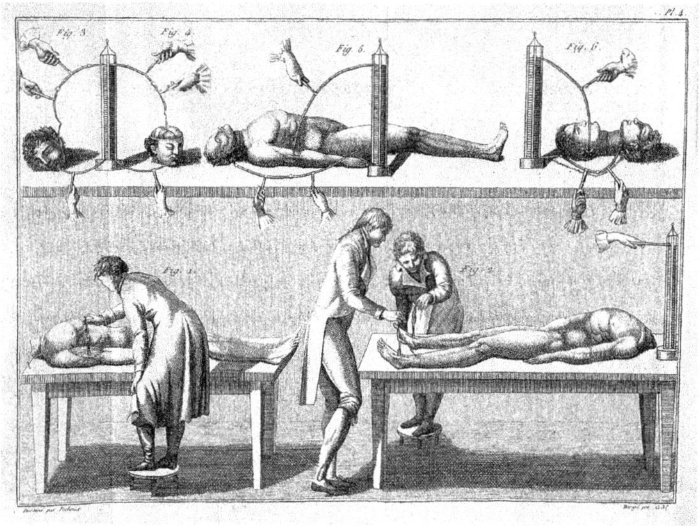

Figure 15.1 Reproduction of plate IV in J. Aldini's "Essai theorique et experimental sur le galvanisme." These drawings by Pecheux depict some of the various procedures that Aldini used to electrify the bodies of corpses. (Courtesy of Creative Commons. https://commons.wikimedia.org/wiki/File:Giovanni_Aldini,_Essai_theorique...sur_le_galvanisme_Wellcome_L0029560.jpg)

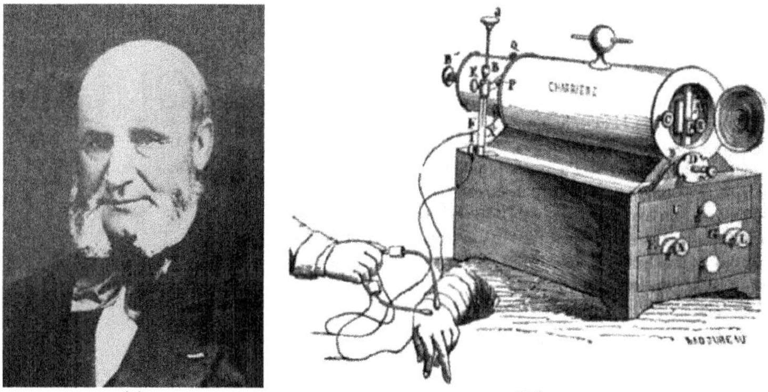

(a) (b)

Figure 15.2 (a) Guillaume Benjamin Amand Duchenne. (Courtesy of Creative Commons. https://commons.wikimedia.org/wiki/Category:Guillaume_Duchenne#/media/File:Guillaume_Benjamin_Amand_Duchenne.jpg) (b) Duchenne's apparatus for the Electrical stimulation and analysis of patients. (Courtesy of creative commons. https://commons.wikimedia.org/wiki/Category:Guillaume_Duchenne#/media/File:DuchenneAppareil.jpg)

public experiments performed by Aldini were the inspiration behind Mary Shelley's *Frankenstein*.

Perhaps inspired by the earlier works of Galvani and Aldini, Guillaume Benjamin Amand Duchenne, a French neurologist, began experimenting with a new technique termed *electropuncture* in the 1850s (Figure 15.2). He was particularly interested in how the muscle and nerves of the body react to electrical stimulation and how the nerves and muscles of the face can convey emotions. He was also a leading physician specializing in neurological disorders, particularly muscle paralysis. He documented the first accounts of several types of muscular atrophy and paralysis caused by nerve disorders, and in the 1860s, he described severe progressive muscle weakness in 13 young boys, a condition that was later named Duchenne muscular dystrophy. Some of the earliest clinical investigations into the use of electricity for clinical diagnoses was carried out by Duchenne, and he developed and refined electrodiagnosis and electrotherapeutic techniques that are used to this day for the diagnosis and treatment of nerve and muscle disorders. His best-known writings are *De l'électrisation localisée* (1855) and *Physiologie des movements* (1867).

Not long after Duchenne's work was published, John McWilliam, a Scottish physiologist, published an article in the *British Medical Journal* in 1889 describing the use of electrical stimulation of the heart to "excite rhythmic contraction" (McWilliam 1889). He went on to describe the use of electrical stimulation to regulate pacing: "a regular series of induction shocks (for example, sixty or seventy per minute) gives a regular series of heartbeat at the same rate." McWilliam had performed extensive studies of the hearts of numerous animals, experimenting with the use of electrical stimulation to revive and control contraction. These studies formed the background and impetus for the development of the artificial pacemaker. Dr. Mark Lidwill, an anesthetist practicing in Sydney, Australia, gave a presentation at the Australasian Medical Congress in Sydney 1929, where he presented a device for stimulating the heart and reported the resuscitation of a baby using this device at the meeting (Lidwell 1929). Lidwill worked at the University of Sydney with Dr. Booth to design a number of machines, one of which could, through application of a conducting needle electrode, set a pacemaker rate from about 80 to 120 pulses per minute. This undoubtedly was the predecessor of today's artificial pacemaker (Mond et al. 1982).

These early observations and experiments "sparked" the curiosity of the public, medical community, and scientists alike and have provided the basis for modern electrophysiological techniques that have been developed for the diagnosis (e.g., the electroencephalogram and electrocardiogram) and treatment (e.g., the cardiac pacemaker) of human ailments. The field of cell biology and tissue engineering has also recognized the benefits of electrical stimulation in controlling cellular behavior. Recently developed conducting polymers are facilitating the development of tissue regeneration platforms that can deliver electrical stimulation to cells and tissues to accelerate regeneration and influence cellular function. Progress in this area will lead to new and exciting developments in the field of tissue regeneration, as well as in the medical bionic field, where electricity can be used to interface between medical implants and the human body.

15.3 MYOGENIC CELL CYCLING AND FUNCTION: POTENTIAL FOR ELECTRICAL INFLUENCE OF MYOGENIC BEHAVIOR

The influence of electrical stimulation on myogenic behavior is best known as the force that facilitates and dictates muscle contraction, as demonstrated by Galvani, Aldini, Duchenne, and McWilliam. In the body, skeletal muscle contractile activity is controlled by the nerves where an action potential travels down the nerve to the nerve ending, resulting in the release of acetylcholine at the neuromuscular junction. This neurotransmitter binds to acetylcholine receptors on the sarcolemma of the muscle, resulting in the activation of sodium–potassium channels on the sarcolemma of muscle fibers, causing an influx of sodium ions. This influx causes an action potential that travels to the t-tubules and sarcoplasmic reticulum, where it induces calcium release to the cytoplasm. Calcium then binds to troponin on the tropomyosin complex and induces actin myosin filament sliding, resulting in muscle contraction. Direct electrical stimulation of skeletal muscle mimics the action potential induced by acetylcholine release, resulting in a release of calcium from the sarcoplasmic reticulum, leading to contraction. As such, the use of electrical stimulation for influencing myogenic behavior both *in vivo* and *in vitro* has great potential in medicine for the treatment of neuromuscular disorders, paralysis, and many other disorders involving skeletal, cardiac, and smooth muscles.

15.3.1 ELECTROTHERAPY, THE EVIDENCE IN MUSCULAR TISSUES

Perhaps the best evidence for the benefits of electrical stimulation on muscle can be seen in the use of transcutaneous electrical nerve stimulation (TENS) and electrical muscle stimulation (EMS) machines for both therapeutic and training purposes. These machines act by stimulation of the nerve (TENS) or muscle (EMS) through the skin and have been used for a number of therapeutic purposes, including the relief of pain (Omura 1987; Bokkon et al. 2011; Heidland et al. 2013). The use of TENS- and EMS-mediated electrical stimulation has also gained popularity as an approach for muscle therapy, as well as for muscle training and conditioning (Hartsell 1986; Trimble and Enoka 1991; Cotter et al. 1991; Quittan et al. 1999; Hirose et al. 2013; Boonyarom et al. 2009). Studies have also shown that the use of electrical stimulation can help reduce the effects of muscle atrophy and denervation in rabbits, rats, and human subjects (Crameri et al. 2000; Hirose et al. 2013; Cotter et al. 1991; Dow et al. 2005; Ashley et al. 2008).

There has been some investigation into the mechanisms by which this effect is mediated at the macroscopic level. Electrical stimulation has been shown to influence myofiber distribution and type (Windisch et al. 1998; Bergh et al. 2010), as well improve or increase the capillary density and blood flow in muscle tissues (Hudlicka et al. 1994; Mathieu-Costello et al. 1996). These changes are reminiscent of changes seen with exercise training and electrical stimulation has been shown to elicit positive effects in critically ill patients (Karatzanos et al. 2012; Routsi et al. 2010; Vivodtzev et al. 2012; Banerjee 2010). These changes at the macroscopic level point to cellular and molecular responses that underlie muscle adaption to increased "exercise loading" mediated by electrical stimulation. In the case of muscle atrophy, it is thought that use of the muscle, through artificial means, prevents changes associated with disuse. The cardiac pacemaker, one of the best-known bionic devices, was developed to provide electrical stimulation directly to the cardiac muscle in order to maintain appropriate function. Pacemaker cells are responsible for coordinating cardiac contraction; however, in a number of patients with conduction defects (and other cardiac defects), this role needs to be fulfilled artificially. The pacemaker uses electrical stimulation to promote timed contraction of cardiac muscle, regulating the beating of the heart. In contrast to the use of EMS in skeletal muscle, the pacemaker is used to regulate normal function rather than to provide a therapeutic function.

Despite fundamental similarities in the contraction-based mechanisms in both skeletal and cardiac muscle, the structures of skeletal muscle and cardiac muscle differ markedly.

In skeletal muscle, the contractile muscle fibers comprise fused myoblasts that form a long, syncytial multinucleated muscle fibers, whereas in the heart, the contractile muscle comprises cardiomyocytes that form an array of cells connected through gap junctions that facilitate transmission of electrical impulses. Skeletal muscle myoblasts (muscle precursor cells) and cardiac myocytes (precursors of the heart) reside in both tissues and share some similarities; in fact, skeletal muscle myoblasts have been shown to be able to electrically couple with cardiac cells through gap junctions and may represent a viable cell source for cardiac remodeling and engineering (Treskes et al. 2015; Siepe et al. 2006, 2007).

As well as being a regulator of muscle contraction in cardiac and skeletal muscles, electrical stimulation has been shown to influence the differentiation of stem cells to cardiomyocytes, as well as the maturation or differentiation of myocytes. The fact that stem cells, myoblasts, and myocytes respond to electrical stimulation and electrical fields has led investigators to explore the mechanisms behind the effect of electrical stimulation on these cells in order to gain deeper insight into development, tissue regeneration, and repair. Not only is electrical stimulation useful for diagnoses of neuromuscular ailments (as developed by Duchenne) or maintenance of regular activity (as in the pacemaker), but also electrical stimulation is now being explored as a means to drive regeneration in muscle tissues.

15.3.2 EFFECTS OF ELECTRICAL STIMULATION AT THE CELLULAR LEVEL

Cardiac and skeletal muscle tissues comprise a number of cellular types, including mature myofibers or cardiomyocytes, progenitor cells (myoblasts in skeletal muscle or myocytes in cardiac muscle), and the cells of associated vasculature and interstitial tissues. Whilst electrical stimulation has the potential to influence the majority of the cell types present in muscles through ion flux (including myoblasts, myocytes, and vasculature), the immediate effect of electrical stimulation in skeletal muscle is seen macroscopically in the muscle fibers or mature cardiac muscle through induced contraction. In rat skeletal muscle, chronic electrical stimulation can induce changes in the status of muscle fiber type profile, changing fast-type muscle fibers to slow-type muscle fibers (Windisch et al. 1998; Putman et al. 2001); however, fiber type changes have been reported to be related to the frequency of the stimulation, with higher-frequency stimulation showing a slow to fast transition (Boonyarom et al. 2009; Radisic et al. 2005). This effect has also been observed in rabbits (Kubis et al. 2002). Studies in humans have also demonstrated an effect of electrical stimulation predominantly on type II fast muscle fibers (Cabric et al. 1988). Although electrical stimulation has been reported to increase the metabolism and maximal force of muscles in humans, the associated fiber type changes can be somewhat variable, suggesting that there are other factors influencing changes in fiber type, other than contractile activity alone (Minetto et al. 2013; Rochester et al. 1995; Dal Corso et al. 2007). Despite some variation in the effects of electrical stimulation in human fiber type changes in the literature, electrical stimulation has been used to reduce muscle wasting in immobile patients (Dirks et al. 2015), as well to help maintain muscle function in denervated and aging muscles (Elzinga et al. 2015; Zealear et al. 2014; Dow et al. 2005).

As well as fiber type changes and associated changes to metabolism, the regeneration and differentiation of muscle tissue can also be accelerated through electrical stimulation, particularly through effects on myoblasts and myocytes (Flaibani et al. 2009; Cotter et al. 1991; Lewis et al. 1997). Electrical stimulation has been shown to influence myoblast behavior *ex vivo*, as well as *in vivo*. *In vitro* studies have shown that the application of electrical stimulation can increase myoblast proliferation, as well as myoblast differentiation (Quigley et al. 2012; Flaibani et al. 2009; Zemianek et al. 2013). Electrical stimulation has also been used to increase the rate of "development" or muscle maturation in three-dimensional (3D) tissue-engineered myogenic constructs (Serena et al. 2008; Hosseini et al. 2012; Park et al. 2008; Langelaan et al. 2011; Lee et al. 2015). A recent study by Ito et al. (2014) also demonstrates that the use of electrical pulse stimulation results in a 4.5-fold increase in twitch force in engineered skeletal muscle constructs, suggesting enhanced maturation and function.

Similar studies have been carried out with cardiomyocytes to generate cardiac muscle constructs *ex vivo*. In cardiac muscle, electrical stimulation can stimulate the differentiation of myocytes to form mature gap junctions. *In vitro* studies have shown that the application of electrical signals mimicking those in the heart can lead to increased myocyte alignment and coupling, increasing the amplitude of synchronous contractions in constructs (Radisic et al. 2004). Interestingly, in these studies the cardiomyocytes in electrically stimulated constructs showed a cellular ultrastructure resembling that of mature cardiac tissue; this was not seen for nonstimulated constructs. Investigations into the effects of electrical stimulation in the heart itself are scarce, as the only effaceable electrical stimulation that can be directly applied *in situ* is that mimicking the role of pacemaker cells, but it is likely that the electrical activity maintained by an "artificial" pacemaker is playing a role in preserving functional cardiac muscle architecture. These studies strongly suggest that electrical stimulation is eliciting an effect on muscle precursors that may be independent of contractile-induced activity of mature muscle.

15.3.3 EFFECTS OF ELECTRICAL STIMULATION IN MUSCLE AT THE MOLECULAR LEVEL

Investigations into the molecular changes seen in electrically stimulated muscle, both *in vivo* and *in vitro*, have uncovered interesting changes in gene expression that may provide therapeutic avenues for the treatment of neuromuscular and cardiac disorders. Electrical stimulation of skeletal muscle *in vivo* has been shown to induce numerous gene expression changes in genes associated with apoptosis and signal transduction, amongst others (Ono et al. 2010). *In vitro* and *in vivo* experiments have shown that electrical stimulation of myoblasts leads to significant gene expression changes in genes associated with myoblast differentiation and metabolism, such as myosin (Carraro et al. 1986; Goldspink et al. 1991), MyoD (Kawahara et al. 2006; Xing et al. 2015), acetylcholine receptor (O'Reilly et al. 2003), neuronal nitric oxide synthase (Reiser et al. 1997), mitochondrial genes (Williams et al. 1987; Petrie et al. 2015; Putman et al. 2004), and genes involved in vascular remodeling and muscle morphogenesis (Wu et al. 2013).

Electrical stimulation of cardiac muscle and cardiomyocytes also induces changes in gene expression. Similarly to skeletal muscle myoblasts, changes in the expression of myosin genes are seen in cardiomyocytes with electrical stimulation (McDonough and Glembotski 1992; Kujala et al. 2012). A number of studies, including microarray studies, on electrically stimulated cardiomyocytes suggest that key genes involved in metabolism are also affected by electrical stimulation (Martherus et al. 2010; Xia et al. 1997), independent of contraction. In addition, changes in the expression of genes in the NFAT3 and GATA4 pathways (Xia et al. 2000), as well as phosphorylation of AMPKinase (Kakinuma et al. 2004), have been reported.

These reports of gene expression changes, both *in situ* in skeletal and cardiac muscles and *in vitro* in cultured myocytes or myoblasts in two-dimensional (2D) or 3D culture, point to a number of potential mechanisms whereby electrical stimulation mediates its effects. There is the obvious effect of increased contraction *in vivo* where adaptions to increased "exercise" loading are elicited, but there are also documented gene expression changes in isolated (i.e., not in physical contact with mature contractile muscle) myoblasts and myocytes, suggesting that the effects of electrical stimulation are not solely related to contractility, or to contractile cells influencing the activity of neighboring progenitors through signaling. This is further corroborated by the fact that electrical stimulation can also influence the fate of embryonic stem cells (Yamada et al. 2007; Serena et al. 2009).

15.4 ELECTRICAL FIELDS, ELECTRICAL STIMULATION, STEM CELLS, AND MUSCLE

An electric field is an "action at a distance" type of force that is generated by charged objects that can influence agents of the opposite or similar charge. When electrical current or charge is applied in many biological applications, an electrical field is produced that influences the cells and tissues in the vicinity. In biological systems, application of electrical stimulation or fields results in the movement of ions, such as calcium; this ion flux can have a significant effect on cells and tissues. Ion flux and electric fields are also generated endogenously by tissues and organisms, for example, the fields produced by the movement of blood around the body (Trivedi et al. 2013) and the local fields produced in the brain around neural activity (Frohlich and McCormick 2010). These fields can have an effect on the tissues around them, and there has been evidence of the existence of feedback loops between cellular activity and endogenous electric fields (Frohlich and McCormick 2010; Trivedi et al. 2013).

Electric fields play an important role in wound healing, as well as during embryonic development (Nuccitelli 2003; Hotary and Robinson 1994). Observations that embryos produce electric fields were reported as far back as the 1940s (Romanoff 1944). Measurements in chick embryos have revealed that voltages of approximately 5–20 mV/mm can be generated (Hotary and Robinson 1992). Since this time, it has been generally accepted that electric fields influence cellular behavior and are of paramount importance during the development of the embryo and in wound regeneration in animals and in adult human tissues (Becker and Murray 1970; Alvarez et al. 1983; Sebastian et al. 2011). Many *in vitro* experiments demonstrate that various cell types migrate in response to electric fields (a process termed *galvanotaxis*), in most cases to the cathode, and that endogenous electric fields influence development (Shi and Borgens 1995; Hotary and Robinson 1990, 1992, 1994; Nishimura et al. 1996; Zhao et al. 1996; Li et al. 2015; for a recent review, see Funk 2015). Disruptions to electric fields in developing chick embryos have been shown to result in defects in the neural tube, notochord, and somite formation, particularly in the tail region (Hotary and Robinson 1992). In terms of muscle progenitors, electrical stimulation and fields can influence myoblast migration and proliferation (Quigley et al. 2012; Lee et al. 2015; Wan et al. 2016), as well as the proliferation and migration of other stem cell types (Li et al. 2014; J. Zhang et al. 2011; Li et al. 2012; Gu et al. 2015; Jahanshahi et al. 2013). Early studies of embryonic *Xenopus* myoblasts in 1981 by Hinkle et al. (1981) demonstrated that myoblasts align to small electric fields. Since then, there have been a number of studies demonstrating the effect of electrical fields and stimulation on myoblast and cardiomyocyte differentiation and function

(Serena et al. 2008; van der Schaft et al. 2013; Leon-Salas et al. 2013; Quigley et al. 2012; Zhao 2009). These studies provide evidence that electrical stimulation *in vivo* may elicit changes in stimulated tissues by influencing progenitor cell migration.

It is established that electrical stimulation of skeletal muscle can enhance muscle and cardiac maturation, myoblast proliferation, and development of functional cardiac and muscle tissues; however, one of the obstacles in muscle tissue engineering is how to deliver electrical stimulation *in vitro* and *in vivo* to target tissues. Conducting polymers are excellent candidates for use in scaffolds designed to enhance muscle regeneration and differentiation both *ex vivo* and *in vivo*. These polymers are generally softer than conventional metal electrodes such as platinum and gold, and also can be doped with biomolecules that can be released from the polymer to further enhance tissue development.

15.5 CONDUCTING POLYMERS: ELECTRICALLY ACTIVE SUBSTRATES FOR TISSUE ENGINEERING

A limiting factor of many scaffolds used for tissue engineering is that they are electrically insulating and do not reflect the native conductivity of electrically active tissues such as cardiac muscle. However, there has been a rapid increase in the availability of biocompatible conductive polymers for potential use in 3D scaffolds. Electroactive scaffolds are of particular relevance for the engineering of electrically active tissues such as skeletal and cardiac muscle, as well as for neural tissues. In clinical applications where electrical stimulation is applied to influence muscle behavior *ex vivo* and *in vivo*, conventional electrodes are generally metallic. However, there are a wide range of conductive polymeric materials available that present alternative electrodes. The advantage of polymer-based electrodes is that they are more processable than their metal counterparts (Kotaki et al. 2006; Prabhakaran et al. 2011a; Jalili et al. 2013; Esrafilzadeh et al. 2013), can be "doped" with biomolecules for controlled release of drugs (Wadhwa et al. 2006; Boehler and Asplund 2014; Justin et al. 2012; Sirivisoot et al. 2011; Esrafilzadeh et al. 2013) and biological molecules (Thompson et al. 2006, 2011; Esrafilzadeh et al. 2013; Luo et al. 2011; Sirivisoot et al. 2011; Justin et al. 2012; Thompson et al. 2011), and can be surface modified with peptides and other biological molecules, such as neurotrophins, to confer desirable biological qualities (Lee et al. 2009; Molino et al. 2013; Gomez and Schmidt 2007; Nickels and Schmidt 2013). In addition, conducting polymers show material characteristics more favorable for cellular growth and tissue tolerance than metals; in particular, the modulus of conducting polymers more closely matched the modulus of tissues than hard metal electrodes. This is especially relevant to implantable electrodes where hard metals can cause tissue damage. A range of conducting polymers have become available over the past few decades, including organic conducting polymers, carbon nanotubes (CNTs), and more recently, graphene. All these polymers have found applications in tissue engineering for the delivery of electrical stimulation to cells in both 2D and in 3D.

15.5.1 ORGANIC CONDUCTING POLYMERS

Organic conducting polymers, such as polypyrrole, polyaniline (PANi), and conducting polymers from the thiophene family, including poly(3,4-ethylenedioxythiophene) (PEDOT), have been used for the growth and electrical stimulation of muscle cells, as well as other cell types. In general, these polymers show very good *in vivo* biocompatibility (Ferraz et al. 2012; X. Wang et al. 2004; Jiang et al. 2002; Z. Wang et al. 2004; Asplund et al. 2009; Qazi et al. 2014), although *in vivo* testing has been relatively limited to date, and excellent *in vitro* compatibility allowing for cellular adhesion, proliferation, and differentiation (Forcelli et al. 2012; Gilmore et al. 2009; Sudwilai et al. 2014). One of the advantages of using organic conducting polymers is the ability to dope them with biological molecules to influence tissue or cellular response, such as extracellular matrix molecules, cytokines, or peptides. Polypyrrole doped with biologically active molecules, such as hyaluronic acid and chondroitin sulfate, can be used to influence the proliferation and differentiation of primary skeletal muscle myoblasts (Gilmore et al. 2009). Also, polymer surfaces can be nanostructured through electrospinning and can be wet spun into fibers to produce orientated scaffolds to facilitate organized cell growth, in particular to align myofibers on anisotropic scaffolds (Jalili et al. 2013; Esrafilzadeh et al. 2013; Prabhakaran et al. 2011a,b; Breukers et al. 2010; Quigley et al. 2013). In addition, conducting polymers can be combined with hydrogels to generate highly conductive soft materials that can be used to generate high-resolution structures (Pan et al. 2012). "Fluffy" 3D polypyrrole structures can also be generated using electrospinning to facilitate the proliferation of cardiac myocyte cultures (Jin et al. 2012). Further to these studies, nanostructured polypyrrole substrates have been used to deliver electrical stimulation to primary muscle myoblasts, significantly affecting their proliferation as well as differentiation *in vitro* (Quigley et al. 2012). Similarly, cardiac myoblasts can be grown and electrically stimulated on polypyrrole surfaces, including small electrodes, to induce synchronized contractions (Nishizawa et al. 2007; You et al. 2011). The proliferation of vascular smooth muscle cells can also be inhibited by growth on polypyrrole doped with heparin; in addition, polypyrrole has been used as a substrate for the stimulation of smooth muscle cells resulting in increased proliferation of the cells (stimulated at 5 Hz frequency) and increased expression of contractile proteins (Rowlands and Cooper-White 2008; Stewart et al. 2012).

Other organic conducting polymers, such as PANi, have been used for the growth and electrical stimulation of myoblasts and cardiac myocytes (Bidez et al. 2006; Jun et al. 2009; You et al. 2011; Hsiao et al. 2013). The use of electrically active scaffolds that incorporate conducting polymers such as PANi has been shown to increase the electrical coupling of cardiac myocytes and to apply electrical stimulation to synchronize the beating of cardiac cultures (Hsiao et al. 2013). In addition, increased expression of the gap junction protein connexin 43 was seen on scaffolds containing PANi (You et al. 2011). PEDOT has also been used on microelectrode arrays to induce electrically stimulated contraction in differentiated C2C12 muscle cells (Nagamine et al. 2011).

Organic conducting polymers have been successfully blended with other materials to successfully create conductive hydrogel and polymer scaffolds that could potentially be used to create 3D tissue constructs for *in vivo* applications. Nanostructured PANi blends have been developed for the growth and differentiation of primary myoblasts and potentially for cardiomyocyte growth. Myoblasts grown on electrospun PANi-PCL (Figure 15.3) and PANi-poly(L-lactide-co-epsilon-caprolactone) (PLCL) blends show increased differentiation, dependent on nanofiber alignment and PANi content (Ku et al. 2012; Chen et al. 2013; Jun et al. 2009; Jeong et al. 2008). PEDOT has also been used to create PEDOT/PU hydrogels that are highly conductive and are able to support the adhesion and proliferation of C2C12 cells, as well as the growth

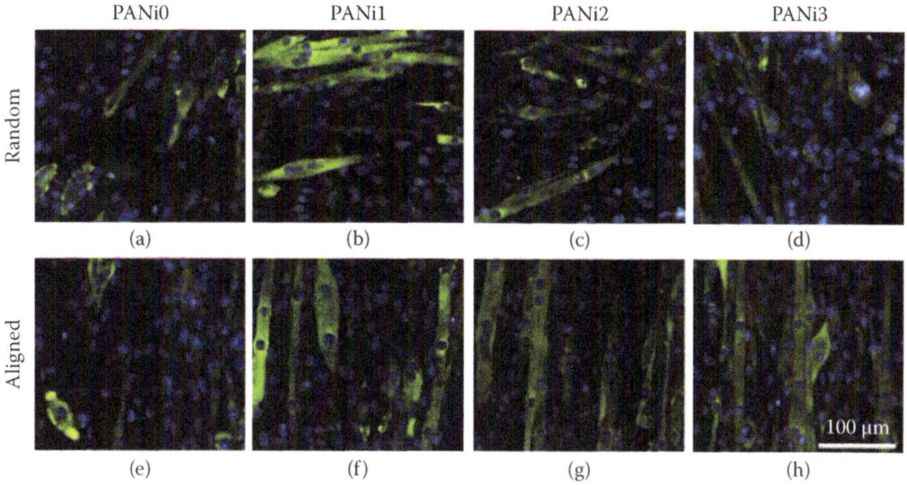

Figure 15.3 Myofibers cultured on PANi-PCL blend electrospun nanofibers. Randomly aligned fibers with increasing PANi loading (a–d) show random orientation of myofibers. Anisotopic fibers with increased PANi loading show aligned myofibers (e–h). Increased PANi loading in the polymer blend does not adversely affect myotube formation. (Adapted from Ku, S. H., et al., *Biomaterials*, 33 [26], 6098–6104, 2012.)

and differentiation of neural cells (Figure 15.4) (Sasaki et al. 2014). In addition, polythiophene (poly(3-thiopheneacetic acid)) has been used to create a conductive hydrogel that supports the adhesion and proliferation of primary myoblasts (Mawad et al. 2012). Blending conductive polymers with hydrogels or modifying conductive polymers to form single-component hydrogel-conducting polymer monomers (Mawad et al. 2012) facilitates the use of conducting polymers for the synthesis of 3D structures for tissue regeneration. The use of hydrogels allows tailoring of properties such as the cross-linking density to meet the specific physical properties of tissues, such as modulus.

Organic conducting polymers present a versatile conductive material that can be easily processed, modified with biological components, and blended or modified with other polymers to form conductive, bioactive scaffolds. The unique properties of this class of conducting polymer make them excellent candidates for use in tissue regeneration, as well as for use in biomedical devices designed to interface with tissues. More recently, CNTs and graphene polymers have also been developed for use in tissue engineering and present an alternative to organic conducting polymers.

15.5.2 CARBON NANOTUBES

CNTs have also been used in scaffolds to create an electroactive environment for the growth and stimulation of cells, including myoblasts and myofibers (Kam et al. 2009; Ahadian et al. 2014a; Nick et al. 2012; MacDonald et al. 2008; Huang et al. 2012; Cellot et al. 2009). CNTs have been used to create hybrid hydrogels for the growth and electrical stimulation of muscle cells (Ostrovidov et al. 2014; Ahadian et al. 2014a; Ramon-Azcon et al. 2013), as well as other cell types, including neural cells (de Asis et al. 2010; Mooney et al. 2012; Ostrovidov et al. 2014; He et al. 2011). Apart from providing an avenue for electrical stimulation (and sensing), an electroactive environment is a more

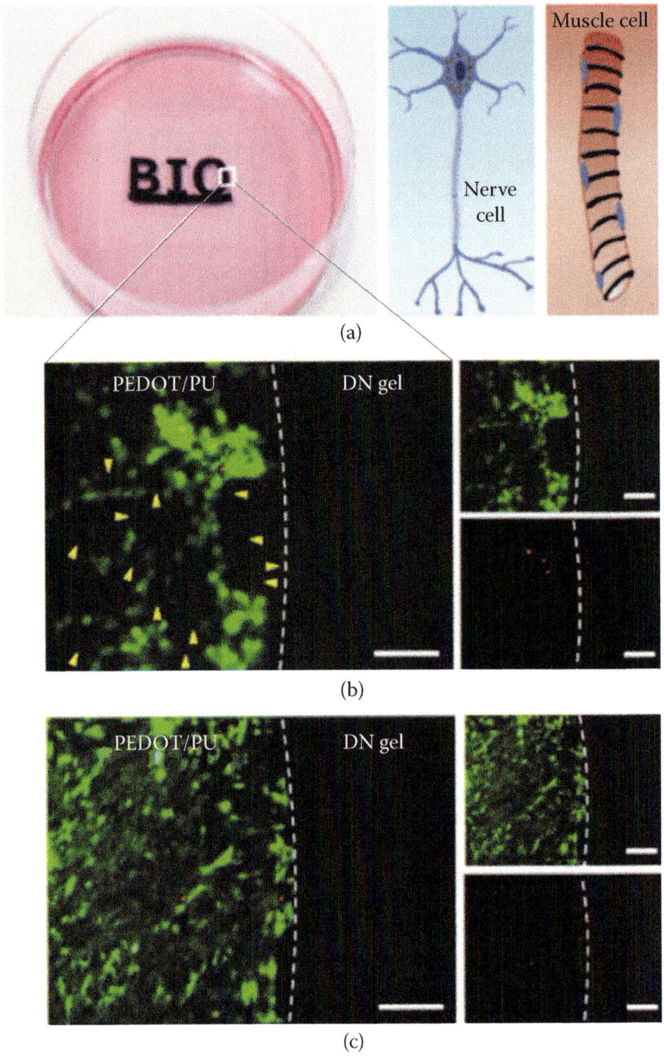

Figure 15.4 Printed electrically conductive PEDOT-PU hydrogels. (a) PEDOT-PU printed gel on a flexible plastic film. (b) Neural cells grown on the printed polymer. (c) Muscle (C2C12) cells grown on the printed polymer. Note the cells only grow on the PEDOT-PU polymer and not on a double network (DN) hydrogel. (Adapted from Sasaki, M., et al., *Adv. Healthc are Mater.*, 3 [11], 1919–1927, 2014.)

conducive environment for the growth of electrically active cells, particularly cardiomyocytes, as discussed in previous sections. Bioactive electroactive cardiac patches have been created using CNT incorporated photo-cross-linkable gelatin methacrylate (GelMA) hydrogels with neonatal rat cardiomyocytes seeded onto the hydrogel surfaces (Shin et al. 2013). The resulting cardiac constructs showed three times higher spontaneous synchronous beating rates and 85% lower excitation threshold than those cultured on GelMA hydrogels without CNT. This indicates that the electrically conductive and nanofibrous networks formed by CNTs are key characteristics leading to improved cardiac cell

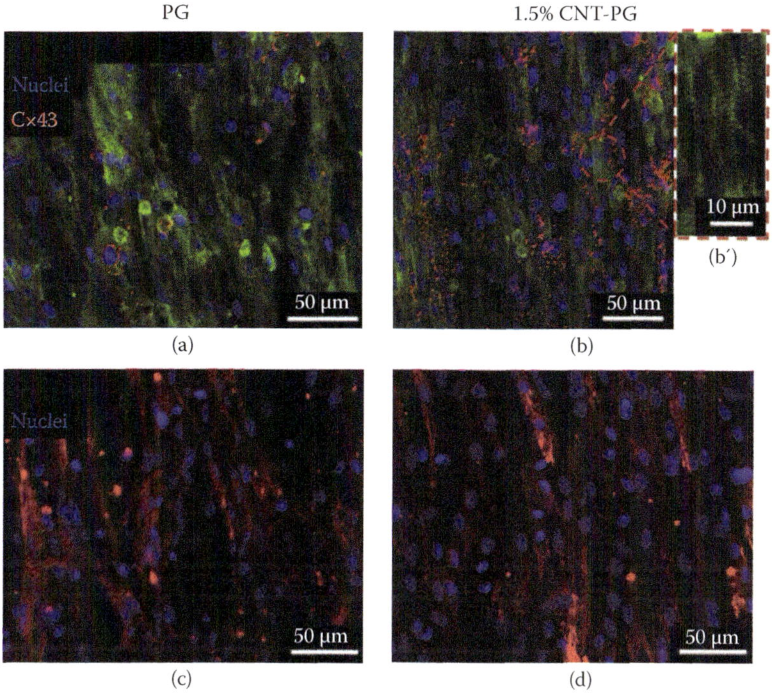

Figure 15.5 Cardiomyocytes grown on gelatin (a and c) and gelatin–CNT blended (b and d) scaffolds. (a and b) Sarcomeric alpha-actinin (green), Cx43 (red), and DAPI (blue). (c and d) Troponin I (red) and DAPI (blue) staining after 7 days of culture. Organized sarcomeres with higher Cx43 expression can be seen in myocytes grown on 1.5% CNT–gelatin (b and d) than on pure gelatin scaffolds (a and c). A high magnification image (b′) shows interconnected sarcomeric structures perpendicular to the direction of the nanofibers. (Adapted from Kharaziha, M., et al., *Biomaterials*, 35 [26], 7346–7354, 2014.)

adhesion, organization, and cell-to-cell coupling. CNTs have also been blended with gelatin for engineering cardiac constructs (Figure 15.5). Similar to the results found on CNT-GelMA hydrogels, CNT gelatin scaffolds showed superior mechanical properties and the presence of CNTs enhanced cardiomyocyte beating (Kharaziha et al. 2014).

CNTs have also been used to generate electroactive skeletal muscle scaffolds. These scaffolds have been used to enhance myoblast differentiation via electrical stimulation (Ostrovidov et al. 2014). Similar to studies by Shin, CNTs have been blended with GelMA to create electrically conductive hydrogels for skeletal muscle myoblast growth and stimulation. Ahadian and colleagues (2014b) further refined this blend of polymers to produce vertically aligned CNTs within the gel via dielectrophoresis. These hydrogels were used to successfully differentiate myoblasts and apply electrical stimulation to myoblast populations, which significantly increased the expression of a number of genes associated with myotube maturation, including alpha-actinin and myosin heavy-chain isoforms. Interestingly, levels of gene expression were highest in gels with aligned fibers versus randomly aligned fibers. CNTs have also been blended with polyurethane and have been used to deliver electrical stimulation to skeletal muscle myoblast cultures, significantly increasing the level of myoblast differentiation (Sirivisoot and Harrison 2011).

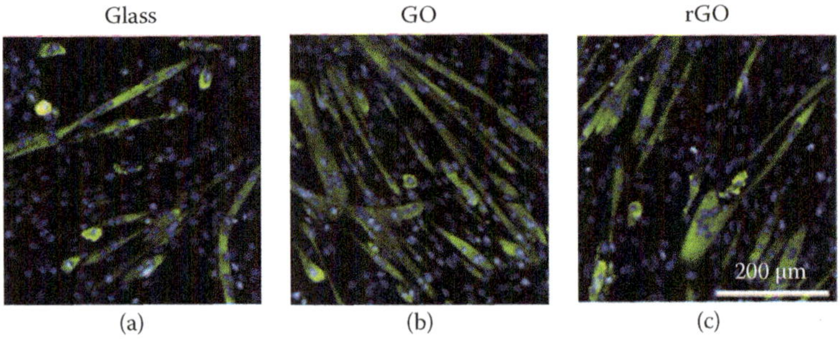

Figure 15.6 Differentiation of skeletal muscle myoblasts to myotubes on (a) glass, (b) graphene oxide (GO), and (c) reduced graphene oxide (rGO). Significantly higher levels of fusion and MHC-positive myotubes were seen on GO and rGO surfaces than on glass controls. (Adapted from Ku, S. H., and Park, C. B., *Biomaterials*, 34 [8], 2017–2023, 2013.)

The incorporation of CNTs into gelatin, and potentially other biomaterials, has great potential for the creation of multifunctional cardiac and skeletal muscle scaffolds for both therapeutic purposes and *in vitro* studies. These hybrid materials could also be used for neuron and other muscle cells (e.g., smooth muscle) to create tissue constructs with improved organization, electroactivity, and mechanical integrity. However, there have been concerns regarding the biosafety of CNTs. A number of studies have shown that the length, purity, and solubility of single- and multiwalled CNTs can have a significant bearing on their toxicity (Liu et al. 2012; Nayak et al. 2010; Haniu et al. 2014). Some studies have shown that by modification, nanotube toxicity can be reduced (X. Zhang et al. 2011; Li et al. 2010). As with any biomaterial, exhaustive testing of long-term biocompatibility and tolerance must be undertaken before use in humans can be considered; however, CNTs show great promise as a material for tissue engineering skeletal muscle and cardiac tissues.

15.5.3 GRAPHENE

Compared with other conducting polymers, graphene is a relatively new player in the biomaterial field. However, it is a versatile conducting polymer with very high conductivity, the ability to be modified, a large surface area, which makes it ideal for sensing applications, and low inflammatory response *in vivo* (Liu et al. 2015). However, in terms of toxicity and biocompatibility, the route of administration, dosage, structure, and chemical nature of graphene all can have a bearing on the biological response (for a recent review on graphene toxicity, see Sydlik et al. 2015). For example, the functionalization or reduction of graphene oxide can reduce toxicity in *in vitro* and *in vivo* studies (Sasidharan et al. 2011). Similar to organic polymers and CNTs, graphene can be processed into different topographical substrates (Segerstrom et al. 2011; Yu et al. 2014) and can be modified with peptides (Wang et al. 2014; Guo et al. 2012). Graphene has been used to create electrodes for electrical stimulation (Heo et al. 2011; Ahadian et al. 2014; Li et al. 2013), as well as for the sensing of biomarkers such as dopamine and myoglobin (Kumar and Prakash 2014; Kumar et al. 2015).

To date, the majority of publications concerning graphene as a tissue engineering platform have focused on bone and neural tissue; however, graphene has great promise as an electroscaffold for muscle tissues due to its inherent conductive properties.

Graphene and graphene oxide have been used as substrates or as composites with other materials, such as organic conducting polymers and hydrogels, for the growth and differentiation of myoblasts (Ku and Park 2013; Bajaj et al. 2014; Chaudhuri et al. 2015; Shin et al. 2015, Ahadian et al. 2014). These substrates have also been used for the electrical stimulation of cells, including myoblasts, to influence cell differentiation (Ahadian et al. 2014b). In these studies, electrical stimulation by graphene resulted in a significant increase in the expression of a number of genes associated with muscle differentiation, as well as corresponding increases in average myotube length and surface area (Ahadian et al. 2014b).

Graphene's excellent conduction properties make it an excellent candidate for the synthesis of sensitive electrodes designed to record the electrical activity of cells. This area has mainly focused on the design of electrodes for neural recordings, but some studies have investigated the use of this material for recording the activity of cardiac cells or myocytes (Cohen-Karni et al. 2010). Cardiac myocytes have also been grown successfully on graphene substrates (Kim et al. 2013). Graphene has also been used for the differentiation of mesenchymal (Park et al. 2014, 2015) and embryonic (Lee et al. 2014) stem cells to cardiac myocytes. Mesenchymal stem cells grown in the presence of reduced graphene oxide (rGO) flakes were shown to have elevated levels of vascular endothelial growth factor (VEGF), basic fibroblast growth factor (FGF), and hepatocyte growth factor (HGF), factors involved in the induction of angiogenesis (Bhang et al. 2009), and the same study found that rGO also led to increased expression of connexin 43, implying enhanced gap junction formation. Remarkably, the implantation of mesenchymal stem cells grown in culture with rGO flakes led to enhanced cardiac repair in a mouse model of cardiac infarction, resulting in significant improvements in cardiac function, including fractional shortening and increased ejection fractions. Its authors conclude that rGO flakes imparted their effect by their increased affinity for extracellular matrix molecules, as well as their electrical conductivity, which enhanced coupling between cells. At the time that this review was written, no reports of electrical stimulation of cardiac myocytes could be found. However, graphene oxide and its derivatives present an alternative to other conducting polymers for the electrical stimulation of cardiac and skeletal muscle cells.

15.6 CONCLUDING REMARKS

As conducting polymers and fabrication techniques become more sophisticated, scaffolds for tissue regeneration will be better able to suit the requirements of specific tissues, such as skeletal and cardiac muscle. The integration of electrically conductive polymers into bioscaffolds presents an exciting opportunity to integrate some previously overlooked properties of electrically active tissues into new generation scaffolds. This will enable the appropriate use of electrical stimulation and electrical fields to accelerate the regeneration process. In addition, the use of conducting polymers will also facilitate the integration of sensors into scaffolds to allow feedback on tissue performance. The use of these polymers to support cell growth and function heralds a new generation of bioactive, intelligent scaffolds that not only facilitate the physical tissue requirements of tissues, such as modulus, but also allow the electrical stimulation of tissue to regulate and stimulate tissue regeneration and activity.

REFERENCES

Ahadian, S., J. Ramon-Azcon, M. Estili, et al. 2014a. Hybrid hydrogels containing vertically aligned carbon nanotubes with anisotropic electrical conductivity for muscle myofiber fabrication. *Sci Rep* 4: 4271.

Ahadian, S., J. Ramon-Azcon, H. Chang, et al. 2014b. Electrically regulated differentiation of skeletal muscle cells on ultrathin graphene-based films. *RSC Adv* 4 (19): 9534–41.

Alvarez, O. M., P. M. Mertz, R. V. Smerbeck, and W. H. Eaglstein. 1983. The healing of superficial skin wounds is stimulated by external electrical current. *J Invest Dermatol* 81 (2): 144–8.

Ashley, Z., H. Sutherland, M. F. Russold, H. Lanmuller, W. Mayr, J. C. Jarvis, and S. Salmons. 2008. Therapeutic stimulation of denervated muscles: The influence of pattern. *Muscle Nerve* 38 (1): 875–86.

Asplund, M., E. Thaning, J. Lundberg, A. C. Sandberg-Nordqvist, B. Kostyszyn, O. Inganas, and H. von Holst. 2009. Toxicity evaluation of PEDOT/biomolecular composites intended for neural communication electrodes. *Biomed Mater* 4 (4): 045009.

Bajaj, P., J. A. Rivera, D. Marchwiany, V. Solovyeva, and R. Bashir. 2014. Graphene-based patterning and differentiation of C2C12 myoblasts. *Adv Healthcare Mater* 3 (7): 995–1000.

Banerjee, P. 2010. Electrical muscle stimulation for chronic heart failure: An alternative tool for exercise training? *Curr Heart Fail Rep* 7 (2): 52–8.

Becker, R. O., and D. G. Murray. 1970. The electrical control system regulating fracture healing in amphibians. *Clin Orthop Relat Res* 73: 169–98.

Bergh, A., H. Nordlof, and B. Essen-Gustavsson. 2010. Evaluation of neuromuscular electrical stimulation on fibre characteristics and oxidative capacity in equine skeletal muscles. *Equine Vet J Suppl* 42 (38): 671–5.

Bhang, S. H., S. W. Cho, J. M. Lim, et al. 2009. Locally delivered growth factor enhances the angiogenic efficacy of adipose-derived stromal cells transplanted to ischemic limbs. *Stem Cells* 27 (8): 1976–86.

Bidez, P. R., 3rd, S. Li, A. G. MacDiarmid, E. C. Venancio, Y. Wei, and P. I. Lelkes. 2006. Polyaniline, an electroactive polymer, supports adhesion and proliferation of cardiac myoblasts. *J Biomater Sci Polym Ed* 17 (1–2): 199–212.

Boehler, C., and M. Asplund. 2014. A detailed insight into drug delivery from PEDOT based on analytical methods: Effects and side effects. *J Biomed Mater Res A* 103 (3): 1200–7.

Bokkon, I., A. Till, F. Grass, and A. Erdofi Szabo. 2011. Phantom pain reduction by low-frequency and low-intensity electromagnetic fields. *Electromagn Biol Med* 30 (3): 115–27.

Boonyarom, O., N. Kozuka, K. Matsuyama, and S. Murakami. 2009. Effect of electrical stimulation to prevent muscle atrophy on morphologic and histologic properties of hindlimb suspended rat hindlimb muscles. *Am J Phys Med Rehabil* 88 (9): 719–26.

Breukers, R. D., K. J. Gilmore, M. Kita, K. K. Wagner, M. J. Higgins, S. E. Moulton, G. M. Clark, D. L. Officer, R. M. Kapsa, and G. G. Wallace. 2010. Creating conductive structures for cell growth: Growth and alignment of myogenic cell types on polythiophenes. *J Biomed Mater Res A* 95 (1): 256–68.

Cabric, M., H. J. Appell, and A. Resic. 1988. Fine structural changes in electrostimulated human skeletal muscle. Evidence for predominant effects on fast muscle fibres. *Eur J Appl Physiol Occup Physiol* 57 (1): 1–5.

Carraro, U., C. Catani, S. Belluco, M. Cantini, and L. Marchioro. 1986. Slow-like electrostimulation switches on slow myosin in denervated fast muscle. *Exp Neurol* 94 (3): 537–53.

Cellot, G., E. Cilia, S. Cipollone, et al. 2009. Carbon nanotubes might improve neuronal performance by favouring electrical shortcuts. *Nat Nanotechnol* 4 (2): 126–33.

Chaudhuri, B., D. Bhadra, L. Moroni, and K. Pramanik. 2015. Myoblast differentiation of human mesenchymal stem cells on graphene oxide and electrospun graphene oxide-polymer composite fibrous meshes: Importance of graphene oxide conductivity and dielectric constant on their biocompatibility. *Biofabrication* 7 (1): 015009.

Chen, M. C., Y. C. Sun, and Y. H. Chen. 2013. Electrically conductive nanofibers with highly oriented structures and their potential application in skeletal muscle tissue engineering. *Acta Biomater* 9 (3): 5562–72.

Cohen-Karni, T., Q. Qing, Q. Li, Y. Fang, and C. M. Lieber. 2010. Graphene and nanowire transistors for cellular interfaces and electrical recording. *Nano Lett* 10 (3): 1098–102.

Cotter, M. A., N. E. Cameron, J. A. Barry, and M. C. Pattullo. 1991. Chronic stimulation accelerates functional recovery of immobilized soleus muscles of the rabbit. *Exp Physiol* 76 (2): 201–12.

Crameri, R. M., A. R. Weston, S. Rutkowski, J. W. Middleton, G. M. Davis, and J. R. Sutton. 2000. Effects of electrical stimulation leg training during the acute phase of spinal cord injury: A pilot study. *Eur J Appl Physiol* 83 (4–5): 409–15.

Dal Corso, S., L. Napolis, C. Malaguti, et al. 2007. Skeletal muscle structure and function in response to electrical stimulation in moderately impaired COPD patients. *Respir Med* 101 (6): 1236–43.

de Asis, E. D., Jr., J. Leung, S. Wood, and C. V. Nguyen. 2010. Empirical study of unipolar and bipolar configurations using high resolution single multi-walled carbon nanotube electrodes for electrophysiological probing of electrically excitable cells. *Nanotechnology* 21 (12): 125101.

Dirks, M. L., D. Hansen, A. Van Assche, P. Dendale, and L. J. Van Loon. 2015. Neuromuscular electrical stimulation prevents muscle wasting in critically ill comatose patients. *Clin Sci (Lond)* 128 (6): 357–65.

Dow, D. E., R. G. Dennis, and J. A. Faulkner. 2005. Electrical stimulation attenuates denervation and age-related atrophy in extensor digitorum longus muscles of old rats. *J Gerontol A Biol Sci Med Sci* 60 (4): 416–24.

Elzinga, K., N. Tyreman, A. Ladak, B. Savaryn, J. Olson, and T. Gordon. 2015. Brief electrical stimulation improves nerve regeneration after delayed repair in Sprague Dawley rats. *Exp Neurol* 269: 142–53.

Esrafilzadeh, D., J. M. Razal, S. E. Moulton, E. M. Stewart, and G. G. Wallace. 2013. Multifunctional conducting fibres with electrically controlled release of ciprofloxacin. *J Control Release* 169 (3): 313–20.

Ferraz, N., M. Stromme, B. Fellstrom, S. Pradhan, L. Nyholm, and A. Mihranyan. 2012. In vitro and in vivo toxicity of rinsed and aged nanocellulose-polypyrrole composites. *J Biomed Mater Res A* 100 (8): 2128–38.

Flaibani, M., L. Boldrin, E. Cimetta, M. Piccoli, P. De Coppi, and N. Elvassore. 2009. Muscle differentiation and myotubes alignment is influenced by micropatterned surfaces and exogenous electrical stimulation. *Tissue Eng Part A* 15 (9): 2447–57.

Forcelli, P. A., C. T. Sweeney, A. D. Kammerich, B. C. Lee, L. H. Rubinson, Y. P. Kayinamura, K. Gale, and J. F. Rubinson. 2012. Histocompatibility and in vivo signal throughput for PEDOT, PEDOP, P3MT, and polycarbazole electrodes. *J Biomed Mater Res A* 100 (12): 3455–62.

Frohlich, F., and D. A. McCormick. 2010. Endogenous electric fields may guide neocortical network activity. *Neuron* 67 (1): 129–43.

Funk, R. H. 2015. Endogenous electric fields as guiding cue for cell migration. *Front Physiol* 6: 143.

Gilmore, K. J., M. Kita, Y. Han, A. Gelmi, M. J. Higgins, S. E. Moulton, G. M. Clark, R. Kapsa, and G. G. Wallace. 2009. Skeletal muscle cell proliferation and differentiation on polypyrrole substrates doped with extracellular matrix components. *Biomaterials* 30 (29): 5292–304.

Goldspink, G., A. Scutt, J. Martindale, T. Jaenicke, L. Turay, and G. F. Gerlach. 1991. Stretch and force generation induce rapid hypertrophy and myosin isoform gene switching in adult skeletal muscle. *Biochem Soc Trans* 19 (2): 368–73.

Gomez, N., and C. E. Schmidt. 2007. Nerve growth factor-immobilized polypyrrole: Bioactive electrically conducting polymer for enhanced neurite extension. *J Biomed Mater Res A* 81 (1): 135–49.

Gu, X., J. Fu, J. Bai, C. Zhang, J. Wang, and W. Pan. 2015. Low-frequency electrical stimulation induces the proliferation and differentiation of peripheral blood stem cells into Schwann cells. *Am J Med Sci* 349 (2): 157–61.

Guo, C. X., S. R. Ng, S. Y. Khoo, X. Zheng, P. Chen, and C. M. Li. 2012. RGD-peptide functionalized graphene biomimetic live-cell sensor for real-time detection of nitric oxide molecules. *ACS Nano* 6 (8): 6944–51.

Haniu, H., N. Saito, Y. Matsuda, et al. 2014. Biological responses according to the shape and size of carbon nanotubes in BEAS-2B and MESO-1 cells. *Int J Nanomed* 9: 1979–90.
Hartsell, H. D. 1986. Electrical muscle stimulation and isometric exercise effects on selected quadriceps parameters*. *J Orthop Sports Phys Ther* 8 (4): 203–9.
He, L., D. Lin, Y. Wang, Y. Xiao, and J. Che. 2011. Electroactive SWNT/PEGDA hybrid hydrogel coating for bio-electrode interface. *Colloids Surf B Biointerfaces* 87 (2): 273–9.
Heidland, A., G. Fazeli, A. Klassen, K. Sebekova, H. Hennemann, U. Bahner, and B. Di Iorio. 2013. Neuromuscular electrostimulation techniques: Historical aspects and current possibilities in treatment of pain and muscle waisting. *Clin Nephrol* 79 (Suppl. 1): S12–23.
Heo, C., J. Yoo, S. Lee, A. Jo, S. Jung, H. Yoo, Y. H. Lee, and M. Suh. 2011. The control of neural cell-to-cell interactions through non-contact electrical field stimulation using graphene electrodes. *Biomaterials* 32 (1): 19–27.
Hinkle, L., C. D. McCaig, and K. R. Robinson. 1981. The direction of growth of differentiating neurones and myoblasts from frog embryos in an applied electric field. *J Physiol* 314: 121–35.
Hirose, T., T. Shiozaki, K. Shimizu, T. Mouri, K. Noguchi, M. Ohnishi, and T. Shimazu. 2013. The effect of electrical muscle stimulation on the prevention of disuse muscle atrophy in patients with consciousness disturbance in the intensive care unit. *J Crit Care* 28 (4): 536.e1–7.
Holt, H. E. 1936. Galvani and the pre-Galvanian electrophysiologists. *Ann Sci* 1 (2): 157–172.
Hosseini, V., S. Ahadian, S. Ostrovidov, G. Camci-Unal, S. Chen, H. Kaji, M. Ramalingam, and A. Khademhosseini. 2012. Engineered contractile skeletal muscle tissue on a microgrooved methacrylated gelatin substrate. *Tissue Eng Part A* 18 (23–24): 2453–65.
Hotary, K. B., and K. R. Robinson. 1990. Endogenous electrical currents and the resultant voltage gradients in the chick embryo. *Dev Biol* 140 (1): 149–60.
Hotary, K. B., and K. R. Robinson. 1992. Evidence of a role for endogenous electrical fields in chick embryo development. *Development* 114 (4): 985–96.
Hotary, K. B., and K. R. Robinson. 1994. Endogenous electrical currents and voltage gradients in *Xenopus* embryos and the consequences of their disruption. *Dev Biol* 166 (2): 789–800.
Howes, G. J. 1984. The phylogenetic relationships of the electric catfish family Malapteruridae (Teleostei: Siluroidei). *J Nat Hist* 19: 37–67.
Hsiao, C. W., M. Y. Bai, Y. Chang, M. F. Chung, T. Y. Lee, C. T. Wu, B. Maiti, Z. X. Liao, R. K. Li, and H. W. Sung. 2013. Electrical coupling of isolated cardiomyocyte clusters grown on aligned conductive nanofibrous meshes for their synchronized beating. *Biomaterials* 34 (4): 1063–72.
Huang, Y. J., H. C. Wu, N. H. Tai, and T. W. Wang. 2012. Carbon nanotube rope with electrical stimulation promotes the differentiation and maturity of neural stem cells. *Small* 8 (18): 2869–77.
Hudlicka, O., M. D. Brown, S. Egginton, and J. M. Dawson. 1994. Effect of long-term electrical stimulation on vascular supply and fatigue in chronically ischemic muscles. *J Appl Physiol (1985)* 77 (3): 1317–24.
Ito, A., Y. Yamamoto, M. Sato, K. Ikeda, M. Yamamoto, H. Fujita, E. Nagamori, Y. Kawabe, and M. Kamihira. 2014. Induction of functional tissue-engineered skeletal muscle constructs by defined electrical stimulation. *Sci Rep* 4: 4781.
Jahanshahi, A., L. Schonfeld, M. L. Janssen, S. Hescham, E. Kocabicak, H. W. Steinbusch, J. J.van Overbeeke, and Y. Temel. 2013. Electrical stimulation of the motor cortex enhances progenitor cell migration in the adult rat brain. *Exp Brain Res* 231 (2): 165–77.
Jalili, R., J. M. Razal, and G. G. Wallace. 2013. Wet-spinning of PEDOT:PSS/functionalized-SWNTs composite: A facile route toward production of strong and highly conducting multifunctional fibers. *Sci Rep* 3: 3438.
Jeong, S. I., I. D. Jun, M. J. Choi, Y. C. Nho, Y. M. Lee, and H. Shin. 2008. Development of electroactive and elastic nanofibers that contain polyaniline and poly(L-lactide-co-epsilon-caprolactone) for the control of cell adhesion. *Macromol Biosci* 8 (7): 627–37.
Jiang, X., Y. Marois, A. Traore, D. Tessier, L. E. Dao, R. Guidoin, and Z. Zhang. 2002. Tissue reaction to polypyrrole-coated polyester fabrics: An in vivo study in rats. *Tissue Eng* 8 (4): 635–647.

Jin, L., T. Wang, Z. -Q. Feng, M. Zhu, M. K. Leach, Y. I. Naim, and Q. Jiang. 2012. Fabrication and characterization of a novel fluffy polypyrrole fibrous scaffold designed for 3D cell culture. *J Mater Chem* 22 (35): 18321–6.

Jun, I., S. Jeong, and H. Shin. 2009. The stimulation of myoblast differentiation by electrically conductive sub-micron fibers. *Biomaterials* 30 (11): 2038–47.

Justin, G. A., S. Zhu, T. R. Nicholson 3rd, J. Maskrod, J. Mbugua, M. Chase, J. H. Jung, and R. M. Mercado. 2012. On-demand controlled release of anti-inflammatory and analgesic drugs from conducting polymer films to aid in wound healing. *Conf Proc IEEE Eng Med Biol Soc* 2012: 1206–9.

Kakinuma, Y., Y. Zhang, M. Ando, T. Sugiura, and T. Sato. 2004. Effect of electrical modification of cardiomyocytes on transcriptional activity through 5′-AMP-activated protein kinase. *J Cardiovasc Pharmacol* 44 (Suppl. 1): S435–8.

Kam, N. W., E. Jan, and N. A. Kotov. 2009. Electrical stimulation of neural stem cells mediated by humanized carbon nanotube composite made with extracellular matrix protein. *Nano Lett* 9 (1): 273–8.

Kane, K., and A. Taub. 1975. A history of local analgesia. *Pain* 1: 125–38.

Karatzanos, E., V. Gerovasili, D. Zervakis, et al. 2012. Electrical muscle stimulation: An effective form of exercise and early mobilization to preserve muscle strength in critically ill patients. *Crit Care Res Pract* 2012: 432752.

Kawahara, Y., K. Yamaoka, M. Iwata, et al. 2006. Novel electrical stimulation sets the cultured myoblast contractile function to 'on'. *Pathobiology* 73 (6): 288–94.

Kellaway, P. 1946. The part played by the electric fish in the early history of bioelectricity and electrotherapy. *Bull Hist Med* 20: 112–37.

Kharaziha, M., S. R. Shin, M. Nikkhah, S. N. Topkaya, N. Masoumi, N. Annabi, M. R. Dokmeci, and A. Khademhosseini. 2014. Tough and flexible CNT-polymeric hybrid scaffolds for engineering cardiac constructs. *Biomaterials* 35 (26): 7346–54.

Kim, T., Y. H. Kahng, T. Lee, K. Lee, and H. Kim do. 2013. Graphene films show stable cell attachment and biocompatibility with electrogenic primary cardiac cells. *Mol Cells* 36 (6): 577–82.

Kotaki, M., X. M. Liu, and C. He. 2006. Optical properties of electrospun nanofibers of conducting polymer-based blends. *J Nanosci Nanotechnol* 6 (12): 3997–4000.

Ku, S. H., S. H. Lee, and C. B. Park. 2012. Synergic effects of nanofiber alignment and electroactivity on myoblast differentiation. *Biomaterials* 33 (26): 6098–104.

Ku, S. H., and C. B. Park. 2013. Myoblast differentiation on graphene oxide. *Biomaterials* 34 (8): 2017–23.

Kubis, H. P., R. J. Scheibe, J. D. Meissner, G. Hornung, and G. Gros. 2002. Fast-to-slow transformation and nuclear import/export kinetics of the transcription factor NFATc1 during electrostimulation of rabbit muscle cells in culture. *J Physiol* 541 (Pt 3): 835–47.

Kujala, K., A. Ahola, M. Pekkanen-Mattila, L. Ikonen, E. Kerkela, J. Hyttinen, and K. Aalto-Setala. 2012. Electrical field stimulation with a novel platform: Effect on cardiomyocyte gene expression but not on orientation. *Int J Biomed Sci* 8 (2): 109–20.

Kumar, A., and R. Prakash. 2014. Graphene sheets modified with polyindole for electro-chemical detection of dopamine. *J Nanosci Nanotechnol* 14 (3): 2501–6.

Kumar, V., M. Shorie, A. K. Ganguli, and P. Sabherwal. 2015. Graphene-CNT nanohybrid aptasensor for label free detection of cardiac biomarker myoglobin. *Biosens Bioelectron* 72: 56–60.

Langelaan, M. L., K. J. Boonen, K. Y. Rosaria-Chak, D. W. van der Schaft, M. J. Post, and F. P. Baaijens. 2011. Advanced maturation by electrical stimulation: Differences in response between C2C12 and primary muscle progenitor cells. *J Tissue Eng Regen Med* 5 (7): 529–39.

Lee, J. H., W. Y. Jeon, H. H. Kim, E. J. Lee, and H. W. Kim. 2015. Electrical stimulation by enzymatic biofuel cell to promote proliferation, migration and differentiation of muscle precursor cells. *Biomaterials* 53: 358–69.

Lee, J. Y., J. W. Lee, and C. E. Schmidt. 2009. Neuroactive conducting scaffolds: Nerve growth factor conjugation on active ester-functionalized polypyrrole. *J R Soc Interface* 6 (38): 801–10.

Lee, T. J., S. Park, S. H. Bhang, J. K. Yoon, I. Jo, G. J. Jeong, B. H. Hong, and B. S. Kim. 2014. Graphene enhances the cardiomyogenic differentiation of human embryonic stem cells. *Biochem Biophys Res Commun* 452 (1): 174–80.

Leon-Salas, W. D., H. Rizk, C. Mo, N. Weisleder, L. Brotto, E. Abreu, and M. Brotto. 2013. A dual mode pulsed electro-magnetic cell stimulator produces acceleration of myogenic differentiation. *Recent Pat Biotechnol* 7 (1): 71–81.

Lewis, D. M., W. S. al-Amood, and H. Schmalbruch. 1997. Effects of long-term phasic electrical stimulation on denervated soleus muscle: Guinea-pig contrasted with rat. *J Muscle Res Cell Motil* 18 (5): 573–86.

Li, J., F. Yang, G. Guo, D. Yang, J. Long, D. Fu, J. Lu, and C. Wang. 2010. Preparation of biocompatible multi-walled carbon nanotubes as potential tracers for sentinel lymph nodes. *Polym Int* 59 (2): 169–174.

Li, L., W. Gu, J. Du, et al. 2012. Electric fields guide migration of epidermal stem cells and promote skin wound healing. *Wound Repair Regen* 20 (6): 840–51.

Li, N., Q. Zhang, S. Gao, Q. Song, R. Huang, L. Wang, L. Liu, J. Dai, M. Tang, and G. Cheng. 2013. Three-dimensional graphene foam as a biocompatible and conductive scaffold for neural stem cells. *J Sci Rep* 3: 1604.

Li, Y., P.-S. Wang, G. Lucas, R. Li, and L. Yao. 2015. ARP2/3 complex is required for directional migration of neural stem cell-derived oligodendrocyte precursors in electric fields. *Stem Cell Res Ther* 6: 41.

Li, Y., M. Weiss, and L. Yao. 2014. Directed migration of embryonic stem cell-derived neural cells in an applied electric field. *Stem Cell Rev* 10 (5): 653–62.

Lidwell, M. C. 1929. *Cardiac Disease in Relation to Anaesthesia, Transactions of the Third Session.* Sydney: Australian Medical Congress (British Medical Association).

Liu, D., L. Wang, Z. Wang, and A. Cuschieri. 2012. Different cellular response mechanisms contribute to the length-dependent cytotoxicity of multi-walled carbon nanotubes. *Nanoscale Res Lett* 7 (1): 361.

Liu, J.-H., T. Wang, H. Wang, Y. Gu, Y. Xu, H. Tang, G. Jia, and Y. Liu. 2015. Biocompatibility of graphene oxide intravenously administrated in mice-effects of dose, size and exposure protocols. *Toxicol Res* 4 (1): 83–91.

Luo, X., C. Matranga, S. Tan, N. Alba, and X. T. Cui. 2011. Carbon nanotube nanoreservior for controlled release of anti-inflammatory dexamethasone. *Biomaterials* 32 (26): 6316–23.

MacDonald, R. A., C. M. Voge, M. Kariolis, and J. P. Stegemann. 2008. Carbon nanotubes increase the electrical conductivity of fibroblast-seeded collagen hydrogels. *Acta Biomater* 4 (6): 1583–92.

Martherus, R. S., S. J. Vanherle, E. D. Timmer, V. A. Zeijlemaker, J. L. Broers, H. J. Smeets, J. P. Geraedts, and T. A. Ayoubi. 2010. Electrical signals affect the cardiomyocyte transcriptome independently of contraction. *Physiol Genomics* 42A (4): 283–9.

Mathieu-Costello, O., P. J. Agey, L. Wu, J. Hang, and T. H. Adair. 1996. Capillary-to-fiber surface ratio in rat fast-twitch hindlimb muscles after chronic electrical stimulation. *J Appl Physiol (1985)* 80 (3): 904–9.

Mawad, D., E. Stewart, D. L. Officer, T. Romeo, P. Wagner, K. Wagner, and G. G. Wallace. 2012. A single component conducting polymer hydrogel as a scaffold for tissue engineering. *Adv Funct Mater* 22 (13): 2692–9.

McDonough, P. M., and C. C. Glembotski. 1992. Induction of atrial natriuretic factor and myosin light chain-2 gene expression in cultured ventricular myocytes by electrical stimulation of contraction. *J Biol Chem* 267 (17): 11665–8.

McWilliam, J. A. 1889. Electrical stimulation of the heart in man. *Br Med J* 1 (1468): 348–50.

Minetto, M. A., A. Botter, O. Bottinelli, D. Miotti, R. Bottinelli, and G. D'Antona. 2013. Variability in muscle adaptation to electrical stimulation. *Int J Sports Med* 34 (6): 544–53.

Molino, P. J., B. Zhang, G. G. Wallace, and T. W. Hanks. 2013. Surface modification of polypyrrole/biopolymer composites for controlled protein and cellular adhesion. *Biofouling* 29 (10): 1155–67.

Mond, H. G., J. G. Sloman, and H. E. Edwards.1982. The first pacemaker. *Pacing Clin Electrophysiol* 5: 278–82.

Mooney, E., J. N. Mackle, D. J. Blond, E. O'Cearbhaill, G. Shaw, W. J. Blau, F. P. Barry, V. Barron, and J. M. Murphy. 2012. The electrical stimulation of carbon nanotubes to provide a cardiomimetic cue to MSCs. *Biomaterials* 33 (26): 6132–9.

Nagamine, K., T. Kawashima, S. Sekine, Y. Ido, M. Kanzaki, and M. Nishizawa. 2011. Spatiotemporally controlled contraction of micropatterned skeletal muscle cells on a hydrogel sheet. *Lab Chip* 11 (3): 513–7.

Nayak, T. R., P. C. Leow, P.-L. R. Ee, T. Arockiadoss, S. Ramaprabhu, and G. Pastorin. 2010. Crucial parameters responsible for carbon nanotubes toxicity. *Curr Nanosci* 6 (2): 141–54.

Nick, C., R. Joshi, J. J. Schneider, and C. Thielemann. 2012. Three-dimensional carbon nanotube electrodes for extracellular recording of cardiac myocytes. *Biointerphases* 7 (1–4): 58.

Nickels, J. D., and C. E. Schmidt. 2013. Surface modification of the conducting polymer, polypyrrole, via affinity peptide. *J Biomed Mater Res A* 101 (5): 1464–71.

Nishimura, K. Y., R. R. Isseroff, and R. Nuccitelli. 1996. Human keratinocytes migrate to the negative pole in direct current electric fields comparable to those measured in mammalian wounds. *J Cell Sci* 109 (Pt 1): 199–207.

Nishizawa, M., H. Nozaki, H. Kaji, T. Kitazume, N. Kobayashi, T. Ishibashi, and T. Abe. 2007. Electrodeposition of anchored polypyrrole film on microelectrodes and stimulation of cultured cardiac myocytes. *Biomaterials* 28 (8): 1480–5.

Nuccitelli, R. 2003. A role for endogenous electric fields in wound healing. *Curr Top Dev Biol* 58: 1–26.

O'Reilly, C., D. Pette, and K. Ohlendieck. 2003. Increased expression of the nicotinic acetylcholine receptor in stimulated muscle. *Biochem Biophys Res Commun* 300 (2): 585–91.

Omura, Y. 1987. Basic electrical parameters for safe and effective electro-therapeutics [electro-acupuncture, TES, TENMS (or TEMS), TENS and electro-magnetic field stimulation with or without drug field] for pain, neuromuscular skeletal problems, and circulatory disturbances. *Acupunct Electrother Res* 12 (3–4): 201–25.

Ono, T., K. Maekawa, W. Sonoyama, S. Kojima, T. Tanaka, G. T. Clark, and T. Kuboki. 2010. Gene expression profile of mouse masseter muscle after repetitive electrical stimulation. *J Prosthodont Res* 54 (1): 36–41.

Ostrovidov, S., X. Shi, L. Zhang, et al. 2014. Myotube formation on gelatin nanofibers—Multi-walled carbon nanotubes hybrid scaffolds. *Biomaterials* 35 (24): 6268–77.

Pan, L., G. Yu, D. Zhai, et al. 2012. Hierarchical nanostructured conducting polymer hydrogel with high electrochemical activity. *Proc Natl Acad Sci U S A* 109 (24): 9287–92.

Parent, A. 2004. Giovanni Aldini: From animal electricity to human brain stimulation. *Can J Neurol Sci* 31 (4): 576–84.

Park, H., R. Bhalla, R. Saigal, M. Radisic, N. Watson, R. Langer, and G. Vunjak-Novakovic. 2008. Effects of electrical stimulation in C2C12 muscle constructs. *J Tissue Eng Regen Med* 2 (5): 279–87.

Park, J., Y. S. Kim, S. Ryu, W. S. Kang, S. Park, J. Han, H. C. Jeong, B. H. Hong, Y. Ahn, and B.-S. Kim. 2015. Graphene potentiates the myocardial repair efficacy of mesenchymal stem cells by stimulating the expression of angiogenic growth factors and gap junction protein. *Adv Funct Mater* 25 (17): 2590–600.

Park, J., S. Park, S. Ryu, et al. 2014. Graphene-regulated cardiomyogenic differentiation process of mesenchymal stem cells by enhancing the expression of extracellular matrix proteins and cell signaling molecules. *Adv Healthcare Mater* 3 (2): 176–81.

Petrie, M., M. Suneja, and R. K. Shields. 2015. Low-frequency stimulation regulates metabolic gene expression in paralyzed muscle. *J Appl Physiol (1985)* 118 (6): 723–31.

Prabhakaran, M. P., L. Ghasemi-Mobarakeh, G. Jin, and S. Ramakrishna. 2011a. Electrospun conducting polymer nanofibers and electrical stimulation of nerve stem cells. *J Biosci Bioeng* 112 (5): 501–7.

Prabhakaran, M. P., L. Ghasemi-Mobarakeh, and S. Ramakrishna. 2011b. Electrospun composite nanofibers for tissue regeneration. *J Nanosci Nanotechnol* 11 (4): 3039–57.

Putman, C. T., W. T. Dixon, J. A. Pearcey, I. M. Maclean, M. J. Jendral, M. Kiricsi, G. K. Murdoch, and D. Pette. 2004. Chronic low-frequency stimulation upregulates uncoupling protein-3 in transforming rat fast-twitch skeletal muscle. *Am J Physiol Regul Integr Comp Physiol* 287 (6): R1419–26.

Putman, C. T., K. R. Sultan, T. Wassmer, J. A. Bamford, D. Skorjanc, and D. Pette. 2001. Fiber-type transitions and satellite cell activation in low-frequency-stimulated muscles of young and aging rats. *J Gerontol A Biol Sci Med Sci* 56 (12): B510–19.

Qazi, T. H., R. Rai, and A. R. Boccaccini. 2014. Tissue engineering of electrically responsive tissues using polyaniline based polymers: A review. *Biomaterials* 35 (33): 9068–86.

Quigley, A. F., K. Wagner, M. Kita, et al. 2013. In vitro growth and differentiation of primary myoblasts on thiophene based conducting polymers. *Biomater Sci* 1 (9): 983–95.

Quigley, A. F., J. M. Razal, M. Kita, et al. 2012. Electrical stimulation of myoblast proliferation and differentiation on aligned nanostructured conductive polymer platforms. *Adv Healthcare Mater* 1 (6): 801–8.

Quittan, M., A. Sochor, G. F. Wiesinger, J. Kollmitzer, B. Sturm, R. Pacher, and W. Mayr. 1999. Strength improvement of knee extensor muscles in patients with chronic heart failure by neuromuscular electrical stimulation. *Artif Organs* 23 (5): 432–5.

Radisic, M., W. Deen, R. Langer, and G. Vunjak-Novakovic. 2005. Mathematical model of oxygen distribution in engineered cardiac tissue with parallel channel array perfused with culture medium containing oxygen carriers. *Am J Physiol Heart Circ Physiol* 288 (3): H1278–89.

Radisic, M., H. Park, H. Shing, T. Consi, F. J. Schoen, R. Langer, L. E. Freed, and G. Vunjak-Novakovic. 2004. Functional assembly of engineered myocardium by electrical stimulation of cardiac myocytes cultured on scaffolds. *Proc Natl Acad Sci U S A* 101 (52): 18129–34.

Ramon-Azcon, J., S. Ahadian, M. Estili, et al. 2013. Dielectrophoretically aligned carbon nanotubes to control electrical and mechanical properties of hydrogels to fabricate contractile muscle myofibers. *Adv Mater* 25 (29): 4028–34.

Reiser, P. J., W. O. Kline, and P. L. Vaghy. 1997. Induction of neuronal type nitric oxide synthase in skeletal muscle by chronic electrical stimulation in vivo. *J Appl Physiol (1985)* 82 (4): 1250–5.

Rochester, L., M. J. Barron, C. S. Chandler, R. A. Sutton, S. Miller, and M. A. Johnson. 1995. Influence of electrical stimulation of the tibialis anterior muscle in paraplegic subjects. 2. Morphological and histochemical properties. *Paraplegia* 33 (9): 514–22.

Romanoff, A. L. 1944. Bio-electric potentials and vital activities of the egg. *Biodynamica* 4: 329–58.

Routsi, C., V. Gerovasili, I. Vasileiadis, E. Karatzanos, T. Pitsolis, E. Tripodaki, V. Markaki, D. Zervakis, and S. Nanas. 2010. Electrical muscle stimulation prevents critical illness polyneuromyopathy: A randomized parallel intervention trial. *Crit Care* 14 (2): R74.

Rowlands, A. S., and J. J. Cooper-White. 2008. Directing phenotype of vascular smooth muscle cells using electrically stimulated conducting polymer. *Biomaterials* 29 (34): 4510–20.

Sasaki, M., B. C. Karikkineth, K. Nagamine, H. Kaji, K. Torimitsu, and M. Nishizawa. 2014. Highly conductive stretchable and biocompatible electrode-hydrogel hybrids for advanced tissue engineering. *Adv Healthcare Mater* 3 (11): 1919–27.

Sasidharan, A., L. S. Panchakarla, P. Chandran, D. Menon, S. Nair, C. N. Rao, and M. Koyakutty. 2011. Differential nano-bio interactions and toxicity effects of pristine versus functionalized graphene. *Nanoscale* 3 (6): 2461–4.

Sebastian, A., F. Syed, D. Perry, V. Balamurugan, J. Colthurst, I. H. Chaudhry, and A. Bayat. 2011. Acceleration of cutaneous healing by electrical stimulation: Degenerate electrical waveform down-regulates inflammation, up-regulates angiogenesis and advances remodeling in temporal punch biopsies in a human volunteer study. *Wound Repair Regen* 19 (6): 693–708.

Segerstrom, S., G. Sandborgh-Englund, and E. I. Ruyter. 2011. Biological and physicochemical properties of carbon-graphite fibre-reinforced polymers intended for implant suprastructures. *Eur J Oral Sci* 119 (3): 246–52.

Serena, E., E. Figallo, N. Tandon, C. Cannizzaro, S. Gerecht, N. Elvassore, and G. Vunjak-Novakovic. 2009. Electrical stimulation of human embryonic stem cells: Cardiac differentiation and the generation of reactive oxygen species. *Exp Cell Res* 315 (20): 3611–19.

Serena, E., M. Flaibani, S. Carnio, L. Boldrin, L. Vitiello, P. De Coppi, and N. Elvassore. 2008. Electrophysiologic stimulation improves myogenic potential of muscle precursor cells grown in a 3D collagen scaffold. *Neurol Res* 30 (2): 207–14.

Shi, R. Y., and R. B. Borgens. 1995. 3-Dimensional gradients of voltage during development of the nervous-system as invisible coordinates for the establishment of embryonic pattern. *Developmental Dynamics* 202 (2): 101–14.

Shin, S. R., S. M. Jung, M. Zalabany, et al. 2013. Carbon-nanotube-embedded hydrogel sheets for engineering cardiac constructs and bioactuators. *ACS Nano* 7 (3): 2369–80.

Shin, Y. C., J. H. Lee, L. H. Jin, M. J. Kim, Y. J. Kim, J. K. Hyun, T. G. Jung, S. W. Hong, and D. W. Han. 2015. Stimulated myoblast differentiation on graphene oxide-impregnated PLGA-collagen hybrid fibre matrices. *J Nanobiotechnol* 13: 21.

Siepe, M., M. N. Giraud, E. Liljensten, U. Nydegger, P. Menasche, T. Carrel, and H. T. Tevaearai. 2007. Construction of skeletal myoblast-based polyurethane scaffolds for myocardial repair. *Artif Organs* 31 (6): 425–33.

Siepe, M., M. N. Giraud, M. Pavlovic, C. Receputo, F. Beyersdorf, P. Menasche, T. Carrel, and H. T. Tevaearai. 2006. Myoblast-seeded biodegradable scaffolds to prevent post-myocardial infarction evolution toward heart failure. *J Thorac Cardiovasc Surg* 132 (1): 124–31.

Sirivisoot, S., and B. S. Harrison. 2011. Skeletal myotube formation enhanced by electrospun polyurethane carbon nanotube scaffolds. *Int J Nanomedicine* 6: 2483–97.

Sirivisoot, S., R. Pareta, and T. J. Webster. 2011. Electrically controlled drug release from nanostructured polypyrrole coated on titanium. *Nanotechnology* 22 (8): 085101.

Sleigh, C. 1998. Life, death and galvanism. *Stud Hist Philos Sci Biol Biomed Sci* 29 (2): 219–48.

Stewart, E. M., X. Liu, G. M. Clark, R. M. Kapsa, and G. G. Wallace. 2012. Inhibition of smooth muscle cell adhesion and proliferation on heparin-doped polypyrrole. *Acta Biomater* 8 (1): 194–200.

Sudwilai, T., J. J. Ng, C. Boonkrai, N. Israsena, S. Chuangchote, and P. Supaphol. 2014. Polypyrrole-coated electrospun poly(lactic acid) fibrous scaffold: Effects of coating on electrical conductivity and neural cell growth. *J Biomater Sci Polym Ed* 25 (12): 1240–52.

Sydlik, S. A., S. Jhunjhunwala, M. J. Webber, D. G. Anderson, and R. Langer. 2015. In vivo compatibility of graphene oxide with differing oxidation states. *ACS Nano* 9 (4): 3866–74.

Thompson, B. C., S. E. Moulton, J. Ding, R. Richardson, A. Cameron, S. O'Leary, G. G. Wallace, and G. M. Clark. 2006. Optimising the incorporation and release of a neurotrophic factor using conducting polypyrrole. *J Control Release* 116 (3): 285–94.

Thompson, B. C., S. E. Moulton, R. T. Richardson, and G. G. Wallace. 2011. Effect of the dopant anion in polypyrrole on nerve growth and release of a neurotrophic protein. *Biomaterials* 32 (15): 3822–31.

Treskes, P., K. Neef, S. Perumal Srinivasan, et al. 2015. Preconditioning of skeletal myoblast-based engineered tissue constructs enables functional coupling to myocardium in vivo. *J Thorac Cardiovasc Surg* 149 (1): 348–56.

Trimble, M. H., and R. M. Enoka. 1991. Mechanisms underlying the training effects associated with neuromuscular electrical stimulation. *Phys Ther* 71 (4): 273–80; discussion 280–2.

Trivedi, D. P., K. J. Hallock, and P. R. Bergethon. 2013. Electric fields caused by blood flow modulate vascular endothelial electrophysiology and nitric oxide production. *Bioelectromagnetics* 34 (1): 22–30.

van der Schaft, D. W., A. C.van Spreeuwel, K. J. Boonen, M. L. Langelaan, C. V. Bouten, and F. P. Baaijens. 2013. Engineering skeletal muscle tissues from murine myoblast progenitor cells and application of electrical stimulation. *J Vis Exp* (73): e4267.

Vivodtzev, I., R. Debigare, P. Gagnon, V. Mainguy, D. Saey, A. Dube, M. E. Pare, M. Belanger, and F. Maltais. 2012. Functional and muscular effects of neuromuscular electrical stimulation in patients with severe COPD: A randomized clinical trial. *Chest* 141 (3): 716–25.

Wadhwa, R., C. F. Lagenaur, and X. T. Cui. 2006. Electrochemically controlled release of dexamethasone from conducting polymer polypyrrole coated electrode. *J Control Release* 110 (3): 531–41.
Walker, W. C. 1937. Animal electricity before Galvani. *Ann Sci* 2 (1): 84–113.
Wan, Q., S. S. Yeung, K. K. Cheung, S. W. Au, W. W. Lam, Y. H. Li, Z. Q. Dai, and E. W. Yeung. 2016. Optimizing electrical stimulation for promoting satellite cell proliferation in muscle disuse atrophy. *Am J Phys Med Rehabil* 95 (1): 28–38.
Wang, C., B. Chen, M. Zou, and G. Cheng. 2014. Cyclic RGD-modified chitosan/graphene oxide polymers for drug delivery and cellular imaging. *Colloids Surf B Biointerfaces* 122: 332–40.
Wang, X., X. Gu, C. Yuan, S. Chen, P. Zhang, T. Zhang, J. Yao, F. Chen, and G. Chen. 2004. Evaluation of biocompatibility of polypyrrole in vitro and in vivo. *J Biomed Mater Res A* 68 (3): 411–22.
Wang, Z., C. Roberge, L. H. Dao, Y. Wan, G. Shi, M. Rouabhia, R. Guidoin, and Z. Zhang. 2004. In vivo evaluation of a novel electrically conductive polypyrrole/poly(D,L-lactide) composite and polypyrrole-coated poly(D,L-lactide-co-glycolide) membranes. *J Biomed Mater Res A* 70 (1): 28–38.
Williams, R. S., M. Garcia-Moll, J. Mellor, S. Salmons, and W. Harlan. 1987. Adaptation of skeletal muscle to increased contractile activity. Expression nuclear genes encoding mitochondrial proteins. *J Biol Chem* 262 (6): 2764–7.
Windisch, A., K. Gundersen, M. J. Szabolcs, H. Gruber, and T. Lomo. 1998. Fast to slow transformation of denervated and electrically stimulated rat muscle. *J Physiol* 510 (Pt 2): 623–32.
Wu, C. H. 1984. Electric fish and the discovery of animal electricity: The mystery of the electric fish motivated research into electricity and was instrumental in the emergence of electrophysiology. *Am Sci* 72 (6): 598–607.
Wu, Y., L. Collier, W. Qin, G. Creasey, W. A. Bauman, J. Jarvis, and C. Cardozo. 2013. Electrical stimulation modulates Wnt signaling and regulates genes for the motor endplate and calcium binding in muscle of rats with spinal cord transection. *BMC Neurosci* 14: 81.
Xia, Y., L. M. Buja, R. C. Scarpulla, and J. B. McMillin. 1997. Electrical stimulation of neonatal cardiomyocytes results in the sequential activation of nuclear genes governing mitochondrial proliferation and differentiation. *Proc Natl Acad Sci USA* 94 (21): 11399–404.
Xia, Y., J. B. McMillin, A. Lewis, M. Moore, W. G. Zhu, R. S. Williams, and R. E. Kellems. 2000. Electrical stimulation of neonatal cardiac myocytes activates the NFAT3 and GATA4 pathways and up-regulates the adenylosuccinate synthetase 1 gene. *J Biol Chem* 275 (3): 1855–63.
Xing, H., M. Zhou, P. Assinck, and N. Liu. 2015. Electrical stimulation influences satellite cell differentiation after sciatic nerve crush injury in rats. *Muscle Nerve* 51 (3): 400–11.
Yamada, M., K. Tanemura, S. Okada, et al. 2007. Electrical stimulation modulates fate determination of differentiating embryonic stem cells. *Stem Cells* 25 (3): 562–570.
You, J.-O., M. Rafat, G. J. C. Ye, and D. T. Auguste. 2011. Nanoengineering the heart: Conductive scaffolds enhance connexin 43 expression. *Nano Lett* 11 (9): 3643–8.
Yu, D., K. Goh, H. Wang, L. Wei, W. Jiang, Q. Zhang, L. Dai, and Y. Chen. 2014. Scalable synthesis of hierarchically structured carbon nanotube-graphene fibres for capacitive energy storage. *Nat Nanotechnol* 9 (7): 555–62.
Zealear, D. L., R. Mainthia, Y. Li, I. Kunibe, A. Katada, C. Billante, and K. Nomura. 2014. Stimulation of denervated muscle promotes selective reinnervation, prevents synkinesis, and restores function. *Laryngoscope* 124 (5): E180–7.
Zemianek, J. M., S. Lee, and T. B. Shea. 2013. Acceleration of myofiber formation in culture by a digitized synaptic signal. *Tissue Eng Part A* 19 (23–24): 2693–702.
Zhang, J., M. Calafiore, Q. Zeng, X. Zhang, Y. Huang, R. A. Li, W. Deng, and M. Zhao. 2011. Electrically guiding migration of human induced pluripotent stem cells. *Stem Cell Rev* 7 (4): 987–96.

Zhang, X., Y. Zhu, J. Li, Z. Zhu, J. Li, W. Li, and Q. Huang. 2011. Tuning the cellular uptake and cytotoxicity of carbon nanotubes by surface hydroxylation. *J Nanopart Res* 13 (12): 6941–52.
Zhao, M., A. Agius-Fernandez, J. V. Forrester, and C. D. McCaig. 1996. Directed migration of corneal epithelial sheets in physiological electric fields. *Invest Ophthalmol Vis Sci* 37 (13): 2548–58.
Zhao, Y. 2009. Investigating electrical field-affected skeletal myogenesis using a microfabricated electrode array. *Sens Actuators A Phys* 154 (2): 281–7.

16

The response of endothelial cells to endogenous bioelectric fields

Peter R. Bergethon
Boston University School of Medicine
Boston, Massachusetts
and
Pfizer Neuroscience and Pain Research Unit, Pfizer, Inc.,
Cambridge, Massachusetts

Contents

16.1 INTRODUCTION

The endothelial cell is a mesenchymally derived cell that lines all vascular and lymphatic vessels in the organism. Endothelial cells are a highly adaptive cell capable of forming blood vessels throughout the life of the organism from embryogenesis through adult response to stress and injury. Their properties enable a facile responsiveness to the local environment throughout life and include the ability to

- Proliferate and migrate
- Change shape, arrangement, and orientation
- Form new tubular structures able to carry blood in response to local tissue conditions
- Self-assemble cellular elements into vascular tissues (muscular arteries, arterioles, capillaries, sinusoids, venules, and veins) appropriate to local need
- Form a command–control system for the metering of the blood supply and the elements transported to and from tissues by the blood

Endothelial cells are involved in three fundamental processes:

1. Control of blood flow to tissues
2. Control of leukocytes and material transfer between blood and tissue
3. Vasculogenesis and angiogenesis

Using this broad schema, this introduction first describes what is understood about the properties of the endothelial cells and the endothelium without recourse to consideration of the possible role of bioelectric forces and signals. Following this introduction, each of these domains is revisited for the observations contributed by the work in bioelectric research.

16.1.1 FUNCTIONAL ANATOMY, NORMAL FUNCTION, AND PHYSIOLOGY OF THE VASCULATURE

In its simplest form, every blood vessel comprises at least an endothelial cell and an associated basal limiting membrane. As noted above, the endothelial cell is derived from the mesenchymal layer of the embryo and more precisely from the lateral plate mesoderm (Gilbert 2000). The initial embryonic process of forming blood vessels is called *vasculogenesis*, with the remodeling and subsequent extension of vessels from existing ones, *angiogenesis*. In the human embryo, vasculogenesis occurs on approximately day 18 (Larsen 1998). Vasculogenesis occurs after mesodermal cells from the lateral plate mesoderm are exposed to high concentrations of bone morphogenic protein (BMP) and FGF2 (basic fibroblast growth factor), which cause differentiation of the mesoderm cells into hemangioblasts (Gilbert 2000; Larsen 1998; Gerecht-Nir et al. 2004). The hemangioblast is the precursor of both the pluripotential hematopoietic stem cells, from which all blood cells will be derived, and the angioblast that is the precursor cell of the endothelial cells. Upon formation, the hemangioblasts cluster together into "blood islands," with those cells in the center destined to become hematopoietic stem cells and those in the outer layer the angioblasts. When the blood islands are exposed to other growth factors, including vascular endothelial growth factor (VEGF), the angioblasts are induced to proliferate and form endothelial tubes that form the basis of the primary capillary network. A second growth factor, angiopoietin-1 (Ang1), enables interaction between the endothelial cells and the pericytes, cells of macrophage lineage that have smooth muscle cell (SMC)–like behaviors and associate and surround the endothelial cells. Both the endothelial cells and pericytes are then surrounded by the basal membrane.

As the networks grow, they merge to eventually connect with the developing heart and vitelline veins that deliver blood and nutrients to the embryonic heart.

The primary capillary networks that develop via vasculogenesis in the embryo are modified, extended, and pruned by angiogenesis so that the functional requirements for each tissue in terms of energy, material and oxygen delivery, and waste removal are provided. In the normal mature state, no cell will be more than 50–100 microns away from a blood vessel (Alberts et al. 2002). The physical processes by which nutrient and waste exchange occurs in the vascular system require an intricate balance between large-diameter vessels under high pressure, smaller low-pressure capillary networks, and the control of blood flow as the demand of various tissues varies on a minute-to-minute basis. The mature organism develops a rich array of vessels, all of which are modifications of the basic vessel developed during vasculogenesis. With the exception of the capillary networks, virtually all blood vessels, including arteries and veins, end up comprising three layers:

1. The tunica intima, which is the endothelial cell layer surrounded by its basal lamina. In the small capillaries, there may be pericytes included in an inconsistent pattern.
2. The tunica media is formed by SMCs that are recruited by the endothelial cells to surround the intima. These SMCs produce an extracellular matrix (ECM) composed of glycosaminoglycans and the connective proteins elastin and collagen, which provide elastic compliance and tensile strength to the vessel. The SMCs both provide the structural integrity needed by any vessel larger than a capillary and impart the capacity of the vessel to constrict and dilate in response to endothelial stimuli such as nitric oxide (NO), and also to neurogenic stimuli and hormonal stimuli such as epinephrine.
3. Finally, the outer layer of the vessel is a connective tissue layer, the tunica externa.

An important unifying principle throughout the vascular system is that it is the endothelial cell that largely determines and controls the structure of a given vessel—recruiting and influencing the pericytes or SMCs into varying proliferative or synthetic states that determine the overall structure and function of the vessel segment.

In the mature vessel, the endothelium plays a central role in the sensing of physical, chemical, and biological stimuli from the blood-facing intravascular side and transforming these stimuli into control signals that influence the acute and chronic behavior of the perivascular connective tissues and vascular smooth muscle cells (vSMCs). The traditional view of the normal biology of the vascular system involves a complex interaction between endothelial, vSMCs and other more transient cells, the ECM, blood, and mechanical forces (Schwartz et al. 1991). Recently, electrical forces associated with blood flow have been implicated in this normal biology, and this is a focus of this chapter. Endothelial cells sense a variety of chemical factors carried in the bloodstream and locally (e.g., hormones and platelet-derived factors, estradiol, VEGF, cytokines, oxidized LDL, ADP, bradykinin, acetylcholine, and histamine) and respond to these chemical factors to increase or decrease vSMC tone by inducing their depolarization or hyperpolarization with subsequent vascular constriction or relaxation (Shaul 2002). Although the primary endothelial cell–derived relaxing factor in vascular biology is NO, other endothelial cell hyperpolarizing factors (ECHFs) are likely, including possible electrogenic coupling between endothelial cells and SMCs in some vascular tissues (Cohen and Vanhoutte 1995).

Endothelial cells are the primary producers of NO, a well-documented vasodilator. NO is synthesized by endothelial nitric oxide synthase (eNOS), from the substrate L-arginine.

eNOS has been shown to associate with the plasma membrane of endothelial cells. There, it interacts with a number of receptor families, modulating the production of NO via calcium- and phosphorylation-dependent pathways. After the stimulation of a G protein–coupled receptor (e.g., bradykinin or acetylcholine), inositol trisphosphate (IP_3) is generated and binds to the IP_3 receptor on the endoplasmic reticulum. The intracellular calcium level increases, due to the emptying of the endoplasmic reticulum and the concomitant flux of extracellular calcium into the cell. This elevated intracellular calcium then binds calmodulin ($[Ca^{2+}]_i$–calmodulin), activating an electron transport chain that facilitates the above eNOS reaction, which converts L-arginine into NO and L-citrullene (Becker et al. 2006).

Endothelial cells are responsive to the mechanical forces of blood flow in a process called mechanotransduction. Blood flow causes mechanical stress (shearing, tensile, and compressive) on the blood vessel components, including the endothelial and SMCs and the matrix, with many aspects of normative and pathobiology responding to these mechanical forces (Li et al. 2005). Shear stress is the tangential mechanical force imparted on a surface by fluid flow. In the context of vascular physiology, shear stress has also been clearly shown to hyperpolarize endothelial cells and induce the production of NO. Shear stress influences the levels of important endothelium-derived vasoactive signaling mediators: the vasodilator NO and the vasocontrictor endothelin-17. Experimentally, static, pulsatile, reciprocal, and turbulent flow–generated shearing forces have been shown to variably affect vascular endothelial cell (VEC) proliferation, apoptosis, migration, endothelial permeability, and alignment (Li et al. 2005; Malek et al. 1993, 1996). Like most stimuli that act at the endothelial cell surface, mechanotransduction acts to release endothelium-dependent vasodilators by increasing intracellular Ca^{2+}. This increase occurs from either Ca^{2+} influx or the IP_3-mediated release of Ca^{2+} from intracellular stores. The rise in intracellular Ca^{2+} activates eNOS, producing NO and activating Ca^{2+}-dependent K^+ channels, leading to hyperpolarization of the VEC. More Ca^{2+} flows into the cell, thus sustaining production of these vasoactive substances (Adams et al. 1989).

The existence of the process of mechanotransduction has spurred interest in finding candidate mechanosensors capable of performing this transfer of information from the global environment of blood flow to the endothelial cells' internal machinery.

Two flow-sensitive ion channels have been elucidated by the work of Gautam et al. (2006), Barakat and colleagues (1999, 2003, 2010), and Hoger et al. (2002). The inward-rectifying KIR 2.1 potassium channel has been shown to respond to fluid flow. K^+ channels have been shown to be equally sensitive to oscillatory flow, with physiological frequencies of 0.5 and 1 Hz, as they are to steady flow. The other flow-sensitive channel is an outward-rectifying chloride channel. Outward-rectifying Cl^- channels are equally sensitive to nonreversing pulsatile flow as they are to steady flow; they are insensitive to oscillatory flow with frequencies higher than 5 Hz.

The most immediate consequence of ion channel activation by flow is a change in membrane potential. For steady-state flow, the membrane potential initially hyperpolarizes and then is reversed by depolarization within 35–160 s of flow onset. The hyperpolarization is due to the activation of the KIR 2.1 channel, whereas the depolarization is due to the activation of the Cl^- channel. Channel-blocking studies performed on these two channels lead to several conclusions. First, activation of K^+ and Cl^- channels is sufficient to account for the membrane potential changes induced by steady flow on endothelial cells. Second, both K^+ and Cl^- channels are activated rapidly upon flow initiation, and the two channels are activated independently of one another. Finally, the membrane hyperpolarization precedes depolarization, which suggests that K^+ channels are activated more rapidly than Cl^- channels. This is in spite of the fact that both channels are activated simultaneously and that the electrochemical driving force is larger for Cl^- than for K^+. Therefore, it can be assumed that both channels are activated immediately, but the K^+ current attains its peak faster than the Cl^- current.

Endothelial hyperpolarization has been shown to increase NO synthesis by endothelial cells, by influencing the Ca^{2+} flux across the membrane. The relationship between membrane potential and Ca^{2+} flux was developed by Nilius (1991), who showed that the endothelial membrane hyperpolarization induces an inward current for Ca^{2+}, thus increasing the bioavailability of Ca^{+2} to form the $[Ca^{2+}]_i$–calmodulin complex needed to activate the NO production by endothelial cell.

16.1.2 CONTROL OF TRANSPORT ACROSS THE ENDOTHELIAL BOUNDARY

The organization of endothelial cells as a physical network is essential to the barrier function of the vascular system. An extensive and well-reviewed literature (Yuan and Rigor 2010) has been developed exploring both the normal physiology and the pathobiology of the highly dynamic response of the endothelial structure as the control system that regulates solvent, solute, and cells across the endothelial system. This system includes endothelial cytoskeleton, cell–cell junction complexes (gap, tight, and adherins junctions), and cell attachments to the ECM and basement membrane (via integrins). Transport across the endothelial cell barriers is controlled by flow between the cells (paracellular or intercellular transport) and across the cells themselves (transcellular pathways). Each vascular assembly across various organs differs, with the highest degree of barrier function characterized by the blood–brain barrier. Here, endothelial cells, pericytes, and astrocytes form a highly sophisticated system of interlocking cells and processes that control transport to an exquisite degree. Barrier function in the area of endothelial function is studied in bioelectric manipulations to date, but may be of much greater importance as future study recognizes the important role played in these interfacial regions by bioelectrochemical forces.

16.1.3 ANGIOGENESIS IS THE PROCESS THAT LEADS TO THE CREATION OF NEW AND REPAIR OF DAMAGED BLOOD VESSELS

In normal growth and development, the vascular system must extend itself to keep up with the increasing size of the growing animal. This is an evolutionary solution to the process of increasing scale of the organism. This requires the capacity for cells to proliferate, migrate, and self-organize into the hollow cylinder that is the organized and functional geometry of a blood vessel. When damaged, blood vessels require the capacity to restore the blood supply by remodeling and repair of the vasculature to a stable geometry. This remodeling requires integration of a fourth process, the coordinated breakdown of the existing basal lamina and ECM to allow endothelial extension beyond the parent vasculature. The overall repair process is dynamic, requiring a precise control to maintain the complex balance between the dynamics of topological growth and functional delivery of blood supply to active tissues. A great deal of research over the last decades has provided a reasonably detailed picture of many aspects of angiogenesis (Folkmann 2007; Folkmann and Klagsbrun 1987; Alberts et al. 2002).

During nonstressed adult life, the endothelium is relatively stable, responding within normative homeostatic boundaries to local tissue demands for blood supply by controlling the blood flow via dilation or constriction of the blood vessel, as already discussed. Under these conditions, the dynamics of the endothelium is local and stable. This is matched by a substantive longevity of each cell. The turnover of undisrupted endothelial cells is very slow, with cellular lifetimes in rodent models on the order of months in liver and lung to years in brain and muscle (Alberts et al. 2002). However, the endothelial cell is capable of rapid return to a proliferative and migratory state when injury or programmed tissue damage (such as endometrial replacement during menstruation) occurs. If the endothelial lining is destroyed in a large vessel resulting in a denuded and exposed medial layer, the adjoining endothelial cells will immediately begin migration into the damaged region and shift to a proliferative state with replication times of 24–28 h. Following migration and proliferation, these new endothelial cells will also respond to the dynamic forces associated with the resolving disruption by orienting and changing shape in order to restore the overall topological structure of the original vessel. In the case where the endothelial cell repair occurs in a region of laminar blood flow, the cell will respond to the mechanical shearing forces of the local blood flow by elongating from a cobblestone geometry to become spindle shaped. The cells will also orient to the shearing field with the long axis of the spindle-shaped cell parallel to the gradient of the field (Malek et al. 1993, 1996).

Angiogenesis is the response by the endothelial cell to a demand by neighboring tissue to establish new blood flow patterns that will support the metabolic demands of that tissue. The fundamental signal that correlates these metabolic demands is the synthesis and secretion of VEGF by the tissue inviting endothelial cell invasion. The translation of the VEGF gene into mRNA and transcription of the mRNA into the protein is a process initiated by local tissue hypoxia, thus clearly relating the metabolic demand state of the tissue. The oxygen tension within tissues is sensed by the hypoxia-inducible factor (HIF) system. When intracellular levels of HIF-1 rise, which occurs normally only during hypoxia, HIF-1 associates with gene promoters or the HIF response element (HRE) region in the nucleus to upregulate a variety of genes, enabling short-term and long-term adaptation to hypoxia, including VEGF, erythropoietin, and glycolytic enzymes that produce ATP independent of oxidative phosphorylation. The VEGF produced diffuses through the stressed tissues and activates the endothelial cells nearby. The response of the endothelial cell to VEGF is fourfold:

1. The responding cell secretes proteases that digest the local basal lamina of the parent vessel in which it is currently found.
2. The endothelial cell migrates down the gradient of the VEGF toward the signal source.
3. The cells switch to a proliferative state and begin to replicate.
4. The endothelial cells elongate and develop extended internal vacuoles that eventually link intercellularly to form tubes that become connected endothelial lined vessels that extend from the parent vessel into the region being invaded.

Once angiogenesis connects the parent blood supply to the tissue under hypoxic stress, blood, oxygen, and other nutrients are brought to the stressed tissue, and with oxygen present, HIF-1 is now rapidly degraded and the genes promoted by the HREs are shut down. The gradient of VEGF falls toward zero and the endothelial cells return to a nonangiogenic state.

This sketch of the systemic properties of endothelium can now be adorned with observations and experiments that extend the understanding of endothelium to include bioelectric forces.

16.2 ELECTRIC FIELD EFFECTS ON THE VASCULAR SYSTEM

There is an extensive literature that explores the impact of electromagnetic forces (EMFs) on biological systems, and this corpus can generally be separated into two groups:

1. Response of cells to EMFs consistent with those endogenously present and of physiological properties
2. Response of cells to imposed EMFs with the possibility of manipulation of the system by supraphysiological doses of EMF

In the mature vascular system, endothelial cells are exposed to two conditions that generate endogenous local electrical fields:

1. The first of these local fields is that associated with injury to tissue, creating endogenous fields in the range of several hundreds of volts per meter. These "wound" potentials and the response of cells to them have been extensively reviewed elsewhere in this volume and will be only summarized here as they relate to the "injury" responses of endothelial cells as described above: proliferation, migration, orientation, and angiogenesis.
2. The second of these local fields to which vascular tissues are exposed are the electric fields generated by the flow of blood through the vessel itself. These fields, the electrokinetic vascular streaming potential (EVSP), are of much smaller magnitude than the wound potentials, that is, on the order of several volts per meter. While there is a small direct electrical current component, they vary most by the frequency of the pulsatile nature of the beating heart responsible for generating the flow of blood. These fields occur concurrently with the shearing forces that influence endothelial cells through mechanotransduction. In addition, the vSMCs that are recruited by the endothelial cells to form the integrated blood vessel are also exposed to the EVSP and have been shown to be influenced by these local electric fields. The largest part of our discussion focuses on this relatively new area of bioelectric research and its implications.

While the focus of this chapter is on the physiological fields to which endothelial cells are exposed, there is some work that has been done looking especially at the susceptibility of the barrier function of endothelial cells to nonphysiological environmental electromagnetic fields. We will note some of this work in the interest of completeness.

16.2.1 ELECTROKINETIC VASCULAR STREAMING POTENTIAL AS A SOURCE OF VASCULAR BIOELECTRICITY

Blood is a complex aqueous mixture that contains an electrolyte solution, colloidal particles (e.g., proteins and lipoproteins), and formed cellular elements. Each of these solutes carries charge though the largest concentration of charged particles comprising most of the ionic strength of the mixture are the ionic electrolytes (Bergethon 2010a). In the bulk electrolyte solution, the charge is evenly distributed so that electroneutrality is maintained. The endothelial wall on the inside of a blood vessel is a complex nonhomogenous surface of net negative charge in constant contact with the blood. A charge-containing solution that comes in contact with a charged surface will be instantly reorganized to achieve local electroneutrality near the surface (Bergethon 2010b). Regions of preferred charge accumulation occur in distinct layers as a result of the competing forces of electric field interactions and thermal randomizing forces acting between the solution and the surface. An immobile layer of charge forms at the surface (largely Na^+ ions in the vasculature); this layer is known as the "inner Helmholtz layer." In general, this immobile layer of charge does not fully provide local electroneutrality, and a second layer of charge forms extending from the inner Helmholtz plane into the bulk solution until electroneutrality is achieved. This second layer can be described by several mathematical and physical models but is generally appreciated to be more loosely held, and therefore is a "diffuse" layer of charge. It is diffuse because a balance exists between the force of electrical attraction of ions and the outer boundary of the inner Helmholtz layer and the randomizing effects of thermal or Brownian motion. This diffuse layer is properly called the "outer Helmholtz layer," although some workers refer to it as the "Gouy-Chapman layer." This entire region is known in physical electrochemistry as the "interphase." The boundary between the inner Helmholtz layer and the outer Helmholtz layer (called the outer Helmholtz plane [OHP]) is a point at which an electric potential exists, but one that can be overcome by other forces applied tangentially to the OHP. Therefore, the electric potential at this boundary is defined as the zeta potential (ζ). The zeta potential as a phenomenological coefficient captures structural information about the surface and the electrical interactions with the components secured in the inner Helmholtz layer. If a mechanical force applied to the bulk solvent causes the fluid to flow, the charge in the diffuse layer is torn away from the attractive electrical potential (quantified as ζ) and a "streaming current" is created. The charge separation resulting from this streaming current creates a countervoltage attempting to restore the loss of electroneutrality. This counter-EMF is termed the "streaming potential." The streaming current and potential are one of the four electrokinetic or electromechanical phenomena seen when charged electrolytes are induced to move past surfaces or colloidal particles: the others are electro-osmosis, electrophoresis, and the Dorn effect (for details, see Bergethon 2010b).

Blood flow induced by the pumping of the heart therefore produces an electrical force that acts within and across the blood vessel. Although the phenomenon of the streaming potential has been known since Quincke described it in 1859, its potential biological role in the vascular system was demonstrated only recently (Bergethon 1991; Sawyer et al. 1966). Sawyer et al. (1966) investigated the role of electrokinetic processes as a factor in hemostasis and measured the streaming potential in live animals and established the magnitude of the EVSP between 0.7 and 3 V m^{-1}. A simple model for field calculation was developed previously (Bergethon 1991) based on the classical derivation of the EVSP as described by the Helmholtz–Smoluchowski equation with a correction for pulsatile flow, $f(Y_a)$: $E_s = [\zeta\varepsilon\varepsilon_0 P/\eta\kappa]\,(f(Y_a))$ where E_s is the streaming potential (volts), ζ is the zeta potential (volts) in the vessel, ε is the dielectric constant of the electrolyte, ε_0 is the permittivity of free space, P is the effective systolic pressure in N m^{-2}, η is the electrolyte viscosity in kg m^{-1} s^{-1}, and κ is conductivity in S m^{-1}. The modeled field, E_s, is time varying with a grossly sinusoidal form.

In vivo, the EVSP is produced by mechanical events that also generate shear and normal tensile mechanical forces in the blood vessel; therefore, an essential feature that must be considered in all discussions of either vascular mechanotransduction or the EVSP is the reality that these two forces coexist and that endothelial cells especially are subject to their mutual influence. Recent experiments demonstrate that these two forces can be separated and studied in isolation (Trivedi and Bergethon 2012). Typical shear experiments can be performed in a flow chamber in which monolayers of cultured bovine aortic endothelial cells (BAECs) are exposed to standard physiological shearing forces (0.35–2.0 N m^{-2}). The streaming potential can be measured across the flow apparatus via a high-impedance voltmeter, and then using the principle of superposition, electrical current can be injected across the flow chamber to neutralize the generated streaming potential (Figure 16.1). Thus, three conditions can be experimentally created and differentiated:

1. *Fluid flow* refers to an environment where both shear stress and the streaming potential are present.
2. *Pure shear stress* refers to an environment where mechanical shear is present, but the streaming potential was neutralized.
3. *Streaming potential* refers to an environment in which a modeled streaming potential is applied with no mechanical shearing force present.

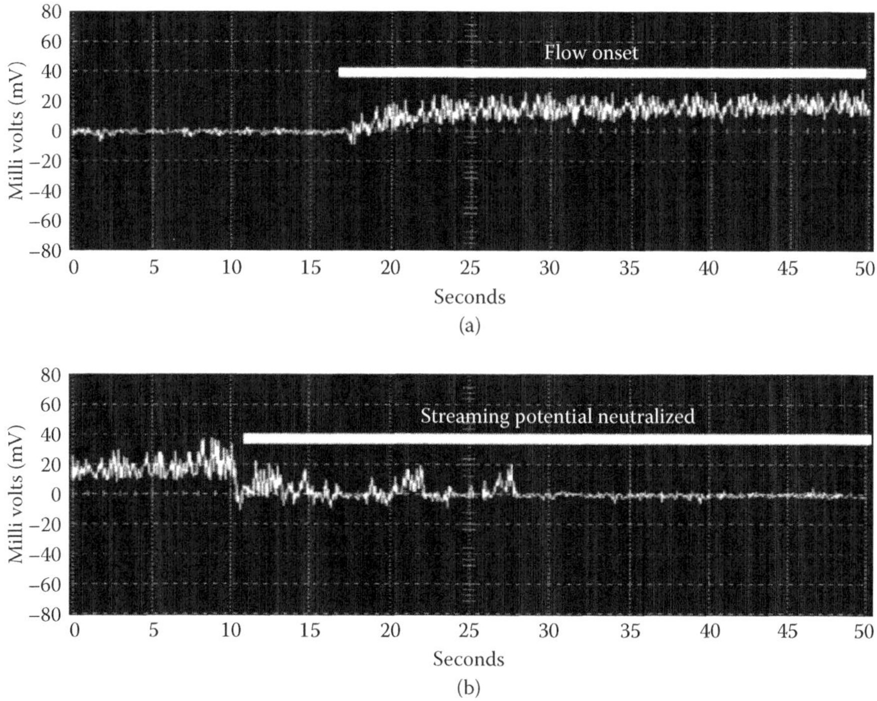

Figure 16.1 Measurement and neutralization of the streaming potential in the flow chamber. (a) The onset of flow coincides with the emergence of an approximately 20 mV streaming potential. (b) Superposition of a counter-EMF across the flow chamber; it neutralizes the observed streaming potential. (From Trivedi, D. P., et al., *Bioelectromagnetics*, 34, 22–30, 2013. With permission.)

It is worth emphasizing that in the literature, all previously published mechanotransduction studies have been performed solely under the fluid flow environment. These experimental conditions have for the first time allowed a pure shear stress and streaming potential environment to be differentiated and studied independently.

16.2.2 EFFECT OF FLOW ON MEMBRANE POTENTIAL

It is well established that BAECs respond to the onset of fluid flow by hyperpolarizing in the first 100 s, and then depolarizing to a membrane potential slightly higher than baseline (Voets et al. 1996; Lieu et al. 2004). Further, this characteristic electrophysiological response to flow onset can be wholly attributed to the activation of two membrane ion channels: an inward-rectifying K^+ channel, KIR 2.1, causing the initial hyperpolarization and an outward-rectifying chloride channel causing the late depolarization back toward baseline (Figure 16.2a). Blockade of KIR 2.1 by Cs^{2+} produces only depolarization in response to flow onset because only the chloride channel is available following activation. Conversely, blockade of the chloride channel by 5-nitro-2-(3-phenylpropylamino)-benzoic acid (NPPB) leaves only the potassium channel response to flow onset, resulting in a sustained hyperpolarization.

These observations provided an ideal test to test the hypothesis that the mechanical shear force and EVSP could be shown to be acting independently. Combining the ability to create fluid flow, pure shear, and streaming potential states with a pharmacological approach in which specific channel-blocking agents were added to the bathing solution while monitoring the transmembrane potential under shear conditions provided an experimental model in which double dissociation of the two polarizing currents could be established. Figure 16.2b shows data generated under fluid flow conditions at shear stress levels of 2.0 N m^{-2}. Figure 16.2c shows the same fluid flow conditions with pharmacological blockade of the chloride channel by NPPB, revealing the hyperpolarizing currents generated when only the KIR is activated by the conditions of fluid flow. Figure 16.2c shows shearing force activation of only the chloride channel under fluid flow conditions. In Figure 16.2c, the KIR channel is blocked with Cs^{2+} and no hyperpolarization is observed. These results replicate previously described transmembrane potential effects of flow onset in the presence of the channel-blocking agents. Thus, BAEC membrane potential responds to flow onset with two independent current flows: a rapidly responding potassium channel that hyperpolarizes the membrane, and a slow-responding chloride channel that gradually depolarizes the membrane back toward baseline (Trivedi and Bergethon 2012; Trivedi 2013).

When a pure shear stress environment is created by neutralization of the measured streaming potential, a prolonged hyperpolarization without the subsequent depolarization back to baseline is the result (Figure 16.3a). The hyperpolarization varies proportionally to the degree of shear force to which the BAECs are exposed, consistent with the interpretation that the KIR channel is modulated by the magnitude of the shearing force and is the channel responding to the mechanical shearing force. Figure 16.3a confirms that the hyperpolarization seen under pure shear stress conditions is due to the KIR channel since 140 mM Cs^{2+} blockade under these conditions abrogates the hyperpolarizing wave. The absence of the depolarizing current generated by the slow-acting chloride channel under these conditions provides support that the chloride channel is the EVSP-sensitive component. This is largely confirmed when under streaming potential conditions the depolarization of the BAECs (seen in fluid flow and KIR blockade [Figure 16.2b]) can be almost completely abrogated by NPPB treatment (Figure 16.3b).

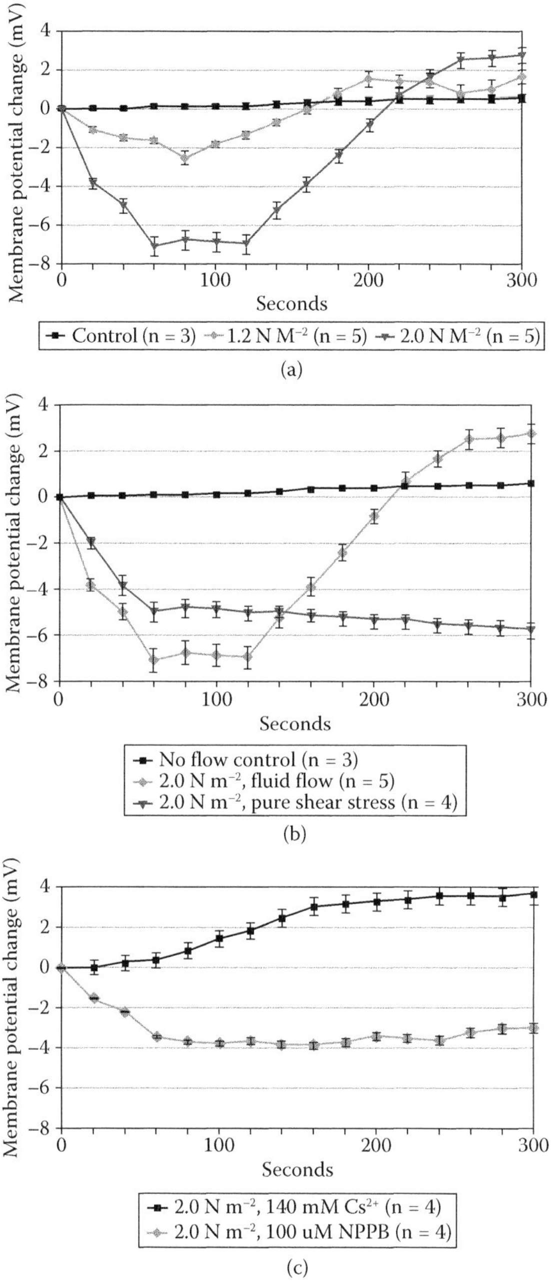

Figure 16.2 (a) BAEC membrane potential change seen at varying shear force conditions in standard fluid flow conditions as reported in the literature. (b) Comparison of BAEC membrane potential changes seen in no-flow, standard fluid flow, and pure shear conditions of 2.0 Nm^{-2}. (c) Each component of the standard fluid flow separated by cesium and NPPB treatments. *n* is the number of independent experiments with pooled cells (usually eight or nine) combined to form each point.

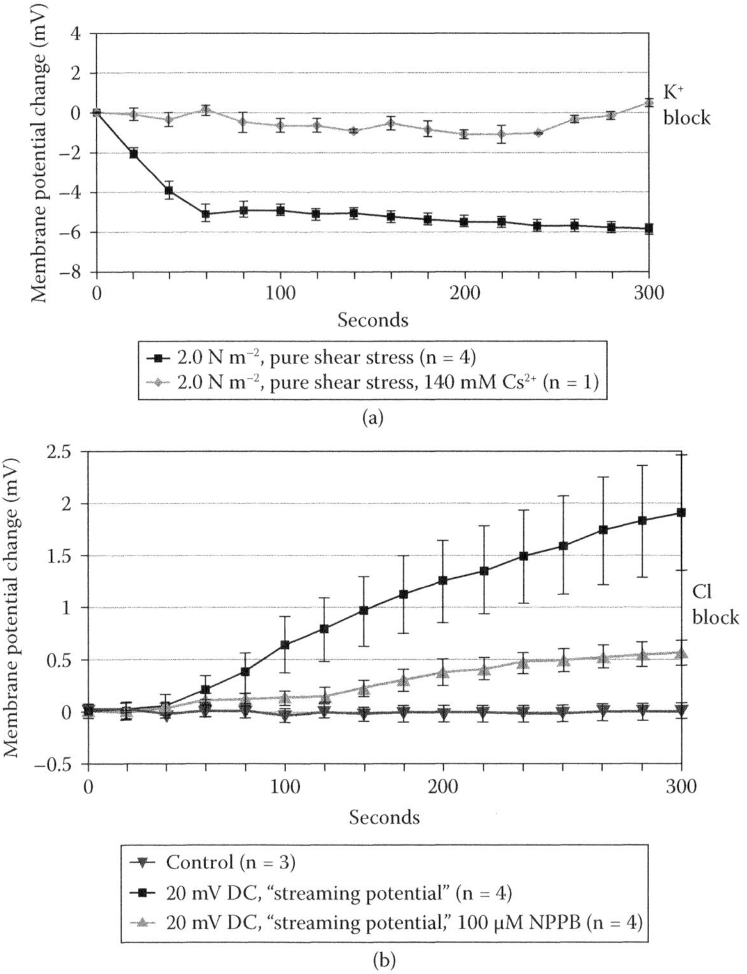

Figure 16.3 (a) Under pure shear conditions (flow and EVSP neutralized), the K^+-dependent current is eliminated with a Cs^{+2} blockade of the K^+ channel. (b) Under streaming potential conditions, the depolarization caused by the electrical field is almost completely eliminated by the NPPB blockade of the Cl^- channel. *n* is the number of independent experiments with pooled cells (usually eight or nine) combined to form each point.

These experiments have therefore largely confirmed that the concomitant physical forces present when endothelial cells are exposed to fluid flow environments are detected by the cells differentially. The lack of complete abrogation of the depolarizing current and experimental fact that the magnitude of the pure shear stress 2.0 N m^{-2} curve seen in Figure 16.3 is significantly less than the fluid flow environment curve (Figure 16.2) suggests that other channels in the endothelial cell may also be responding to these forces in a more complex fashion. This complexity is consistent with work done in the mechanotransduction field, and as will be discussed shortly, there are clearly other molecular signaling systems in the endothelial cell relating to NO production that are responsive to the bioelectric signal with EVSP parameters. These data show that the field of mechanotransduction in endothelial cells is incomplete unless bioelectric fields such as the EVSP are specifically controlled for, and the full story requires substantial research before it is completely elucidated.

16.2.3 DISSECTION OF AN ENDOTHELIAL CELL NITRIC OXIDE RESPONSE TO FLUID FLOW, PURE SHEAR STRESS, AND STREAMING CONDITIONS

It is well established that the application of shearing stress to the endothelial cell causes production of NO (Davies 1995). The degree of shearing stress is proportional to the concentration of NO produced and secreted by the endothelium. As is now appreciated, this well-established relationship has not taken into account the presence of the bioelectrics of the streaming potential. Our laboratory performed experiments in which the conditions of pure shear stress and streaming potential were compared. Figure 16.4 shows the results of a series of experiments in which the NO production is measured from BAECs exposed to varying degrees of shear force (0.35–2.0 N m^{-2}) under standard fluid flow conditions. When the streaming potential is neutralized at 2.0 N m^{-2}, creating conditions of pure shear, the NO response is potentiated with an approximate doubling of the NO signal with the EVSP signal removed. Using a mathematical treatment developed for understanding the early and late phases of the ATP-induced NO signal in BAECs (Trivedi et al. 2013), the NO response was shown to result from stimulation in the early phase when the hyperpolarizing conditions are likely to influence Ca^{+2} current. These findings make sense in the context of the prolonged hyperpolarization seen by preventing the activation of the slow chloride channel in this type of pure shear stimulation. The signals controlling NO production following its initiation include the transmembrane potential that, if held in a hyperpolarized state, leads to increased Ca^{+2} ion flux into the cell, with a resulting increase in the production of NO.

Importantly, experiments in which BAECs are exposed only to EVSP parameter fields with no other concomitant stimulus such as mechanical shear do not elicit NO production. Therefore, the streaming potential, although not a primary signal for initiating NO production in endothelium, in the presence of a mechanical shearing force interacts to upregulate NO production as a complex response to fluid flow (Trivedi et al. 2013; Trivedi and Bergethon 2012; Trivedi 2013).

More detailed study of the response by cultured monolayers of endothelial cells to EVSP parameterized fields alone shows effects on the membrane potential and a complex interaction with the signaling cascade, leading to NO production (Trivedi et al. 2013). Our laboratory has studied the electrophysiological and NO production response of confluent monolayers of BAECs to EVSP parameterized electric fields using fluorescent indicators (bis-(1,3-dibutylbarbituric acid)trimethine [DiBAC4(3)]) for membrane potential and diaminofluorescein-2-diacetate(2-(3,6-diacetyloxy-4,5-diamino-9H-xanthen-9-yl)-benzoic acid) (DAF-2DA) for NO. The EVSP parameters to which a BAEC would be exposed were calculated after Bergethon (1991), taking into account the heart rate and systemic blood pressure of the cow (Olsen 1971). These were estimated physiological bounds of normo-, hypo-, and hypertension (100–300 mmHg systolic: 1.67–6.6 V m^{-1}) and brady- through tachycardia (0.5–2 Hz) in the bovine animal. Direct current (DC) and alternating current (AC) components were studied independently. BAECs consistently respond by depolarization to these physiological parameterized fields, although in a complex fashion (Figure 16.5). DC fields in the physiological range tested showed small

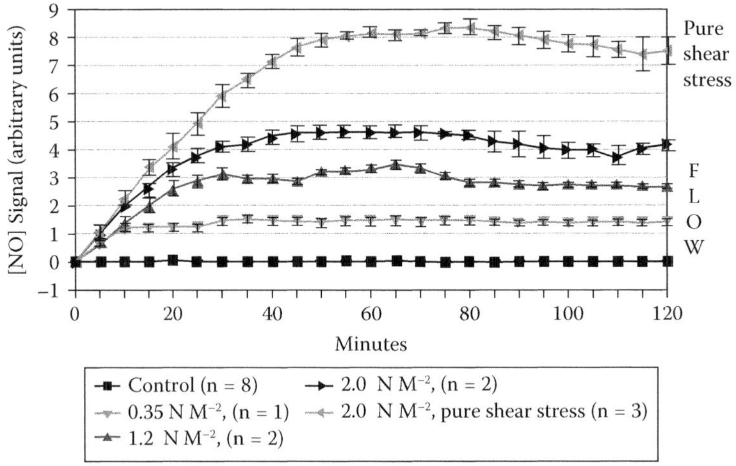

Figure 16.4 NO production at various standard fluid flow conditions with varying shear stresses. Top curve shows a pure shear stress condition of 2.0 N M^{-2} with streaming potential neutralized. *n* is the number of independent experiments with pooled cells (usually eight or nine) combined to form each point.

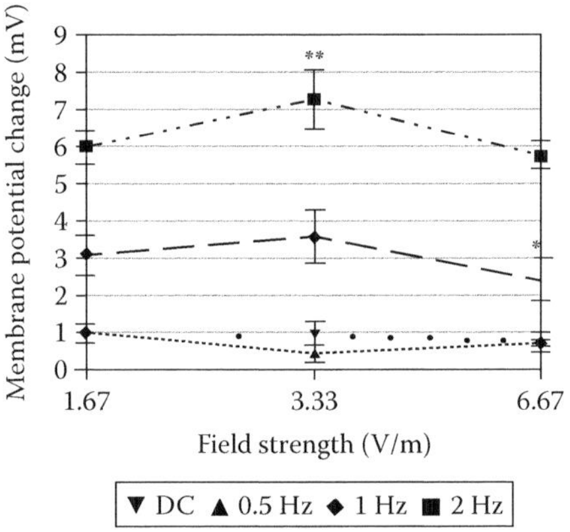

Figure 16.5 EVSP-induced frequency-dependent depolarization. (a) Direct current ($n = 6$) and 0.5 Hz ($n = 6$) stimulation both show a ~1 mV depolarization, with no significant difference at the three field strengths (1.67 V m^{-1}, $p = 0.35$; 3.33 V m^{-1}, $p = 0.08$; 6.67 V m^{-1}, $p = 0.42$). At 1 Hz stimulation ($n = 9$), a ~3 mV depolarization occurs (*), with a lower depolarization at 6.67 V m^{-1} (vs. 1.67 V m^{-1}, $p = 0.01$; vs. 3.33 V m^{-1}, $p = 0.02$). At 2 Hz stimulation ($n = 9$), a ~6 mV depolarization occurs (**), with a higher depolarization at 3.33 V m^{-1} (vs. 1.67 V m^{-1}, $p = 0.001$; vs. 6.67 V m^{-1}, $p = 0.0001$).

depolarizations that were statistically differentiable from controls, suggesting that that this effect was not artifact. A dependence on field strength under these small and physiologically limited DC potentials is not seen. There is a much more robust and complicated depolarization response of endothelial cells on field strength when periodic fields are applied. BAECs show a threshold effect at 0.5 Hz with a logarithmic dependence between the membrane depolarization and the electric field magnitude for a given frequency. Given the evidence that the EVSP fields are likely acting at least by inactivating the slow chloride channels and the degree of depolarization seen under the combination of periodic fields and varying magnitudes, it is likely that other channels are contributing to these effects, but this physiology has yet to be elucidated.

The EVSP fields do not activate NO production in BAECs alone. Since the effect of the electric fields causes depolarization and NO synthesis is proportional to membrane hyperpolarization, this might not be surprising. The expectation in this simple model, then, would be that EVSP fields might attenuate the effect when a physiological NO stimulant such as ATP is applied to the cell system. Experiments testing this hypothesized model showed that NO production in BAECs actually increased with a field strength and frequency dependency (Trivedi et al. 2013). The kinetics of NO production following ATP stimulation of endothelial cells can be separated into an early phase that is logarithmic and a later phase that is exponential. Each of these phases can be separated by varying the Ca^{2+} conditions of the extracellular fluid or blockade of L- and T-type calcium channels with Ni^{2+}. The early phase, which is associated with activation of the NOS pathway by mobilization of intracellular calcium, is independent of calcium channel blockade by Ni^{2+}. Still, the presence of extracellular calcium is required for normal endothelial cell responses consistent with the role of calcium exchange between intracellular and extracellular compartments in normal endothelial cell NOS function. The later exponential phase that is thought to be mediated by extracellular calcium flux into the cell is blocked with both calcium channel blockade and removal of extracellular calcium. The EVSP fields have been shown to impact these early and later phases of the NOS pathway:

- In the presence of 1.1 μM extracellular calcium, the EVSP potentiates the early-logarithmic phase in a manner proportional to the field strength. The late-exponential phase is not influenced by the EVSP field.
- When the L- and T-type calcium channels are blocked with Ni^{+2}, the early-logarithmic-phase potentiation is eliminated (suggesting that the signal above is due to EVSP interaction with these types of channels). A later phase signal appears with EVSP stimulation that is proportional to field strength (suggesting an interaction of the EVSP field independent of these calcium channels in this kinetic range).
- When calcium is removed from the extracellular fluid, no effect is seen on the early-logarithmic phase, but a strong field strength and frequency dependence on the exponential phase of the NO production are found (suggesting an interaction with intracellular stores and recycling of intracellular calcium under these conditions).

There is not sufficient experimental evidence to identify the full complement of biochemical components with which the EVSP field is interacting in the endothelial cell at this point. These experiments do suggest, however, interactions with L- and T-type calcium channels, the slow chloride channel, and potentially intracellular calcium turnover components, perhaps in the calmodulin or IP_3 signaling pathways. Although substantial experimental work remains to be done to elucidate the interactions of the EVSP with endothelial cells in just these pathways, the clear physiological significance of this endogenous electrical field can be imputed.

The probable mechanism linking EVSP to cellular effects is the coupling of applied periodic electrical fields to membrane-associated processes. This effect in enzymes has been established both theoretically and experimentally (Cole 1972; Jaffe 1977; Poo 1981; Serpersu and Tsong 1984; Astumian and Robertson 1989). This mechanism, electroconformational coupling, is applied to the conformational transitions in a macromolecule with intramolecular separations of charge or altered dipole moments that respond to changes in the electrical field across the membrane. The principle of electroconformational coupling is that a chemical reaction, even when catalyzed by an enzyme, may proceed at a relatively negligible rate until the field-induced conformational change activates the enzyme and substantially increases the k_{cat}. Enzymatic electroconformational coupling is generally predicted to be frequency specific, and experimental evidence supports these predictions. Importantly, while studies of the altered activity of the membrane-associated Na^+ and K^+-ATPase in AC fields have been shown to have a frequency dependence (Serpersu and Tsong 1984; Astumian and Robertson 1989), a voltage dependence of the effect is not always clearly demonstrated above a certain threshold (Serpersu and Tsong 1984). These properties of voltage and frequency dependence are seen in the EVSP effects on vascular endothelial and SMCs (*vide infra*).

16.2.4 BIOELECTRICITY, ANGIOGENESIS, AND THE ENDOTHELIAL RESPONSE TO INJURY

Most of the work demonstrating endothelial response to electrical fields associated with injury potentials and implicated in angiogenesis shows field strengths nearly two orders of magnitude higher than those associated with the EVSP. Like the EVSP experimental literature, these experiments explore field parameters based on endogenous empirical measures, and thus have physiological constraints associated with them. Much of this research has been recently reviewed (Wang et al. 2011), and only a brief survey of this field of endothelial research is offered here.

16.2.4.1 Proliferation

Applied electrical fields alter the proliferative rate of a wide variety of cells, including endothelial cells. Blackiston and coworkers (2009) have argued that this is due to signaling through the membrane potential. They note that when applied electric fields cause hyperpolarizing currents in human endothelial cells, cell division is arrested. Wang and coworkers (2003, 2011) have explored some of the relatively high field conditions for manipulating endothelial cell proliferation and found that proliferation of human umbilical vein endothelial cells (HUVECs) was unaffected by low field strengths (50–100 V m^{-1}), but field strengths of 200 V m^{-1} reduced proliferation by 50%. These effects seemed to be mediated through cell cycle control that arrested the cell cycle progression at the G1/S transition, possibly through the decreased expression of cyclin E, which would prevent passage of the cells through G1. In addition, electric field exposure increases the expression of p27kip1, a molecule that inhibits the cyclin E–Cdk2 complex.

16.2.4.2 Migration and orientation

Many cells have been shown to orient themselves and migrate influenced by an applied electrical field vector (Borgens et al. 1981, 1986). This phenomenon is called electrotaxis or galvanotaxis. Endothelial cells have been shown to migrate differentially depending on tissue source, that is, cathodically if they are BAECs and anodically if they are HUVECs (Bai et al. 2004). The migration of endothelial cells along the gradient of an applied electrical field is reported to be field strength dependent with maximal differential effect apparent between 150 and 200 V m^{-1}. Some endothelial cells will respond to field strengths as low as 100 V m^{-1}, and almost all are reported to show electrotaxis once field strengths of 500 V m^{-1} are applied (Bai et al. 2004). These same workers have shown that endothelial cells will orient their long axis perpendicular to the direction of the applied electric field (Zhao et al. 2004; Bai et al. 2004) and will elongate over time to maximize the amount of cell membrane exposed to this perpendicular field component. Interestingly, within the electrical field, the endothelial cells will change from a cobblestone, random orientation to orient to the field as described. In this fashion, the endothelial cell responds to the electrical field with an orthogonal orientation, compared with the way endothelial cells respond to the mechanical shearing force field.

16.2.4.3 Angiogenesis

The central role of VEGF and HIF-1α in angiogenesis was described earlier. Within this context, it is probably not surprising that *in vivo* stimulation of tissues that caused them to become metabolically active induces angiogenesis. This effect is well established in the literature (Hudlicka and Brown 2009; Hang et al. 1995). Much more surprising is the observation that in muscle tissue, stimulation below that needed for physiological contraction also induces VEGF mRNA transcription and protein production. This was associated with increases in physiologically relevant parameters such as increased blood flow and capillary density (Kanno et al. 1999; Lindermann et al. 2000). The stimuli used to achieve these effects were on the order of 50 Hz and 0.1 V. Low-level square-wave stimulation (1 ms duty cycle per 100 μA, 2 Hz) in the epidural space over ischemic rat cerebral cortex has been shown to increase pertinent growth factors in the brain tissue (glia-derived neurotrophic factor [GDNF], brain-derived neurotrophic factor [BDNF], and VEGF) (Baba et al. 2009). These same workers showed evidence that angiogenesis was increased in the region of the electrically stimulated ischemic cortex over control and nonstimulated cortex.

Zhao and coworkers (2004) have shown that DC fields of 150 V m^{-1} induced increased VEGF secretion from HUVEC, even in the absence of other cells. This group has also shown that the stimulation acted via VEGF receptor activation, and its downstream effects were mediated through PI3K-Akt and Rho-ROCK signaling pathways, resulting in reorganization of the actin cytoskeleton. The ability of electrical stimulation to cause VEGF production is not limited to endothelial cells and has been shown in muscle (Kanno et al. 1999), osteoblasts (Kim et al. 2006), and mouse embryonic stem cells (Sauer et al. 2005). All these stimulation paradigms have been on the order of hundreds of volts per meter. The strong relationship between HIF-1 and VEGF production in the context of electrical stimulation of 500 V m^{-1} has been described by Sauer and coworkers (2005). They were able to demonstrate that when 4-day-old embryonic bodies were exposed for 60 s to 500 V m^{-1} fields, HIF-1α increased. These effects are likely attributed to the production of local reactive oxygen species or ischemic conditions since the effect of the electrical fields on HIF-1α levels could be abolished by antioxidant treatment or inhibition of NAD(P)H oxidase.

16.2.5 BIOELECTRIC CONTROL OF BLOOD TISSUE TRAFFICKING

While literature exists exploring the effects of EMFs on the tight endothelial barrier system of the blood–brain barrier, virtually all the studies done are in a high-frequency field, and most are focused on the effects of microwave, MRI gradients, or MRI-associated radio frequency stimulation (Nittby et al. 2008; Stam 2010). The effects on endothelial cell permeability are variable, and the literature at this point has no clear direction in spite of the clear social concern for environmental EMF on the blood–brain barrier. A few studies of low-frequency (50 Hz) fields have been done (Gulturk et al. 2010), but these are not physiological bioelectric fields either. This area of bioelectric effect on the endothelium is fraught with methodological challenges and will bear watching into the future to see if there are significant nonthermal effects demonstrable with more research effort.

16.3 CLINICAL IMPLICATIONS: A SYSTEMS APPROACH INTEGRATING FORCES, HOMEOSTASIS, AND PATHOBIOLOGY

This chapter has emphasized the systemic nature of the endothelial cell. It is clear that the endothelium has a central organizing role as an interface between the blood supply and the local tissues, and as such, it responds to chemical, mechanical, electrical, and structural signals. Whether in growth and development, normative physiological response, or pathobiology, this systems view is essential to understanding the overall role of bioelectric forces in the vasculature. We conclude our discussions by first being more specific about the bioelectric influence of the EVSP on vSMCs, and then the disruptions of the entire endothelial vascular system in the pathobiology of the vascular system.

16.3.1 ACUTE AND CHRONIC EXPOSURE TO EVSP PARAMETER FIELDS ALTER vSMC BEHAVIOR

The endothelial cell is integrated with the other major cell in the vasculature, the vSMC, and the bioelectric forces of the EVSP have been shown to influence both these cells (Bergethon 1991; Bergethon et al. 2013). vSMCs are present in the tunica media of the blood vessel and are exposed to EVSP electrical fields. Sinusoidal fields of extremely low frequency (ELF) consistent with natural heart rates and field magnitudes constrained by natural blood pressures, blood viscosity and conductivity, and tissue zeta potential have been shown to have biological effects on vSMCs. Acute exposure of SMCs to 2 Hz fields with magnitudes from 0.75 to 6.67 V m^{-1} causes a steady-state depolarization of vSMC cells by approximately 10 mV. Frequency effects support an electroconformational coupling mechanism as the basis of the interaction. The presence of a 2 Hz 1.25 V m^{-1} field was also shown to alter the electrophysiological response of SMCs treated with the vasoconstrictor 5-hydroxytryptamine (5-HT) by significantly increasing the degree of membrane depolarization with concurrent exposure (Bergethon 1991). Chronic application of isolated EVSP fields on vSMCs has biologically relevant impact on the electrophysiology, electropharmocology, proliferation, elastin accumulation, and ultrastructural morphology of the vSMC (Bergethon et al. 2013). We now consider the integrated vascular system in the context of pathobiology.

16.3.2 CARDIOVASCULAR DISEASE ETIOLOGY

Cardiovascular diseases (CVDs) are the leading cause of death worldwide (WHO 2011). Myocardial infarction (MI) alone is responsible for 20% of all deaths in the United States. An estimated 17.3 million people died from CVD worldwide in 2008, representing 30% of all global deaths in that year. Roughly half were due to coronary artery disease, and a third were due to stroke. The problem is becoming particularly prevalent in developing countries. More than 80% of the world's CVD deaths occurred in the developing world. By 2030, it is projected that ~23.6 million people worldwide will die from CVDs.

Atherosclerosis is the major contributory pathobiology to CVD, and the hallmark of atherosclerosis is the atheromatous plaque (Chrissobolis and Sobey, 2016). The elements, forces, and systemic interactions that cause the formation of atheromatous plaques and their progression from intimal cushions through to ruptured complicated plaques are not all known (Eisenstein 1992). Our modern model of atherogenesis has advanced to some degree since von Rokitansky proposed the "encrustation hypothesis" in 1852. Still, the overall picture is based on an interaction between the cells comprising the vessel itself (the endothelial and SMCs) and some form of lipid deposition and dysregulation following either chemical infiltration or response to injury and inflammation (Ross and Glomset 1976).

The pathological process in atherogenesis is a dysregulation of the normal complex interrelationships between endothelial cells, SMCs, the ECM, and the biophysical environment of the blood vessel. During normal life, the vascular tree undergoes a growth and development phase, followed by a more mature functional period, as has been discussed for the endothelial cells. This is well reflected as well in the vSMCs, which are sometimes characterized as being proliferative during early vascular development; during this phase, they actively proliferate and also synthesize the ECM components, including large amounts of elastin, collagen, glycoproteins, and glycosaminoglycans. These cells then enter a mature phase, sometimes referred to as "contractile," in which their mitogenicity and protein synthetization are much attenuated and in which the terminally differentiated SMC expresses cytoskeletal marker proteins, such as α-actin and SMC myosin (Thyberg 1996). In this contractile phase, SMC respond with vasoconstriction and dilation to a variety of regulatory mechanisms, which are usually classified as neurogenic influences; chemical regulation from blood-derived sources; autoregulatory, an effect now understood to reflect largely the secretion of NO by endothelial cells as a reflection of intravascular pressure and mechanical forces; and metabolic regulation, that is, responsiveness to the demands of the local metabolic environment itself. However, these sole cellular components of the vascular media monitor the local environment (both chemical and physical) and, under the correct combination of influences, will revert to the more immature proliferative state. It is this potential for reversion to this proliferative state with its increased accumulation of ECM proteins and overabundance of smooth cells that is associated with subintimal thickening and atheromatous growth. Atheromatous plaques develop primarily in elastic arteries (e.g., the aorta, carotid, and iliac arteries) and large and medium-sized muscular arteries (e.g., coronary and popliteal arteries). Symptomatic CVD most often involves the arteries supplying the heart, brain, kidneys, and lower extremities. MI, stroke, aortic aneurysms, and peripheral vascular disease are the most common consequences.

The response to injury hypothesis considers atheromatous plaque formation to be a chronic inflammatory response of the arterial wall initiated by injury to the endothelial cells, which are the primary constituent of the luminal surface (tunica intima). Chronic and repetitive endothelial injury induced in experimental animals by mechanical denudation, hemodynamic forces, immune complex deposition, irradiation, and chemicals causes intimal thickening and, in the presence of high-lipid diets, atheroma formation. The specific cause of endothelial dysfunction in early atherosclerosis is still widely debated. The two most important determinants of endothelial alterations are thought to be hemodynamic disturbances and hypercholesterolemia. In support of the hemodynamic argument, plaques are often found at the ostia of vessels arising from the aorta, branch points, and along the posterior wall of the abdominal aorta, all areas of disturbed flow patterns. Areas of disturbed, turbulent flow and low shear stress are prone to atherosclerosis, while those with smooth, laminar flow seem protected. The normal laminar flow typically encountered in lesion-protected areas of the arterial vasculature blocks inflammatory mechanisms that mediate endothelial dysfunction and apoptosis of endothelial cells.

16.4 CONCLUSIONS

This chapter has discussed research demonstrating complex and physiologically important effects on BAEC from electrical fields at the levels associated with vascular blood flow in the mammalian vasculature. The EVSP is clearly a potential force in the pathobiology of atherosclerosis since it has a regulatory function in both VEC and vSMC. An important role for EVSP-type ELF fields as a control factor in endothelial cell function and vascular biology is thus well supported through both the biological and biophysical mechanisms, and the multiple specific targets of action remain to be elucidated.

Some speculation on the clinical implications of the EVSP on endothelial cells and the larger vascular context might be entertained. The amplitude of the EVSP field varies proportionally to blood pressure, the rate of pulsatile flow and flow rate vary as a complex function of metabolic demand, and the zeta potential reflects the integrity of the vascular interface with the blood. The EVSP can change sign, magnitude, and even waveform when the vessel surface is altered as occurs in endothelial injury, in atherosclerotic plaque formation, and as the blood pressure and pulse rate rises and falls. A dissociation of the shear force and the EVSP may play a role in vascular endothelial cell activity in areas of turbulence (e.g., following vessel narrowing from atheromatous plaque) since in these regions the shearing force falls dramatically while the EVSP would remain, leading to a region of SMC depolarization that may act to depolarize and cause vasoconstriction of local SMC. In addition, the pulsatile flow that leads to a varying periodic field is attenuated in regions of turbulence, leading to a lower frequency of the EVSP. The homeostasis between the shearing force and EVSP may be altered in areas of turbulence with a reduction in NO production because of reduced shear and reduced EVSP local frequency. In addition, the ELF fields of EVSP character affect both the vascular endothelial cell and the SMC independently. It is intriguing that the long-term alteration of EVSP might change endothelial response just as it alters ECM production in the vSMCs. Since part of the process of mechanotransduction will depend on the connection to ECM (secreted by vSMC) by laminin and other ECM components, the chronic effect of EVSP in altering the ECM physical chemistry and physics can alter the interaction of the mechanical and electrical forces leading to endothelial response. The complex interdependence and possibilities of vascular cell interactions comodulated by ELF electrical fields open further questions about vascular homeostasis and pathobiology. All these factors are important considerations in a new vascular biophysics and biochemistry of the blood vessel that should now include ELF electrical fields such as the EVSP. The opportunities for understanding and intervention in vascular biology open a whole new field of medical therapy. Only research engagement in this fertile new field of vascular biology will hold back progress.

REFERENCES

Adams, D. J., Barakeh, J., Laskey, R., and Van Breemen, C. Ion channels and regulation of intracellular calcium in vascular endothelial cells. *FASEB J* 3 (1989): 2389–2400.

Alberts, B., Johnson, A., Lewis, J., et al. *Blood Vessels and Endothelial Cells in Molecular Biology of the Cell.* 4th ed. New York: Garland Science, 2002.

Astumian, R. D., and Robertson, B. Nonlinear effect of an oscillating electric field on membrane proteins. *J Chem Phys* 91 (1989): 4891.

Baba, T., Kameda, M., Yasuhara, T., et al. Electrical stimulation of the cerebral cortex exerts anti-apoptotic, angiogenic, and anti-inflammatory effects in ischemic stroke rats through phosphoinositide 3-kinase/Akt signaling pathway. *Stroke* 40 (2009): 598–605.

Bai, H., McCaig, C. D., Forrester, J. V., and Zhao, M. Direct current electric fields induce distinct preangiogenic responses in microvascular and macrovascular cells. *Arterioscler Thromb Vasc Biol* 24 (2004): 1234–1239.

Barakat, A. I., and Gojova, A. Role of ion channels in cellular mechanotransduction—Lessons from the vascular endothelium. In *Cellular Mechanotransduction.* Cambridge: Cambridge University Press, 2010, chapter 6.

Barakat, A. I., Leaver, E. V., Pappone, P. A., and Davies, P. F. A flow-activated chloride-selective membrane current in vascular endothelial cells. *Circ Res* 85 (1999): 820–828.

Barakat, A. I., and Lieu, D. K. Differential responsiveness of vascular endothelial cells to different types of fluid mechanical shear stress. *Cell Biochem Biophys* 38 (2003): 323–343.

Becker, W. M., Kleinsmith, L. J., and Hardin, J. Signal transduction mechanisms: Messengers and receptors. In *The World of the Cells, 6th edition.* San Francisco: Pearson Education, Benjamin Cummings, 2005, pp. 405–406.

Bergethon, P. R. Altered electrophysiologic and pharmacologic responses to smooth muscle cells on exposure to electric fields generated by blood flow. *Biophys J* 60 (1991): 588–595.

Bergethon, P. R. The electrified interphase. In *The Physical Basis of Biochemistry: The Foundations of Molecular Biophysics.* New York: Springer, 2010a, pp. 583–601.

Bergethon, P. R. Dynamic bioelectrochemistry—Charge transfer in biological systems. In *The Physical Basis of Biochemistry: The Foundations of Molecular Biophysics.* New York: Springer, 2010b, pp. 713–737.

Bergethon, P. R., Kindler, D. D., Hallock, K., et al. Continuous exposure to low amplitude extremely low frequency electric fields characterizing the vascular streaming potential alter elastin accumulation in vascular smooth muscle cells. *Bioelectromagnetics* 34 (2013): 358–365.

Blackiston, D. J., McLaughlin, K. A., and Levin, M. Bioelectric controls of cell proliferation: Ion channels, membrane voltage and the cell cycle. *Cell Cycle* 8 (2009): 3527–3536.

Borgens, R. P., Blight, A. R., and Murphy, D. J. Axonal regeneration in spinal cord injury: A perspective and new technique. *J Comp Neurol* 250 (1986): 157.

Borgens, R. B., Roederer, E., and Cohen, H. J. Enhanced spinal cord regeneration in lamprey by applied electric fields. *Science* 213 (1981): 611.

Chrissobolis, S., and Sobey, C. G. Vascular biology and atherosclerosis of cerebral vessels. In *Stroke: Pathophysiology, Diagnosis, and Management.* (ed. Grotha, J. C., et al.,) 6th edition, London: Elsevier, 2016, pp. 3–12.

Cohen, R. A., and Vanhoutte, P. M. Endothelium-dependent hyperpolarization, beyond nitric oxide and cyclic GMP. *Circulation* 92 (1995): 3337–3349.

Cole, K. S. *Membranes, Ions and Impulses.* Berkeley, CA: University of California Press, 1972.

Davies, P. Flow-mediated endothelial mechanotransduction. *Physiol Rev* 75 (1995): 519–560.

Eisenstein, R. Vascular extracellular tissue and atherosclerosis. *Artery* 5 (1992): 207.

Folkmann, J. Is angiogenesis an organizing principle in biology and medicine? *J Pediatr Surg* 42 (2007): 1–11.

Folkmann, J., and Klagsbrun, M. Angiogenic factors. *Science* 235 (1987): 442–447.

Gautam, M., Shen, Y., Thirkill, T. L., Douglas, G. C., and Barakat, A. I. Flow-activated chloride channels in vascular endothelium—Shear stress sensitivity, desensitization dynamics and physiological implications. *J Biol Chem* 281 (2006): 36492–36500.

Gerecht-Nir, S., Sivan Osenberg, S., Nevo, O., et al. Vascular development in early human embryos and in teratomas derived from human embryonic stem cells. *Biol Reprod* 71 (2004): 2029–2036.

Gilbert, S. F. *Developmental Biology*. 6th ed. Sunderland, MA: Sinauer Associates, 2000.

Gulturk, S., Demirkazik, A., Kosar, I., et al. Effect of exposure to 50 Hz magnetic field with or without insulin on blood-brain barrier permeability in streptozotocin-induced diabetic rats. *Bioelectromagnetics* 31 (2010): 262–269.

Hang, J., Kong, L., Gu, J. W., and Adair, T. H. VEGF gene expression is up-regulated in electrically stimulated rat skeletal muscle. *Am J Physiol* 269 (1995): H1827–H1831.

Hoger, J. H., Ilyin, V. I., Forsyth, S., and Hoger, A. Shear stress regulates the endothelial KIR 2.1 ion channel. *Proc Natl Acad Sci U S A* 99 (2002): 7780–7785.

Hudlicka, O., and Brown, M. D. Adaptation of skeletal muscle microvasculature to increase or decreased blood flow: Role of shear stress, nitric oxide and vascular endothelial growth factor. *J Vasc Res* 46 (2009): 504–512.

Jaffe, L. F. Electrophoresis along cell membranes. *Nature* 265 (1977): 600–602.

Kanno, S., Oda, N., Abe, M., et al. Establishment of a simple and practical procedure applicable to therapeutic angiogenesis. *Circulation* 99 (1999): 2682–2687.

Kim, I. S., Song, J. K., Hang, Y. L., et al. Biphasic electric current stimulates proliferation and induces VEGF production in osteoblasts. *Biochim Biophys Acta* 1783 (2006): 907–916.

Larsen, W. J. *Essentials of Human Embryology*. 2nd ed. New York: Churchill Livingstone, 1998.

Li, Y. J., Haga, J. H., and Chien, S. Molecular basis of the effects of shear stress on vascular endothelial cells. *J Biomech* 38 (2005): 1949–1971.

Lieu, D. K., Pappone, P. A., and Barakat, A. I. Differential membrane potential and ion current responses to different types of shear stress in vascular endothelial cells. *Am J Physiol Cell Physiol* 286 (2004): C1367–C1375.

Lindermann, J. R., Kloehn, M. R., and Greene, A. S. Development of an implantable muscle stimulator: Measurement of stimulated angiogenesis and post-stimulus vessel regression. *Microcirculation* 7 (2000): 119–128.

Malek, A. M., Greene, A. L., and Izumo, S. Regulation of endothelin 1 gene by fluid shear stress is transcriptionally mediated and independent of protein kinase C and cAMP. *PNAS* 90 (1993): 5999–6003.

Malek, A. M., and Izumo, S. Mechanism of endothelial cell shape change and cytoskeletal remodeling in response to fluid shear stress. *J Cell Sci* 109 (1996): 713–726.

Nilius, B. Regulation of transmembrane calcium fluxes in endothelium. *Physiology* 6 (1991): 110–114.

Nittby, H., Grafstro, G., and Eberhardt, J. L. Review: Radiofrequency and extremely low-frequency electromagnetic field effects on the blood brain barrier. *Electromagn Biol Med* 27 (2008): 103–126.

Olsen, J. D. Periodic elevations of blood pressure of cattle during exposure to subzero environmental temperatures. *Int J Biometeor* 15 (1971): 225.

Poo, M. M. In situ electrophoresis of membrane components. *Ann Rev Biophys Bioeng* 10 (1981): 245–276.

Quincke, G. T'ber die Fortfujung materielle Theilchen durch stromende Elektrictat. *Pogg Ann* 113 (1859): 513.

Ross, R., and Glomset, J. The pathogenesis of atherosclerosis. *N Engl J Med* 295 (1976): 369–377.

Sauer, H., Bekhite, M. M., Heschler, J., and Wartenberg, M. Redox control of angiogenic factors and CD31-positive vessel-like structures in mouse embryonic stem cells after direct current electrical field stimulation. *Exp Cell Res* 304 (2005): 380–390.

Sawyer, P. N., Himmelfarb, E., Lustrin, I., and Ziskind, H. Measurement of streaming potentials of mammalian blood vessels, aorta and vena cava, in vivo. *Biophys J* 6 (1966): 641–651.

Schwartz, C. J., Valente, A. J., Sprague, E. A., Kelley, J. L., and Nerem, R. M. The pathogenesis of atherosclerosis: An overview. *Clin Cardiol* 14 (1991): I1–I16.

Serpersu, E. H., and Tsong, T. Y. Activation of electogenic Rb+ transport of (Na,K)-ATPase by an electric field. *J Biol Chem* 259 (1984): 7155–7162.

Shaul, P. W. Regulation of endothelial nitric oxide synthase: Location, location, location. *Annu Rev Physiol* 64 (2002): 749–774.

Stam, R. Review: Electromagnetic fields and the blood–brain barrier. *Brain Res Rev* 65 (2010): 80–89.

Thyberg, J. Differentiated properties and proliferation of arterial smooth muscle cells in culture. *Int Rev Cytol* 169 (1996): 183–265.

Trivedi, D. P. The effects of streaming potential modeled electrical fields on bovine aortic endothelial cells. Boston University, Pro Quest Dissertations Publishing, 3536986, 2013. http://search.proquest.com/docview/1321125711.

Trivedi, D. P., and Bergethon, P. R. Revising the model of vascular hemostasis: Mechanotransduction is opposed by a novel electrical force. *Ann Neurol* 72 (2012): S3.

Trivedi, D. P., Hallock, K. J., and Bergethon, P. R. Electric fields caused by blood flow modulate vascular endothelial electrophysiology and nitric oxide production. *Bioelectromagnetics* 34 (2013): 22–30.

Voets, T., Droogmans, G., and Nilius, B. Membrane currents and the resting membrane potential in cultured pulmonary artery endothelial cells. *J Physiol* 497 (1996): 95–107.

Wang, E., Yin, Y., Bai, H., Reid, B., Zhao, Z., and Zhao, M. Electrical control of angiogenesis. In *The Physiology of Bioelectricity in Development, Tissue Regeneration and Cancer*, ed. C. E. Pullar. Boca Raton, FL: CRC Press, 2011, pp. 155–175.

Wang, E., Yin, Y., Zhao, M., Forrester, J. V., and McCraig, C. D. Physiological electric fields control the G1/S phase cell cycle checkpoint to inhibit endothelial cell proliferation. *FASEB J* 17 (2003): 458–60.

WHO (World Health Organization). *The Atlas of Heart Disease and Stroke: Cardiovascular Disease Fact Sheet*. CH. no. 317. Geneva: WHO, 2011.

Yuan, S. Y., and Rigor, R. R. *Regulation of Endothelial Barrier Function*. San Rafael, CA: Morgan & Claypool Life Sciences, 2010.

Zhao, M., Bai, H., Wang, E., et al. Electrical stimulation directly induces pre-angiogenic responses in vascular endothelial cells by signaling through VEGF receptors. *J Cell Sci* 117 (2004): 397–405.

17 The role of electrical field on stem cells *in vitro*

Miina Björninen
University of Wollongong
Wollongong, New South Wales, Australia

Suvi Haimi
University of Twente
Enschede, The Netherlands
and
University of Tampere
Tampere, Finland

Michael J. Higgins
University of Wollongong
Wollongong, New South Wales, Australia

Jeremy M. Crook
University of Wollongong
Wollongong, New South Wales, Australia

Contents

17.1 INTRODUCTION

Endogenous direct current electric fields (DC-EFs) occur *in vivo* in the form of transepithelial cellular potentials (TEPs) or neuronal field potentials (Mycielska and Djamgoz 2004). Moreover, tissues and bodily fluids take part in creating EFs, for instance, in the form of piezoelectric phenomena in certain tissues comprise mainly collagen (Fukada and Yasuda 1964) or as electrolyte movement or electrolyte gradients (Levin 2007). DC-EFs therefore have an essential role in controlling nerves and muscles, but also in nonexcitable tissues. Although produced in most tissues of the body, DC-EFs are particularly important during embryonic and fetal development, as well as wound healing and tissue regeneration (Cameron et al. 1993; Mycielska and Djamgoz 2004; Zhao 2009). Currents measured at injury sites are typically on the order of 100 $\mu A/cm^2$, leading to DC voltage gradients of ~10 mV/mm (given a resistivity of ~1000 Ω/cm for soft tissues) (McCaig et al. 2005). Severed newt limbs are characterized by large currents at the site of tissue regeneration and a voltage drop of 60 mV/mm within the first 125 μm of extracellular space (Borgens et al. 1977), while a steady DC-EF of 140 mV/mm persists at wound edges of human epithelium (Barker et al. 1982). Moreover, disturbances to environmental EFs cause aberrant development (Robinson 1985; Cameron et al. 1993).

While there is growing interest in applying DC-EF to offset and reverse the polarity of injury-induced potentials for promoting tissue regeneration (Borgens et al. 1981; McCaig et al. 2005), endogenous DC-EFs are three to four orders of magnitude lower than EFs (1–2 V/mm) applied in "classical" electrophysiology to depolarize neurons and fire action potentials. Notwithstanding, the phenomena of endogenous DC-EF in physiological processes of development and regeneration are consistent with the more recent discovery that EFs influence and are important to stem cell behavior *in vivo*, supporting organ development, tissue repair, and homeostasis. Accordingly, electrical stimulation (ES) is increasingly being used as a tool to control stem cell behavior and function, including both self-renewal and differentiation, for basic research and translational application.

In vitro ES delivery systems can be divided into direct and indirect stimulation (Figure 17.1). Whereas direct stimulation requires electrodes to be in contact with the cell culture, electrodes are placed outside the cell culture for indirect stimulation. Indirect stimulation can be further classified as capacitive stimulation, whereby homogenous EFs are generated across the cell culture area, and inductive stimulation, involving the production of electromagnetic fields (EMFs). Whether direct or indirect, both stimulation platforms can employ different ES waveforms, such as sinusoidal, sawtooth, and square waves, ranging from monophasic (DC) to biphasic (alternating current [AC]) waveforms, and from continuous to pulsed waveforms or more complex pulse bursts (Balint et al. 2013). The simplest ES is DC voltage, which is applied by batteries (Pelto et al. 2013b).

A further important consideration for applying ES for stem cell manipulation is the use of conventional two-dimensional (2-D) flat-bed culture or advanced three-dimensional (3-D) culture. 2-D culture requires fewer cells and is more amenable to observing cell morphology and migration effects of ES in a simple and well-characterized environment. However, 2-D culture has limited potential for modeling the natural cell environment or stem cell niche conditions. 3-D culture is thought to more closely

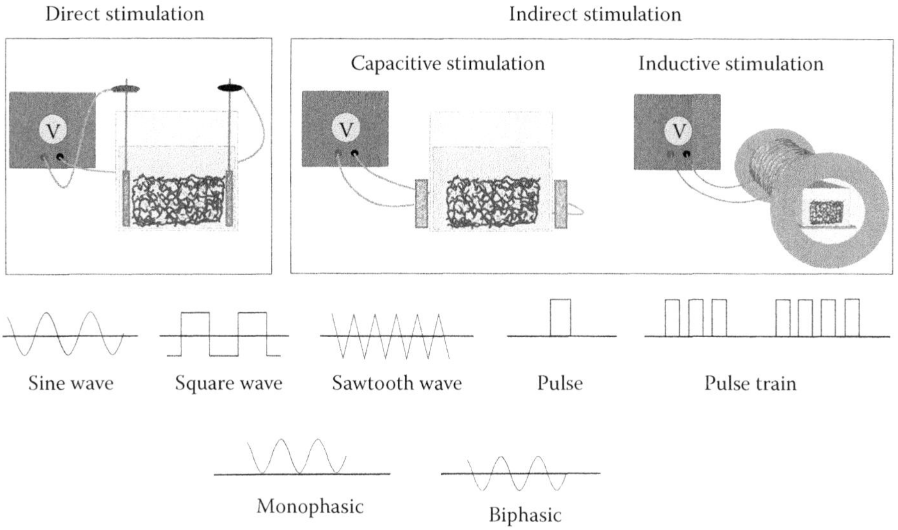

Figure 17.1 *In vitro* ES delivery systems and commonly applied waveforms.

recapitulate the *in vivo* state, including ES-induced epigenetic and genetic regulation of the stem cell state. For example, differential gene expression of chemokines between 2-D and 3-D cultures has been reported (Hwang et al. 2012), with 3-D culture resembling *in vivo* conditions. ES in 3-D culture can be conductive cell substrate or suspension based (Serena et al. 2009; Jaatinen et al. 2015).

This chapter provides an overview of the different effects and methods of *in vitro* ES on stem cell culture for basic and applied research. Topics covered include the effects of ES on stem cell behavior and function, including self-renewal and differentiation, and the application of ES for stem cell research and translation. Methods canvassed for stem cell ES include direct stimulation using integrated electrode cell culture systems, indirect stimulation by capacitive coupling versus inductive stimulation, stimulation using microelectrode arrays (MEAs), and the general requirements for safe and sustainable ES of stem cell cultures.

17.2 CELLULAR AND MOLECULAR EFFECTS OF ES

Given the occurrence of endogenous EFs, it is not surprising that ES has been found to influence stem cell maintenance and differentiation *in vitro*, the latter exemplified by cardiomyocyte induction from human embryonic stem cells (Chan et al. 2013), osteogenic differentiation from human mesenchymal stem cells (MSCs) (McCullen et al. 2010), and neuronal induction from human neural stem cells (NSCs) (Pires et al. 2015; Stewart et al. 2015).

While insight to the mechanisms of action of ES is limited, research is expanding toward better understanding of regulated signaling pathways underlying cellular mechanics such as cytoskeletal and membrane mechanics, behaviors such as migration and proliferation, cell adhesion, and lineage specificity, as well as extracellular signaling via extracellular matrix (ECM). In the latter case, movement and adsorption of extracellular proteins are purportedly influenced by ES, including cell membrane receptor binding (Adey 1993). Notwithstanding, much attention has been given to calcium signaling, whereby ES has been demonstrated to influence intracellular calcium dynamics (Hammerick et al. 2010a; McCullen et al. 2010), which in turn affects cell proliferation, motility, differentiation (Yamada et al. 2007; McCullen et al. 2010), and other cellular processes. These important cell effects are consistent with the myriad biochemical effects of calcium ions (Ca^{2+}), including allosteric regulatory effects on numerous cytoplasmic enzymes and proteins and changes in gene transcription.

ES has been shown to increase intracellular calcium concentration via Ca^{2+} translocation through cell membrane voltage-gated calcium channels (Figure 17.2) (Brighton et al. 2001). An intracellular calcium increase correlates with increasing EF in human adipose stem cells (ASCs) (McCullen et al. 2010). The increase appears to be rapid following initiation of ES. For example, cytosolic free calcium increases and cAMP decreases after 30 s of EF initiation in mouse ASCs (Hammerick et al. 2010a). Earlier studies demonstrate that different EFs act via various mechanisms for coupling DC-EF to calcium signaling for bone marrow stromal cells (bMSCs) and osteoblasts. As such, different EF magnitudes and frequencies activate different mechanisms to increase cytosolic Ca^{2+}, including Ca^{2+} release from intracellular stores mediated by phospholipase C activation or opening of ion channels such as the membrane Ca^{2+} pump and Na^{+}-Ca^{2+} exchanger (Figure 17.2) (Khatib et al. 2004; Sun et al. 2007).

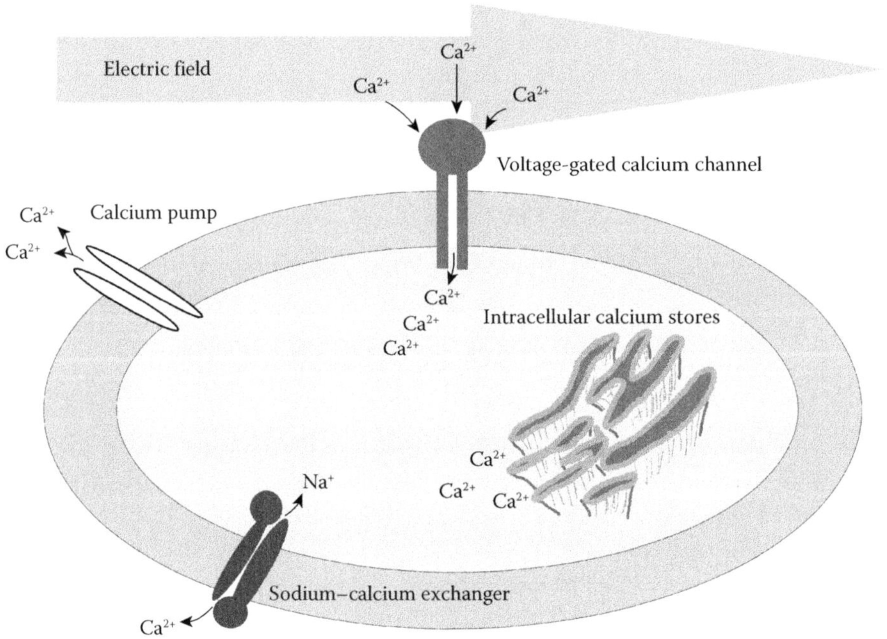

Figure 17.2 *In vitro* ES effects on intracellular calcium dynamics. ES influences intracellular calcium concentration via Ca^{2+} translocation through cell membrane voltage-gated calcium channels, Ca^{2+} release from intracellular stores, and opening of ion channels, such as the membrane Ca^{2+} pump and Na^{+}-Ca^{2+} exchanger.

17.2.1 STEM CELL SELF-RENEWAL

Stem cell self-renewal is the process of cell division to make more stem cells, retaining the undifferentiated state. Therefore, cells remain in the cell cycle, maintaining their multipotency or pluripotency. Self-renewal programs involve signaling pathways that balance proto-oncogenes that promote self-renewal, gate-keeping tumor suppressors that constrain self-renewal, and care-taking tumor suppressors that preserve genomic integrity. Consistent with the role of endogenous EFs in cell homeostasis of tissues, these cell-intrinsic mechanisms of self-renewal are regulated by multiple ES waveforms, including DC, degenerate wave, and capacitive coupling, supported by enhanced proliferation of stem cells, with a decreased proportion of cells in the G0/G1 cell cycle phase and an increased proportion of cells at phases S and G2/M (Griffin et al. 2011; Hernandez-Bule et al. 2014). Moreover, biphasic electrical currents (ECs) with current densities of 1–30 $\mu A/cm^2$ for 200 µs at 100 Hz have been used to increase fetal NSC proliferation corresponding to enhanced production of neurospheres (Chan et al. 2013). Studies of adipose-derived MSCs suggest that the effect of ES on proliferation does not compromise differentiation potential, with MSCs able to be induced to adipogenic, chondrogenic, and osteogenic cell lineage (Hernandez-Bule et al. 2014).

The mechanisms responsible for coupling ES with cell proliferation remain largely undetermined, although several intracellular pathways have been implicated. Mitogen-activated protein kinase (MAPK) cascades have been shown to play a role in ES of human MSC proliferation, namely, classical extracellular signal-regulated kinase (ERK) and p38 kinase (Kim et al. 2009). MAPK pathways transmit, amplify, and assimilate signals from many different stimuli to produce an appropriate physiological response. In the case of MSCs, downstream vascular endothelial growth factor (VEGF) has been implicated in ES-enhanced proliferation, which is blocked by p38 kinase and ERK inhibitors, as well as calcium channel blockers (Kim et al. 2009). Interestingly, Kim et al. report that a longer duration (250 µs vs. 25 µs) of biphasic EC ES with lower amplitude (1.5 $\mu A/cm^2$ vs. 15 $\mu A/cm^2$) at a frequency of 100 Hz promotes more proliferation.

Although cell proliferation is complex and ES mechanisms are poorly characterized, the implication of ERK and p38 signaling in ES regulation, together with demonstrated MAPK cascades in cell proliferation, suggests further downstream processes incorporating translocation of activated ERKs and p38 to the nucleus and transcription factor transactivation to promote mitosis.

Demonstrated substrates of ERKs include 90 KDa ribosomal S6 Kinase (p90rsk, also known as MAPKAP-K1), cytosolic phospholipase A2, and microtubule-associated proteins (MAPs) such as MAP-1, MAP-2, and Tau (Zhang and Liu 2002). ERKs can therefore influence gene expression via p90rsk whose substrates include cell cycle–related transcription factors cAMP response element binding protein (CREB), estrogen receptor alpha (ERα), c-Fos, and glycogen synthase kinase 3 (GSK3). Thus, akin to the effects of other stimuli, such as mitogenic stimulation by cytokines, ES likely regulates cell proliferation via known signaling pathways of mitogenic stimulation, including G_1/S transition and S phase progression. This is consistent with ES being sensed through pathways concerned with mechanotransduction and chemotaxis, in addition to the involvement of growth factor receptors (e.g., VEGF, epidermal growth factor [EGF], and fibroblast growth factor [FGF]) in ES transduction (Balint et al. 2013).

Mammalian p38 kinases are activated by cellular stress arising from, for example, ultraviolet (UV) radiation, heat shock, osmotic stress, and inflammation (including proinflammatory cytokines), as well as protein synthesis inhibitors and various mitogens (Thornton and Rincon 2009). Not surprisingly, like ES, they all impact on cell proliferation, although not all in the same way. For example, whereas UV radiation elicits cell cycle arrest and apoptosis, mitogens trigger mitosis. Consistent with the former, P38 has classically been associated with apoptosis and cell cycle arrest at G_1/S (Thornton and Rincon 2009). Although there is no compelling evidence for a direct effect of p38 inducing cell proliferation, in some instances (e.g., following DNA damage), it can increase cell survival (Thornton and Rincon 2009). While it remains to be determined, blocking ES-enhanced MSC proliferation by p38 inhibition is indicative of another potentially novel "nonclassical" function of p38 in cell cycling. Further studies are undoubtedly necessary.

17.2.2 STEM CELL DIFFERENTIATION

ES is increasingly recognized as a useful *in vitro* technique to enhance stem cell differentiation. By far the majority of studies to date relate to MSC differentiation for osteogenic and cardiac cell induction, although recent reports extend to pluripotent and other stem cell types (Table 17.1).

While there are undoubtedly lineage-specific (e.g., nonexcitable vs. excitable cell types, or further specialized neuronal, myocytic, bone, etc., cell subtypes) and stem cell type–specific mechanisms of ES, in addition to differential electrocoupling between the various modalities of applied ES (e.g., indirect vs. direct ES, intensity, and duration), in any case, there are myriad cooperative and competing cellular and subcellular processes concomitantly affected by other cues of the cell microenvironment (e.g., culture media components and supplements, and mechanical cues of substrates). For example, calcium dynamics (see Section 17.2) regulates human MSC differentiation and is altered by ES and other physical stimuli, such as substrate stiffness.

Therefore, comprehensive characterization of a cell culture system and cellular biochemistry during ES will be important to attaining, understanding, standardizing, and replicating desired differentiation effects. By doing so, ES may be used as a routine and relatively noninvasive approach to directing stem cell fate to pure and germane populations of somatic cells for *in vitro* R&D and translation, such as early-phase drug and toxicity testing.

17.2.3 CELL MIGRATION

A cellular effect of DC-EFs is galvanotaxis, which is directional migration toward a cathode or anode. *In vitro* DC-EFs comparable in magnitude to those occurring naturally *in vivo* produce galvanotaxis in a range of cultured cells (Mycielska and Djamgoz 2004). Murine ASCs migrate toward the cathode in DC fields of physiologic strength (Hammerick et al. 2010). EFs also induce murine ASCs to position perpendicularly to the field vector and exhibit a transitory increase in cytosolic calcium. Similar to cell proliferation (see above), their galvanotactic response likely involves chemotactic signaling pathways (Hammerick et al. 2010a).

EF directed migration has also been shown for bone marrow MSCs, although predominantly to the anode at a low threshold and with a physiological EF of ~25 mV/mm (Zhao 2009). Migratory response was enhanced with increasing EF. The response to

Table 17.1 **Studies of the effects of direct ES on stem cells**

STEM CELL APPLICATION	REFERENCE	ES TYPE AND PARAMETERS	CELL EFFECT(S) OF ES	CELL TYPE	2-D/3-D AND SUBSTRATE
Bone	Sun et al. (2007)	EF: 0.1 V/cm, 1 Hz sinusoidal AC	• Proliferation threefold • ALP activity • Mineralization	bMSCs	2-D: Material unspecified
	McCullen et al. (2010)	EF: 1, 3, or 5 V/cm, sinusoidal 1 Hz, 4 h/day for 14 days	• Proliferation (5 V/cm) • Mineralization (1 V/cm)	Human ASCs	2-D: Interdigitated gold electrodes on glass
	Kim et al. (2009)	EC: Pulsed biphasic electric current (BEC), 100 Hz, 250 μs pulse duration with 1.5 $\mu A/cm^2$ or 25 μs with 15 $\mu A/cm^2$, 24 h/day	• Proliferation (250 and 25 μs) • VEGF production (250 μs) • BMP-2 production (250 μs)	bMSCs	2-D: Au
	Griffin et al. (2011)	DW: 15 Hz (62.5 ms pulse width) and 160 mV peak-to-peak CC: 10 mV/mm DC: 75 mV/mm	• Proliferation (DW) • MMP-2 and MT1-MMP expression (CC, DW) • Cell invasion to collagen (CC, DW)	Human bMSCs	2-D: Glass
	Hammerick, et al. (2010a)	EF: Pulsed, 6 V/cm, 50 Hz, 6 h/day	• Gene expression of ALP, osteopontin, RUNX-2, Coll I	Mouse ASCs	2-D: TCP
	Hu et al. (2014)	EC: 0.035–3.5 V/cm for 2, 4, and 12 h	• ECM calcium deposition (0.35 V/cm)	Rat bMSC	2-D: PPy films

(*Continued*)

Table 17.1 *(Continued)* **Studies of the effects of direct ES on stem cells**

STEM CELL APPLICATION	REFERENCE	ES TYPE AND PARAMETERS	CELL EFFECT(S) OF ES	CELL TYPE	2-D/3-D AND SUBSTRATE
	Hronik-Tupaj et al. (2011)	EF: 20 mV/cm, 60 kHz, 40 min/day	• Upregulation of ALP and Coll I, and stress response marker heat shock protein 27 • Enhanced reduced nicotinamide adenine dinucleotide (NADH), flavin adenine dinucleotide (FAD), dinucleotide lipofuscin, and calcium deposition	Human MSCs	2-D: Glass
	Björninen et al. (2014)	EC: Pulsed BEC, ±0.2 V, 2.5 ms pulse width, 100 Hz	• Proliferation • Mineralization	Human ASCs	2-D: PPy films containing the dopant chondroitin sulfate
	Creecy et al. (2013)	EC: Sinusoidal, 10 Hz, 10 or 40 μA, for 6 h/day	• Gene expression of RUNX-2 (10 μA), osteocalcin (10 and 40 μA)	Human bMSCs	3-D: Coll I hydrogel
Neural	Gu et al. (2015)	20 Hz, 100 ms, 3 V for 1 h	• Proliferation • Phosphorylated ERK 1/2, cyclin D1 and CDK4 protein levels	Rat peripheral blood stem cells	2-D: Poly-L-lysine-treated glass
	Pires et al. (2015)	100 Hz pulsed DC, 1 V with 10 ms pulses	• Neurons • Neurite length	NSCs	2-D: PEDOT–polystyrene sulfonate

(Continued)

Table 17.1 *(Continued)* **Studies of the effects of direct ES on stem cells**

STEM CELL APPLICATION	REFERENCE	ES TYPE AND PARAMETERS	CELL EFFECT(S) OF ES	CELL TYPE	2-D/3-D AND SUBSTRATE
	Chang et al. (2011)	EC: Calculated current density of 4 or 8 μA/cm^2 for 200 μs at 100 Hz	• Proliferation • Expression of βIII-tubulin (4 and 8 μA), neural nucleus marker (1.33 and 4 μA), and mature neuron marker (4 μA)	Fetal NSCs	2-D: ITO
	Matsumoto et al. (2013)	Rectangular, 10 Hz, 100 mV, 2 ms, 30 min on days 0, 1, 2 of differentiation	• Expression of neurogenin2	Mouse bMSC	Not specified
	Jaatinen et al. (2015)	EF: Pulse 5 s on, 20 s off; the EF was 35 mV/mm in 1 mA, 53 mV/mm in 1.5 mA, and 155 mV/mm in 1 mA	• Expression of βIII-tubulin (1.5 mA) and mature neuron marker (1 and 1.5 mA) • Gene expression of βIII-tubulin (1 and 1.5 mA) and mature neuron marker (1 mA)	Human ASCs	Cell suspension
	Thrivikraman et al. (2014)	EC: DC 1 mV to 2 V for 10 min/day	• βIII-tubulin	Human MSCs	2-D: PANI films
	Yamada et al. (2007)	EF: 0, 5 (0.01 A), 10 (0.57 A), and 20 V (1.46 A), 1 train of 5 pulses (950 ms interpulse interval)	• Expression of βIII-tubulin	ESCs and embryoid bodies	2-D: TCP

(Continued)

Table 17.1 *(Continued)* **Studies of the effects of direct ES on stem cells**

STEM CELL APPLICATION	REFERENCE	ES TYPE AND PARAMETERS	CELL EFFECT(S) OF ES	CELL TYPE	2-D/3-D AND SUBSTRATE
	Lei et al. (2014)	EC: 40 and 80 mV	• Protein expression of MAP-2 and glial fibrillary acidic protein (80 mV)	NSC/neural progenitor cells	2-D: ITO
	Huang et al. (2012)	EC: 5 mV, 0.5 mA, 25 ms intermittent stimulation	• Gene expression of gamma-enolase, methyl CpG binding protein 2 • Protein expression of MAP-2	NSCs	3-D: CNTs
	Prabhakaran et al. (2011)	100 mV/mm for 60 min	• Neurite outgrowth	NSCs	3-D: Electrospun PLLA/PANI fibers
	Kobelt et al. (2014)	0.53 or 1.83 V/m 10 min/day for 2 days	• Neurite outgrowth • Mature neurite morphology • Expression of βIII-tubulin, neuronal nuclei • Better organized F-actin	Adult neural stem progenitor cell	2-D: Glass
	Stewart et al. (2015)	Charged-balanced biphasic current pulses at 250 Hz ± 1 mA current amplitude, 100 μs pulse width, 20 μs open-circuit interphase gap and 3.78 ms short-circuit phase between pulses	• Total neurite length • Expression of βIII-tubulin	Human fetal neural stem/progenitor cell	2-D: PPy containing the anionic dopant dodecylbenzenesulfonate

(Continued)

Table 17.1 *(Continued)* **Studies of the effects of direct ES on stem cells**

STEM CELL APPLICATION	REFERENCE	ES TYPE AND PARAMETERS	CELL EFFECT(S) OF ES	CELL TYPE	2-D/3-D AND SUBSTRATE
Cardiac	Pavesi et al. (2014)	Monophasic (8 V, 2 ms, 1 Hz) and biphasic (+4 V, 1 ms and −4 V, 1 ms; 1 Hz)	• Biphasic stimulation induced the protein expression of connexin 43 • Gene expression profile more closely related to that of neonatal cadiomyocytes, particularly for biphasic stimulation	ASCs	2-D: PDMS
	Mooney et al. (2012)	0.15 V/cm, 2 ms pulse, 1 Hz	• Cell reorientation and elongation • Expression of cardiac myosin heavy chain, Nkx2.5, GATA-4, cardiac troponin, connexin 43	Human MSCs	3-D: CNT- based PLA nanofiber scaffolds
	Crowder et al. (2013)	500 V/m and 5 ms pulse width at 1 Hz	• Gene expression: Myosin heavy chain (PCL-CNT) • Colocalization of alpha-myosin heavy chain and F-actin (PCL) • Cell–cell contacts (PCL)	Human MSCs	3-D: Electrospun scaffolds of PCL with or without CNT
	Serena et al. (2009)	1 V/mm either 1 or 90 s	• ROS production	Embryoid bodies	3-D: Suspension culture

(Continued)

Table 17.1 *(Continued)* **Studies of the effects of direct ES on stem cells**

STEM CELL APPLICATION	REFERENCE	ES TYPE AND PARAMETERS	CELL EFFECT(S) OF ES	CELL TYPE	2-D/3-D AND SUBSTRATE
	Wu et al. (2009)	Field strengths ranging from 250 to 750 V/m, applied for 60 s	• ROS production	Mouse ES cell-derived embryoid bodies	Cell suspension
	Llucia-Valldeperas et al. (2014)	2 ms monophasic square-wave pulses of 25 mV/cm at 1 Hz for 7 and 14 days	• Cardioregeneration capability • Gene expression: GATA-4 early transcription factor, myocyte enhancer factor 2A	Cardiomyocyte progenitor cells	2-D: PDMS
	Kim et al. (2013)	0, 1, and 3 h at 1.5 V/1.8 cm with a biphasic square pulse (5 ms) at 5 Hz frequency	• Focal adhesion kinase activation for cell adhesion and survival	Sca-1(+) cardiac stem cells	2-D
	Nunes et al. (2013)	Rectangular, biphasic, 1 ms, 3–4 V/cm	• Myofibril ultrastructural organization • Conduction velocity • Improved both electrophysiological and $Ca2^{+}$ handling properties	Human ESC-derived and iPSC-derived cardiac tissues	3-D: "Biowire" (sterile surgical suture in type I collagen gels)
	Wang et al. (2013)	20% strain simulation, 5 V ES, and 20% strain + 5 V ES	• Cell density	Rat MSCs	3-D: Decellularised porcine myocardium

(Continued)

Table 17.1 *(Continued)* **Studies of the effects of direct ES on stem cells**

STEM CELL APPLICATION	REFERENCE	ES TYPE AND PARAMETERS	CELL EFFECT(S) OF ES	CELL TYPE	2-D/3-D AND SUBSTRATE
Angiogenesis	Sauer et al. (2005)	A single electrical field pulse with field strengths ranging from 125 to 750 V/m and a duration of 60 s	• ROS • Protein expression of hypoxia-inducible factor-1a and VEGF	Mouse ESCs	2-D
General	Hernandez-Bule et al. (2014)	5-min pulses of 448 kHz sine wave current, 50 μA/mm^2, separated by 4 h interpulse lapses, along a total period of 48 h	• Proliferation	Human ASCs	2-D: Petri dish
	Hwang et al. (2012)	2-D: 1.5 μA/cm^2 amplitude, 250 ms duration, and 100 pulses/s 3-D: 125 ms duration with 100 pulses/s at varying amplitudes of 10, 20, or 40 μA/cm^2	• BMP-2, insulin-like growth factor, and VEGF, chemokines (CXCL2, interleukin [IL]-8), and chemokine receptors (CXCR4 and IL-8RB)	Human bMSCs	2-D: Au 3-D: Collagen sponge

Note: ALP, alkaline phosphatase; BMP, bone morphogenic protein; CC, capacitive coupling; CNT, carbon nanotube; EC, electric current; Coll I, collagen type I; DW, degenerate wave; ESC, embryonic stem cell; GATA-4, GATA binding protein 4; NSC, neural stem cell; PCL, poly(ε-caprolactone); PLA, polylactide; PLLA, poly-L-lactide; rGO, reduced graphene oxide; ROS, reactive oxygen species; TCP, tissue culture plate.

prolonged EF application persisted for a minimum of 8 h and was passage dependent, with reduced directed migration at higher passages (e.g., 7–10).

Comparison between DC, capacitive coupling, pulsed electromagnetic field (PEMF), and degenerate wave demonstrated DC increased cellular migration in a scratch-wound assay. While all ES waveforms enhanced expression of migratory genes, DC induced the greatest level of expression (Griffin et al. 2011).

Electrotaxis of human NSCs is time and voltage dependent (Feng et al. 2012), where NSCs derived from human embryonic stem cells have been shown to undergo directional migration toward a cathode using small applied DC-EFs, as low as 16 mV/mm (Feng et al. 2012). Reversal of the field polarity reversed migration. Increasing field strength correlated with directedness and migration distance to the cathode.

Like other cell behaviors, the mechanisms underlying ES of cell migration are poorly understood, although assembly of actin filaments, cytoskeletal tension, anterior redistribution of Golgi apparatus, and polarization of membrane receptors are associated phenomena (Balint et al. 2013). Not surprisingly, potential intracellular pathways include the calcium–calmodulin ($Ca[^{2+}]$-CaM) pathway also implicated in neuronal migration in the developing central nervous system (Balint et al. 2013; Kobayashi et al. 2015). Moreover, canonical WNt/GSK3β signaling has recently been implicated in DC-EF regulation of rodent neural progenitor cell migration, with electrotaxis reduced by pharmacological inhibition of Wnt/GSK3β signaling (Liu et al. 2015). Reduced electrotactic response was associated with downregulation of GSK3β phosphorylation, β-catenin activation, and CLASP2 expression. RNA interference of GSK3β also reduced electrotactic response, together with decreased β-catenin activity and GLASP2 expression (Liu et al. 2015). Interestingly, ES applied in the physiological DC-EF range of 30 mV/mm increases the expression of N-cadherin and β-catenin, promoting cell–cell adhesion and chain migration in neuroblast and mouse neurosphere cultures (Cao et al. 2015). Furthermore, the EF effect on chain migration in 3-D cultures of subventricular zone involves upregulation of P2Y purinoceptor 1 (P2Y1) and downstream activation of protein kinase C via N-cadherin and β-catenin (McCaig et al. 2005).

17.3 METHODS OF *IN VITRO* ELECTRICAL STIMULATION

17.3.1 DIRECT STIMULATION PLATFORMS

Direct ES systems are simple and require a relatively small amount of space; hence, they are commonly used for *in vitro* EF stimulation. However, this type of ES may be biased due to the formation of a capacitive electrolyte layer at the electrodes, which are directly in contact with cell culture media (Balint et al. 2013; Merrill et al. 2005). In addition, formation of faradic products (reduced or oxidized chemical species), corrosion of the electrodes, changes in pH, and reduced levels of oxygen are the most characteristic disadvantages (Balint et al 2013; Merrill et al. 2005; Ercan 2010). DC applied in direct stimulation platforms causes major stress on electrodes due to formation of irreversible faradic products and cytotoxic reactions, such as hydrolysis (Merrill et al. 2005; Hronik-Tupaj and Kaplan 2012). This can be prevented by using salt bridges, which act like batteries by transferring DC through a series of redox reactions occurring in the salt bridge (Figure 17.3) (Hammerick et al. 2010a; Hronik-Tupaj and Kaplan 2012). High voltage (>80 V) is

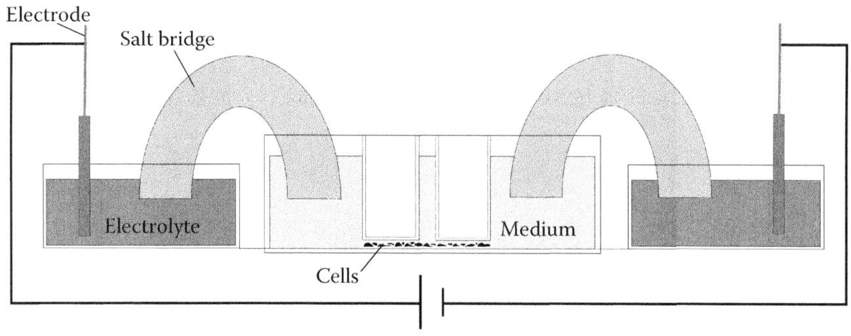

Figure 17.3 ES chamber with salt bridge configuration. Salt bridges are ideal for DC-EF stimulation *in vitro*.

required from the electrodes, as they need to overcome the high resistance of the salt brides (Jennings et al. 2008). Silver–silver chloride electrodes are commonly used with salt bridges (usually connected by immersing them in saline) due to their biocompatibility and routine use in electrophysiology studies (Hronik-Tupaj and Kaplan 2012).

Harmful electrophysiological effects can also be minimized by choosing biphasic stimulation patterns (Merrill et al. 2005; Kim et al. 2009). This allows safer use of simpler and smaller electrode systems that exploit noble metals, metal oxides, ceramics, and graphene-based materials as electrode materials. From noble metals, platinum (Pt), titanium (Ti), and their alloys are commonly used electrode materials due to their stability, corrosion resistance, and biocompatibility (Norlin et al. 2002; Hosseini et al. 2012). Pt is one of the most used electrode materials in biological applications; however, very low charge injection capacity limits its use (Zhao et al. 2011), especially in neural applications where high injection capacities are needed (Musa et al. 2011; Depan and Misra 2014).

Other noble metals that are less used for *in vitro* stimulation include silver (Prabhakaran et al. 2011) and gold (Au). However, silver can have cytotoxic effects in cell culture conditions (Zhang et al. 2014). Stainless steel has also been used for monophasic and biphasic pulses (Pavesi et al. 2014) and even in DC-EF stimulation (Thrivikraman et al. 2014). Titanium nitride (TiN) is a commonly used ceramic material for ES in tissue engineering (Serena et al. 2009; Aryan et al. 2011). From metal oxides, iridium oxide and its mixtures with other conductive materials are commonly used in neural applications due to a high electrical performance (Weiland et al. 2002; Negi et al. 2010; Moral-Vico et al. 2014). Carbon electrodes are also used in many *in vitro* applications, as they possess superior charge injection capacity over TiN and high biocompatibility (Radisic et al. 2004; Tandon et al. 2006, 2009). As a drawback, carbon is a very brittle material and therefore needs to be handled carefully. Electrodes have also been used simultaneously as chemical inductors by Jaatinen et al. (2015), who exploited copper-coated electrodes for neural differentiation of ASCs. Copper was gradually released during ES and entered cells via electrolysis.

17.3.1.1 Electric field stimulation

The term *EF* is potentially misleading in the case of direct stimulation systems, as EC is also evidently involved in the form of electrolyte movement between the working and counterelectrode. The strength of EF can be altered by adjusting the potential difference between the electrodes or by the conductivity of the cell culture medium—the lower the conductivity, the higher the EF. Low-conductivity media (low ionic content) commonly used in the electroporation of cells are often exploited in EF stimulation, as the electrode potentials can be kept relatively low (Serena et al. 2009; Silve et al. 2015).

In conventional direct EF stimulation, the distance between the electrodes is relatively long, requiring use of high potentials that may damage cells, and the stimulation may also be highly heterogenic across the stimulated area. Therefore, interdigitated electrodes have been used to create better-quantified ES that requires low voltages and is relevant to physiological feature sizes (McCullen et al. 2010).

17.3.1.2 Integrated electrode culture systems

Another way to deliver direct ES is by integrated electrode culture systems, which widely vary in configurations. This type of stimulation platform brings EC in close contact with the cell membrane; hence, it is often referred to as EC stimulation (Kim et al. 2009, 2011; Mooney et al. 2012; Pelto et al. 2013a). Yet in many of the cases, EFs are also involved. This technique requires a suitable electrically conductive substrate material that encourages cell attachment to its surface. There are a few different approaches for setting up an integrated stimulation system. In so-called two-electrode systems (Figure 17.4a–c), cells are directly seeded onto the working electrode and counterelectrode is in contact with the cells via medium (Richardson et al. 2007; Kim et al. 2009, 2011; Hwang et al. 2012; Björninen et al. 2014; Stewart et al. 2015). This leads to electrical charging of the working electrode due to relatively low conductivity of the cell culture medium. In another approach presented in Figure 17.4d, EC is generated along a conductive cell substrate, which is placed in between the electrodes (Mooney et al. 2012; Pelto et al. 2013a; Hu et al. 2014). In the third approach, electrodes are connected with the conductive substrate outside the cell culture chamber (Zhang et al. 2013; Hu et al. 2014; Pires et al. 2015). The before-mentioned ES configurations act as examples of how the performance of the substrate material can be altered to serve a certain purpose. In the configurations presented in Figure 17.4, the cell substrates act as working electrodes and the counterelectrodes become strongly polarized due to relatively low conductivity of cell culture media. This creates EFs between the electrodes that are subsequently perceived

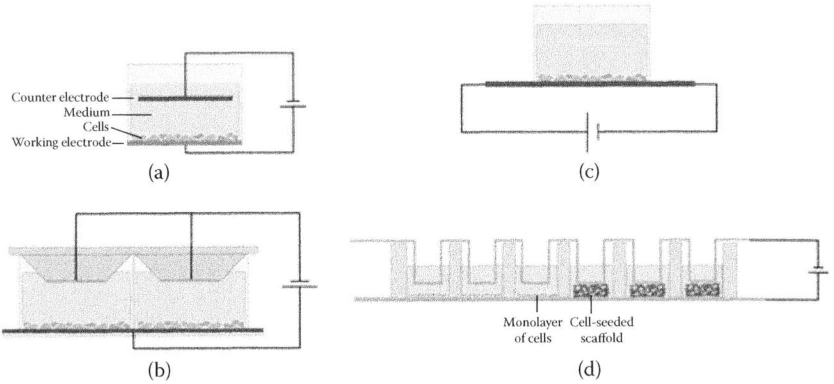

Figure 17.4 Integrated two-electrode stimulation systems. (a) Counterelectrode is fully embedded into cell culture medium. (b) Counterelectrodes are integrated in the chamber lid. (c) Electrodes are attached to conductive cell substrate outside the cell culture well. (d) Counterelectrode is molded to fit in six-well plates.

by the cells. Figure 17.4d shows examples of configurations where the cell substrates, given that they are more conductive than the cell culture media, conduct EC from one electrode to another, not allowing a strong EF to be generated between the electrodes, unlike in the cases presented in Figure 17.4a–c. This of course depends on the conductivity of the substrate. The systems presented in Figure 17.4a–c allow use of substrates of poor conductivity, whereas conductivity of the cell substrate should be higher than that of cell culture media in systems presented in Figure 17.4d in order to serve the purpose as a conductor. It should also be noted that the system presented in Figure 17.4d allows use of DC, as demonstrated by Zhang et al. (2013). DCs of 20–250 μA were applied through the PPy–chitosan films for 4–16 h/day. When it comes to the use of conductive 3-D scaffolds, they can be either squeezed between the electrodes, as presented in Figure 17.4d, or in contact with the working electrode only, which results in stronger polarization of the scaffold upon ES.

Au is commonly used as a conductive substrate in 2-D ES studies (Park et al. 2008; Kim et al. 2009; McCullen et al. 2010). Au is highly conductive, biocompatible, and can be fabricated into different patterns or morphologies on a base material, such as glass (Brunetti et al. 2010; McCullen et al. 2010). In addition, metal oxides, such as indium tin oxide (ITO), have been used in studies conducted in 2-D platforms (Hsiao et al. 2013; Lei et al. 2014).

Carbon-based materials and conductive polymers (CPs) have gained interest in the recent decade, as they can be made biocompatible, at least partially degradable, and they can be used to deliver EC in 3-D structures (Mooney et al. 2012; Li et al. 2013; Pelto et al. 2013a; Gao and Duan 2015). Conductive (or conjugated) polymers, such as commonly used polypyrrole (PPy) or poly(3,4-ethylenedioxythiophene) (PEDOT), go through redox reactions upon AC stimulation when they are being placed as a working electrode (Figure 17.3a) (Björninen et al. 2014; Stewart et al. 2015; Boehler and Asplund 2015). This allows an induction of EF in the close proximity of cells (Meng et al. 2011; Hu et al. 2014; Thrivikraman et al. 2014). CP properties can easily be tailored by doping them with counterions that can be small ions, such as chloride ion, or large biomolecules, such as chondroitin sulfate (Gilmore et al. 2009). The dopant largely determines not only the surface topography or biocompatibility, but also the electrical activity of the CP (Gilmore et al. 2009). Recently, different biomolecules have been explored for ES of stem cells, especially in neural applications (Richardson et al. 2007; Stewart et al. 2015; Pires et al. 2015), but also for ES of MSCs (Björninen et al. 2014; Serra Moreno et al. 2014).

Stewart et al. (2015) provides an example of monitoring PPy performance during ES of NSCs. Biphasic EC was applied, which allowed monitoring of the potentials generated in the circuit. Oxidation of the films occurs during the positive potential and reduction during negative potential. V_a relates to changes in electrolyte, and V_p relates to changes at the electrode surface. V_p describes the stability of the coating.

Redox reactions occurring in CPs during ES can be further exploited in actuation and drug release applications (Thompson et al. 2011; Hsiao et al. 2013; Lei et al. 2014). Regarding carbon-based materials, carbon nanotubes (Mooney et al. 2012; Crowder et al. 2013), graphene (Li et al. 2013), and reduced graphene oxide (Hsiao et al. 2013) have been exploited as conductive substrates either alone or together with CPs.

17.3.2 INDIRECT STIMULATION PLATFORMS

From indirect stimulation platforms, inductive stimulation is much more commonly used for *in vitro* stimulation than capacitive coupling. In capacitive coupling, high voltages are required between the electrodes, limiting the use of this stimulation type (Hess 2012). However, a few *in vitro* capacitive coupling systems have been described. Griffin et al. (2011) tested a system where electrodes were placed below and above bMSCs seeded to a culture dish (Griffin et al. 2011), whereas Ahadian et al. (2012) reported the use of interdigitated noncontactless electrodes for myoblast stimulation. One of the most common applications of capacitive coupling is for treatments of bone fractures and spinal fusions (Kahanovitz 2002; Griffin and Bayat 2011).

Another form of indirect ES is inductive stimulation, which creates EMFs in the cell culture. EMF is a combination of electric and magnetic fields that can induce each other in certain conditions (Saliev et al. 2014). EMF is generated by a coil (Figure 17.1, see "Inductive stimulation") that surrounds the cell culture system (Kang et al. 2013). Alternatively, a Helmoltz configuration, where coils are placed on both sides of the cell culture chamber, can be used to create a uniform magnetic field across the cell culture area (Schuderer et al. 2004; Ma et al. 2014; Yong et al. 2014). The disadvantage of EMF is that the current in the coils produces heat, which needs to be considered in the stimulation system design (Schuderer et al. 2004).

17.3.3 MICROELECTRODE ARRAYS

An MEA is a spatiotemporal measurement system that enables simultaneous ES and recording of individual cells or whole tissues (Spira and Hai 2013; Kim et al. 2014). It is mostly used for measuring electrical activity of excitable cells, such as neural or cardiac cells *in vivo* and *in vitro* (Spira and Hai 2013). MEA is noninvasive and can be used for stimulating and recording cells even for several weeks (Kim et al. 2014). Cells adhere to electrodes—often individual cells contacting each electrode. The normal pattern of planar electrodes is 8×8 or 6×10. The same electrodes can either record extracellular action potentials or deliver charge to electrically stimulate cells. MEA systems widely vary in configurations, for instance, from planar to 3-D (Kim et al. 2014).

In order to achieve a high signal-to-noise ratio (5:1 or higher), MEA materials should have a low impedance (Obien et al. 2015). Materials typically used for electrodes are Au, TiN, Pt, stainless steel, aluminum, and iridium oxide (Obien et al. 2015).

Insulators are an essential part of MEAs, as they passivate the conductor lines leading to the electrodes. Inorganic silicon-based materials or polymers, such as polydimethylsiloxane (PDMS), are used as insulators (Kim et al. 2014).

A useful application for ES of nonexcitable cells is the combination of MEA and fluorescence microscopy. This can be useful, for instance, in observing calcium influx in the cell in response to ES (Nitzan et al. 2011).

17.4 SAFE LIMITS OF ELECTRICAL STIMULATION

In the case of direct ES, the safety of ES arises from the fact that cells and electrodes should not be damaged. This often requires compromises, as the optimal stimulation pattern for cells can be damaging to the electrodes, leading to irreversible faradic reactions at the electrode surface. This can be cytotoxic to the cells and corrosive to the electrodes. Reversible and irreversible faradic reactions at the electrodes are extensively reviewed elsewhere (Merrill et al. 2005). Cell damage under ES originates from the overstimulation of the cells. This threshold is dependent on charge density, frequency, and waveform; hence, it needs to be evaluated in each separate study.

17.4.1 ELECTRODES

In the case of direct stimulation, electrode material plays a crucial role, as stated earlier. The choice of electrode material should be evaluated in each separate case to suit the demands of the ES parameters. Charge injection capacity can be increased by increasing the roughness, and hence the active surface area, of the electrodes. This can be done, for instance, by sputter coating the electrode surface with the electrode material, or coating it with CPs or carbon nanotubes that possess a high surface area due to high porosity (Aregueta-Robles et al. 2014). For example, the CP poly(3,4-ethylenedioxythiophene) (PEDOT) has a high charge injection limit (15 mC/cm^2) and wide potential limit window compared with metallic materials (Wallace et al. 2012). Due to 3-D microtopography and porosity, the surface area of CPs is much greater than that of conventional metal electrodes, and thus leads to a higher charge density and lower impedance. The charge injection mechanism for organic CP electrode materials is more advantageous for biological applications than metals; redox reactions occurring within organic CPs result in electronic current being converted to ionic current. This electronic-to-ionic conversion of current is seemingly more compatible with living cells that also utilize ionic currents.

In some instances, microelectrodes have been isolated from the cell culture, resulting in switching from direct stimulation to capacitive stimulation, with the latter requiring lower voltages (Ahadian et al. 2012). This has been exploited in electrophoresis applications (Park et al. 2009; Shafiee et al. 2009). Ahadian et al. (2012) reported a sharp increase in pH electrode corrosion when AC-EF (voltage 10 V, frequency 1 Hz, and duration 10 ms) was applied via an interdigitated array (IDA) of Pt in direct contact with cell culture medium. No pH changes were observed when IDA-Pt was separated from culture medium by glass coverslip.

17.4.2 WAVEFORM

Monophasic waveforms are purportedly more damaging than biphasic waveforms due to greater formation of electrochemical products (Lilly et al. 1955; Mortimer et al. 1970). However, electrodes used with charge-balanced waveforms are likely to corrode at lower potentials than monophasic waveforms due to damaging positive potentials during anodic phase (Merrill et al. 2005). Therefore, charge-imbalanced waveforms might serve better in maintaining charge balance depending on the electrode system (Merrill et al. 2005). Accordingly, even though charge-balanced waveforms reverse the faradic products generated by the cathodic or anodic phase, applied potentials need to be lower than those with the monophasic waveform. In addition, pH changes under biphasic stimulation have been reported to range from 4 to 10 in close (1 mm) proximity of the electrodes (Ballestrasse et al. 1985). Pavesi et al. (2014) compared monophasic (8 V, 2 ms, 1 Hz) and biphasic (+4 V, 1 ms and −4 V, 1 ms; 1 Hz) pulses for ASCs and found a greater effect of the biphasic waveform in inducing the relocalization of cardiomyogenic marker connexin 43 at the cell membrane (Pavesi et al. 2014).

One should also bear in mind that short EF pulses (~1 μs to 100 ms) with higher than 1 kV/cm EF cause disruption of mammalian cell membrane, leading to electroporation—formation of temporarily existing pores in the cell membrane (Fox et al. 2006). For instance, electroporation for transfection purposes using an electroporation microdevice can be done with a potential difference of 1 V. Even though most of the ES for regenerative purposes is done with much lower potential differences, short pulses with relatively high potential may compromise cell viability.

17.4.3 ELECTRICAL STIMULATION DOSE

Safe limits for EF or EC are highly dependent on the ES parameters, configuration, cell culture conditions, and cell type. Therefore, it should be noted that published methods can only be used as a directional guideline for setting up an ES experiment. The safe limits of EF have been tested by Thrivikraman et al. (2014) and McCullen et al. (2010) in experiments conducted with MSCs. McCullen et al. reported cell death at 100 V/cm of sinusoidal, 1 Hz ES. Thrivikraman et al. seeded cells on conductive polyaniline (PANI) films fabricated with different conductivities and applied DC-EF across the cell culture. This was, as a minimum, partially EC stimulation, depending on the films' conductivity. Cell death was reported at 0.5 mV/cm with the most conductive films, and at 1 V/cm with the majority of other film types.

Stimulation duration also plays a crucial role in the viability of cells. Hu et al. (2014) applied a DC-EF 0.035 and 0.35 V/cm through PPy films that were seeded with rat bMSCs and reported decreased viability when stimulation time was increased from 2 h and 4 h to 12 h.

17.5 SUMMARY

While understanding the mechanisms of action of ES on stem cell behavior and function is in its infancy, research is rapidly expanding to better discern the underlying processes of stem cell regulation. As such, ES is increasingly being recognized as a powerful and perhaps vitally important tool for stem cell R&D, especially when attempting to recapitulate the cues of the cell microenvironment *in vivo* for germane and therefore meaningful cell and tissue modeling. Conversely, exploiting ES to induce abnormal cell behavior is potentially equally important, with transgressions helping to identify what the normal and abnormal (in the case of disease) rules are and how they are imposed. As a result, ES is being used to illuminate the fundamentals of stem cell biology, such as cell signaling, growth, death, differentiation, and tissue formation, as well as uncovering opportunities for advanced therapeutics, such as electroceuticals (Famm et al. 2013). The advent of new molecular tools, such as light-controlled receptors, and other nanoscale probing characterization techniques is also critical for bringing to reality EF-based therapeutics by elucidating the molecular effects of DC-EF in controlling stem cell processes. With the advent of additive fabrication and 3-D printing, the availability of suitable, easy-to-build (e.g., 3-D electrode devices) or ready-to-use, and cost-effective ES platforms and device technology will make ES more accessible than ever before toward being universally applied in stem cell research laboratories as a routine and standard procedure.

REFERENCES

Adey, W. R. (1993). Biological effects of electromagnetic fields. *J Cell Biochem* 51 (4): 410–416.

Ahadian, S., J. Ramon-Azcon, S. Ostrovidov, G. Camci-Unal, V. Hosseini, H. Kaji, K. Ino, H. Shiku, A. Khademhosseini, and T. Matsue. (2012). Interdigitated array of Pt electrodes for electrical stimulation and engineering of aligned muscle tissue. *Lab Chip* 12 (18): 3491–3503.

Aregueta-Robles, U. A., A. J. Woolley, L. A. Poole-Warren, N. H. Lovell, and R. A. Green. (2014). Organic electrode coatings for next-generation neural interfaces. *Front Neuroeng* 7: 15.

Aryan, N. P., M. I. Asad, C. Brendler, S. Kibbel, G. Heusel, and A. Rothermel. (2011). In vitro study of titanium nitride electrodes for neural stimulation. *Conf Proc IEEE Eng Med Biol Soc* 2011: 2866–2869.

Balint, R., N. J. Cassidy, and S. H. Cartmell. (2013). Electrical stimulation: A novel tool for tissue engineering. *Tissue Eng Part B Rev* 19 (1): 48–57.

Ballestrasse, C. L., R. T. Ruggeri, and T. R. Beck. (1985). Calculations of the pH changes produced in body tissue by a spherical stimulation electrode. *Ann Biomed Eng* 13 (5): 405–424.

Barker, A. T., L. F. Jaffe, and J. W. Vanable Jr. (1982). The glabrous epidermis of cavies contains a powerful battery. *Am J Physiol* 242 (3): R358–R366.

Björninen, M., A. Siljander, J. Pelto, J. Hyttinen, M. Kellomäki, S. Miettinen, R. Seppänen, and S. Haimi. (2014). Comparison of chondroitin sulfate and hyaluronic acid doped conductive polypyrrole films for adipose stem cells. *Ann Biomed Eng* 42 (9): 1889–1900.

Boehler, C., and M. Asplund. (2015). A detailed insight into drug delivery from PEDOT based on analytical methods: Effects and side effects. *J Biomed Mater Res A* 103 (3): 1200–1207.

Borgens, R. B., E. Roederer, and M. J. Cohen. (1981). Enhanced spinal cord regeneration in lamprey by applied electric fields. *Science* 213 (4508): 611–617.

Borgens, R. B., J. W. Vanable Jr., and L. F. Jaffe. (1977). Bioelectricity and regeneration: Large currents leave the stumps of regenerating newt limbs. *Proc Natl Acad Sci U S A* 74 (10): 4528–4532.

Brighton, C. T., W. Wang, R. Seldes, G. Zhang, and S. R. Pollack. (2001). Signal transduction in electrically stimulated bone cells. *J Bone Joint Surg Am* 83-A (10): 1514–1523.

Brunetti, V., G. Maiorano, L. Rizzello, B. Sorce, S. Sabella, R. Cingolani, and P. P. Pompa. (2010). Neurons sense nanoscale roughness with nanometer sensitivity. *Proc Natl Acad Sci U S A* 107 (14): 6264–6269.

Cameron, I. L., W. E. Hardman, W. D. Winters, S. Zimmerman, and A. M. Zimmerman. (1993). Environmental magnetic fields: Influences on early embryogenesis. *J Cell Biochem* 51 (4): 417–425.

Cao, L., J. Pu, R. H. Scott, J. Ching, and C. D. McCaig. (2015). Physiological electrical signals promote chain migration of neuroblasts by up-regulating P2Y1 purinergic receptors and enhancing cell adhesion. *Stem Cell Rev* 11 (1): 75–86.

Chan, Y. C., S. Ting, Y. K. Lee, K. M. Ng, J. Zhang, Z. Chen, C. W. Siu, S. K. Oh, and H. F. Tse. (2013). Electrical stimulation promotes maturation of cardiomyocytes derived from human embryonic stem cells. *J Cardiovasc Transl Res* 6 (6): 989–999.

Chang, K.-A., J. W. Kim, J. A. Kim, S. Lee, S. Kim, W. H. Suh, H.-S. Kim, S. Kwon, S. J. Kim, and Y.-H. Suh. (2011). Biphasic electrical currents stimulation promotes both proliferation and differentiation of fetal neural stem cells. *PLoS One* 6 (4): e18738.

Creecy, C. M., C. F. O'Neill, B. P. Arulanandam, V. L. Sylvia, C. S. Navara, and R. Bizios. (2013). Mesenchymal stem cell osteodifferentiation in response to alternating electric current. *Tissue Eng Part A* 19 (3–4): 467–474.

Crowder, S. W., Y. Liang, R. Rath, A. M. Park, S. Maltais, P. N. Pintauro, W. Hofmeister, C. C. Lim, X. Wang, and H.-J. Sung. (2013). Poly(ε-caprolactone)-carbon nanotube composite scaffolds for enhanced cardiac differentiation of human mesenchymal stem cells. *Nanomedicine (London, England)* 8 (11): 1763–1776.

Depan, D., and R. D. K. Misra. (2014). The development, characterization, and cellular response of a novel electroactive nanostructured composite for electrical stimulation of neural cells. *Biomater Sci* 2 (12): 1727–1739.

Ercan, B., and T. J. Webster. (2010). The effect of biphasic electrical stimulation on osteoblast function at anodized nanotubular titanium surfaces. *Biomaterials.* 31 (13): 3684–3693.

Famm, K., B. Litt, K. J. Tracey, E. S. Boyden, and M. Slaoui. (2013). Drug discovery: A jump-start for electroceuticals. *Nature* 496 (7444): 159–161.

Feng, J. F., J. Liu, X. Z. Zhang, L. Zhang, J. Y. Jiang, J. Nolta, and M. Zhao. (2012). Guided migration of neural stem cells derived from human embryonic stem cells by an electric field. *Stem Cells* 30 (2): 349–355.

Fox, M. B., D. C. Esveld, A. Valero, R. Luttge, H. C. Mastwijk, P. V. Bartels, A.van den Berg, and R. M. Boom. (2006). Electroporation of cells in microfluidic devices: A review. *Anal Bioanal Chem* 385 (3): 474–485.

Fukada, E., and I. Yasuda. (1964). Piezoelectric effects in collagen. *Jpn J Appl Phys* 3 (2): 117.

Gao, H., and H. Duan. (2015). 2D and 3D graphene materials: Preparation and bioelectrochemical applications. *Biosens Bioelectron* 65: 404–419.

Gilmore, K. J., M. Kita, Y. Han, A. Gelmi, M. J. Higgins, S. E. Moulton, G. M. Clark, R. Kapsa, and G. G. Wallace. (2009). Skeletal muscle cell proliferation and differentiation on polypyrrole substrates doped with extracellular matrix components. *Biomaterials* 30 (29): 5292–5304.

Griffin, M., and A. Bayat. (2011). Electrical stimulation in bone healing: Critical analysis by evaluating levels of evidence. *Eplasty* 11: e34.

Griffin, M., S. A. Iqbal, A. Sebastian, J. Colthurst, and A. Bayat. (2011). Degenerate wave and capacitive coupling increase human MSC invasion and proliferation while reducing cytotoxicity in an in vitro wound healing model. *PLoS One* 6 (8): e23404.

Gu, X., J. Fu, J. Bai, C. Zhang, J. Wang, and W. Pan. (2015). Low-frequency electrical stimulation induces the proliferation and differentiation of peripheral blood stem cells into Schwann cells. *Am J Med Sci* 349 (2): 157–161.

Hammerick, K. E., A. W. James, Z. Huang, F. B. Prinz, and M. T. Longaker. (2010a). Pulsed direct current electric fields enhance osteogenesis in adipose-derived stromal cells. *Tissue Eng Part A* 16 (3): 917–931.

Hammerick, K. E., M. T. Longaker, and F. B. Prinz. (2010b). In vitro effects of direct current electric fields on adipose-derived stromal cells. *Biochem Biophys Res Commun* 397 (1): 12–17.

Hernandez-Bule, M. L., C. L. Paino, M. A. Trillo, and A. Ubeda. (2014). Electric stimulation at 448 kHz promotes proliferation of human mesenchymal stem cells. *Cell Physiol Biochem* 34 (5): 1741–1755.

Hess, R., Jaeschke, A. (2012). Synergistic effect of defined artificial extracellular matrices and pulsed electric fields on osteogenic differentiation of human MSCs. *Biomaterials.* 33 (35): 8975–8985.

Hosseini, V., S. Ahadian, S. Ostrovidov, G. Camci-Unal, S. Chen, H. Kaji, M. Ramalingam, and A. Khademhosseini. (2012). Engineered contractile skeletal muscle tissue on a microgrooved methacrylated gelatin substrate. *Tissue Eng Part A* 18 (23–24): 2453–2465.

Hronik-Tupaj, M., and D. L. Kaplan. (2012). A review of the responses of two- and three-dimensional engineered tissues to electric fields. *Tissue Eng Part B Rev* 18 (3): 167–180.

Hronik-Tupaj, M., W. L. Rice, M. Cronin-Golomb, D. L. Kaplan, and I. Georgakoudi. (2011). Osteoblastic differentiation and stress response of human mesenchymal stem cells exposed to alternating current electric fields. *Biomed Eng Online* 10: 9.

Hsiao, Y.-S., C.-W. Kuo, and P. Chen. (2013). Multifunctional graphene–PEDOT microelectrodes for on-chip manipulation of human mesenchymal stem cells. *Adv Funct Mater* 23 (37): 4649–4656.

Hu, W.-W., Y.-T. Hsu, Y.-C. Cheng, C. Li, R.-C. Ruaan, C.-C. Chien, C.-A. Chung, and C.-W. Tsao. (2014). Electrical stimulation to promote osteogenesis using conductive polypyrrole films. *Mater Sci Eng C* 37: 28–36.

Huang, Y.-J., H.-C. Wu, N.-H. Tai, and T.-W. Wang. (2012). Carbon nanotube rope with electrical stimulation promotes the differentiation and maturity of neural stem cells. *Small* 8 (18): 2869–2877.

Hwang, S. J., Y. M. Song, T. H. Cho, R. Y. Kim, T. H. Lee, S. J. Kim, Y. K. Seo, and I. S. Kim. (2012). The implications of the response of human mesenchymal stromal cells in three-dimensional culture to electrical stimulation for tissue regeneration. *Tissue Eng Part A* 18 (3–4): 432–445.

Jaatinen, L., S. Salemi, S. Miettinen, J. Hyttinen, and D. Eberli. (2015). The combination of electric current and copper promotes neuronal differentiation of adipose-derived stem cells. *Ann Biomed Eng* 43 (4): 1014–1023.

Jennings, J., D. Chen, and D. Feldman. (2008). Transcriptional response of dermal fibroblasts in direct current electric fields. *Bioelectromagnetics* 29 (5): 394–405.

Kahanovitz, N. (2002). Electrical stimulation of spinal fusion: A scientific and clinical update. *Spine J* 2 (2): 145–150.

Kang, K. S., J. M. Hong, J. A. Kang, J.-W. Rhie, Y. H. Jeong, and D.-W. Cho. (2013). Regulation of osteogenic differentiation of human adipose-derived stem cells by controlling electromagnetic field conditions. *Exp Mol Med* 45 (1): e6.

Khatib, L., D. E. Golan, and M. Cho. (2004). Physiologic electrical stimulation provokes intracellular calcium increase mediated by phospholipase C activation in human osteoblasts. *FASEB J* 18 (15): 1903–1905.

Kim, I. S., Y. M. Song, T. H. Cho, H. Pan, T. H. Lee, S. J. Kim, and S. J. Hwang. (2011). Biphasic electrical targeting plays a significant role in Schwann cell activation. *Tissue Eng Part A* 17 (9–10): 1327–1340.

Kim, I. S., J. K. Song, Y. M. Song, T. H. Cho, T. H. Lee, S. S. Lim, S. J. Kim, and S. J. Hwang. (2009). Novel effect of biphasic electric current on in vitro osteogenesis and cytokine production in human mesenchymal stromal cells. *Tissue Eng Part A* 15 (9): 2411–2422.

Kim, R., S. Joo, H. Jung, N. Hong, and Y. Nam. (2014). Recent trends in microelectrode array technology for in vitro neural interface platform. *Biomed Eng Lett* 4 (2): 129–141.

Kim, S. W., H. W. Kim, W. Huang, M. Okada, J. A. Welge, Y. Wang, and M. Ashraf. (2013). Cardiac stem cells with electrical stimulation improve ischaemic heart function through regulation of connective tissue growth factor and miR-378. *Cardiovasc Res* 100 (2): 241–251.

Kobayashi, H., S. Saragai, A. Naito, K. Ichio, D. Kawauchi, and F. Murakami. (2015). Calm1 signaling pathway is essential for the migration of mouse precerebellar neurons. *Development* 142 (2): 375–384.

Kobelt, L., A. Wilkinson, A. McCormick, R. Willits, and N. Leipzig. (2014). Short duration electrical stimulation to enhance neurite outgrowth and maturation of adult neural stem progenitor cells. *Ann Biomed Eng* 42 (10): 2164–2176.

Lei, K. F., I. C. Lee, Y.-C. Liu, and Y.-C. Wu. (2014). Successful differentiation of neural stem/progenitor cells cultured on electrically adjustable indium tin oxide (ITO) surface. *Langmuir* 30 (47): 14241–14249.

Levin, M. (2007). Large-scale biophysics: Ion flows and regeneration. *Trends Cell Biol* 17 (6): 261–270.

Li, N., Q. Zhang, S. Gao, Q. Song, R. Huang, L. Wang, L. Liu, J. Dai, M. Tang, and G. Cheng. (2013). Three-dimensional graphene foam as a biocompatible and conductive scaffold for neural stem cells. *Sci Rep* 3: 1604.

Lilly, J. C., J. R. Hughes, E. C. Alvord Jr., and T. W. Galkin. (1955). Brief, noninjurious electric waveform for stimulation of the brain. *Science* 121 (3144): 468–469.

Liu, J., B. Zhu, G. Zhang, J. Wang, W. Tian, G. Ju, X. Wei, and B. Song. (2015). Electric signals regulate directional migration of ventral midbrain derived dopaminergic neural progenitor cells via Wnt/GSK3beta signaling. *Exp Neurol* 263: 113–121.

Llucia-Valldeperas, A., B. Sanchez, C. Soler-Botija, C. Galvez-Monton, S. Roura, C. Prat-Vidal, I. Perea-Gil, J. Rosell-Ferrer, R. Bragos, and A. Bayes-Genis. (2014). Physiological conditioning by electric field stimulation promotes cardiomyogenic gene expression in human cardiomyocyte progenitor cells. *Stem Cell Res Ther* 5 (4): 93.

Ma, Q., P. Deng, G. Zhu, et al. (2014). Extremely low-frequency electromagnetic fields affect transcript levels of neuronal differentiation-related genes in embryonic neural stem cells. *PLoS One* 9 (3): e90041.

Matsumoto, M., T. Imura, T. Fukazawa, Y. Sun, M. Takeda, T. Kajiume, Y. Kawahara, and L. Yuge. (2013). Electrical stimulation enhances neurogenin2 expression through β-catenin signaling pathway of mouse bone marrow stromal cells and intensifies the effect of cell transplantation on brain injury. *Neurosci Lett* 533: 71–76.

McCaig, C. D., A. M. Rajnicek, B. Song, and M. Zhao. (2005). Controlling cell behavior electrically: Current views and future potential. *Physiol Rev* 85 (3): 943–978.

McCullen, S. D., J. P. McQuilling, R. M. Grossfeld, J. L. Lubischer, L. I. Clarke, and E. G. Loboa. (2010). Application of low-frequency alternating current electric fields via interdigitated electrodes: Effects on cellular viability, cytoplasmic calcium, and osteogenic differentiation of human adipose-derived stem cells. *Tissue Eng Part C Methods* 16 (6): 1377–1386.

Meng, S., Z. Zhang, and M. Rouabhia. (2011). Accelerated osteoblast mineralization on a conductive substrate by multiple electrical stimulation. *J Bone Miner Metab* 29 (5): 535–544.

Merrill, D. R., M. Bikson, and J. G. R. Jefferys. (2005). Electrical stimulation of excitable tissue: Design of efficacious and safe protocols. *J Neurosci Methods* 141 (2): 171–198.

Mooney, E., J. N. Mackle, D. J. Blond, E. O'Cearbhaill, G. Shaw, W. J. Blau, F. P. Barry, V. Barron, and J. M. Murphy. (2012). The electrical stimulation of carbon nanotubes to provide a cardiomimetic cue to MSCs. *Biomaterials* 33 (26): 6132–6139.

Moral-Vico, J., S. Sánchez-Redondo, M. P. Lichtenstein, C. Suñol, and N. Casañ-Pastor. (2014). Nanocomposites of iridium oxide and conducting polymers as electroactive phases in biological media. *Acta Biomater* 10 (5): 2177–2186.

Mortimer, J. T., C. N. Shealy, and C. Wheeler. (1970). Experimental nondestructive electrical stimulation of the brain and spinal cord. *J Neurosurg* 32 (5): 553–559.

Musa, S., D. R. Rand, C. Bartic, W. Eberle, B. Nuttin, and G. Borghs. (2011). Coulometric detection of irreversible electrochemical reactions occurring at Pt microelectrodes used for neural stimulation. *Anal Chem* 83 (11): 4012–4022.

Mycielska, M. E., and M. B. Djamgoz. (2004). Cellular mechanisms of direct-current electric field effects: Galvanotaxis and metastatic disease. *J Cell Sci* 117 (Pt 9): 1631–1639.

Negi, S., R. Bhandari, L. Rieth, R. Van Wagenen, and F. Solzbacher. (2010). Neural electrode degradation from continuous electrical stimulation: Comparison of sputtered and activated iridium oxide. *J Neurosci Methods* 186 (1): 8–17.

Nitzan, H., S.-I. Mark, and H. Yael. (2011). Optical validation of in vitro extra-cellular neuronal recordings. *J Neural Eng* 8 (5): 056008.

Norlin, A., J. Pan, and C. Leygraf. (2002). Investigation of interfacial capacitance of Pt, Ti and TiN coated electrodes by electrochemical impedance spectroscopy. *Biomol Eng* 19 (2–6): 67–71.

Nunes, S. S., J. W. Miklas, J. Liu, et al. (2013). Biowire: A platform for maturation of human pluripotent stem cell-derived cardiomyocytes. *Nat Methods* 10 (8): 781–787.

Obien, M. E. J., K. Deligkaris, T. Bullmann, D. J. Bakkum, and U. Frey. (2015). Revealing neuronal function through microelectrode array recordings. *Front Neurosci* 8: 423.

Park, J. S., K. Park, H. T. Moon, D. G. Woo, H. N. Yang, and K.-H. Park. (2008). Electrical pulsed stimulation of surfaces homogeneously coated with gold nanoparticles to induce neurite outgrowth of PC12 cells. *Langmuir* 25 (1): 451–457.

Park, K., H.-J. Suk, D. Akin, and R. Bashir. (2009). Dielectrophoresis-based cell manipulation using electrodes on a reusable printed circuit board. *Lab Chip* 9 (15): 2224–2229.

Pavesi, A., M. Soncini, A. Zamperone, S. Pietronave, E. Medico, A. Redaelli, M. Prat, and G. B. Fiore. (2014). Electrical conditioning of adipose-derived stem cells in a multi-chamber culture platform. *Biotechnol Bioeng* 111 (7): 1452–1463.

Pelto, J., M. Björninen, A. Pälli, E. Talvitie, J. Hyttinen, B. Mannerström, R. Suuronen Seppanen, M. Kellomäki, S. Miettinen, and S. Haimi. (2013a). Novel polypyrrole-coated polylactide scaffolds enhance adipose stem cell proliferation and early osteogenic differentiation. *Tissue Eng Part A* 19 (7–8): 882–892.

Pelto, J. M., S. P. Haimi, A. S. Siljander, S. S. Miettinen, K. M. Tappura, M. J. Higgins, and G. G. Wallace. (2013b). Surface properties and interaction forces of biopolymer-doped conductive polypyrrole surfaces by atomic force microscopy. *Langmuir* 29 (20): 6099–6108.

Pires, F., Q. Ferreira, C. A. Rodrigues, J. Morgado, and F. C. Ferreira. (2015). Neural stem cell differentiation by electrical stimulation using a cross-linked PEDOT substrate: Expanding the use of biocompatible conjugated conductive polymers for neural tissue engineering. *Biochim Biophys Acta* 1850 (6): 1158–1168.

Prabhakaran, M. P., L. Ghasemi-Mobarakeh, G. Jin, and S. Ramakrishna. (2011). Electrospun conducting polymer nanofibers and electrical stimulation of nerve stem cells. *J Biosci Bioeng* 112 (5): 501–507.

Radisic, M., H. Park, H. Shing, T. Consi, F. J. Schoen, R. Langer, L. E. Freed, and G. Vunjak-Novakovic. (2004). Functional assembly of engineered myocardium by electrical stimulation of cardiac myocytes cultured on scaffolds. *Proc Natl Acad Sci U S A* 101 (52): 18129–18134.

Richardson, R. T., B. Thompson, S. Moulton, C. Newbold, M. G. Lum, A. Cameron, G. Wallace, R. Kapsa, G. Clark, and S. O'Leary. (2007). The effect of polypyrrole with incorporated neurotrophin-3 on the promotion of neurite outgrowth from auditory neurons. *Biomaterials* 28 (3): 513–523.

Robinson, K. R. (1985). The responses of cells to electrical fields: A review. *J Cell Biol* 101 (6): 2023–2027.

Saliev, T., Z. Mustapova, G. Kulsharova, D. Bulanin, and S. Mikhalovsky. (2014). Therapeutic potential of electromagnetic fields for tissue engineering and wound healing. *Cell Prolif* 47 (6): 485–493.

Sauer, H., M. M. Bekhite, J. Hescheler, and M. Wartenberg. (2005). Redox control of angiogenic factors and CD31-positive vessel-like structures in mouse embryonic stem cells after direct current electrical field stimulation. *Exp Cell Res* 304 (2): 380–390.

Schuderer, J., W. Oesch, N. Felber, D. Spät, and N. Kuster. (2004). In vitro exposure apparatus for ELF magnetic fields. *Bioelectromagnetics* 25 (8): 582–591.

Serena, E., E. Figallo, N. Tandon, C. Cannizzaro, S. Gerecht, N. Elvassore, and G. Vunjak-Novakovic. (2009). Electrical stimulation of human embryonic stem cells: Cardiac differentiation and the generation of reactive oxygen species. *Exp Cell Res* 315 (20): 3611–3619.

Serra Moreno, J., M. G. Sabbieti, D. Agas, L. Marchetti, and S. Panero. (2014). Polysaccharides immobilized in polypyrrole matrices are able to induce osteogenic differentiation in mouse mesenchymal stem cells. *J Tissue Eng Regen Med* 8 (12): 989–999.

Shafiee, H., J. Caldwell, M. Sano, and R. Davalos. (2009). Contactless dielectrophoresis: A new technique for cell manipulation. *Biomed Microdevices* 11 (5): 997–1006.

Silve, A., I. Leray, M. Leguèbe, C. Poignard, and L. M. Mir. (2015). Cell membrane permeabilization by 12-ns electric pulses: Not a purely dielectric, but a charge-dependent phenomenon. *Bioelectrochemistry* 106 (Pt. B): 369–378.

Spira, M. E., and A. Hai. (2013). Multi-electrode array technologies for neuroscience and cardiology. *Nat Nano* 8 (2): 83–94.

Stewart, E., N. R. Kobayashi, M. J. Higgins, A. F. Quigley, S. Jamali, S. E. Moulton, R. M. Kapsa, G. G. Wallace, and J. M. Crook. (2015). Electrical stimulation using conductive polymer polypyrrole promotes differentiation of human neural stem cells: A biocompatible platform for translational neural tissue engineering. *Tissue Eng Part C Methods* 21 (4): 385–393.

Sun, S., Y. Liu, S. Lipsky, and M. Cho. (2007). Physical manipulation of calcium oscillations facilitates osteodifferentiation of human mesenchymal stem cells. *FASEB J* 21 (7): 1472–1480.

Tandon, N., C. Cannizzaro, P. H. Chao, R. Maidhof, A. Marsano, H. T. Au, M. Radisic, and G. Vunjak-Novakovic. (2009). Electrical stimulation systems for cardiac tissue engineering. *Nat Protoc* 4 (2): 155–173.

Tandon, N., C. Cannizzaro, E. Figallo, J. Voldman, and G. Vunjak-Novakovic. (2006). Characterization of electrical stimulation electrodes for cardiac tissue engineering. *Conf Proc IEEE Eng Med Biol Soc* 1: 845–848.

Thompson, B. C., S. E. Moulton, R. T. Richardson, and G. G. Wallace. (2011). Effect of the dopant anion in polypyrrole on nerve growth and release of a neurotrophic protein. *Biomaterials* 32 (15): 3822–3831.

Thornton, T. M., and M. Rincon. (2009). Non-classical p38 map kinase functions: Cell cycle checkpoints and survival. *Int J Biol Sci* 5 (1): 44–51.

Thrivikraman, G., G. Madras, and B. Basu. (2014). Intermittent electrical stimuli for guidance of human mesenchymal stem cell lineage commitment towards neural-like cells on electroconductive substrates. *Biomaterials* 35 (24): 6219–6235.

Wallace, G. G., S. Moulton, R. M. Kapsa, and M. J. Higgins. (2012). *Organic Bionics*. Hoboken, NJ: Wiley.

Wang, B., G. Wang, F. To, J. R. Butler, A. Claude, R. M. McLaughlin, L. N. Williams, A. L. de Jongh Curry, and J. Liao. (2013). Myocardial scaffold-based cardiac tissue engineering: Application of coordinated mechanical and electrical stimulations. *Langmuir* 29 (35): 11109–11117.

Weiland, J. D., D. J. Anderson, and M. S. Humayun. (2002). In vitro electrical properties for iridium oxide versus titanium nitride stimulating electrodes. *IEEE Trans Biomed Eng* 49 (12): 1574–1579.

Wu, C. X., J. Y. Ma, K. Lin, and Y. K. Shi. (2009). Differentiation of bone marrow mesenchymal stem cells into cardiomyocytes in rats following low-tension pulse electric stimulation: In the hope of verifying effects of electricity factors in development of the myocardium. *J Clin Rehabil Tissue Eng Res* 13 (40): 7859–7864.

Yamada, M., K. Tanemura, S. Okada, et al. (2007). Electrical stimulation modulates fate determination of differentiating embryonic stem cells. *Stem Cells* 25 (3): 562–570.

Yong, Y., Z. D. Ming, L. Feng, Z. W. Chun, and W. Hua. (2014). Electromagnetic fields promote osteogenesis of rat mesenchymal stem cells through the PKA and ERK1/2 pathways. *J Tissue Eng Regen Med* 10 (10): E537–E545. DOI: 10.1002/term.1864.

Zhang, J., K. G. Neoh, X. Hu, E.-T. Kang, and W. Wang. (2013). Combined effects of direct current stimulation and immobilized BMP-2 for enhancement of osteogenesis. *Biotechnol Bioeng* 110 (5): 1466–1475.

Zhang, T., L. Wang, Q. Chen, and C. Chen. (2014). Cytotoxic potential of silver nanoparticles. *Yonsei Med J* 55 (2): 283–291.

Zhang, W., and H. T. Liu. (2002). MAPK signal pathways in the regulation of cell proliferation in mammalian cells. *Cell Res* 12 (1): 9–18.

Zhao, M. (2009). Electrical fields in wound healing—An overriding signal that directs cell migration. *Semin Cell Dev Biol* 20 (6): 674–682.

Zhao, Z., C. Watt, A. Karystinou, A. J. Roelofs, C. D. McCaig, I. R. Gibson, and C. De Bari. (2011). Directed migration of human bone marrow mesenchymal stem cells in a physiological direct current electric field. *Eur Cell Mater* 22: 344–358.

18 Effects of electrical stimulation on cutaneous wound healing: Evidence from *in vitro* studies and clinical trials

Sara Ud-Din and Ardeshir Bayat
Medicine and Health Manchester, University of Manchester, United Kingdom

Contents

18.1 INTRODUCTION

An effective means of promoting wound healing is electrical stimulation (ES) (Szuminsky et al. 1994). ES is defined as the application of an electrical current through electrodes placed either within the wound itself or on the periwound skin (Isseroff and Dahle 2012). An important parameter of ES in wound healing is the type of applied polarity, which may affect protein synthesis, cell migration, growth of bacteria, galvanotaxis, inflammation, and the processes of bioelectric events of injury (Szuminsky et al. 1994). ES has been used for a number of clinical applications, such as pain management and wound healing, including chronic and acute wounds (Thakral et al. 2013). There are a range of treatment strategies available for wound management, including both invasive and noninvasive modalities, which are outlined in Figure 18.1. The use of ES for the treatment of both acute and chronic wounds has gained prominence in the literature.

Several different modalities and waveforms of electricity have been described, including direct current (DC), alternating current (AC), high-voltage pulsed current (HVPC), and low-intensity direct current (LIDC) (Szuminsky et al. 1994). One of the most familiar types of ES is transcutaneous electrical nerve stimulation (TENS), which has been used frequently for pain control (Szuminsky et al. 1994). Additionally, frequency rhythmic electrical modulation systems (FREMS) is also a form of transcutaneous electrotherapy using ES that varies the pulse, frequency, duration, and voltage (Thakral et al. 2013). Recently, a biofeedback device named the Fenzian system, where its waveform was deciphered to look like degenerate waves (DWs), has been used in the treatment of acute cutaneous wounds (Fernandez-Chimeno et al. 2004). ES devices have varying voltages, currents, modes, and lengths of time of application. The decisions regarding the mode of application are compounded by choices of frequency and duration of pulses, the size of the current applied, and the selection of appropriate electrodes (Fernandez-Chimeno et al. 2004).

The rationale for use of ES is the potential at the epidermis, which is known as transepithelial potential (TEP) and varies between 10 and 60 mV (average, 23 mV) (Foulds and Barker 1983). The TEP is caused by the concentration of negative

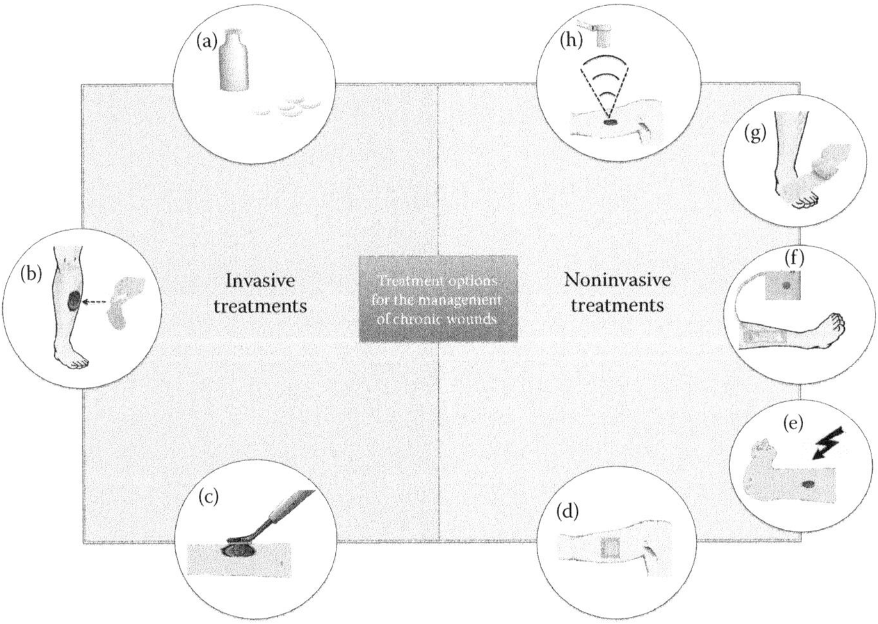

Figure 18.1 Diagram to demonstrate some of the available treatment strategies for the management of chronic wounds under the categories invasive and noninvasive. Invasive treatments include (a) antibiotic therapy, (b) skin substitute therapy, and (c) debridement. Noninvasive treatments include (d) wound dressings, (e) electrical stimulation, (f) negative pressure wound therapy, (g) compression bandaging, and (h) ultrasound.

chlorine ions at the surface and positive sodium and potassium ions in the tissues. During wounding, this epithelial seal is broken. The TEP collapses to 0 mV at the wound edge and ions begin to leak out, establishing an "injury current" (10–100 μA) accompanied by its electric field (EF), 140 mV/mm, which rapidly falls to 0 mV/mm only 2–3 mm away from the wound edge (Foulds and Barker 1983). Depending on the EF direction, healthy cells will be attracted or repelled and the speed and direction of their motility (electrotaxis) will be influenced by this (McCaig et al. 2005; Cheng et al. 1995).

18.2 MODES OF ES

The main types of ES currents are DC, AC, and pulsed current (PC) (Figure 18.2).

18.2.1 ALTERNATING CURRENT

AC is the continuous bidirectional flow of charged particles in which a change in direction of flow occurs at least once every second (Kloth 2005). As AC is not used clinically for wound healing, it is not discussed in this chapter.

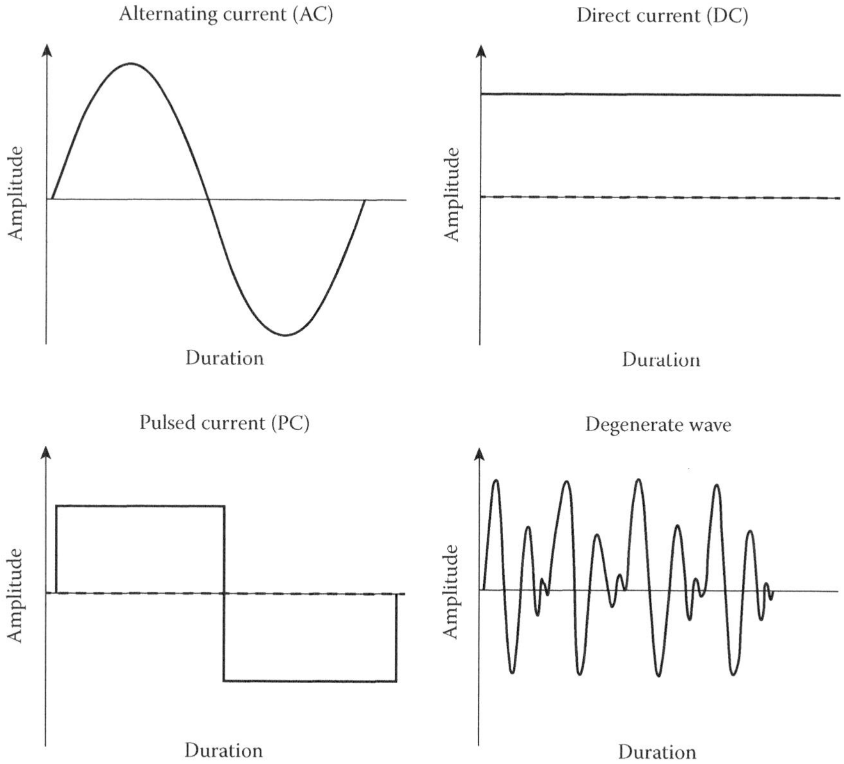

Figure 18.2 Illustrations showing one example of each of the main electrical waveforms available for the treatment of wounds, including alternating current, direct current, and pulsed current. There are other subtypes of each of these waveforms.

18.2.2 DIRECT CURRENT

DC is the continuous, unidirectional flow of charged particles for 1 s or longer (Baker et al. 1997). In the tissues, the direction of DC flow is determined by the polarity, with negatively charged ions moving toward the anode and positively charged ions moving toward the cathode (Baker et al. 1997). The electrode polarity remains constant until it is changed manually on the ES device. Continuous DC has no pulses and therefore no waveform. When DC is used to treat wounds, less than 1.0 mA should be delivered to the electrodes.

18.2.3 PULSED CURRENT

PC is the unidirectional or bidirectional flow of charged particles, in which each pulse is separated by a longer period of no current flow (Kloth 2005; Gentzkow et al. 1993). PC can have two waveforms: monophasic or biphasic. A monophasic pulse represents a very brief movement of electrons or ions away from the isoelectric line, returning to the zero line after a finite period of time (less than 1.0 s) (Gentzkow et al. 1993). The biphasic PC waveform also represents a very brief duration of movement of electrons or ions. However, in this case, the pulse is bidirectional and can be asymmetric or symmetric about the isoelectric line.

Currently, there is substantial evidence that supports the effectiveness of ES for cutaneous wound healing. This chapter focuses on the effects of ES on cutaneous wound healing, with particular emphasis on tissue oxygenation, angiogenesis, antibacterial effects, cell migration, and protein synthesis, utilizing clinical trials and *in vitro* evidence (Table 18.1, Figure 18.3).

18.3 EFFECTS OF ES ON CUTANEOUS WOUND HEALING

18.3.1 TISSUE OXYGENATION

There is growing evidence from human subject studies that ES facilitates a temporary increase in local tissue oxygen tension (Peters et al. 1998; Clover et al. 2003). It is commonly recognized that cells involved in tissue repair require oxygen to function most efficiently. Peters et al. (1998) evaluated diabetic patients with impaired vascular function treated with ES delivered through a silver mesh sock for 60 min on two consecutive days. They showed that they did not have the delayed response to an increase in $PtcO_2$ (Peters et al. 1998). This result suggests that ES increases cutaneous oxygen saturation secondary to increasing local perfusion in diabetic subjects.

In a study by Clover et al. (2003), they noted a significant increase in capillary density in patients with peripheral artery disease after 3 and 6 weeks of TENS treatment and 4 weeks posttreatment. Microvessel density was determined by microscope visualization of capillaries. Perfusion was also determined by transcutaneous oxygen tension and provided measurement of skin oxygen supply in superficial vessels. Transcutaneous oxygen measurements were significantly greater in the treatment group at 3 and 6 weeks of treatment and 4 weeks posttreatment (Clover et al. 2003).

Table 18.1 **Summary of the literature pertaining to the effects of electrical stimulation on wound healing, including tissue oxygenation, angiogenesis, antibacterial effects, cell migration, and protein synthesis**

AUTHOR	NO. OF PATIENTS	ES TYPE	DURATION	EFFECTS
Tissue oxygenation				
Dogden et al.	N = 10 diabetic and N = 20 normal	Monophasic via cathode and biphasic via cathode	30-min session	Increased $PtcO_2$ after ES in normal subjects and no increase in diabetic patients
Peters et al.	N = 19	Subsensory ES	Four 60-min periods over 2 days	No delayed response post ES
Clover et al.	N = 36 with peripheral vascular disease	Subcontractile ES	Three 60-min sessions over 6 weeks	Increased microvessel density and tissue perfusion
Angiogenesis				
Greenberg et al.	Burn wounds in pig skin (number not stated)	Monophasic pulsed ES with varying polarities	30 min daily	Neovascularity was seen on day 10 following negative versus positive polarity
Mertz et al.	Pig wounds (number not stated)	Monophasic pulsed current	Two 30-min sessions for 7 days	Enhanced epithelialization following both cathodal and anodal ES
Junger et al.	N = 15 with venous leg ulcers	Monophasic pulsed current	30 min daily for 38 days	Improved capillary density, increased oxygen tension and tissue perfusion
Cramp et al.	N = 30	TENS	15-min session	Increase in blood perfusion
Gilcreast et al.	N = 132 diabetic patients	High-voltage pulsed current	60-min session	Increased blood flow
Sherry et al.	N = 20	TENS	2 burst per second	Applied at 25% above motor threshold increased blood flow

(Continued)

Table 18.1 *(Continued)* **Summary of the literature pertaining to the effects of electrical stimulation on wound healing, including tissue oxygenation, angiogenesis, antibacterial effects, cell migration, and protein synthesis**

AUTHOR	NO. OF PATIENTS	ES TYPE	DURATION	EFFECTS
Nolan et al.	N = 1	TENS	20-min session	Increased blood flow
Wilkstrom et al.	N = 9	TENS	60-min sessions on 3 occasions	Stimulated peripheral circulation
Ud-Din et al.	N = 20	DW stimulation (clinical)	4 sessions average 20 min	Increased blood flow and hemoglobin levels
Sebastian et al.	N = 20	DW stimulation (*in vitro*)	4 sessions average 20 min	Upregulated angiogenesis
Antibacterial effects				
Wolcott et al.	N/A	Cathodal DC	Microampere levels over several days	Wounds were pathogen free following treatment
Rowley	N/A	AC and cathodal DC	Microampere and milliampere levels—duration not stated	Significant bacteriostatic effect on *in vitro* growth rates of *Escherichia coli*
Guffey et al.	N/A	HVPC and DC	30-min session	Anodal and cathodal continuous DC inhibited *Staphylococcus aureus* growth
Kincaid and Lavoie	N/A	HVPC	250 V dosage at cathode, 250 V dosage at both anode and cathode	pH changes noted at these doses
Szuminsky et al.	N/A	HVPC	500 V	Bacterial destruction observed

(Continued)

Table 18.1 *(Continued)* **Summary of the literature pertaining to the effects of electrical stimulation on wound healing, including tissue oxygenation, angiogenesis, antibacterial effects, cell migration, and protein synthesis**

AUTHOR	NO. OF PATIENTS	ES TYPE	DURATION	EFFECTS
Cell migration				
Mertz et al.	N/A	Monophasic pulsed current in ovine model	Two 30-min sessions for 7 days	Greater epithelialization when treated with the cathode on day 0 and the anode from days 1 to 7
Eberhardt et al.	N/A Human skin	Exogenous ES	Not stated	Greater number of neutrophils
Protein synthesis				
Bourguignon et al.	N/A Human fibroblasts	HVPC	50 and 75 V an 100 pps	Maximum synthesis occurred at these parameters
Bourguignon et al.	N/A	HVPC	50 and 75 V and 100 pps	Increase in Ca^{2+} uptake and upregulation of insulin receptors
Bassett and Herrmann	N/A Fibroblast cultures	Electrostatic field	1000 V/cm	20% increase in DNA and collagen synthesis after 14 days
Cheng et al.	N/A Rat skin	Microampere levels of DC	10 to 1000 microampere for 2 h *in vitro*	Concentration of adenosine triphosphate was increased by 5 times following ES and increased amino acid uptake thereby obtaining maximum effect on protein synthesis
Falanga et al.	N/A Human fibroblasts	DC	100 V and 100 pps	Upregulates receptors for TGF-β on human dermal fibroblasts

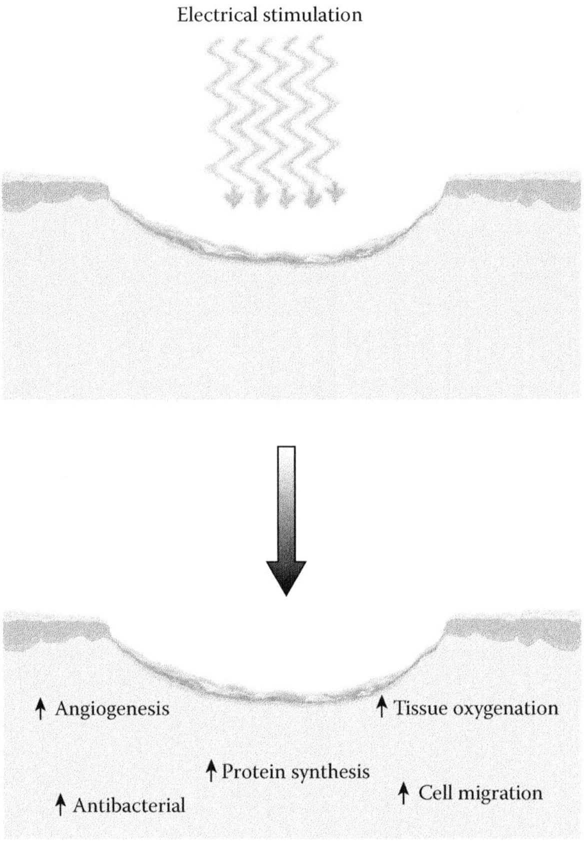

Figure 18.3 Illustration to demonstrate that when electrical stimulation is applied to a wound, angiogenesis is enhanced, tissue oxygenation is increased, protein synthesis is improved, and cell migration is enhanced, and this provides antibacterial effects.

18.3.2 WOUND ANGIOGENESIS

Clinical studies have reported increased blood flow secondary to increasing capillary density in human wounds treated with ES (Mertz et al. 1993; Junger et al. 1997). Mertz et al. (1993) showed enhanced epithelialization in pig wounds following both cathodal and anodal stimulation by 20% compared with wounds treated with either positive or negative polarity alone (Mertz et al. 1993).

Junger et al. (1997) demonstrated a mean increase of 43.5% in capillary density in venous leg ulcers in 15 patients (Junger et al. 1997). Monophasic PC was applied for 30 min daily for a mean of 38 days. The parameters were 140 μs pulse duration with an average current of 630 μA at 128 pps or 315 μA at 64 pps. Capillary density, which was observed by light microscopy, improved from a baseline of 8.05 capillaries/mm^2 to 11.55 capillaries/mm^2 poststimulation ($p < 0.039$). $PtcO_2$ was also measured in the periwound skin prior to and following the ES treatments. Oxygen tension was shown to increase from 13.5 to 24.7 mmHg, and skin perfusion increased as determined by laser Doppler fluxmetry.

Cramp et al. (2000) reported increased laser Doppler blood flow in a double-blinded study of healthy subjects with the application of TENS. There was a significant increase in blood flow in the low-intensity TENS group compared with the control and high-frequency TENS groups at 3, 6, 9, 12, and 15 min after the start of treatment. Gilcreast et al. (1998) evaluated perfusion in 132 diabetic subjects that were nontobacco users, before and after ES. A group of subjects demonstrated a significant increase in transcutaneous oxygen measurement (27%, $n = 35$). Responders were older and more likely to have neuropathy, higher blood glucose levels (glycated hemoglobin 9%), and good perfusion to the forefoot (toe blood pressure 70 mmHg). TENS is considered to be one of the most common therapeutic modalities used in clinical practice for the relief of chronic and acute pain (Liebano et al. 2011). Some authors have observed that in addition to its analgesic effects, TENS can also alter skin temperature and increase blood flow (Atalay and Yilmaz 2009). This observation has led to various studies investigating the effect on the peripheral vascular system and how this facilitates tissue repair (Sherry et al. 2001). Some studies have shown that TENS significantly increases skin temperature with low-frequency (2–4 Hz) (Kaada et al. 1984) and high-frequency (75–100 Hz) TENS (Nolan et al. 1993), and in local blood flow (Kaada et al. 1984). Other studies suggested that when applied at the same intensity, low-frequency TENS enhanced blood flow levels more than high-frequency TENS (Cramp et al. 2000; Wikstrom et al. 1999).

A clinical trial by Ud-Din et al. (2012) was conducted involving multiple temporal punch biopsies treated with biofeedback ES and demonstrated increased blood flow and hemoglobin levels in acute cutaneous wounds (on day 14 postwounding) created in 20 human volunteers, compared with controls who had not received ES. The biofeedback ES used is a transcutaneous low-intensity device that detects changes in skin impedance. This device forms part of a biofeedback link with the individual's normal physiological repair. Using a concentric electrode, the device detects the skin's electrical impedance and adjusts the outgoing microcurrent electrical biofeedback impulses (Colthurst and Giddings 2007). The device delivers 0.004 mA, 20–80 V, and has a frequency default of 60 Hz and impulses, which last approximately 6/100 of a second. This treatment modality accelerated the rate of cutaneous wound healing in all cases, as evidenced by gene and protein studies showing upregulated angiogenesis and downregulated inflammation (Sebastian et al. 2011).

18.3.3 ANTIBACTERIAL EFFECTS

Bacterial load and infection are thought to be important factors in chronic wounds and delayed healing (Edwards and Harding 2004; Halbert et al. 1992; Madsen et al. 1996). The antibacterial effects of ES have been studied both *in vitro* and *in vivo*, and results have indicated that ES may impose a bacteriostatic or bactericidal effect on microbes that commonly infect wounds. Halbert et al. (1992) took bacterial cultures from 83 limbs and showed an association between delayed wound healing and higher bacterial counts in leg ulcers. Compared with noncolonized ulcers, colonized ulcers had longer duration and larger size and took a longer time to heal (Halbert et al. 1992).

Wolcott et al. (1969) observed that human wounds initially colonized with *Pseudomonas* or *Proteus* organisms were pathogen-free following several days of treatment with microampere levels of cathodal DC. In a study by Rowley (1972), milliampere levels of AC and cathodal DC delivered through platinum electrodes showed that *in vitro* growth rates of *Escherichia coli* demonstrated a significant bacteriostatic effect with DC, while they were not affected by AC. Additionally, numerous studies have established that antibacterial activity occurred in the presence of silver cations deposited *in vivo* or *in vitro* by low levels of DC (Marino et al. 1984, 1985; Deitch et al. 1983; Colmano et al. 1980; Thibodeau et al. 1978; Alvarez et al. 1983).

Further studies have compared *in vitro* antibacterial effects of HVPC and DC and found that HVPC applied at 50–800 mA and 100 pps for 30 min had no inhibitory effects on *Staphylococcus aureus*, whereas both anodal and cathodal continuous DC applied at 1, 5, and 10 mA did inhibit *S. aureus* growth (Guffey and Asmussen 1989). Electrochemical pH changes have not been shown to occur at the anode or cathode when HVPC is applied to human tissues for 30 min (Newton and Karselis 1983). However, when Kincaid and Lavoie (1989) evaluated antibacterial effects of HVPC *in vitro*, they observed pH changes only at the cathode at a dosage of 500 V and at both the anode and cathode at 250 V. Szuminsky et al. (1994) aimed to identify the mechanisms by which HVPC applied at 500 V causes bacterial destruction *in vitro*. They were unable to determine whether the effect was due to the direct action of the current on the organisms, electrophoretic recruitment of antimicrobial factors, local heat generation, or pH changes. Although these studies showed positive effects *in vitro*, the high voltages used would not be tolerated if applied to human wound subjects.

Kloth (2005) suggested that since there are many silver-impregnated dressings available that deliver silver ions to wounds to reduce the bacterial problem, the efficacy of these dressings may be enhanced by actively repelling the silver ions into the wound with anodal DC.

18.3.4 CELL MIGRATION

Several studies have shown that cells involved in wound healing migrate toward the anode or cathode of an EF (known as galvanotaxis) delivered into the cell cultures (Yang et al. 1984; Nishimura et al. 1996; Sheridan et al. 1996). Macrophages, which are important during the inflammatory wound healing phase, migrate toward the anode (Yang et al. 1984), and neutrophils migrate toward both the anode and cathode (Nishimura et al. 1996; Sheridan et al. 1996). There has been considerable support for the notion that fibroblasts migrate toward the cathode (Bourguignon and Bourguignon 1987; Bourguignon et al. 1986). Exogenously applied electrical fields of the same magnitude as in mammalian wounds direct the migration of human keratinocytes toward the cathode (Bourguignon et al. 1986). Mertz et al. (1993) used monophasic PC in an ovine model to assess epidermal cell migration macroscopically for 7 days following two 30-min sessions. They demonstrated that wounds treated with the cathode on day 0, followed by the anode on days 1–7, demonstrated 20% greater epithelialization than wounds treated with either positive or negative polarity alone. Eberhardt et al. (1986) investigated the effects of exogenous ES on cell composition in human skin and found that 69% of 500 cells counted 6 h post-ES were neutrophils, compared with 45% found for control wounds. The authors suggested that the difference in neutrophil percentage was due to the galvanotaxic effect by the currents.

A potential mechanism has been suggested for ES facilitating cell migration *in vivo* by galvanotaxis (Kloth 2005). A cell may detect an EF by the movement of proteins within the plasma membrane. An example is that epidermal growth factor receptors have been shown to move to the cathode side of keratinocytes exposed to a DC EF (Fang et al. 1999).

18.3.5 PROTEIN SYNTHESIS

Bourguignon et al. used healthy human fibroblasts in cell cultures with HVPC (Bourguignon and Bourguignon 1987; Bourguignon et al. 1986). They stated that the fibroblasts were induced to increase their rate of DNA and protein synthesis. Maximum synthesis occurred with parameters of 50 and 75 V and 100 pulses per second (pps), where the cells were close to the cathode. Voltages over 250 V inhibited both protein and DNA synthesis. A further study by Bourguignon et al. (1989) reported an increase in Ca^{2+} uptake and upregulation of insulin receptors on the fibroblast membrane during the second minute of stimulation.

A study by Cheng et al. (1982) investigated the effects of microampere levels of DC stimulation on rat skin. They showed that the concentration of adenosine triphosphate was increased by five times following application of DC (parameters: 10–1000 μA applied to skin strips 0.5 mm thick at 500 μA for 2 h *in vitro*). Additionally, they noted that 100–500 μA of DC increased amino acid uptake 30%–40% above control levels, and that 50 μA was required to obtain a maximum stimulation effect on protein synthesis. Falanga et al. (1987) have shown that ES upregulates receptors for TGF-β on human dermal fibroblasts. Transforming growth factor-β (TGF-β) is known to play an important role in collagen synthesis. The fibroblasts were exposed to 100 V and 100 pps and had receptor levels of TGF-β that were six times greater than the control fibroblasts.

It has been suggested that the possible mechanisms by which ES enhances tissue healing include the opening of voltage-sensitive calcium channels in the fibroblast plasma membrane (Kloth 2005). Therefore, the upregulation of TGF-β receptors on the cell surface may cause increased rates of collagen and DNA synthesis; thus, fibroblasts are stimulated to proliferate.

A number of clinical and *in vitro* studies have revealed the effects of ES on healing processes, demonstrating increased collagen deposition, enhanced angiogenesis, increased local perfusion, and accelerated wound healing. ES has also been shown to play a role in bacteriostatic and bactericidal effects demonstrated both *in vitro* and *in vivo*. There has been documentation of changes in multiple wound-related phenomena in experimental wounds in response to ES, including increased ATP and DNA synthesis, galvanotaxic directional migration of epithelial cells, decreasing edema, and increased angiogenesis. Despite variations in the type of current, duration, and dosing of ES, the majority of trials showed increased perfusion, angiogenesis, and antibacterial effects.

18.4 CONCLUSIONS

ES has been shown to have beneficial effects on wound healing. It is important to determine which therapies to use given the differences of modalities available, as well as cost and time invested. Among the options of advanced therapies available, ES offers an alternative that has been shown to have a positive effect at different stages of wound healing in many preclinical studies. ES decreases bacterial infection, increases local perfusion, and accelerates wound healing, therefore addressing these pivotal factors in wound complications. ES offers a unique treatment option to heal complicated wounds and improve surgical wound results. ES is a simple, inexpensive intervention to improve wound healing. Rigorous clinical trials are needed to help understand the dosing, timing, and type of ES to be used.

REFERENCES

Alvarez, O., Mertz, P., Smerbeck, R., et al. The healing of superficial skin wounds is stimulated by external electrical current. *Journal of Investigative Dermatology* 81 (1983): 144–8.

Atalay, C., Yilmaz, K. B. The effect of transcutaneous electrical nerve stimulation on postmastectomy skin flap necrosis. *Breast Cancer Research and Treatment* 117 (2009): 611–14.

Baker, L. L., Chanbers, R., Demuth, S. K., Villar, F. Effects of electrical stimulation on wound healing in patients with diabetic ulcers. *Diabetes Care* 20 (1997): 405–12.

Bourguignon, G., Bourguignon, L. Electric stimulation of protein and DNA synthesis in human fibroblast. *Federation of American Societies for Experimental Biology* 1 (1987): 398–402.

Bourguignon, G., Bourguignon, M., Khorshed, A., et al. Effect of high voltage pulsed galvanic stimulation on human fibroblasts in cell culture. *Journal of Cell Biology* 103 (1986): 344a.

Bourguignon, G., Wenche, J., Bourguignon, L. Electric stimulation of human fibroblasts causes an increase in Ca^{2+} influx and the exposure of additional insulin receptors. *Journal of Cell Physiology* 140 (1989): 397–85.

Cheng, K., Tarjan, P. P., Oliveira-Gandia, M. F., et al. An occlusive dressing can sustain natural electrical potential of wounds. *Journal of Investigative Dermatology* 104 (1995): 662–5.

Cheng, N., Van Hoof, H., Bock, E., et al. The effects of electric currents on ATP generation, protein synthesis, and membrane transport in rat skin. *Clinical Orthopaedics* 171 (1982): 264–72.

Clover, A. J., McCarthy, M. J., Hodgkinson, K., Bell, P. R., Brindle, N. P. Noninvasive augmentation of microvessel number in patients with peripheral vascular disease. *Journal of Vascular Surgery* 38 (2003): 1309–12.

Colmano, G., Edwards, S., Barranco, S. Activation of antibacterial silver coatings on surgical implants by direct current: Preliminary studies in rabbits. *American Journal of Veterinary Research* 41 (1980): 964–6.

Colthurst, J., Giddings, P. A retrospective case note review of the Fenzian electrostimulation system: A novel non-invasive, non-pharmacological treatment. *The Pain Clinic* 19 (2007): 7–14.

Cramp, A. F., Gilsenan, C., Lowe, A. S., Walsh, D. M. The effect of high- and low-frequency transcutaneous electrical nerve stimulation upon cutaneous blood flow and skin temperature in healthy subjects. *Clinical Physiology* 20 (2000): 150–7.

Deitch, E., Marino, A., Gillespie, T., et al. Silver nylon: A new antimicrobial agent. *Antimicrobial Agents Chemotherapy* 23 (1983): 356–9.

Eberhardt, A., Szczypiorski, P., Korytowski, G. Effect of transcutaneous electrostimulation on the cell composition of skin exudate. *Acta Physiologica Polonica* 37 (1986): 41–6.

Edwards, R., Harding, K. G. Bacteria and wound healing. *Current Opinion in Infectious Diseases* 17 (2004): 91–6.

Falanga, V., Bourguignon, G., Bourguignon, L. Electrical stimulation increases the expression of fibroblast receptors for transforming growth factor-beta. *Journal of Investigative Dermatology* 88 (1987): 488–92.

Fang, K., Ionides, E., Oster, G., et al. Epidermal growth factor receptor relocalization and kinase activity are necessary for directional migration of keratinocytes in DC electric fields. *Journal of Cell Science* 112 (1999): 1967–78.

Fernandez-Chimeno, M., Houghton, P. E., Holey, L. Electrical stimulation for chronic wounds. *Cochrane Database of Systematic Reviews* 1 (2004): CD004550.

Foulds, I. S., Barker, A. T. Human skin battery potentials and their possible role in wound healing. *British Journal of Dermatology* 109 (1983): 515–22.

Gentzkow, G. D., Alon, G., Taler, G. A., Eltorai, I. M., Montroy, R. E. Healing of refractory stage III and IV pressure ulcers by a new electrical stimulation device. *Wounds* 5 (1993): 160–72.

Gilcreast, D. M., Stotts, N. A., Froelicher, E. S., Baker, L. L., Moss, K. M. Effect of electrical stimulation on foot skin perfusion in persons with or at risk for diabetic foot ulcers. *Wound Repair and Regeneration* 6 (1998): 434–41.

Guffey, J., Asmussen, M. In vitro bactericidal effects of high voltage pulsed current versus direct current against *Staphylococcus aureus. Journal of Clinical Electrophysiology* 1 (1989): 5–9.
Halbert, A. R., Stacey, M. C., Rohr, J. B., Jopp-McKay, A. The effect of bacterial colonization on venous ulcer healing. *Australasian Journal of Dermatology* 33 (1992): 75–80.
Isseroff, R. R., Dahle, S. E. Electrical stimulation therapy and wound healing: Where are we now?*Advances in Wound Care* 1 (2012): 238–243.
Junger, M., Zuder, D., Steins, A., et al. Treatment of venous ulcers with low frequency pulsed current (Dermapulse): Effects on cutaneous microcirculation. *Der Hautartz* 18 (1997): 879–903.
Kaada, B., Olsen, E., Eielsen, O. In search of mediators of skin vasodilation induced by transcutaneous nerve stimulation. III. Increase in plasma VIP in normal subjects and in Raynaud's disease. *General Pharmacology* 15 (1984): 107–13.
Kincaid, C., Lavoie, K. Inhibition of bacterial growth in vitro following stimulation with high voltage, monophasic pulsed current. *Physical Therapy* 69 (1989): 651–5.
Kloth, L. C. Electrical stimulation for wound healing: A review of evidence from in vitro studies, animal experiments, and clinical trials. *International Journal of Lower Extremity Wounds* 4 (2005): 23–44.
Liebano, R. E., Rakel, B., Vance, C. G., Walsh, D. M., Sluka, K. A. An investigation of the development of analgesic tolerance to TENS in humans. *Pain* 152 (2011): 335–42.
Madsen, S. M., Westh, H., Danielsen, L., Rosdahl, V. T. Bacterial colonization and healing of venous leg ulcers. *Acta Pathologica, Microbiologica et Immunologica Scandinavica* 104 (1996): 895–9.
Marino, A., Deitch, E., Albright, J. Electric silver antisepsis. *IEEE Transactions on Biomedical Engineering* 32 (1985): 336–7.
Marino, A. A., Deitch, E. A., Malakanok, V., Albright, J. A., Specian, R. D. Electrical augmentation of the antimicrobial activity of silver-nylon fabrics. *Journal of Biological Physics* 12 (1984): 93–8.
McCaig, C. D., Rajnicek, A. M., Song, B., Zhao, M. Controlling cell behavior electrically: Current views and future potential. *Physiology Reviews* 85 (2005): 943–78.
Mertz, P., Davis, S., Cazzaniga, A., et al. Electrical stimulation: Acceleration of soft tissue repair by varying the polarity. *Wounds* 5 (1993): 153–9.
Newton, R., Karselis, T. Skin pH following high voltage pulsed galvanic stimulation. *Physical Therapy* 63 (1983): 1593–6.
Nishimura, K., Isseroff, R., Nuccitelli, R. Human keratinocytes migrate to the negative pole in direct current electric fields comparable to those measured in mammalian wounds. *Journal of Cell Science* 109 (1996): 199–207.
Nolan, M. F., Hartsfield, J. K., Witters, D. M., Wason, P. J. Failure of transcutaneous electrical nerve stimulation in the conventional and burst modes to alter digital skin temperature. *Archives of Physical and Medical Rehabilitation* 74 (1993): 182–7.
Peters, E., Armstrong, D., Wunderlich, R., et al. The benefit of electrical stimulation to enhance perfusion in persons with diabetes mellitus. *Journal of Foot and Ankle Surgery* 37 (1998): 396–400.
Rowley, B. Electrical current effects on *E. coli* growth rates. *Proceedings of the Society for Experimental Biology and Medicine* 139 (1972): 929–34.
Sebastian, A., Syed, F., Perry, D., et al. Acceleration of cutaneous healing by electrical stimulation: Degenerate electrical waveform down-regulates inflammation, up-regulates angiogenesis and advances remodelling in temporal punch biopsies in a human volunteer study. *Wound Repair and Regeneration* 19 (2011): 693–708.
Sheridan, D., Isseroff, R., Nuccitelli, R. Imposition of a physiologic DC electric field alters the migratory response of human keratinocytes on extracellular matrix molecules. *Journal of Investigative Dermatology* 106 (1996): 642–6.
Sherry, J. E., Oehrlein, K. M., Hegge, K. S., Morgan, B. J. Effect of burst-mode transcutaneous electrical nerve stimulation on peripheral vascular resistance. *Physical Therapy* 81 (2001): 1183–91.

Szuminsky, N. J., Albers, A. C., Unger, P., Eddy, J. G. Effect of narrow, pulsed high voltages on bacterial viability. *Physical Therapy* 74 (1994): 660–7.

Thakral, G., Lafontaine, J., Najafi, B., Talal, T. K., Kim, P., Lavery, L. A. Electrical stimulation to accelerate wound healing. *Diabetic Foot and Ankle* 4 (2013): 22081.

Thibodeau, E., Handelman, S., Marquis, R. Inhibition and killing of oral bacteria by silver ions generated with low-intensity direct current. *Journal of Dental Research* 57 (1978): 922–6.

Ud-Din, S., Perry, D., Giddings, P., et al. Electrical stimulation increases blood flow and haemoglobin levels in acute cutaneous wounds without affecting wound closure time: Evidenced by non-invasive assessment of temporal biopsy wounds in human volunteers. *Experimental Dermatology* 21 (2012): 758–64.

Wikstrom, S. O., Svedman, P., Svensson, H., et al. Effect of transcutaneous nerve stimulation on microcirculation in intact skin and blister wounds in healthy volunteers. *Scandinavian Journal of Plastic and Reconstructive Surgery Hand Surgery* 33 (1999): 195–201.

Wolcott, L., Wheeler, P., Hardwicke, H., et al. Accelerated healing of skin ulcers by electrotherapy: Preliminary clinical results. *Southern Medical Journal* 62 (1969): 795–801.

Yang, W., Onuma, E., Hui, S. Response of C3H/10T1/2 fibroblasts to an external steady electric field stimulation. *Experimental Cell Research* 155 (1984): 92–7.

19 Effect of electrical stimulation on bone healing

Michelle Griffin
Centre for Nanotechnology and Regenerative Medicine
London, United Kingdom

Ardeshir Bayat
Medicine and Health
Manchester, University of Manchester, United Kingdom

Contents

19.1 INTRODUCTION

Nonunions and delayed unions pose a challenging problem to the orthopedic community to manage. Electrical stimulation has been shown to be an effective and noninvasive method for enhancing bone healing and treating fracture nonunions in the clinical setting. The positive effect of electrical stimulation on bone healing has caused great research interest into understanding how to apply electrical fields to fracture sites and the mode by which it works. Typically, electrical stimulation can be applied using direct current (DC), capacitive coupling (CC), and inductive coupling (IC) to improve bone healing. In this chapter, we report on the molecular methods by which each mode of electrical stimulation has been shown to behave and stimulate bone healing. In addition, this chapter reviews the best-available clinical evidence for the use of electrical stimulation in bone healing.

19.2 CLINICAL NEED FOR ELECTRICAL STIMULATION

Bone is a dynamic organ that undergoes a process of regeneration following injury to achieve healing (Ryaby 1998). Bone healing is a multistage process that is highly orchestrated in response to injury (Ryaby 1998). The immediate inflammatory response leads to the recruitment of stem cells and their differentiation into chondrocytes to produce cartilage and osteoblasts to form bone (Ai-Aql et al. 2008). After the cartilage matrix is formed, it is mineralized to form bone, followed by the resorption of the mineralized cartilage (Ai-Aql et al. 2008). Further bone remodeling takes place in which the initial bony callus is formed to restore the anatomical structure and support mechanical loads (Ai-Aql et al. 2008). Unfortunately, in 5%–10% of fractures the healing will be disrupted, resulting in a delayed union or nonunion (Ryaby 1998). In the United States, nonunions and delayed unions pose a huge economic burden, with an estimated 6 million fractures per year (Ryaby 1998). For the patient, nonunions lead to a loss of function and significant pain, as well as the increased costs and reduced quality of life (Giannoudis et al. 2007). Many factors have been shown to contribute to nonunions, including fracture displacement, severity of the fracture, infection, and the speed of the initial management (Giannoudis et al. 2007).

The gold standard treatment of care for nonunions or delayed unions can involve both surgical interventions and nonsurgical treatment (Giannoudis et al. 2007). Cast immobilization and surgical treatment involving external fixation, plating, and internal intramedullary nail fixation are all techniques to facilitate fracture healing (Giannoudis et al. 2007). Due to the significant problems nonunions can cause the patient and surgeon, extensive research has been carried out to create and ascertain the effectiveness of adjunctive therapies to improve bone healing. One treatment that has received considerable attention is electrical stimulation. Clinically, electrical stimulation has been applied to bone fractures to promote healing by DC, CC, or IC techniques (Griffin et al. 2011a).

19.3 HISTORY OF ELECTRICAL STIMULATION

Electrical signals have been shown to play an instructive role in many cell behaviors, affecting the cell proliferation, differentiation, and migration. Electrical stimulation has also been shown to enhance wound healing and regeneration *in vivo*. The effect of electrical stimulation on bone healing has been a topic for centuries. The realization that bone had electrical signals was proposed by Iwao Yasuada's work on piezoelectricity in the 1950s (Fukada and Yasuda 1957). Fukada and Yasuda (1957) observed that when bone was emerged in acid for 3 weeks to remove the apatite between the collagen fibers, electrical signals were still produced. Bassett and Becker (1962) further confirmed that mechanical deformation also causes electrical stimuli, which can promote bone formation. They determined that the amplitude of electrical potentials is dependent on the rate of bone loading (Bassett and Becker 1962). The formation of bone by electrical stimulation has been shown to be caused by the upregulation and downregulation of signaling molecules at the cellular level. The clinical success of electric and electromagnetic fields has caused extensive research over the years to understand how electrical stimulation improves bone healing.

19.4 MODES OF ELECTRICAL STIMULATION USED FOR BONE HEALING

Electric and electromagnetic fields can be generated to bone by several methods, all of which has their own advantages and disadvantages (Figure 19.1). In the 1960s to the 1970s, DC was investigated. Direct electrical current techniques are invasive and involve the implantation of one or more electrodes directly at the fracture or fracture site (Griffin et al. 2011a; Aaron et al. 2004). DC stimulators are used in spinal procedures. Usually, a subcutaneous lithium battery powers the unit for 6 months, and then this needs to be removed during a secondary procedure (Griffin et al. 2011a; Aaron et al. 2004). Such devices are useful because they provide a constant current directly at the fracture site (Griffin et al. 2011a; Aaron et al. 2004). However, such devices have a limited battery life and provide infection risks (Griffin et al. 2011a; Aaron et al. 2004).

Today electrical stimulators are often classified into electromagnetic or ultrasound. Electromagnetic stimulators are then grouped into inductive or CC. CC involves using an external power source of 20–200 kHz, creating electric fields of 1–100 mV/cm at the fracture site (Griffin et al. 2011a; Nelson et al. 2003). Electrodes applied on the skin are connected to an external battery pack. CC is not invasive, but does require patients to change the batteries daily (Griffin et al. 2011a; Nelson et al. 2003). The alternative is IC, otherwise known as pulsed electromagnetic field (PEMF) coupling (Griffin et al. 2011a; Aaron et al. 2004; Nelson et al. 2003). The device is externally applied to the fracture site for up to 10 h per day, in a biphasic and quasi-rectangular waveform, varying in both amplitude and frequency (Griffin et al. 2011a; Aaron et al. 2004; Nelson et al. 2003).

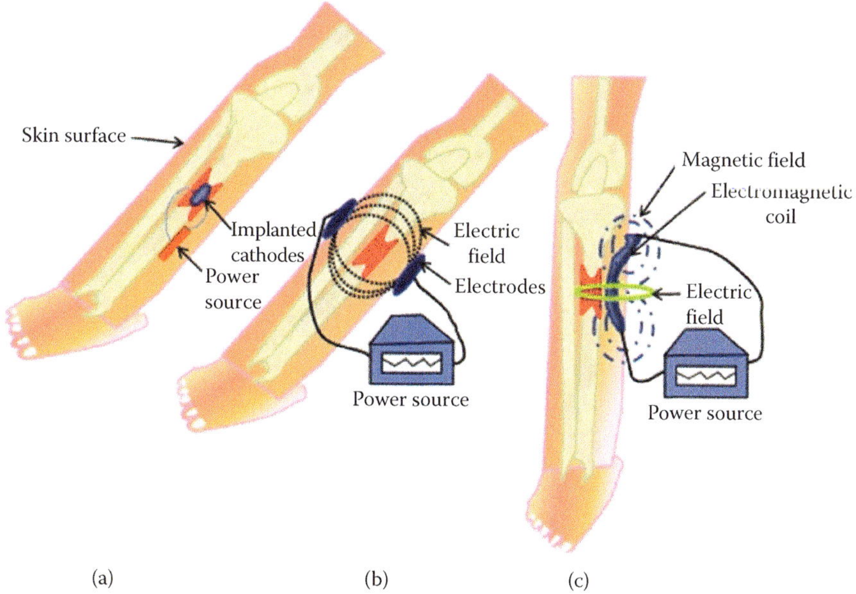

Figure 19.1 The three methods of administering electric stimulation are shown in this diagram. (a) Direct current. (b) Capacitive coupling. (c) Inductive coupling. (From Griffin, M., et al., *Eplasty*, 11, e34, 2011. With permission.)

The coil appropriately sized for the fracture site creates an electric signal and enhances bone formation. As these devices are heavy, these treatments often cause problems with patient compliance (Griffin et al. 2011a; Aaron et al. 2004; Nelson et al. 2003).

More recently, low-intensity pulsed ultrasound (LIPUS) has also become popular, which is a noninvasive method of treatment. LIPUS has received a lot of attention to fully understand how it improves bone physiology and healing, but its effectiveness has remained controversial (Hannemann et al. 2014). The ultrasound used to improve the bone healing is a mechanical energy delivered as an acoustic pressure wave, with higher intensities than diagnostic purposes being shown to improve bone healing (Hannemann et al. 2014).

19.5 HOW CELLS RESPOND TO ELECTRICAL STIMULATION: POSSIBLE MECHANISMS

Each mode of electrical stimulation has been investigated for its effect at the cellular level in causing changes in both signal transduction and gene and protein regulation (Figure 19.2).

19.5.1 DIRECT CURRENT

Few mechanisms of action have been proposed to account for the effect of DC on bone healing. One theory is that DC stimulates osteogenesis by the electrochemical reaction occurring at the cathode ($O_2 + 2H_2O + 4e- \rightarrow 4OH$), creating end products referred to as faradic products (Baranowski and Black 1987; Bassett and Herrmann 1961; Bodamyali et al. 1999; Brighton et al. 1951, 1975; Cho et al. 2001; Steinbeck et al. 1998). The hydroxyl ions (OH) at the cathode have been shown to lower the oxygen concentration and pH (Bodamyali et al. 1999). In turn, this prevents bone resorption and triggers bone formation by the osteoblasts (Bodamyali et al. 1999). Hydrogen peroxide (H_2O_2) has also been documented to be formed at the cathode, which has been shown to enhance osteoclast differentiation and inadvertently trigger bone formation by osteoblasts (Bodamyali et al. 1999; Steinbeck et al. 1998). Hydrogen peroxide has also been shown to cause a stimulatory effect on the secretion of vascular endothelial growth factor (VEGF) by macrophages, which is important for fracture healing to promote angiogenesis (Steinbeck et al. 1998).

Another proposed mechanism by which DC stimulations affect bone healing is by directly influencing osteoblast gene expression. The use of electrical stimulation has also been shown to enhance the osteogenic differentiation of the osteoblasts, demonstrated by the expression of bone morphogenetic protein (BMP)-2, collagen type I, and osteonectin and the phosphorylation of Samd4 (Cao et al. 2013). DC has also been shown to promote bone repair and osteogenesis but, as already highlighted, requires implanted electrical electrodes. The replacement of the metal electrodes with a biodegradable conductive polymer film could overcome this problem. Zhang et al. (2013) manufactured polypyrrole–chitosan films comprising polypyrrole nanoparticles dispersed in a chitosan matrix and found that DC stimulation with 200 μA for 4 h per day enhanced the osteoblast metabolic activity on day 7 by 1.8-fold more than that without DC stimulation. Osteogenic gene expression was also enhanced by the covalent immobilization of BMP-2, and a combination with electrical stimulation resulted in a synergic effect on osteoblast maturation (Zhang et al. 2013). The interactions of electrical stimulation of osteoblasts seeded on scaffold materials have also recently been researched (Gittens et al. 2013;

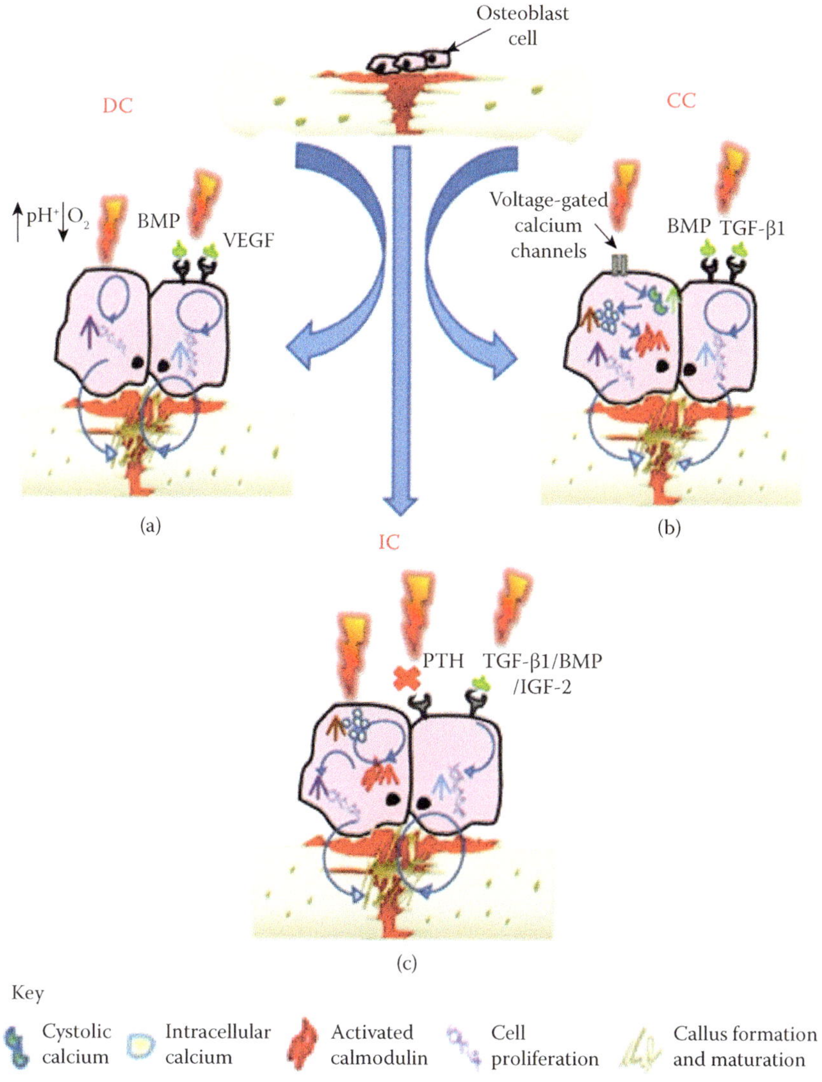

Figure 19.2 Proposed mechanism of action of the different types of electrical stimulation methods. (a) Proposed mechanism for direct current (DC). DC lowers the oxygen level and increases the pH, which causes an increase in osteoblast cell proliferation. This in turn enhances callus formation and maturation, leading to bone healing. All three types of electrical stimulation enhance growth factors. This in turn increases cell proliferation, which enhances callus formation and maturation, leading to bone healing and improved clinical outcome. (b) Proposed mechanism for capacitive coupling (CC). CC causes an increase in cystolic calcium through voltage-gated calcium channels. This then increases intracellular calcium, which in turn enhances activated calmodulin stores. Cell proliferation then increases, which enhances callus formation and maturation, leading to bone healing. (c) Proposed mechanism for inductive coupling (IC). IC causes a direct increase in intracellular calcium, which in turn enhances activated calmodulin stores. Cell proliferation is increased, which enhances callus formation and maturation, leading to bone healing. BMP, bone morphogenetic protein; IGF-2, insulin growth factor 2. PTH, parathyroid hormone; TGF-β1, transforming growth factor beta 1; VEGF, vascular endothelial growth factor. (From Griffin, M., et al., *Eplasty*, 11, e34, 2011. With permission.)

Bodhak et al. 2012). The effect of DC with 5–25 μA applied for 5 days in intervals of 8 h for 15 min duration to human fetal osteoblast (hFOB) cells on titanium surfaces showed enhanced cell adhesion and proliferation (Bodhak et al. 2012).

DC has also been shown to affect the regulation of the extracellular matrix, which in turns causes changes in bone formation and bone healing (Wang et al. 1998). Wang et al. (1998) illustrated the effect of DC on cell culture to understand the effect of electrical stimulation on bone metabolism. DC of 100 $\mu A/cm^2$ increased the calcium ion metabolism. It was found that the intracellular calcium ion concentration increased, which could be key in regulating bone metabolism (Wang et al. 1998). Early studies have shown that DC improves bone formation *in vivo* (Klems and Schleicher 1981; Fredericks et al. 2007; Hellewell and Beljan 1979; Zorlu et al. 1998). Thirty New Zealand white rabbits had an implantable DC stimulator placed across the decorticated transverse processes before placement of an autograft. The mRNA was elevated for BMP-2, BMP-6, and BMP-7. The authors concluded that DC-induced bone formation seen clinically could be mediated by the upregulation of these osteoinductive factors (Fredericks et al. 2007). The effectiveness of DC on bone formation *in vivo* has also been studied and found to improve the bone growth in osteotomy models (Hellewell and Beljan 1979). The effect of DC has also been compared to ultrasound in an *in vivo* animal model. Four groups were formed, each consisting of 16 ultrasound, 16 electrostimulation, 16 ultrasound control, and 16 electrostimulation control animals. It was observed that in the ultrasound and electrostimulation groups, the fracture healing was greater than in the control groups, but there was no difference between ultrasound and electrostimulation groups (Zorlu et al. 1998).

19.5.2 CAPACITIVE COUPLING

Several studies have been carried out to understand the mechanism by which CC stimulation affects bone healing (Brighton et al. 2001; Lorich et al. 1998). CC has been shown to stimulate bone formation by calcium translocations via voltage-gated calcium channels. CC has also been shown to activate calcium voltage-gated channels, triggering an increase in phospholipase A2, which raises prostaglandin E2 synthesis (Brighton et al. 2001; Lorich et al. 1998). Cystolic Ca^{2+} increases intracellular calcium, which in turn enhances calmodulin levels (Brighton et al. 2001; Lorcih et al. 1998). Cellular proliferation by upregulation of the nucleotide synthesis and a wide array of enzymatic proteins is caused by the activated calmodulin, causing bone formation (Brighton et al. 2001; Lorich et al. 1998).

Several studies have also reported that CC causes changes at the cellular gene and protein level to enhance protein and gene expression. Wang et al. (2006) illustrated that postconfluent culture of MC3T3-E1 bone cells exposed to CC enhanced their BMP-2 to BMP-8 expression. It was clear that the optimal capacitively coupled signal (60 kHz, 20 mV/cm at a 50% duty cycle for 24 h) could specifically and selectively upregulate a number of osteoconductive BMPs. Furthermore, as BMPs are known to increase alkaline phosphatase (ALP) activity in MC3T3-E1, ALP activity levels were measured in cell extracts (Wang et al. 2006). The results showed that the levels were increased by nearly 50% ($p < 0.01$) in the stimulated samples (Figure 19.3) (Wang et al. 2006). Another study also demonstrated that MC3T3-E1 cells increased levels of transforming growth factor β1 (TGF-β1) mRNA determined by quantitative reverse transcription–polymerase chain reaction of osteoblastic cells (MC3T3-E1) when exposed to CC stimulation (Zhuang et al. 1997).

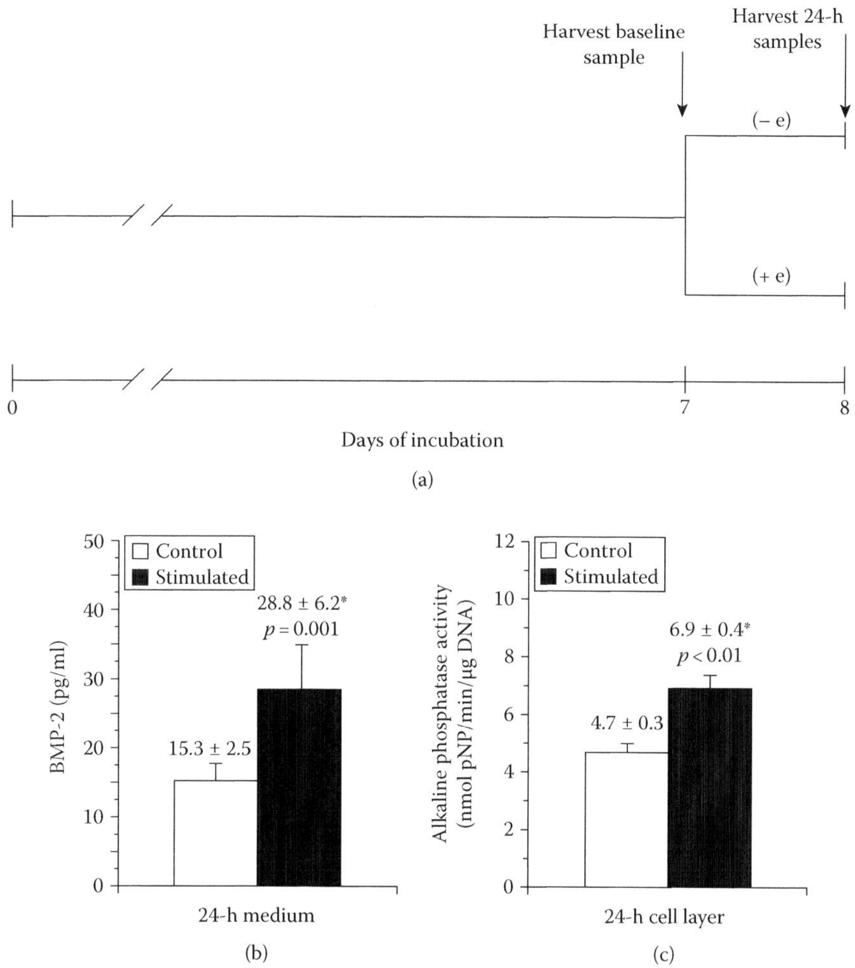

Figure 19.3 Measurement of BMP-2 protein levels and alkaline phosphatase activity in electrically stimulated cultures. (a) Experimental design. Cells were incubated for 7 days before the stimulation period, and baseline cell layer samples were harvested. Incubation was then continued for 24 h in fresh medium in the absence (−e) or presence (+e) of a capacitive coupled signal (60 kHz, 20 mV/cm, and 50% duty cycle), and medium and cell layer samples were harvested. (b) BMP-2 protein was measured and found to be significantly higher in stimulated culture than the unstimulated control. (c) Alkaline phosphatase activity in the cell layer was found to be significantly higher in the stimulated culture than in the unstimulated control. (From Wang, Z., et al., *J. Bone Joint Surg. Am.*, 88, 1053–1065, 2006. With permission.)

Clark et al. (2014) evaluated the gene expression after CC and found that human calvarial osteoblasts of 60 kHz, 20 mV/cm, 50% duty cycle for 2-h duration per day enhanced the expression of selected genes important in fracture healing. The treatment significantly unregulated the mRNA expression of a number of genes, including BMP-2 and -4, and TGF-β1, -β2, and -β3, as well as fibroblast growth factor (FGF)-2, osteocalcin, and ALP (Clark et al. 2014).

The authors of the chapter also compared a novel degenerate sine wave (DW) with CC. In this study, DW was compared with CC and found to enhance the differentiation of the Sa-OS-2 osteoblast-like cells to a greater extent, as found by increased ALP and collagen I gene expression by quantitative real-time–polymerase chain reaction analysis ($p < 0.01$) (Figure 19.4). Furthermore, DW enhanced the mineralization of the cells compared with CC, as shown by an increased transcription of osteocalcin, osteonectin, osteopontin, and bone sialoprotein ($p < 0.05$) (Figure 19.4) (Griffin et al. 2013).

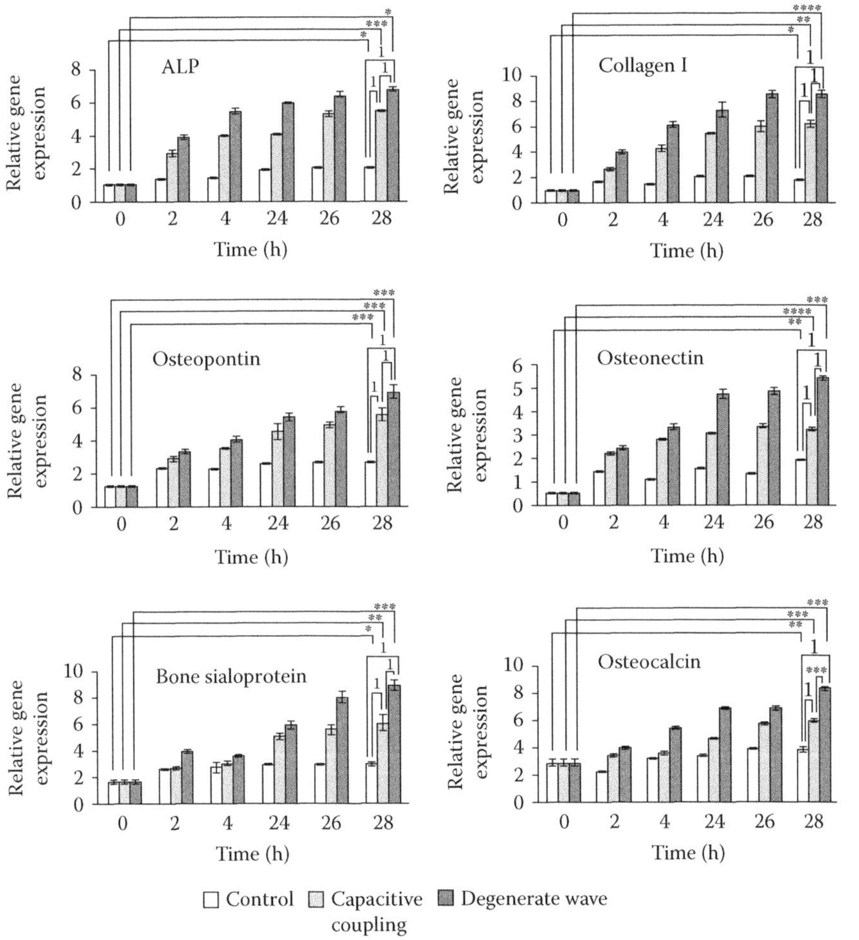

Figure 19.4 Differentiation and mineralization-specific biomarkers in osteoblasts after electrical stimulation. Alkaline phosphatase, collagen I, and mineralization-specific genes such as osteopontin, osteonectin, bone sialoprotein, and osteocalcin reveal that differentiation and mineralization of osteoblasts was greater after degenerate wave (DW) stimulation than capacitive coupling (CC) stimulation by 28 h. *$p < 0.05$; **$p < 0.01$; ***$p < 0.001$; \$$p < 0.0001$. (From Griffin, M., et al., *PLoS One*, 8, e72978, 2013. With permission.)

19.5.3 INDUCTIVE COUPLING

IC has been extensively investigated to fully understand the effect on bone healing and bone formation. At a cellular and a molecular level, IC has been shown to cause an effect on the production of the proteins that regulate gene transcription, affect several membrane receptors, and stimulate osteoblasts to secrete growth factors, including BMPs.

First, IC has been shown to exhibit its effect on bone healing by altering the calcium uptake in bone (Fredericks et al. 2007). IC has been demonstrated to increase the calcium uptake of bone by inactivating its signal to parathyroid hormone (PTH) (Luben et al. 1982; Spadaro et al. 2002). The blocking of the signal prevents the store of cyclic adenosine monophosphate to accumulate (Luben et al. 1982; Spadaro et al. 2002). Another proposed mechanism by which IC affects bone healing is the effect of the electrical stimulation on the activation of intracellular calcium stores. IC stimulation has been shown to increase the intracellular calcium stores of osteoblast cells, activating intracellular calcium stores, which in turn increases activated calmodulin levels, which enhances osteoblast cell proliferation (Fredericks et al. 2007).

Several studies have also shown that IC enhances bone healing by directing regulating gene and protein expression. For example, Tsai et al. (2007) illustrated that pulsed electromagnetic fields on poly(lactic-co-glycolic acid) (PLGA) scaffolds in a bioreactor enhanced proliferation and differentiation when specific parameters were implemented. Osteoblasts were exposed to pulsed stimulation with an average (root mean square) amplitude of either 0.13, 0.24, or 0.32 mT, resulting in an electric field waveform consisting of a single, narrow 300-micron quasi-rectangular pulse with a repetition rate of 7.5 Hz (Tsai et al. 2007). Interestingly, PEMF stimulation at 2 and 8 h at 0.13 mT increased cell number on days 6 and 12 but decreased ALP activity on days 6 and 12. However, 0.32 mT PEMF-treated groups inhibited cell proliferation but increased ALP activity (Tsai et al. 2007). Hronik-Tupak et al. (2011) found that electrical stimulation improves human mesenchymal stem cell (hMSC) osteogenic differentiation for a 20 mV/cm, 60 kHz electric field. PEMF has also been shown to enhance the osteogenic differentiation of human BMSCs by increasing mineralization at the expense of proliferation with upregulation of several osteogenic marker genes in the PEMF-treated group, including BMP-2, TGF-β1, osteoprotegerin, matrix metalloproteinase 1 (MMP-1) and MMP-3, osteocalcin, and bone sialoprotein (Hronik-Tupaj et al. 2011).

When evaluating electrical stimulation on cell behavior, many studies have focused on the effect of stimulation of the osteogenesis of adult stem cells and osteoblast cells. The authors of this chapter have also compared several electrical stimulation waveforms and their effect on bone marrow–derived stem cells (hBMSCs) for their ability to migrate and invade (Griffin et al. 2011b). BMSCs have been shown to migrate toward the site of injury and differentiate into osteoblasts and chondrocytes at the wound site. Several electrical stimulation parameters, including DC, CC, PEMF, and DW, were assessed for their effect on cytotoxicity, proliferation, cell kinetics, and apoptosis in hBMSCs. The electrical stimulation was applied for 3 h a day, up to 5 days. It was observed that DW had the greatest proliferative and least cytotoxic effect compared with the other waveforms (Griffin et al. 2011b). Using a collagen invasion assay, CC and DW caused a greater number of cells to invade than unstimulated cells (Griffin et al. 2011b). All electrical waveforms increased the expression of migration- and invasion-related genes, with DC stimulation showing significantly higher expression of chemokine receptor type 4 (CXCR-4) and stromal cell–derived factor 1 (SDF-1) on days 2, 3, and 5 than the other waveforms (Griffin et al. 2011b). It was clear that different waveforms have differing effects on cellular activity, and further exploration into understanding each waveform is required. Further, understanding of hBMSCs' homing using electrical stimulation would be beneficial to enhancing bone tissue regeneration (Griffin et al. 2011b).

Enhancement of bone formation and bone healing has been demonstrated *in vivo* using PEMF. Several osteotomy *in vivo* models have proved there is a stimulatory effect of PEMF on bone healing (Inoue et al. 2002; Sarker et al. 1993). Mixed-breed dogs were treated with PEMF for 1 h daily for 4 weeks after surgery for 8 weeks (Inoue et al. 2002). In the PEMF group, load bearing was earlier and, at 8 weeks, was greater than in the control group. Using a similar model, it has also been illustrated that PEMF enhanced the bone formation and fracture healing of the osteotomy sites after 9 h per day in an animal rat model. The authors concluded that fractured rat tibia could be enhanced by the application of PEMF, which was clearly observed in the third week (Sarker et al. 1993).

Midura et al. (2005) illustrated that 0.2 mm fibular osteotomies in rats showed greater callus formation and stiffness using the PEMF Physio-Stim® treatment. However, the authors concluded that there was specificity in the relationship between waveform and biological outcomes. The beneficial effect on bone healing was not observed when a different PEMF waveform, Osteo-Stim®, was utilized (Figure 19.5)

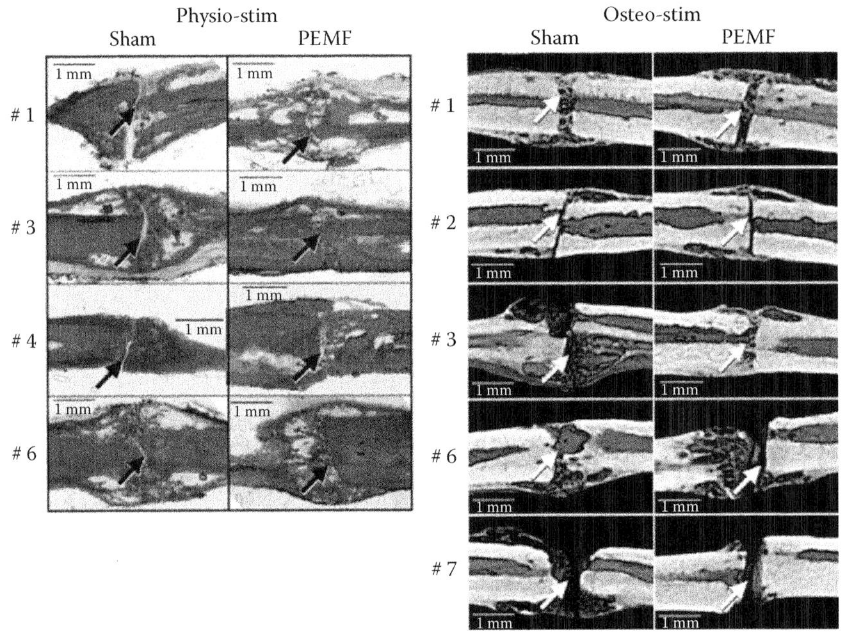

Figure 19.5 Images of the hard callus tissues at 5 weeks of healing surrounding the original osteotomy site. Images of the Physio-Stim specimens are Alizarin Red S–stained, undecalcified thick midsagittal sections, while those of the Osteo-Stim specimens are midsagittal views from high-resolution micro–computed tomography imaging *ex vivo* (20 μm voxel size). The arrow denotes the original osteotomy site in each image. With the exception of rat 4, the PEMF-treated fibulae exhibited better cortical alignments, consolidation of the newly formed cancellous bony callus, and fill of the original osteotomy gap with new bone tissue than their respective sham-treated fibulae. (From Midura, R. J., et al., *J. Orthop. Res.*, 23, 1035–1046, 2005. With permission.)

(Midura et al. 2005). A similar study supported these findings, with significant reduction in the amount of time-dependent bone volume loss in PEMF-treated, distal fibular segments compared with their contralateral sham-treated bones. Osteotomy gap size was also significantly smaller in hind limbs exposed to PEMF than sham controls (Ibiwoye et al. 2004).

PEMF has also been shown to improve distraction osteogenesis in a rabbit tibial leg-lengthening model (Fredericks et al. 2003). Tibiae were tested for torsional strength after 9, 16, and 23 days postdistraction. PEMF-treated tibiae were significantly stronger than sham controls at 9, 16, and 23 days postdistraction (Fredericks et al. 2003).

19.6 CLINICAL EVIDENCE OF ELECTRICAL STIMULATION

Electrical stimulation has been extensively successful in the treatment of nonunions (Bassett et al. 1982), delayed unions (Cundy and Paterson 1990), acute fractures (Wahlström 1984), osteotomies (Mammi et al. 1993), spinal fusions (Kane 1988), and congenital pseudarthroses (Poli et al. 1985). Due to the numerous clinical studies utilizing electrical stimulation, several meta-analyses and systemic reviews have been performed to understand the efficacy of electrical stimulation for bone healing. In the hierarchy of evidence, randomized controlled trials and meta-analyses are considered the best quality, and so we assess those which have been undertaken to assess electrical stimulation for bone healing.

A meta-analysis of electrical stimulation for long-bone healing was carried out in 2008, with 11 studies being identified as suitable for analysis. Due to the methodologies and high heterogeneity between studies, it was difficult to ascertain the impact of electromagnetic stimulation on fracture healing (Mollon et al. 2008). In 2011b, Griffin et al. similarly performed a Cochrane-style meta-analysis to understand the specific effect of electromagnetic stimulation on treating delayed union or nonunion of long-bone fractures in adults. Ten studies were found to be relevant, which included 125 participants (Griffin et al. 2011b). However, only four studies were identified as randomized controlled trials comparing electromagnetic simulation with a sham control (Griffin et al. 2011). Barker et al. (1984) investigated the effect of PEMF on tibial nonunion in 17 participants and evaluated the clinical union at 24 weeks. Fractures in five of the nine patients with PEMF united compared with five of the seven patients with dummy machines (Barker et al. 1984). Scott and King (1994) illustrated the effect of CC or a sham device on 23 patients with long-bone nonunions. A nonunion was defined as the presence of a fracture for 9 months or longer and no clinical or radiological signs of progress for the most recent 3 months (Scott and King 1994). After 6 months, 6 of 11 nonunions treated with an active device had healed, as opposed to only 1 of 12 cases treated with a sham device ($p = 0.02$) (Scott and King 1994). Sharrard (1990) observed the effect of PEMF on 51 participants with delayed union of the tibia and evaluated the radiographical union at 12 weeks. Simonis et al. (2003) demonstrated the effect of PEMF or sham device for 6 months on tibial nonunions existing for 1 year in 34 participants. Patients in the treatment group showed progression to union significantly more often than the controls (16/18 vs. 8/16, respectively, $p = 0.02$) (Simonis et al. 2003). While favoring electromagnetic stimulation, the overall pooled estimate of the effect size of electromagnetic stimulation on healing in the treatment of both delayed union and nonunion was not statistically significant (relative risk [RR] 1.96; 95% confidence interval [CI] 0.86–4.48) (Simonis et al. 2003). The quality of the evidence of reporting in the included studies was poor, with incomplete reporting of follow-up and lack of documentation of excluded participants.

Schmidt-Rohlfing et al. (2011) also carried out a systematic review evaluating both electromagnetic fields and high-frequency electric fields for bone healing. The group identified 14 randomized clinical trials, including a total of 915 patients for bone healing, which identified studies with the primary endpoint as the "rate of bony healing" (Schmidt-Rohlfing et al. 2011). The authors also concluded that due to the heterogeneous physical parameters in the studies, including the frequencies, time course, and flux densities, and in view of the methodological deficits, it was impossible to create any general conclusions (Sharrard 1990). Furthermore, the analysis did not enable the authors to make any recommendations to directly influence patient care (Sharrard 1990).

Hannemann et al. (2011) has recently performed a systematic review and meta-analysis to evaluate the best currently available evidence from randomized controlled trials comparing pulsed electromagnetic fields or LIPUS bone growth stimulation with placebo for acute fractures. The pooled results from 730 patients from 13 trials were analyzed, and no significant difference between PEMF or LIPUS and control was found (Hannemann et al. 2011). The author concluded that there were insufficient data from the randomized trials to fully see the benefit of PEMF or LIPUS for bone growth when used for acute fractures. However, PEMF and LIPUS did significantly shorten the time to radiological union for acute fractures with nonoperative treatment and accelerate the time to clinical union for acute diaphyseal fractures (Hannemann et al. 2011).

It is clear from all systematic reviews carried out to date that evidence from appropriately conducted randomized controlled trials is required to ascertain the true effect of electrical stimulation on bone healing for the clinical setting. Future studies, to improve electrical stimulation for bone healing, are required to have more consistent outcome measures, including time and electrical stimulation parameters. More importantly, outcome measures need to be uniform to be able to compare electrical stimulation studies.

19.7 CHALLENGES OF UNDERSTANDING THE MECHANISM OF ELECTRICAL STIMULATION

Over the last 160 years, our understanding of the cellular consequences of electrical stimulation for bone healing has caused the uptake of electrical simulation for the orthopedic field. It is clear from the experimental studies carried out to date that DC, CC, and IC have all been shown to enhance bone healing due to alterations in cellular behavior. Several mechanisms have been proposed and hypothesized to account for the effect of electrical stimulation on bone healing.

One major drawback in the literature evaluating electrical stimulation is the different electrical fields and systems used to assess cell response. Future studies need to fully understand and define the electrical and magnetic field they are producing to enable studies to be compared. Future studies should be directed at establishing a set of standard parameters of DC, IC, and CC to interpret the effect of electrical on biological interactions. Furthermore, it is clear that studies have failed to elicit which electrical stimulation parameters may be optimal for specific clinical scenarios.

19.8 CONCLUSIONS

In conclusion, extensive research is still ongoing to fully understand the effect of electrical stimulation on bone healing. The widespread use of electrical stimulation to aid bone healing in clinical practice will continue to spark research interest. Investigations have begun to clarify how cells respond to electrical stimuli by alterations in transmembrane signaling and gene expression. Further work is needed to explore the precise mechanism using specific electrical waveforms and parameters to improve the clinical application of electrical stimulation for bone healing.

REFERENCES

Aaron R. K., Ciombor D. M., Simon B. J. Treatment of nonunions with electric and electromagnetic fields. *Clin Orthop* 419 (2004): 21–9.

Ai-Aql Z. S., Alagl A. S., Graves D. T., Gerstenfeld L. C., Einhorn T. A. Molecular mechanisms controlling bone formation during fracture healing and distraction osteogenesis. *J Dent Res* 87 (2008): 107–18.

Baranowski T. J., Black J. The mechanism of faradic stimulation of osteogenesis. In *Mechanistic Approaches to Interactions of Electric and Electromagnetic Fields with Living Systems*, eds. M. Blank, E. Findl. New York: Plenum Press, 1987, 399–416.

Barker A. T., Dixon R. A., Sharrard W. J., Sutcliffe M. L. Pulsed magnetic field therapy for tibial non-union. Interim results of a double-blind trial. *Lancet* 1 (1984): 994–6.

Bassett C. A., Becker R. O. Generation of electric potentials by bone in response to mechanical stress. *Science* 137 (1962): 1063–4.

Bassett C. A., Herrmann I. Influence of oxygen concentration and mechanical factors on differentiation of connective tissues in vitro. *Nature* 190 (1961): 460–1.

Bassett C. A., Mitchell S. N., Schink M. M. Treatment of therapeutically resistant nonunions with bone grafts and pulsing electromagnetic fields. *J Bone Joint Surg Am* 64 (1982): 1214–20.

Bodamyali T., Kanczler J. M., Simon B., et al. Effect of faradic products on direct current-stimulated calvarial organ culture calcium levels. *Biochem Biophys Res Commun* 264 (1999): 657–61.

Bodhak S., Bose S., Kinsel W. C., Bandyopadhyay A. Investigation of in vitro bone cell adhesion and proliferation on Ti using direct current stimulation. *Mater Sci Eng C Mater Biol Appl* 32 (2012): 2163–8.

Brighton C. T., Adler S., Black J., et al. Cathodic oxygen consumption and electrically induced osteogenesis. *Clin Orthop Relat Res* 107 (1975): 277–82.

Brighton C. T., Ray R. D., Soble L. W., et al. In vitro epiphyseal-plate growth in various oxygen tensions. *J Bone Joint Surg Am* 51 (1951): 1383–96.

Brighton C. T., Wang W., Seldes R., et al. Signal transduction in electrically stimulated bone cells. *J Bone Joint Surg Am* 83 (2001): 1514–23.

Cao J., Man Y., Li L. Electrical stimuli improve osteogenic differentiation mediated by aniline pentamer and PLGA nanocomposites. *Biomed Rep* 1 (2013): 428–32.

Clark C. C., Wang W., Brighton C. T. Up-regulation of expression of selected genes in human bone cells with specific capacitively coupled electric fields. *J Orthop Res* 32 (2014): 894–90.

Cho M., Hunt T. K., Hussain M. Z. Hydrogen peroxide stimulates macrophage vascular endothelial growth factor release. *Am J Physiol Heart Circ Physiol* 280 (2001): 2357–63.

Cundy P. J., Paterson D. C. A ten-year review of treatment of delayed union and nonunion with an implanted bone growth stimulator. *Clin Orthop Relat Res* 259 (1990): 216–22.

Fredericks D. C., Piehl D. J., Baker J. T., Abbott J., Nepola J. V. Effects of pulsed electromagnetic field stimulation on distraction osteogenesis in the rabbit tibial leg lengthening model. *J Pediatr Orthop* 23 (2003): 478–83.

Fredericks D. C., Smucker J., Petersen E. B., et al. Effects of direct current electrical stimulation on gene expression of osteopromotive factors in a posterolateral spinal fusion model. *Spine (Phila Pa 1976)* 32 (2007): 174–81.
Fukada E., Yasuda I. On the piezoelectric effect of bone. *J Physical Soc Jpn* 12 (1957): 1158–69.
Giannoudis P. V., Atkins R. Management of long-bone non-unions. *Injury* 38 (2007): S1–2.
Gittens R. A., Olivares-Navarrete R., Rettew R., et al. Electrical polarization of titanium surfaces for the enhancement of osteoblast differentiation. *Bioelectromagnetics* 34 (2013): 599–612.
Griffin M., Bayat A. Electrical stimulation in bone healing: Critical analysis by evaluating levels of evidence. *Eplasty* 11 (2011a): e34.
Griffin M., Iqbal S. A., Sebastian A., Colthurst J., Bayat A. Degenerate wave and capacitive coupling increase human MSC invasion and proliferation while reducing cytotoxicity in an in vitro wound healing model. *PLoS One* 6 (2011b): e23404.
Griffin M., Sebastian A., Colthurst J., Bayat A. Enhancement of differentiation and mineralisation of osteoblast-like cells by degenerate electrical waveform in an in vitro electrical stimulation model compared to capacitive coupling. *PLoS One* 8 (2013): e72978.
Griffin X. L., Costa M. L., Parsons N., Smith N. Electromagnetic field stimulation for treating delayed union or non-union of long bone fractures in adults. *Cochrane Database Syst Rev* 4 (2011): CD008471.
Hannemann P. F., Mommers E. H., Schots J. P., Brink P. R., Poeze M. The effects of low-intensity pulsed ultrasound and pulsed electromagnetic fields bone growth stimulation in acute fractures: A systematic review and meta-analysis of randomized controlled trials. *Arch Orthop Traum Surg* 134 (2014): 1093–106.
Hellewell A. B., Beljan J. R. The effect of a constant direct current on the repair of an experimental osseous defect. *Clin Orthop Relat Res* 142 (1979): 219–22.
Hronik-Tupaj M., Rice W. L., Cronin-Golomb M., Kaplan D. L., Georgakoudi I. Osteoblastic differentiation and stress response of human mesenchymal stem cells exposed to alternating current electric fields. *Biomed Eng Online* 10 (2011): 9.
Ibiwoye M. O., Powell K. A., Grabiner M. D., et al. Bone mass is preserved in a critical-sized osteotomy by low energy pulsed electromagnetic fields as quantitated by in vivo micro-computed tomography. *J Orthop Res* 22 (2004): 1086–93.
Inoue N., Ohnishi I., Chen D., et al. Effect of pulsed electromagnetic fields (PEMF) on late-phase osteotomy gap healing in a canine tibial model. *J Orthop Res* 20 (2002): 1106–14.
Klems H., Schleicher G. Effect of direct current on the healing of cancellous bone defects in animal experiments: Quantitative results [author's translation]. *Z Orthop Ihre Grenzgeb* 119 (1981): 217–21.
Kane W. J. Direct current electrical bone growth stimulation for spinal fusion. *Spine* 13 (1988): 363–5.
Lorich D. G., Brighton C. T., Gupta R., et al. Biochemical pathway mediating the response of bone cells to capacitive coupling. *Clin Orthop Relat Res* 350 (1998): 246–56.
Luben R. A., Cain C., Chen M., et al. Effects of electromagnetic stimulation on bone and bone cells in vitro: Inhibition to parathyroid hormone by low energy low frequency fields. *Proc Natl Acad Sci U S A* 79 (1982): 4180–4.
Mammi G. I., Rocchi R., Cadossi R., et al. The electrical stimulation of tibial osteotomies. Double-blind study. *Clin Orthop Relat Res* 288 (1993): 246–53.
Mollon B., da Silva V., Busse J. W., Einhorn T. A., Bhandari M. Electrical stimulation for long-bone fracture-healing: A meta-analysis of randomized controlled trials. *J Bone Joint Surg Am* 90 (2008): 2322–30.
Midura R. J., Ibiwoye M. O., Powell K. A., et al. Pulsed electromagnetic field treatments enhance the healing of fibular osteotomies. *J Orthop Res* 23 (2005): 1035–46.
Nelson F. R., Brighton C. T., Ryaby J., et al. Use of physical forces in bone healing. *J Am Acad Orthop Surg* 11 (2003): 344–54.
Poli G., Dal Monte A., Cosco F. Treatment of congenital pseudarthrosis with endomedullary nail and low frequency pulsing electromagnetic fields: A controlled study. *J Bioelectr* 4 (1985): 195–209.

Ryaby, J. T. Clinical effects of electromagnetic and electric fields on fracture healing. *Clin Orthop Relat Res* 355 (1998): S205–15.

Sarker A. B., Nashimuddin A. N., Islam K. M., et al. Effect of PEMF on fresh fracture-healing in rat tibia. *Bangladesh Med Res Counc Bull* 19 (1993): 103–12.

Schmidt-Rohlfing B., Silny J., Gavenis K., Heussen N. Electromagnetic fields, electric current and bone healing—What is the evidence? *Z Orthop Unfall* 149 (2011): 265–70.

Scott G., King J. B. A prospective, double-blind trial of electrical capacitive coupling in the treatment of non-union of long bones. *J Bone Joint Surg Am* 76 (1994): 820–6.

Sharrard W. J. A double-blind trial of pulsed electromagnetic fields for delayed union of tibial fractures. *J Bone Joint Surg Br* 72 (1990): 347–55.

Simonis R. B., Parnell E. J., Ray P. S., Peacock J. L. Electrical treatment of tibial non-union: A prospective, randomised, double-blind trial. *Injury* 34 (2003): 357–62.

Spadaro J. A., Bergstrom W. H. In vivo and in vitro effects of a pulsed electromagnetic field on net calcium flux in rat calvarial bone. *Calcif Tissue Int* 70 (2002): 496–502.

Steinbeck M. J., Kim J. K., Trudeau M. J., et al. Involvement of hydrogen peroxide in the differentiation of clonal HD-11EM cells into osteoclast-like cells. *J Cell Physiol* 176 (1998): 574–87.

Tsai M. T., Chang W. H., Chang K., Hou R. J., Wu T. W. Pulsed electromagnetic fields affect osteoblast proliferation and differentiation in bone tissue engineering. *Bioelectromagnetics* 28 (2007): 519–28.

Wahlström O. Stimulation of fracture healing with electromagnetic fields of extremely low frequency (EMF of ELF). *Clin Orthop Relat Res* 186 (1984): 293–30.

Wang Q., Zhong S., Ouyang J., et al. Osteogenesis of electrically stimulated bone cells mediated in part by calcium ions. *Clin Orthop Relat Res* 348 (1998): 259–68.

Wang Z., Clark C. C., Brighton C. T. Up-regulation of bone morphogenetic proteins in cultured murine bone cells with use of specific electric fields. *J Bone Joint Surg Am* 88 (2006): 1053–65.

Zhang J., Neoh K. G., Hu X., Kang E. T., Wang W. Combined effects of direct current stimulation and immobilized BMP-2 for enhancement of osteogenesis. *Biotechnol Bioeng* 110 (2013): 1466–75.

Zhuang H., Wang W., Seldes R. M., Tahernia A. D., Fan H., Brighton C. T. Electrical stimulation induces the level of TGF-beta1 mRNA in osteoblastic cells by a mechanism involving calcium/calmodulin pathway. *Biochem Biophys Res Commun* 237 (1997): 225–9.

Zorlu U., Tercan M., Ozyazgan I., et al. Comparative study of the effect of ultrasound and electrostimulation on bone healing in rats. *Am J Phys Med Rehabil* 77 (1998): 427–32.

Index